ASTROBIOLOGY: FUTURE PERSPECTIVES

ASTROPHYSICS AND SPACE SCIENCE LIBRARY

VOLUME 305

ASTROBIOLOGY: FUTURE PERSPECTIVES

Edited by

PASCALE EHRENFREUND
Leiden Observatory, The Netherlands

WILLIAM IRVINE
University of Massachusetts, Amherst, U.S.A.

TOBY OWEN
University of Hawaii, Honolulu, U.S.A.

LUANN BECKER
University of California, Santa Barbara, U.S.A.

JEN BLANK
Lawrence Livermore National Laboratory, Livermore, U.S.A.

JOHN BRUCATO
Naples Observatory, Naples, Italy

LUIGI COLANGELI
Naples Observatory, Naples, Italy

SYLVIE DERENNE
ENS de Chimie de Paris, Paris, France

ANNE DUTREY
Bordeaux Observatory, Floirac, France

DIDIER DESPOIS
Bordeaux Observatory, Floirac, France

ANTONIO LAZCANO
UNAM, México D.F., México

and

FRANCOIS ROBERT
LEME, Paris, France

KLUWER ACADEMIC PUBLISHERS
DORDRECHT / BOSTON / LONDON

A C.I.P. Catalogue record for this book is available from the Library of Congress.

ISBN 1-4020-2304-9
ISBN 1-4020-2305-7 (e-book)

Published by Kluwer Academic Publishers,
P.O. Box 17, 3300 AA Dordrecht, The Netherlands.

Sold and distributed in North, Central and South America
by Kluwer Academic Publishers,
101 Philip Drive, Norwell, MA 02061, U.S.A.

In all other countries, sold and distributed
by Kluwer Academic Publishers,
P.O. Box 322, 3300 AH Dordrecht, The Netherlands.

Cover image by James Garry; **fastlight** aerospace illustration.

Printed on acid-free paper

Printed in the Netherlands.

TABLE OF CONTENTS

PREFACE

The International Space Science Team: *"Prebiotic matter: From the interstellar medium to the Solar System"* had its "first light" in October 2001 and has since been active in addressing interdisciplinary scientific aspects in support of the new research discipline, Astrobiology. The team is a consortium of 12 scientists, each representing a specific research field crucial to revealing our origin as a consequence of the evolving Universe. The team investigated the conditions in the Solar System, and beyond, that allow nature to assemble the basic organics which play such an important role in the chemical evolution that preceded biological evolution.

Some of the results from discussions held during team meetings were published in October 2002:

"Astrophysical and Astrochemical Insights into the Origin of Life" Reports on Progress in Physics 65, 1427-1487 (2002)
Ehrenfreund, P.; Irvine, W.; Becker, L.; Blank, J.; Brucato, J. R.; Colangeli, L.; Derenne, S.; Despois, D.; Dutrey, A.; Fraaije, H.; Lazcano, A.; Owen, T.; Robert, F.

In order to extend those discussions and go deeper in some important areas, the team organized a workshop at the International Space Science Institute in Bern, Switzerland between April 1-4, 2003 with a title: "Astrobiology - Future Perspectives". To allow for the most fruitful interactions, the workshop was restricted to 30 top experts in the fields that comprise Astrobiology. This book reflects the state-of-the-art concerning selected topics and tries to give a glimpse at the future of exciting research in Astrobiology.

ASTROBIOLOGY, a new exciting interdisciplinary research field, seeks to unravel the origin and evolution of life wherever it might exist in the Universe. The current view of the origin of life on Earth is that it is strongly connected to the origin and evolution of our planet and, indeed, of the Universe as a whole. In order to establish a coherent picture of processes that may have played an important role in the chemical evolution leading to life, we have to understand the evolution of the very early Universe. In particular, we must investigate the formation of the first biogenic elements in stellar interiors and during stellar mass loss and explosions. Recent observations, balloon experiments and space missions such as the Wilkinson – Microwave Anisotropy Probe (WMAP) have refined the timescale of the Universe now known to be ~ 13.7 billion years old and expected to expand forever. The first objects in the Universe capable of ionizing gas formed about 200 million years after the Big Bang. It is generally believed that the elemental composition of the medium out of which the earliest stars and galaxies condensed consisted primarily of H and He. Nonetheless, the most red-shifted quasars, galaxies and Lyα absorbers currently observed all exhibit at least some admixture of heavier elements, as do the most ancient stars in our Milky Way Galaxy. Recent studies of primordial star formation show that in the absence of heavy elements the formation of stars with masses 100 times that of the Sun would have been favoured. Low-mass stars could not have formed before a minimum level of heavy-element enrichment had been reached. This enrichment has an important effect on the fragmentation properties of a gravitationally unstable gas, influencing the fragmentation of cloud clumps into low-mass protostars. The formation and distribution of heavy elements and the formation of low-mass stars contain major open questions in the field of Astrobiology.

In the interstellar medium and circumstellar environments, heavy elements are mixed and complex molecules and dust are formed and continuously modified according to the physical and chemical conditions they experience. New generations of stars and planets arise from agglomeration of dust and gas in interstellar clouds. The last decade has shown an impressive improvement in our understanding of protoplanetary disks and the processes that can form terrestrial and giant planets and the dark worlds at the outer edge of our Solar System—the Kuiper Belt region. These disks quickly become planets in some regions, or form small bodies that can eventually collide with already formed planets in others. Consequently it seems possible that both exogenous and endogenous sources of organic matter could have provided the first building blocks of life on the early Earth and likely merged to create the atmosphere and hydrosphere in which life flourished. In order to develop insights into the origin and development of life, minor Solar System objects (e.g., comets), planetary surface processes, hydrospheres, and atmospheres remain major targets of attention. Impacts and exogenous delivery had both beneficial and destructive effects on the evolution of planetary biospheres; determining the inventories of organic compounds and other volatiles in comets, asteroids, meteorites and interplanetary dust particles is therefore of major importance.

The transition from abiotic organic matter to entities that we define to be "alive" is not yet understood, nor are the specific conditions on the early Earth that must have played a major role in taking that step. The development of multiple processes such as self-replication, autocatalysis, Darwinian molecular selection, storage and transmission of genetic information, molecular stability and reactivity, and membrane formation are among the elementary steps toward molecular evolution and life that need to be further explored. Clues to these past events are encoded in ancient rocks, microfossils, and in the living cells themselves. Morphological, geochemical and isotopic biosignatures in rocks provide crucial records for unraveling the history of primitive life. Recent discoveries of microbial life in environments that are extreme by human standards improve our understanding of where life may exist elsewhere in the Universe.

The search for habitats and signatures of life beyond the Earth includes the exploration of our Solar System and the search for extrasolar planetary systems. Currently, we know of three other intriguing objects in our Solar System in this connection. Mars may have had conditions suitable for the origin of life at the same time this remarkable transition occurred on Earth. Mars may even harbour simple forms of microbial life today, at depth, or in hydrothermal or ice-rich regions. Recent studies of Europa imply the presence of a liquid ocean below a thick ice crust, raising the possibility of a marine biosphere.

Titan exhibits a rich organic chemistry (C and N) in its dense atmosphere and may thus provide clues to the chemical evolution that must precede biology. Farther away still, one of the most remarkable set of discoveries of the last decade has been the detection of numerous extrasolar planets. Continuing development of ground and space-based telescopes is a necessary first step toward revealing whether any of these distant worlds contain life. A spectroscopic detection of abundant oxygen in their atmospheres would be a compelling signal, but it is believed that the Earth had life long before it had significant atmospheric oxygen—other indicators may provide a key to the detection of life outside of our Solar System.

The chapters of this book discuss Astrobiology on the basis of recent developments in relevant fields. Chapter 1 focuses on the current cosmological concept which is manifested by recent observations and data from space missions, and elaborates on the consequences of primordial star formation and the time and location of the synthesis of the first heavy elements. The formation of organic molecules in interstellar and circumstellar environments and their transport to protostellar disks is investigated in Chapter 2. Recent knowledge of the chemistry occurring in protoplanetary disks, the main physical and chemical processes associated with the formation of solar-type stars and their accretion disks, as well as the possible contributions to the organic inventory of primitive Solar System bodies are discussed in Chapter 3. Following, Chapter 4 discusses insights into planet formation. A review of the allotropic forms of carbon in Chapter 5 provides a comprehensive view of the evolution of this element in space and its ability to build complex molecular and macromolecular precursors for life.

An overview of current knowledge about organic molecules in planetary atmospheres is given in Chapter 6, which reports in-depth on the three types of atmospheric environments that can be found in our Solar System, namely, the highly oxidized terrestrial planet atmospheres, the mildly reduced atmospheres of Titan, Pluto and Triton, and the highly reduced atmospheres of the giant planets. Interpretations of telescopic observations show that H_2O-ice is ubiquitous on surfaces throughout the outer Solar System. Additionally, carbon-bearing molecular material is emerging as a major component in the outer Solar System, where that material appears entrained in H_2O-ice in comet nuclei and many planetary satellites, as well as in the more volatile N_2 ice on Triton and Pluto. Chapter 7 discusses laboratory data in relation to observations of organics in the Solar System. By delivering prebiotic molecules to the Earth, comets could have played a role in the early phases of the development of life on our planet. In order to explore this possibility, Chapter 8 presents the most recent assessment of the molecular content of comets.

The recent discovery of a large number of Solar System bodies that orbit the Sun beyond Neptune has identified new possibilities for the study of primordial matter and processes in the early solar nebula. Indeed, Kuiper Belt objects are among the most primitive solid bodies in the Solar System though they are very difficult to study due to their intrinsic faintness and remoteness. Chapter 9 is an overview of present knowledge and future prospects for progress on this subject. In complementary fashion, interplanetary dust particles are among the most pristine materials of the Solar System presently accessible for laboratory analysis. Current progress in the investigation of these tiny particles, along with a description of new innovative techniques, is reported in Chapter 10 and provides important constraints on the evolution of our Solar System and the delivery processes of prebiotic matter to the planetary surfaces.

The environmental conditions under which life developed on the early Earth are unknown, and traces of Earth's earliest history have been recycled by the tectonic activity of our dynamic planet. Nonetheless, in Chapter 11 the formation of the Earth, its primordial atmosphere and the impact record are briefly described to bridge the gap between the formation of terrestrial planets and the ancient fossil record on Earth. The evidence for early life and its initial evolutionary steps on Earth are linked intimately with the geological evolution of the early Earth. While there are no records of the first appearance of life, and the earliest isotopic indications of the existence of organisms fractionating carbon in ~3.8 Ga (billion years) rocks from the Isua greenstone belt in Greenland are tenuous, there are well-preserved microfossils and microbial mats that occur in 3.5-3.3 Ga, early-Archaean sedimentary formations from the Barberton (South Africa) and Pilbara (Australia) greenstone belts. These are described in Chapter 12.

On Earth, organic matter of biological origin is subjected to various alteration processes, dominated by oxidation and/or thermal degradation due to deep burial. Other alteration processes that may have been rampant in the early Earth include impact metamorphism, irradiation and thermal degradation of dissolved organic species. The implications of these transformation processes, all of which affect the resulting molecular and carbon isotope chemistry, for the use of carbon compounds as biosignatures on Earth and other planetary bodies are discussed in Chapter 13. New, sensitive techniques have enabled us, in recent years, to elucidate the nature of extraterrestrial macromolecular material extracted from meteorites; this is of significant importance because the major fraction of carbon in the interstellar medium (in dust, comets, and meteors) appears to be incorporated in such macromolecular networks. The specific characteristics of extraterrestrial, macromolecular carbonaceous material are highlighted through a comparison of carbonaceous meteorites with early-Archean cherts from the Warrawoona Group (Australia), discussed in Chapter 14. The origin of life and remaining unsolved questions, such as prebiotic assembly, energy transduction, the tree of life, lateral gene transfer, and chirality are discussed in Chapter 15. Laboratory experiments on analog and prebiotic precursor material are described in Chapter 16.

Mars has been a central focus of interest in the context of extraterrestrial life. The search for extinct or extant life on Mars is one of the main goals of space missions to the red planet during the next decade. In January 2004, the European MARS-EXPRESS went into orbit, and two NASA Exploration Rovers, Spirit and Opportunity, arrived safely on Mars to pursue land-based investigations of the planet. These missions will test the planet's ancient and current habitability while accumulating enormous quantities of new information. Europa is a future target for the exploration of possible subsurface water and life therein.The search for Life elsewhere in the Solar System and corresponding, relevant space missions to Mars and Europa are summarized in Chapter 17. Accompanying this is a summary of the possible prebiotic organic chemistry active on an immense scale in the atmosphere of Saturn's satellite, Titan, to be explored by the CASSINI-HUYGENS mission starting in mid-2004.

Chapters 18 and 19 describe the efforts in the US and in Europe, respectively, to introduce Astrobiology as a new, valuable scientific discipline for research, education, and public support of science.

Future perspectives and recommendations for a successful exploitation of interdisciplinary research ultilizing astronomical observations, space missions, laboratory and field research, as well as the design of instrumentation, are given in the final Chapter 20 by the ISSI team.

We are grateful to the International Space Science Institute in Bern, Switzerland to have hosted and supported our team over a period of three years, and to the science of gastronomy so nicely practiced in Bern.

Pascale Ehrenfreund, Leiden Observatory, Holland (Teamleader)
(Cosmic dust)
Luann Becker, UC Santa Barbara, USA
(Geochemistry)
Jennifer Blank, Lawrence Livermore National Laboratory, USA
(Geochemistry, Extraterrestrial delivery)
John Brucato, Observatory Napoli, Italy
(Cosmic dust)
Luigi Colangeli, Observatory Napoli, Italy
(Cosmic dust, Comets)
Sylvie Derenne, ENS de Chimie de Paris, France
(Organic chemistry of meteorites)
Didier Despois, Observatory Bordeaux, France
(Comets)
Anne Dutrey, Observatory Bordeaux, France
(Protoplanetary disks)
Bill Irvine, Univ.Massachusetts, USA
(Interstellar chemistry, Comets)
Antonio Lazcano, UNAM, Mexico
(Prebiotic chemistry)
Toby Owen, Univ. Hawaii, USA (Co-chair)
(Solar System)
Francois Robert, Lab. de Mineralogie, France (Co-chair)
(Solar nebula, Meteorites)

Chapter 1

THE SYNTHESIS OF THE ELEMENTS AND THE FORMATION OF STARS

Marco Spaans
Kapteyn Astronomical Institute

1 GENERAL INTRODUCTION

One of the prerequisites for life is the presence of elements such as H, C, O, N, S and P. Those elements, with the exception of hydrogen, the most abundant element in space, were not yet present in the early universe. The hot chemistry in the very early universe, so-called Big Bang nucleosynthesis, only produced very light elements, predominantly hydrogen and helium and traces of deuterium, tritium, lithium and beryllium (c.f. Schramm 1998, 2002). All other chemical elements that occur in terrestrial biochemistry were formed by nucleosynthesis during the course of stellar evolution. It is generally believed that the elemental composition of the medium out of which the earliest stars and galaxies condensed consisted primarily of H and He. However, the most redshifted[1] quasars, galaxies and Ly α absorbers (thought to be intergalactic clouds or protogalaxies) currently observed all exhibit at least some admixture of heavier elements, as do the most ancient stars in our Milky Way. This requires the heavy first elements to have been formed at very early times, during the first five hundred million years of the universe's evolution. In astronomy all elements heavier than He are referred to as metals, and we will use that

[1] The redshift z of an object, that emits a photon at time t, depends on the observed frequency, ν, of that photon at time t_0, and the corresponding laboratory value, ν_0, according to $1 + z = \nu_0/\nu = R(t_0)/R(t)$, with $R(t')$ the size of the universe at time t'. For time one has the relation $t' = T_0(1 + z)^{-1.5}$, where $t' = 0$ corresponds to the big bang, $z = 0$ to the present, and $T_0 \approx 14$ Gyr is the current age of the universe.

P. Ehrenfreund et al. (eds.), Astrobiology: Future Perspectives, 1–16.

nomenclature in this chapter. Consequently metallicity refers to the abundance of such elements normally relative to that of the Sun. Recent studies of primordial star formation show that in the absence of elements heavier than He the formation of stars with 100 times the mass of the Sun would have been strongly favored (Abel, Bryan & Norman 2000), and that low-mass stars could not have formed before a minimum level of metal enrichment had been reached (Silk 1983; Norman & Spaans 1997; Schneider et al. 2002; Bromm et al. 2002; Hirashita & Ferrara 2002; Bromm & Loeb 2003; Cazaux & Spaans 2004). The value of this minimum level is very uncertain, but should be, in solar units, no less than $10^{-5} - 10^{-4}$ to allow a distribution of stellar masses similar to that observed today (the so-called Salpeter-like initial mass function or IMF) larger than 10^{-4} to allow efficient H_2 formation, and be of the order of $10^{-4} - 10^{-2}$ to facilitate efficient cooling by metal lines and complex molecules. The fact that these processes, which are all believed to be crucial in the efficient formation of stars and planets, require a rather modest metallicity suggests that a sharp transition might occur in the early universe from a star-less, pristine environment to one with active star formation and a rich molecular chemistry.

Depending on their mass, some of the first generation stars (known as population III stars or Pop III objects) may have collapsed into very massive black holes (VMBHs) at the end of their lives. Such stars do not contribute to the metal enrichment of the surrounding gas. It then appears that the initial cosmic metal enrichment had to rely on the heavy-element yields from so-called pair unstable supernovae (SNe). Such SN explosions leave no remnants, but do produce metals and, subsequently, dust grains. It should be noted that the dispersion and mixing of these heavy elements in the pristine gas is complicated and depends strongly on the details of SN blast waves as well as on the depth of the gravitational potential well of the primordial galaxy in which the first stars are formed (Madau, Ferrara & Rees 2001). Since the shock waves and radiation produced by the first stars may (temporarily) inhibit the ongoing formation of these primordial galaxies, the origins of the first elements and the large scale evolution of structure in the universe are intimately related.

Metallicity enrichment has important consequences for the fragmentation of a gravitationally unstable gas cloud through the cooling that metals provide. Molecules, that may be formed from metals, are even more efficient coolants. As the intrinsic ability of a gas cloud to cool increases, the mass scale down to which this cloud can fragment into individual or binary proto-stars decreases depending on the ambient equation of state (c.f. Spaans & Silk 2000; Scalo & Biswas 2002). These low-mass proto-stellar environments evolve into low-mass stars, potentially with planetary systems that exist long enough (more than a Gyr) for carbon-based life to develop. Hence, once the universe has been enriched to a minimum metallicity, that may vary strongly with location, a mode of star formation as we still observe in our own Milky Way today develops. The stars formed in this manner are called second and third generation, or population II and I, and constitute the bulk of the stars (likely with planets) in the universe, as observed with, e.g., the Hubble Space Telescope (c.f. Madau et al. 1996).

In the following sections the broad outline given above will be discussed in more detail. Three questions will be taken as crucial to the origins of the elements and the nature of star formation:

I. How are the first elements formed?
II. When (and over what time period) are these first elements formed?
III. How are stars formed after the first elements have been distributed?

2 OBSERVATIONAL CONSIDERATIONS

A prediction of standard cosmology is that the elemental composition of the medium out of which the earliest stars and galaxies condensed consisted primarily of hydrogen and helium (^{4}He) with small admixtures of deuterium, lithium ^{7}Li, and ^{3}He (c.f. Schramm 1998, 2002). The last few years have seen great advances in the discovery of highly redshifted quasars and galaxies and the study of their spectra. High-resolution spectroscopy has simultaneously led to analyses of the chemical composition of gases along the line of sight to distant sources. At redshifts of $z \sim 2 - 4$, Songaila & Cowie (2001) find a pervasive and uniform distribution of ionized carbon and silicon even in relatively tenuous intergalactic clouds. At similar redshifts Songaila & Cowie (1996) also found that triply ionized carbon (C IV) can be detected in 75% of the clouds with column densities $N(\mathrm{HI}) > 3 \times 10^{14}\,\mathrm{cm}^{-2}$, while Ellison et al. (2000) were able to measure C IV lines to a limiting sensitivity $\log N(\mathrm{C\ IV}) = 11.75$ against the bright lensed quasar Q1422+231. This suggests that the ratio C IV / H I remains roughly constant down to column densities of at least $N(\mathrm{HI}) \sim 10^{14}\ \mathrm{cm}^{-2}$ corresponding to roughly the mean density of intergalactic gas at the highest redshifts observed $z \sim 3.5$. It is not yet clear whether these carbon abundances persist into the intergalactic voids. With reasonable assumptions this would mean that the carbon abundance at these redshifts is a factor of 5000 lower than today, i.e., $n_\mathrm{C}/n_\mathrm{H} \sim 10^{-7}$.

Using data on oxygen in various stages of ionization obtained from extreme-ultraviolet absorption lines in Lyman-α absorbers, Telfer et al. (2002) find an oxygen to carbon ratio [O/C] $\sim 0.3 - 1.2$ at comparable redshifts. Triply ionized silicon is also observed in the low density clouds, Implying a silicon to carbon ratio somewhat higher than in the Sun (Songaila & Cowie 2001).

Recent ESO observations of the 0.8 $M_{\odot}$ star HE0107-5240, that lies at a distance of 11 kpc from the sun in the halo of the Milky Way, indicate that this star has an extremely low iron abundance of 200000 times lower than that in the Sun ($[Fe/H] \approx -5.3$, Christlieb et al. 2002). This indicates that stars can be formed at very low metallicities and persist to the present day. The relative metal abundances in extremely metal-poor halo stars are discussed in detail in Oh et al. (2001) and Qian & Wasserburg (2002).

Our galactic neighbors, the Large and Small Magellanic Clouds (LMC and SMC), have gas phase metallicities of 0.25 and 0.05, in units of solar, respectively. While a dwarf galaxy like Zwicky 18 has an even more modest metallicty of $Z/Z_{\odot} \approx 0.02$, with $Z_{\odot}$ the solar value. Conversely, observations of quasars and (clusters of) galaxies at high redshift, $z \sim 4 - 6$, indicate vast amounts of molecular gas and metallicities that are of the order of the Solar value (Omont et al. 1996; Bertoldi et al. 2003; Miley et al. 2004). Such galaxies are actively forming stars, and the quasars are probably powered by a combination of star formation and accretion onto $\sim 10^7 - 10^9 M_{\odot}$ black holes. These systems at high redshifts are representative of the $> 3\sigma$ density peaks in the primodial matter distribution, while the intergalactic observations at somewhat lower redshift trace the bulk of the diffuse baryonic matter.

All in all, the available observational data point toward one conclusion: metals already existed at very early times, but their abundances may vary strongly with location from the present ($z = 0$) to the highest redshifts accessible to modern telescopes ($z \sim 6$).

It has been speculated that unknown circumstances might conceivably contrive to produce some ^{12}C prior to the formation of Pop III stars. This would have consequences for the chemistry of the early universe. Standard cosmological models argue against the existence of heavy elements as admixtures in the primordial brew. However, an observational confirmation or denial of this assumption would provide important insights (Harwit & Spaans 2003).

3 POP III OBJECT FORMATION: THE FIRST STARS

A number of authors (Palla, Salpeter & Stahler 1983; Haiman, Thoul & Loeb 1996; Gnedin & Ostriker 1997) have suggested that a very early generation of Pop III stars could have formed at redshifts ranging out to $z \sim 30$, when the universe was only about 100 million years old. Recent simulations due to Abel, Bryan & Norman (2000, 2002) place the epoch of population III formation at $z \sim 18$. These massive stars eventually evolve into type II supernovae and may produce not only sufficient ionizing radiation to reionize the Universe (which was primarily neutral following the formation of atomic hydrogen at $z \sim 1000$) but may also produce and explosively eject heavy elements. The location in time of this epoch of reionization, expected to be coincident with the formation of the first heavy elements, is poorly constrained observationally. Recent observations with the WMAP instrument indicate, albeit with a large uncertainty, that reionization may already have been occuring at $z \sim 20$ (c.f. Kogut 2003; Bennett et al. 2003). Combined with the Gunn-Peterson trough observations[2] (Fan et al. 2001; Gunn & Peterson 1965) this hints at an epoch of reionization that may be quite extended and patchy, rather than a sharp and smooth transition (c.f. Razoumov et al. 2002; Ciardi et al. 2000).

Pop III stars are formed as gas clouds contract in dark matter halos with masses of the order of $10^{6.5} - 10^{7.0} M_{\odot}$, so-called pre-galactic structures, under the influence of gravity (Scannapieco, Schneider & Ferrara 2003; Abel et al. 2002). These gas clouds are capable of radiating away their thermal and gravitational potential energy through the Lyman α line (around 10^4 K) and through H_2 rotational emission (around 500 K). Molecular hydrogen, which is crucial to allow the gravitational collapse to continue despite the built-up of kinetic motions, is usually formed on dust grains. In the pristine gas at high redshift, no dust particles are available and H_2 is formed in the gas phase through the reaction $H + e^- \rightarrow H^-$ where the necessary electrons are left over from the recombination epoch at $z \sim 1000$ (Seager, Sasselov & Scott 1999), followed by $H^- + H \rightarrow H_2 + e^-$ (Tegmark et al. 1997; Norman & Spaans 1997). This formation route to H_2 is quite slow and results in an abundance of H_2 of roughly 10^{-4}. A similar pathway is available for the formation of HD, where the deuterium abundance is only 10^{-5} with respect to hydrogen as a consequence of big bang nucleosynthesis.

These small trace elements of H_2 and HD are sufficient to cool gas clouds down to about 100–200 K, a factor of a few below the first excitation level of H_2, and to densities of $10^3 - 10^4$ cm^{-3}. Abel et al. (2000) find that less than 1% of the primordial gas in the first small-scale

[2] At a redshift of about 1000 the universe recombines and becomes neutral. This neutral period is called the dark ages and it ends with the re-ionization of the universe by pop III stars and quasars, from $z \sim 30 - 6$. Currently, the intergalactic medium is ionized and hence one expects to find more and more neutral gas as one goes to higher redshift. This neutral gas may be seen in absorption against the first quasars and leads to deep (Gunn-Peterson) absorption troughs.

structures cools and collapses to even higher densities, $n_H > 10^5$ cm^{-3}, sufficient to be actually available for primordial star formation.

The masses of pop III stars are high since the ability of the primordial gas to cool is much smaller than that of a gas enriched in metals. Consequently, the Jeans mass (above which gravitational collapse becomes very likely) is large, up to $10^3 M_\odot$ (Bromm et al. 2002). The latter authors find in their numerical simulations that the above temperatures and densities are insensitive to the initial conditions. Instead, they are related to the microphysics of H_2: the lowest excitation energies of H_2 and HD and the critical density above which the level populations follow a Boltzmann distribution. Hence, the time scale for the formation of pop III stars is typically limited by the cooling time (rather than the free-fall time) for the very low metallicities appropriate for the early universe. Furthermore, the angular momentum of the progenitor cloud (Abel, private communication) is very small, suppressing fragmentation effects. Finally, the absence of metals and dust grains implies that radiation pressure from the proto-star on infalling gas, is negligible (Haardt et al. 2002). Hence, the usual mechanisms in the local universe that prevent the growth of a proto-star to extreme masses are not active in pop III formation at high redshift. Even though H_2 is the simplest molecule known, its collisional excitation by hydrogen atoms at temperatures above about 1000 K depends on long range exchange reactions whose rate coefficients are still uncertain (Tiné, Lepp & Dalgarno 2000). It should be realized that these uncertainties in the quantum chemistry of H_2 are related directly to the time over which gravitational collapse can be continued and thus to the mass of the final pop III object.

If pop III stars form in each other's vicinity then the radiation field of an already formed star may impinge on a nearby collapsing proto-stellar cloud and dissociate the H_2 molecules necessary for cooling (Yoshida et al. 2003). This negative feedback quickly diminishes as the first metals are available for additional cooling. Moreover, the presence of an X-ray background from mini-quasars may lead to positive feedback. Additional free electrons, crucial to the gas phase formation of H_2, are released through ionization of atomic hydrogen and more than offset the increase in the H_2 photo-dissociation rate. This effect appears to be mild (Machacek, Bryan & Abel 2003), although many uncertainties remain as to the formation and number of black holes accreting in the centers of such mini-quasars (Haiman 1999).

All in all, pop III star formation appears to take place over an extended redshift range of $z \approx 10 - 30$ and hence over a period of about 400 million years when the universe was only 100–200 Myr old (Scannapieco et al. 2003). For redshifts of about 5–10, when the universe is no more than 1 Gyr old, one expects stars with heavy element abundances similar to those observed today (Pop I and Pop II) to be formed in significant numbers, coincident with the formation of the first massive galaxies.

It should be noted in this that if population III stars were formed in proto-galactic clouds, much of the newly formed admixture of heavy elements could have remained localized, rather than becoming as dispersed throughout the extragalactic medium as the observations of Songaila & Cowie (2001) indicate. The dispersal of metals has been investigated by, e.g., Ferrara & Tolstoy (2000) and indicates that the energy of a few supernova explosions, $E \sim 10^{51}$ erg, is sufficient to drive the baryonic matter out of a primordial dwarf galaxy with a mass of the order of $10^6 - 10^7 M_\odot$. Hence, if the first elements were formed by pop III stars located in these relatively low mass (compared to present galaxy masses) systems then metals may have been dispersed into the intergalactic medium. These metals can then be incorporated into other forming galaxies, be mixed with the pristine gas, boost the ambient cooling rate and thus facilitate the efficient formation of more stars (positive feedback). These effects have been investigated in detail by Scannapieco et al. (2003) and it is found that the fraction of pop III objects formed as a function of redshift depends heavily on the spatial distribution of metals and is fairly independent of the mean metallicity of the universe.

Finally, for pop III stars one cannot really speak of an IMF, but rather of a preferred formation mass. This mean mass of a pop III star is of the order of 200 $M_\odot$ (Abel et al. 2000) in a range of 100–500 $M_\odot$. It should be noted here that Tumlinson, Venkatesan & Shull (2004) have argued for pop III stars that are less massive than 140 $M_\odot$ and are also in agreement with the relative abundances in extreme metal-poor halo stars mentioned above. In any case, once a very modest amount of metals, $\sim 10^{-4}$ of solar, has been injected by other pop III stars (already at a redshift of ~ 15), a Salpeter-like IMF is obtained (Ferrara 2003).

4 POP III OBJECT EVOLUTION: THE FIRST METALS AND DUST

The evolution of massive, low-metallicity stars is quite uncertain (c.f. Baraffe, Heger & Woosley 2001; Heger et al. 2003). The lifetime of a pop III star, about 10^6 yr, is only one or two orders or magnitude longer than its formation time. Pop III objects are unstable to nuclear-powered radial pulsations on the main sequence, but the growth timescale for these instabilities is much longer than for metal-rich stars and the pulsation probably does not have sufficient time to drive appreciable mass loss in primordial stars (Baraffe, Heger & Woosley 2001).

Heger et al. (2002) find that Pop III stars, with masses of the order of 100–500 $M_{\odot}$, encounter an electron-positron pair instability that causes them to collapse and burn oxygen and sillicon explosively. Pop III stars can explode with energies up to 100 times that of an ordinary core collapse supernova. Stars less massive than 140 $M_{\odot}$ or more massive than 260 $M_{\odot}$ should collapse into VMBHs instead of exploding (the event horizon swallows the entire star). The pair-creation SNe are thus bounded by regions of stellar mass that are nucleosynthetically sterile.

Nucleosynthetic yields of pop III stars were investigated by Heger & Woosley (2002) and Umeda & Nomoto (2003). These authors find that the nucleosynthetic signature of pop III stars strongly depends on the mass of the helium core, $M_{\rm He} \approx 63 - 133 M_{\odot}$ in the pair instability mass range. The He core determines the maximum temperature that is reached during the bounce phase of the supernova. A maximum of 57 $M_{\odot}$ of radioactive ^{56}Ni is produced at the upper-end of the He core mass range. An integration over the pair instability supernova mass range yields a roughly solar distribution of heavy nuclei with an even nuclear charge (e.g., Si, S, Ar), but this distribution is quite deficient in heavy nuclei with odd nuclear charge (e.g., Na, Al, P, V, Mn). This bimodality is caused by the fact that there is no stage of stable post-helium burning that can set the neutron excess. This pattern persists when the nucleosynthetic products of pop III stars in the 12–40 $M_{\odot}$ mass range are included. Furthermore, no elements heavier than zinc are produced in pair instability SNe because of a lack of s- and r-processes. It turns out that the Fe/Si ratio is quite sensitive to whether the upper bound on the IMF lies above 260 $M_{\odot}$ or lies somewhere between 140 and 160 $M_{\odot}$. Note that Umeda & Nomoto (2003) argue that core-collapse (high-energy) SNe in the 20–130 $M_{\odot}$ range better fit the observational data on extreme metal-poor stars (Oh et al. 2001; Qian & Wasserburg 2002) than more massive pop III stars do.

One expects part of the produced heavy elements (Si and C) to be incorporated into (small) dust grains. The time scale of dust formation is uncertain, but it is likely to exceed 50–100 million years (Spitzer 1978). Dust particles can strongly enhance the formation rate of H_2 and provide sites for grain surface chemistry, provided their combined surface area is sufficient. The physics of dust coagulation is quite complicated and the grain size distribution is uncertain, although the distribution of Mathis, Rumpl & Nordsieck (1977) provides a good parameterization for Milky Way dust (Li & Draine 2001; Draine & Lee 1984). It appears that turbulent environments tend to lead to rather compressed aggregrates through collisions (c.f. Dominik & Tielens 1997).

The dust grain properties at high redshift are likely to be quite different from those in the Milky Way. For example, the 2175 Å bump in the Milky Way extinction curve, a feature believed to be caused by carbonaceous material varies greatly in strength between the Milky Way and galaxies like M31, the LMC and the SMC, being completely absent in the latter. The presence of strong radiation fields and shock waves, both also associated with pop III star formation activity, are thought to play an important role in the explanation of this observational fact through their impact on the structure (compact or open/uneven) of these small dust grains (Draine 1990).

Still, for $Z/Z_{\odot} \sim 10^{-3} - 10^{-4}$, a level that should be achieved by the first wave of star formation, one expects dust grain chemistry to play an important role in the chemical composition of the early universe. This holds as long as the dust temperature, that is coupled to the temperature of the cosmic microwave background, is below 40 K, i.e., $z < 15$ to prevent evaporation effects (e.g., Spaans & Silk 2000). Finally, note that a large fraction of dust particles in the present universe are believed to be formed through condensation in the upper atmospheres of mass-losing stars, but that no such mass-loss occurs for pop III stars.

In conclusion, it appears that pop III stars are able to produce the first heavy elements, albeit with a different nucleosynthetic signature compared to later generations of stars. These metals are expected to lead to the formation of dust particles and drive a rich ion-molecule and neutral-neutral chemistry (see also Millar 2004, this volume), already at redshifts in excess of $z \sim 5 - 10$, when the universe was less than 1 Gyr old.

5 POP I AND II STARS

With the first two questions adressed, an overview is given now of our basic understanding of the formation of later generations of stars (Pop II, the oldest observed stars, and Pop I, which have heavy element abundances similar to the Sun). In this, we will refer to the interstellar medium (ISM) as gas that has been enriched during the epoch of pop III star formation to an ambient metallicity of more than $10^{-3} - 10^{-2}$, i.e., a rich ion-molecule and neutral-neutral chemistry takes place (see also Millar 2004, this volume). The formation of pop I and II stars should be quite similar and can be studied very well in our own Milky Way. Given the huge literature on this subject and the many branches of astronomy, physics and chemistry which are involved in interstellar gas processes, it is impossible to discuss all the aspects of the enriched interstellar medium and star formation. The emphasis therefore lies on the richness of processes rather than a complete characterization of each individual one.

The ISM, as it exists after the epoch of pop III star formation, plays a crucial role in the overall thermal and chemical balance of galaxies, like the Milky Way, and the process of star formation. In galaxies, pop I and

II stars form by the contraction and fragmentation of molecular clouds, where H_2 is still important for the overall chemistry, but species like CII, CI, OI, CO and H_2O provide cooling at a level much higher than H_2 can. For the Milky Way these processes have led to a galaxy where 90% of the observable matter is incorporated in stars. The remaining 10% of this mass is in the form of dust and gas and continues the star formation process. For our own Galaxy, the ISM is concentrated in spiral arms and a disk. Less massive, dwarf galaxies do not have this nice spiral structure but do contain large amounts of interstellar material. For galaxies like ellipticals on the other hand, the conversion of interstellar gas into stars seems to have been completed, since their interstellar gas mass is less than 0.1% of the total luminous mass.

5.1 The Multi-Phase ISM

Three different phases can be distinguished in the ISM. Atomic and molecular gas in clouds forms the cold dense phase which occupies not more than $\phi = 3$ percent of the volume in interstellar space, but contains the bulk of the interstellar gas mass. This dense gas is in the form of Spitzer-type HI clouds ($n_H = 30$ cm^{-3}, T = 100 K, $\phi = 2\%$) or in molecular clouds ($n_H > 300$ cm^{-3}, T = 20 K, $\phi = 1\%$). Warm ionized gas ($n_H = 0.3$ cm^{-3}, T = 8000 K), as found in the Reynolds layer and in the (turbulent) boundary layers around denser clouds encompasses roughly $\phi = 30\%$ of the ISM volume and about 10% of its mass. The very diffuse and hot interiors of expanding supernova bubbles ($n_H = 0.003$ cm^{-3}, T = 10^6 K) form the hot phase which represents roughly 70% of the ISM volume and a negligible mass fraction.

In the sixties and seventies, the origin and interrelationship of these phases, and their energy and ionization sources, have been studied by Field et al. (1969) and McKee & Ostriker (1977). It was found that a new stable phase reflects the onset of a new cooling mechanism or the decline of a heating source (c.f. Shull 1987). In this framework, the cold HI clouds and warm ionized medium result from the enhanced cooling by [CII] 158 μm at higher densities and Lyα and [OI] 6300 Å cooling at higher temperatures, respectively. The hot phase reflects the recent input of supernova energy. Molecular clouds are a natural extension of the cold HI clouds when one considers the role of self-gravity in increasing the density of molecular gas.

5.2 Pop I and II Stars: Interaction with the ISM

As stars evolve and reach the end of their main sequence, they develop winds which return a significant part of their stellar mass to the ISM (for low-mass stars, $M < 10M_\odot$) or they end their lifes violently in supernova explosions (massive stars, $M > 10M_{\odot}$). These ejecta contain elements heavier than He, generally referred to as metals, and enrich the ambient interstellar gas. A fraction of these metals are incorporated into dust grains, whereas the rest remains in the gas phase and may find its way into molecules as the stellar winds and supernova blast waves expand and cool into dense shells. These dense shells can in turn become the sites of star formation and continue the evolutionary cycle of the galaxy. The aim of ISM studies is to probe this cycle and to understand its role in the star formation process.

During their main sequence life, stars inject large amounts of radiation energy into the ISM. As they die, their ejecta contain large amounts of kinetic energy which power strong shocks. Shocked gas reaches high temperatures and densities and is the site of bright atomic and molecular emission lines, as well as many chemical processes. The combination of these energy sources leads to a process called feedback. For star formation to continue, i.e., for interstellar gas to cool and condense, the energy returned to the ISM must be disposed off. This occurs through line (atoms and molecules) and continuum processes (dust, plasmas) in which kinetic energy and hard photons are converted into the internal degrees of freedom of ambient gas and radiated at long (radio, millimeter, far-infrared) wavelengths, to which the ISM is mostly transparent.

The abundances of atoms, molecules and dust grains in turn are regulated by the overall chemical balance of the ISM and depend on the ambient enrichment, density, and temperature of interstellar gas clouds. The presence of stellar sources drives ionization and dissociation processes, and provides heating of the gas. In such irradiated regions, many molecular species are formed, and bright cooling lines are emitted in response to the energy input. The global effects of feedback are therefore intimately related to the chemical balance of interstellar gas.

The density, which plays an understandably important role in this chemical balance of interstellar clouds, varies over many orders of magnitude under the influence of gravity and hydrodynamic (hydromagnetic) processes. Stellar ejecta are initially quite diffuse even when they have cooled and condensed. The reason is that their mass by itself is not sufficient to create a self-gravitating complex. As many of these diffuse clouds coalesce, larger molecular clouds are formed, which are self-gravitating and provide sites for efficient star formation. Conversely, the collision of shells rises the ambient density as well and can also lead to self-gravitating structures.

The route from a diffuse cloud to a self-gravitating molecular cloud core may take tens of millions of years. During this time the interstellar gas undergoes strong chemical changes. To understand the process of star formation, one needs to comprehend the combined thermal and chemical balance of diffuse and dense interstellar gas clouds as they make their way from stellar winds to proto-stellar objects.

5.3 Pop I and II Stars: Formation

When a star or a binary (50% of all known stars appear to occur in pairs) is formed through molecular cloud collapse and fragmentation, it needs to shed the angular momentum originally contained in the molecular cloud core. As the core shrinks, its spin rate will become larger and centrifugal forces will cause the system to flatten along the equatorial plane. If no angular momentum is lost from the system, then these centrifugal forces would become larger than the gravitational pull of the proto-stellar system, and the collapse of the proto-star as well as the further accretion of material would be inhibited.

The process of stellar collapse then can be summarized through the following paradigm (Shu 1985). A proto-stellar core collapses inside out and the initial angular momentum of the system produces an accretion disk. This disk transfers mass onto the central proto-star while angular momentum is transferred outward. In general, it appears that the formation of outflowing jets combined with an accretion disk are crucial elements. These processes (in particular for high mass stars) are poorly understood since an adequate description of viscosity in hydromagnetic disks is still lacking, despite important insights into the stability of such systems (Balbus & Hawley 1991). Nevertheless, the presence of a disk, as well as jets, has been observed (Blake 1997) and provides an opportunity for the formation of planetary bodies. Indeed, most of the present angular momentum of the Solar system is contained in the giant planet Jupiter. Conversely, in a binary system a lot of the original angular momentum can be absorbed in the combined orbital motion.

The formation of planets is interesting from the ISM point of view since the initial chemical conditions are provided by the history of the parent molecular cloud. Furthermore, the formation of planetesimals is a coagulation process which must start with the smallest solid bodies present in interstellar gas, i.e., dust grains. The physical structure of proto-planetary disks thus provides important insights into star and planet formation as well as the chemical history of interstellar gas (see also Wuchterl et al. 2004, this volume).

The formation of high mass pop I and II stars proceeds on a time scale of $\sim 10^5$ yr which is roughly two orders of magnitude faster than for low mass stars. Although the intrinsic star formation process is believed to be the same, high mass star forming regions appear to be in more turbulent regions and the construction of detailed models is limited by a proper understanding of these processes (Shu 1997). Furthermore, high mass stars are formed in more crowded regions. This may imply that nearby stars and other cores have a non-negligible influence on the star formation process. In the following, we will give a brief overview of observational results for low-mass YSOs and hot molecular cores associated with massive star formation. We refer the reader to the review by Blake (1997, and references therein) for more details.

5.3.1 Outflows

Outflows are generally observed toward proto-stellar objects (Snell et al. 1980) and are believed to be a necessary consequence of angular momentum shedding. The formation of the jet is a much debated subject (Pelletier & Pudritz 1992; Shu et al. 1995), but imaging and kinematic studies provide a means to observe the structure of the jet at various scales. Combined interferometric and single dish mapping in CS 2-1 and SiO 2-1 have been used to indicate that jets impact on dense, cold gas ($n_H \sim 10^5$ cm^{-3}, $T \sim 20$ K) that is subsequently heated and compressed ($n_H \sim 10^7$cm^{-3}, $T \sim 100$ K). The jets can move at speeds up the 400–500 km s^{-1} and extend over several hundred AU (e.g., NGC 1333). One observes both refractory and volatile (SiO, SO, CS, HCN, H_2 CO, CH_3OH) material. The refractory SiO traces energetic events, and the combined species indicate the importance of dust grain sputtering and mantle desorption, very similar to the hot molecular cores discussed in the next section.

5.3.2 The Envelope

The accreting envelope provides the material which is fed onto the proto-star through the accretion disk. Single dish (sub-)millimeter telescopes probe of the simultaneous presence of infall (accretion) and outflow during the YSO evolution. In the Ophiuchus and Serpens molecular clouds (van Dishoeck et al. 1995; McMullin et al. 1994) one finds distinct kinematic features and enhanced abundances of refractory molecules like SiO and SO_2. In a source like IRAS 4A, kinematic signatures consistent with accretion are revealed and all molecules, including CO, appear to be depleted by factors of 25–50. Clearly, freeze-out of volatile molecules onto dust grains plays a crucial role in these envelopes, with obvious importance to grain-surface chemistry. In less embedded sources like YSOs in Taurus, one finds strong emission of CO and HCO^+. Thesc measurements indicate the presence of scale free density distributions with central condensations of the order of 10^6 cm^{-3}. The abundance of HCO^+ yields a lower limit on the fractional ionization in these envelopes of $\sim 10^{-8}$.

5.3.3 Disks and Hot Molecular Cores

Circumstellar disks regulate the accretion of material onto the central star through a boundary layer, and are believed to be 10–100 AU on theoretical grounds. It is important to constrain this radial extent better, and to determine the surface density and temperature profile of these disks (see also Markwick & Charnley 2004, this volume). Early maps of molecular emission (CO 2-1) from a number of systems indicated that disk radii may extend up to 1000–3000 AU (Sargent 1996). Recent developments in (sub-)millimeter instrumentation now allow high resolution dust emission maps at 2.7 mm to be made (Looney et al. 1997).

Hot molecular cores are small (< 0.1 pc), dense ($n(H_2 \sim 10^6 - 4 \times 10^7$ cm^{-3}), hot ($T \sim 100 - 300$ K), and dark ($A_V \sim 500$ mag) clumps of gas in regions of (massive) star formation. As such, they provide information on the impact of nearby star formation, as well as the chemistry of warm dense regions. Due to illumination by young stars and shock heating they are chemically quite distinct from cold clouds. Their chemistry is discussed in detail by Millar (2004, this volume).

For most of the history of the universe the nucleosynthesis of heavy elements has taken place either quiescently or explosively in Pop I and Pop II stars. This is discussed in Cataldo (2004, this volume) and much more extensively by Matteucci (2003). Hopefully, this chapter has provided the reader with an overview of heavy element production and star formation throughout the history of the universe. Stars and planets are being formed to this very day and much more can be learned from these processes about the origins of life, as subsequent chapters will discuss.

Acknowledgements

The author is greatful to Martin Harwit, Pascale Ehrenfreund, Tom Abel and Arjen Siegers for stimulating discussions.

References

Abel, T.R., Bryan, G. & Norman, M.L. 2000, ApJ, 540, 39
Abel, T.R., Bryan, G. & Norman, M.L. 2002, Science, 295, 93
Aguirre, A., 1999, ApJ, 521, 17
Applegate, J.H., Hogan, C.J. & Scherrer, R.J. 1988, ApJ 329, 572
Balbus, S.A., & Hawley, J.F. 1991, ApJ, 376, 214
Baraffe, I,m Heger, A. & Woosley, 2001, ApJ, 550, 890
Blake, G.A. 1997 in Molecules in Astrophysics: Probes and Processes, ed. E.F. van Dishoeck, (Kluwer, Dordrecht), p. 31
Bennett, C.L., Hill, R.S., Hinshaw, G., Nolta, M.R., Odegard, N., Page, L., et al., 2003, ApJS, 148, 97
Bromm, V. & Loeb, A., 2003, Nature, 425, 812
Bromm, V., Coppi, P.S. & Larson, R.B., 2002, ApJ, 564, 23
Cazaux, S. & Spaans, M., ApJ, 2004, in press

Christlieb, N., Bessell, M.S., Beers, T.C., Gustafsson, B., Korn, A., Barklem, P.S., Karlson, T., Mizuno-Wieder & Rossi, S.A., 2002, Nature, 419, 904
Ciardi, B., Ferrara, A., Governato, F. & Jenkins, A., 2000, MNRAS, 314, 611
Dominik, C. & Tielens, A.G.G.M., 1997, ApJ, 480, 647
Draine, B.T., 1990, in The Evolution of the Interstellar Medium, (Pacific), p. 193
Draine, B.T. & Lee, H.M., 1984, ApJ, 285, 89
Ellison, S.L., Songaila, A., Schaye, J. & Pettini, M. 2000, ApJ 120, 1175
Fan, X., Narayanan, V.K., Lupton, R.H., Srauss, M.A., Knapp, G.R., et al., 2001, AJ 122, 2833
Ferrara, A., 2003, Ap&SS, 284, 415
Ferrara, A. & Tolstoy, 2000, MNRAS, 313, 291
Field, G.B., Goldsmith, D., & Habing, H.J. 1969, ApJ, 155, L49
Gnedin, N.Y. & Ostriker, J.P., 1997, ApJ 486, 581
Gunn J.E. & Peterson, B.A., 1965, ApJ 142, 1633
Haardt, F., Ripamonti, E., Colpi, M. & Ferrara, A., 2002, Ap&SS, 281, 479
Haiman, Z., 1999, Ap&SS, 269, 191
Haiman, Z., Thoul, A.A. & Loeb, A. 1996, ApJ 464, 523
Harwit, M. & Spaans, M., 2003, ApJ, 589, 53
Heger, A. & Woosley, S.E., 2002, ApJ, 567, 532
Heger, A., Fryer, C.L., Woosley, S.E., Langer, N. & Hartmann, D.H., 2003, ApJ, 591, 288
Heger, A., Woosley, S.E., Baraffe, I. & Abel, T., 2002, Conf. on Luminous Celestial Objects, proceedings of MPA/ESO, p. 369
Hirashita, H. & Ferrara, A., 2002, MNRAS, 337, 921
Kogut, A., 2003, NewAR, 47, 977
Li, A. & Draine, B.T., 2001, 550, L213
Looney, L.W., Mundy, L.G., & Welch, W.J. 1997, ApJ, 484, L157
Machacek, M.E., Bryan, G.L. & Abel, T., 2003, MNRAS, 338, 273
Madau, P., Ferrara, A. & Rees, M.J., 2001, ApJ, 555, 92
Mathis, J.S., Rumpl, W. & Nordsieck, K.H., 1977, ApJ, 217, 425
Matteucci, F., 2003, 'The Chemical Evolution of the Galaxy', A&SS Library Vol. 253 (Kluwer)
McKee, C.F., & Ostriker, J.P. 1977, ApJ, 218, 148
McMullin, J.P., Mundy, L.G., Wilking, B., Hezel, T., & Blake, G.A. 1994, ApJ, 424, 222
Miley, G.K., Overzier, R.A., Tsvetanov, Z.I., Bouwers, R.J., Benítez, N., et al., 2004, Nature, 427, 47
Norman, C.A. & Spaans, M., 1997, ApJ, 480, 145
Oh, S.P., Nollett, K.M., Madau, P. & Wasserburg, G.J., 2001, ApJ, 562, L1
Omont, A., Petitjean, P., Guilloteau, S., McMahon, R.G., Solomon, P.M. & Pecontal, E., 1996, Nature, 382, 428
Palla, F., Salpeter, E.E. & Stahler, S.W. 1983, ApJ 271, 632
Pelletier, G., & Pudritz, R.E. 1992, ApJ, 493, 117
Qian, Y.-Z. & Wasserburg, G.J., 2002, ApJ, 567, 515
Razoumov, A.O., Norman, M.L., Abel, T. & Scott, D., 2002, ApJ, 572, 695
Sargent, A.I. 1996, in Disks and Outflows around Young Stars, eds. S.V.W. Beckwith, J. Staude, (Springer, Berlin)
Scalo, J. & Biswas, A., 2002, MNRAS, 332, 769
Scannapieco, E., Schneider, R. & Ferrara, A., 2003, ApJ, 589, 35
Schneider, R., Ferrara, A., Natarajan, P. & Omukai, K., 2002, ApJ, 571, 30
Schramm, D.N., 1998, SSRv, 84, 167

Schramm, D.N., 2002, IAUS 187, edited by K. Nomoto & J.W. Truran, p. 1
Seager, S., Sasselov, D.D. & Scott, D., 1999, ApJ, 523, L1
Shu, F.H. 1997, in Molecules in Astrophysics: Probes and Processes, ed. E.F. van Dishoeck, (Kluwer, Dordrecht), p. 19
Shu, F.H., Najita, J., Ostriker, E.O., & Shang, H. 1995, ApJ, 455, L155
Shu, F.H. 1985 in The Milky Way Galaxy, ed. D. Reidel, (Kluwer, Dordrecht), p. 561
Shull, J.M. 1987, in Interstellar Processes, eds. D.J. Hollenbach & H. Thronson (Kluwer, Dordrecht), p. 225
Tumlinson, J., Venkatesan, A. & Shull, J.M., 2004, astro-ph/0401376
Silk, J., 1983, MNRAS, 205, 705
Snell, R.L., Lorem, R.B., & Plambeck, R.L. 1980, ApJ, 239, L17
Songaila, A. & Cowie, L.L., 1996, AJ 112, 335
Songaila, A. & Cowie, L.L., 2001, in *The Extragalactic Infrared Background and its Cosmological Implications*, M. Harwit & M.G. Hauser, eds., Proceedings of an IAU Symposium held at Manchester, August 15–18, 2000, Astronomical Society of the Pacific.
Spaans, M. & Silk, J., 2000, ApJ, 538, 115
Spitzer, Ly., 1978, 'Physical Processes in the ISM' (Wiley & Sons)
Tegmark, M., Silk, J., Rees, M.J., Blanchard, A., Abel, T. & Palla, F., 1997, ApJ, 474, 1
Telfer, R.C., et al., 2002, ApJ 579, 500
Tiné, S., Lepp, S. & Dalgarno, A., 1998, MmSAI, 69, 345
Umeda, H. & Nomoto, K., 2003, astro-ph/0308029
van Dishoeck, E.F., & Blake, G.A., 1998, ARA&A, 36, 317
van Dishoeck, E.F., Blake, G.A., Jansen, D.J., & Groesbeck, T. 1995, ApJ, 447. 760
Yoshida, N., Abel, T., Hernquist, L. & Sugiyama, N., 2003, ApJ, 592, 645

Chapter 2

ORGANIC MOLECULES IN THE INTERSTELLAR MEDIUM

T J Millar
Department of Physics, UMIST
PO Box 88, Manchester M60 1QD, UK
Tom.Millar@umist.ac.uk

Abstract: This article discusses the formation of organic molecules in interstellar and circumstellar environments and their transport, or not, to protostellar disks.

Keywords: Astrochemistry, interstellar clouds, circumstellar envelopes, proto-planetary disks

1 INTRODUCTION

To date, over 100 molecules have been detected in interstellar and circumstellar regions (Table 2.1). These molecules are seen under a wide range of physical conditions, from low density, diffuse clouds, to dark dust clouds thought to be the precursors of low mass stars, to evolved objects such as AGB stars and protoplanetary nebulae. In addition to these gas-phase molecules, infrared observations of absorption and emission bands have been identified with an organic component of interstellar dust.

Table 2.1 shows that the bulk of interstellar and circumstellar molecules are organic. The list of interstellar molecules contains well-known organic molecules such as acetone and dimethyl ether, in addition to a simple sugar, glycolaldehyde CH_2CHOH, ethylene glycol $HOCH_2CH_2OH$, and the simplest amino acid, glycine, which has now

P. Ehrenfreund et al. (eds.), Astrobiology: Future Perspectives, 17–31.

Table 2.1. Observed interstellar and circumstellar molecules: August 2003.

				Number of atoms			
2	3	4	5	6	7	8	≥ 9
Interstellar inorganic molecules (22)							
H_2	H_2O	NH_3					
OH	H_2S	H_3O^+					
SO	SO_2						
SiO	HNO						
SiS	NH_2						
NH	N_2H^+						
NO	N_2O						
NS	H_3^+						
PN							
HCl							
SO^+							
HF							
Interstellar organic molecules (79)							
CH^+	HCN	H_2CO	HC_3N	CH_3OH	HC_5N	$HCOOCH_3$	HC_7N
CH	HNC	H_2CS	C_4H	CH_3CN	CH_3CCH	CH_3C_3N	CH_3OCH_3
CN	HCO	$HNCO$	CH_2NH	CH_3NC	CH_3NH_2	CH_3COOH	CH_3CH_2OH
CO	OCS	$HNCS$	CH_2CO	CH_3SH	CH_3CHO	H_2C_6	CH_3CH_2CN
CS	HCO^+	c-C_3H	NH_2CN	NH_2CHO	CH_2CHCN	CH_2OHCHO	CH_3C_4H
C_2	HOC^+	l-C_3H	$HOCHO$	HC_2CHO	C_6H		CH_3C_5N
CO^+	HCS^+	C_3N	c-C_3H_2	C_5H	c-C_2H_4O		CH_3COCH_3
	C_2H	C_3O	CH_2CN	H_2CCCC	CH_2CHOH		HC_9N
	C_2O	C_3S	H_2CCC	HC_3NH^+			$HC_{11}N$
	C_2S	H_2CN	$HCCNC$				$OHCH_2CH_2OH$
	CH_2	CH_3	$HNCCC$				NH_2CH_2COOH
	CO_2	C_2H_2	CH_4				
	C_3	$HOCO^+$	H_2COH^+				
		$HCNH^+$					
Circumstellar molecules (23)							
CP	$SiCN$	$HCCN$	C_5	C_2H_4		C_7H	C_8H
SiC	c-SiC_2	c-SiC_3	SiH_4	C_5N			
SiN	$NaCN$		SiC_4				
$NaCl$	$MgCN$						
$AlCl$	$MgNC$						
KCl	$AlNC$						
AlF							
SH							

been probably detected after many years of searches (Kuan et al. 2003a). Searches for other biologically relevant molecules, such as pyrimidine, aziridine, and azulene (Charnley et al. 2001; Dickens et al. 2001; Kuan et al. 2003b) have been unsuccessful. In addition to common organic species found on Earth, the interstellar medium also contains some rather exotic species, several of which were detected first in space. Thes include many carbon-chain molecules such as the cyanopolyynes ($HC_{2n+1}N$, $n = 1 - 5$), the radicals $C_nH, n = 1 - 8$, and the cumulenes ($H_2C_{2n}, n = 2, 3$). Some molecules are 'cyclic' or more accurately, are triangular, such as C_3H, C_3H_2, SiC_3, and ethylene oxide, C_2H_4O, although, to date, only one true ring molecule, benzene, has been detected and in only one source, a post-AGB object, the proto-planetary nebula CRL 618.

One should note that the identifications summarized in Table 2.1 are not a complete inventory of molecules, but are affected by three main selection effects. First, the most sensitive means to detect molecules is to observe rotational transitions at millimetre and submillimetre wavelengths using radio telescopes. Such transitions are allowed only if the molecule posseses a permanent electric dipole moment μ_D, with the strength of the transition proportional to μ_D^2. Thus homonuclear molecules such as O_2 and N_2, together with organics such as methane and ethane, C_2H_6, cannot be observed. Their detection must be by the less sensitive means of observing vibrational transitions at infrared wavelengths or through weak magnetic dipole transitions (for O_2) or through detection of a related species (such as N_2H^+). Second, even when a dipole moment is present, partition function effects make it easier to detect linear molecules compared to non-linear molecules. In essence, for a given energy, a linear molecule has much fewer energy states, and therefore transitions available, than a non-linear molecule. Thus, per transition, non-linear species have weaker transitions and are more difficult to detect. Third, detection of weak transitions is affected by the "interstellar forest", that is blending with lines from other complex species. As a result, it is often difficult to observe enough "clean" transitions of a complex molecule to be absolutely certain as to its identification. This was the issue that prevented the clear detection of glycine for many years.

In this paper, I will review the evidence for such species and discuss their formation and survival in order to investigate whether an interstellar component of organic material survives through to the formation of the Solar System. I will also discuss briefly the *in situ* formation of organic molecules in the protostellar disk out of which the planets formed.

2 ORGANIC CHEMISTRY IN INTERSTELLAR CLOUDS

Interstellar clouds have a variety of physical properties and a variety of chemical compositions. In this section I will discuss the composition of diffuse clouds, dark clouds, and hot molecular cores, with particular emphasis on the gas-phase component.

2.1 Diffuse Clouds

Diffuse clouds are so-called because they are transparent to optical light. Thus, they are also permeated by the interstellar UV radiation field which can photoionize and photodissociate molecules on fairly rapid time scales, typically less than 10,000 years. Such clouds are predominantly atomic, although molecular hydrogen, which self-shields against the dissociating photons, may take up to half of the protons in some cases. Simple diatomics, such as CH and CN are observed to have fractional abundances relative to total hydrogen of about $10^{-7} - 10^{-8}$. Triatomic molecues are difficult to detect, with C_3 being observed in a small number of clouds (Maier et al. 2001) at even lower abundances. As the column density of the cloud increases, the dust grains can extinguish more of the UV so that more complex molecules can be synthesized and survive. In such *translucent* clouds, simple molecules such as C_2H and HCN can be detected, and interferometric studies of molecular line absorption against extra-galactic continuum sources have detected more complex molecules (Liszt & Lucas 2001; Lucas & Liszt 2002). Diffuse gas in external galaxies can also be studied, particularly through use of millimeter absorption line studies (Wiklind & Combes 1997; Wiklind & Combes 1998), and shows some additional complexity in terms of the molecules seen in classical diffuse clouds in the Milky Way. Nevertheless, despite relatively high abundances for the inferred physical conditions, the total fraction of carbon in molecules is very small.

Evidence for molecular complexity in diffuse gas almost entirely rests on observations of optical and infrared absorption bands. The optical bands, the so-called Diffuse Interstellar Bands (DIBs), now number over 200 but are still unidentified despite some 70 years of effort. Their origin, in gas or solids, has been hotly debated with the evidence from ultra-high resolution spectroscopy now favouring the gas-phase, with identifiable sub-structure in some of the stronger bands indicating that the molecules are large (Sarre et al. 1995; Walker et al. 2001). Again, the lack of chemical identification makes it difficult to estimate abundances. For a DIB with equivalent width W at wavelength λ (both in Å), the column density of an absorber having oscillator strength f needed is:

$$N = 1.13 \times 10^{20} W/\lambda^2 f$$

In the ζ Oph diffuse cloud the total hydrogen column density is about $1.3 \times 10^{21}\text{cm}^{-2}$. Thus, to account for a strong narrow DIB at 6000 Å, having $W = 0.1$ Å and $f = 0.001$, typical for a small carbon chain, a fractional abundance of about 3×10^{-7} is required, much larger than that of C_3 in ζ Oph, 2×10^{-9} (Maier et al. 2002). Upper limits of 4×10^{-9} and 2×10^{-10} have been set for the fractional abundances of C_4 and C_5 (Maier et al. 2002). These carbon-chain species have fast photodissociation rates which imply a lifetime of only a few thousand years in diffuse clouds; their low abundances show that gas-phase processes are not very efficient in synthesizing them *in situ*. Since the DIB particles must be much more abundant, it appears that an element of *Darwinism* is at work, that is, the molecular carriers must be photo-resistant to the interstellar radiation field.

In particular, long sight-lines such as those toward the Galactic Center, or Cyg OB2 No. 12, which probe low-density diffuse gas, show the presence of the 3.4 μm absorption band due to C–H stretching modes in -CH_2 and -CH_3 groups in aliphatic species molecules trapped in solid particles. This band is absent from the IR spectra of molecular clouds, where a band at 3.47 μm is seen. The fraction of carbon tied up in this organic material depends on the type of material and on its particular absorption coefficient. However, within some uncertainty, it appears that about 0.1% of carbon is in this form, much larger than the fraction of carbon, excluding CO, in gas phase molecules.

A number of unidentified infrared (UIR) emission bands have been detected in circumstellar and interstellar clouds (Allamandola et al. 1987). The bands are most readily detected in carbon-rich environments irradiated by UV photons. The consensus of opinion is that they are due to fluorescent emission from a class of ring molecules, the polycyclic aromatic hydrocarbons (PAHs) (Sellgren 1984). As with the case of the DIBs, there is as yet no identification of a single PAH molecule, although it is likely that each individual band is created by emission from several PAHs. Again one can get an estimate of abundances by comparing the strengths of the bands to those of terrestrial materials. Some information on the number of carbon atoms per molecule can be deduced from the requirement that the molecules must be able to survive against photodissociation and that the fluorescent emission is the result of a single (UV) photon excitation which raises the particle temperature to effectively 1000 K. This implies that the PAHs molecules contain about 50 carbon atoms (Sellgren 1984), with around 10% of cosmic carbon in this form. A number of questions arise from these observations. Can such a large fraction of carbon be synthesised into complex organics within interstellar clouds through gas-phase processes? How important is photodissociation and photoionization for the destruction of these molecules? Are complex organics made through the processing, through photon or particle irradiation, of molecular ices in diffuse interstellar clouds? If so, what are the processes which remove organics to the gas phase? If the interstellar formation of these species is negligible, can they be made in carbon-rich objects, for example AGB stars, and transported to the general interstellar medium?

2.2 Dark Clouds

Dark clouds, which have densities $n(H_2) \sim 10^4 - 10^5$ cm^{-3} and temperatures of $\sim$ 10 K and effectively exclude interstellar UV photons, contain a much wider variety of gas-phase organic molecules, as evidenced by Table 2.2. Although selection effects ensure that we are not seeing the total inventory of organics, the total amount of gas-phase carbon, again excluding CO, is only 1% of the total available. Could there be a significant gas-phase organic component that we cannot see directly? This is certainly possible although it is noticeable that in the dark cloud TMC-1 the fractional abundances within a particular class of molecule, for example the cyanopolyynes, fall off with increasing chain length. Such a decrease is consistent with theoretical models of gas-phase synthesis, as discussed further in Section 2.3. Note too that an important characteristic of the chemical composition of dark clouds is the predominance of highly unsaturated carbon-chain molecules, the largest of which is $HC_{11}N$. In part, this is the selection effect mentioned previously, in that linear molecules have smaller partition functions than branched chain or ring molecules, so that the intensity per transition is larger. Even so, the non-detection of species such as ethanol, ethyl cyanide, acetone, and dimethyl ether is significant.

2.3 Hot Molecular Cores

Hot molecular cores (HMCs) are small ($\sim$ 0.1 pc diameter), dense ($n(H_2) \sim 10^7 - 10^9$ cm^{-3}) and hot ($T \sim 100 - 300$ K) clumps of gas in

Table 2.2. Fractional molecular abundances in a variety of molecular cloud types.

Species	PKS1413[a]	Diffuse[b]	Translucent[c]	TMC–1[d]	L134N[d]	G34.3hc[e]
CN	1(−9)	2(−8)	–	3(−8)	8(−10)	> 6(−10)
HCO^+	2(−9)	2(−9)	2(−9)	8(−9)	1(−8)	> 2(−9)
HCN	1(−9)	3(−9)	4(−8)	2(−8)	1(−8)	2(−8)
C_2H	–	2(−8)	7(−8)	5(−8)	4(−9)	> 8(−9)
c-C_3H_2	–	1(−9)	4(−8)	3(−8)	1(−9)	–
H_2CO	< 8(− 10)	7(−9)	6(−9)	2(−8)	2(−9)	2(−8)
CH_3OH	–	–	2(−8)	2(−9)	4(−9)	6(−8)
CH_3CN	–	< 4(− 11)	< 1(− 9)	1(−9)	< 1(− 9)	1(−8)
HC_3N	–	–	5(−10)	6(−9)	9(−10)	> 3(−9)

The data in this Table are based on the compilation from Turner (2000). [a]PKS1413+135 is a quasar at z = 0.247 (Wiklind & Combes 1997). [b]Diffuse refers to galactic clouds seen in absorption against extragalactic continuum sources (Lucas & Liszt 1997; Liszt & Lucas 2001). [c]Translucent cloud data from Turner (2000). [d]TMC–1 and L134N are cold dark clouds (Ohishi & Kaifu 1998; Dickens et al. 2000). [e]G34.3hc is a hot molecular core (Millar et al. 1997; Hatchell et al. 1998b). a(-b) = $a \times 10^{-b}$.

regions of massive star formation. They are important astrophysically because they represent an early stage in star formation; they may contain embedded massive protostars, or be the fossils of failed star formation. In general they have very rich molecular spectra at millimetre and sub-millimetre wavelengths and show a preponderence of highly saturated molecules, unlike the cold dark clouds. For example, the only cyanopolyyne detected in HMCs is HC_3N. The fractional abundances of saturated molecules such as H_2O, NH_3, CH_3OH, and H_2S, are typically $10^3 - 10^5$ larger than their abundances in cold clouds. In addition, large saturated molecules are also observed to have very large fractional abundances. Examples include ethanol and dimethyl ether, which have fractional abundances $\sim 10^{-9} - 10^{-7}$, (Ikeda et al. 2001), again much larger than the values of the upper limits, typically $\sim 10^{-11}$, in cold clouds. HMCs have a further property which is unexpected in such warm regions, namely they possess quite large fractionation in deuterium, with typical D/H ratios in molecules being 100 times the cosmic D/H ratio and about an order of magnitude less than in cold clouds. The fractionation of D in interstellar molecules is related to small zero-point energy differences between reactants and products, so that it is a process which is efficient only at low temperatures, that is less than $\sim$ 50 K. Thus the chemical synthesis of fractionated molecules, such as H_2CO, HCN, and CH_3OH, must have occured at lower temperatures than their current kinetic temperatures. This, coupled with the large abundances of saturated molecules, has been interpreted as the result of grain surface chemistry, that is chemistry within the icy mantles of interstellar grains in a cold pre-HMC phase. Subsequent heating of the gas by either embedded or nearby stars evaporates the mantles and returns these molecules to a hot gas phase.

3 ORGANIC CHEMISTRY IN STELLAR ENVELOPES

Organic molecules are also found in the circumstellar envelopes of late-type stars, with carbon-rich (i.e. those with a C/O ratio greater than one) objects having a particularly rich chemistry; more than 50 molecules have been detected in the C-rich AGB star IRC+10216. Rather surprisingly, given the temperatures and pressures that exist at the stellar surface, which ensure that LTE is a good approximation, many of the detected molecules are highly unsaturated carbon chains - indeed many of them, such as the cyanopolyynes, are detected also in the cold dark clouds. AGB stars are also objects in which carbonaceous dust particles are formed and then ejected into the interstellar medium. A detailed review of the formation of molecules and grains in late-type stars has been given by Millar (2003); here we merely sketch the salient points.

Close to the stellar surface of an AGB star, the gas temperature is on the order of 2000 K and LTE holds. The result, in the case of C-rich objects, is the formation of CO, which essentially uses up all of the oxygen atoms. The excess carbon forms stable molecules such as C_2H_2, CS and HCN. Because AGB stars pulsate, the gas in the photosphere experiences regular shocks which heat and compress it. Although the shock gives the gas an outward momentum, it is not sufficient to escape the gravitational attraction of the star and the material eventually falls back towards the stellar surface. The resulting conditions in this excursion allow for a shock chemistry to develop. In particular, the conversion of acetylene to ring molecules such as benzene and derivatives, is possible and can instigate the development of an organic chemistry that leads to the formation of PAH particles (Cau 2002). The pulsation allows the gas to be shocked a number of times, each shock producing more PAHs. At a large enough density, the PAH molecules aggregate and grow to form small carbonaceous grains which are driven by radiation pressure out into the interstellar medium. Collisions between the dust grains and the gas causes an outflow of molecular gas, a mass-loss process, to occur. This mass loss, which may have a typical rate of $10^{-5} M_\odot\ yr^{-1}$, has a typical velocity of about $15\,kms^{-1}$ and forms a circumstellar envelope (CSE) which may extend, in terms of detectable CO emission, up to 10^{18} cm from the star. The combination of LTE and shock chemistry ensures that fairly stable molecules are injected into the CSE. As the material flows away it meets the incoming interstellar UV radiation field - the central star has a negligible UV field - which photodissociates and photoionises these parent molecules. The result is that a rich photochemistry ensues in the outer CSE. Eventually all species which the interstellar UV photons can destroy will be returned to atoms and atomic ions and will be ejected to the interstellar medium in this form. In this degradation, many interesting and exotic organic molecules can be formed. Figs. 2.1 and 2.2 show the radial distributions of the photodissociation products of C_2H_2 and of the cyanopolyyne species, respectively. The ethynyl radical, C_2H, the primary daughter product of C_2H_2 photodissociation, plays a significant role in the sequential synthesis of the cyanopolyyne chain, through reactions such as:

$$C_2H + HC_{2n-1}N \longrightarrow HC_{2n+1}N + H$$

Chemistry is important if, once formed, the time for reactive loss is comparable with that of the radial expansion of the gas in the CSE. Close to the photosphere, where the external field cannot penetrate, the reaction

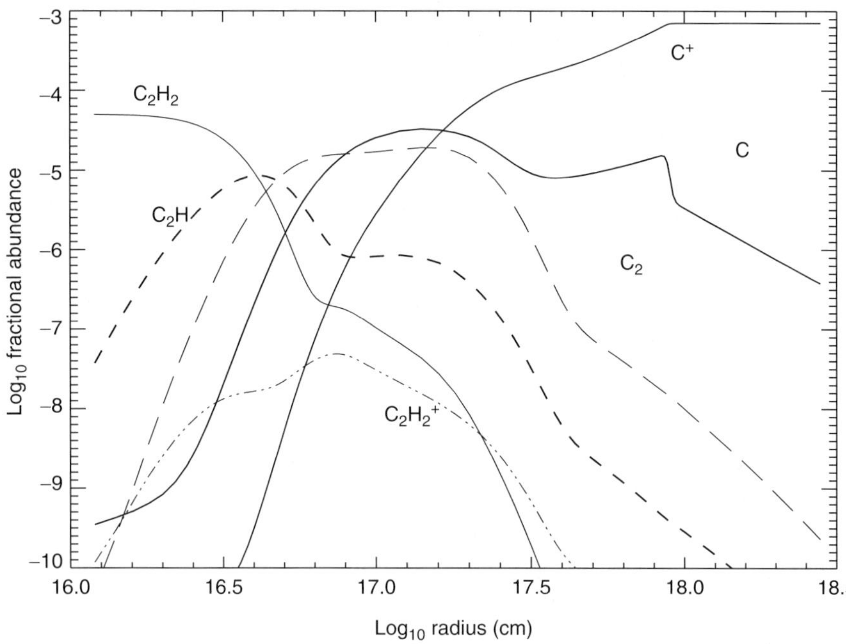

Figure 2.1. The radial distributions of acetylene and its daughter species are shown. The formation of molecular shells of C_2H and C_2 is evident. The parameters used are representative of IRC+10216 (after Millar et al. 2000).

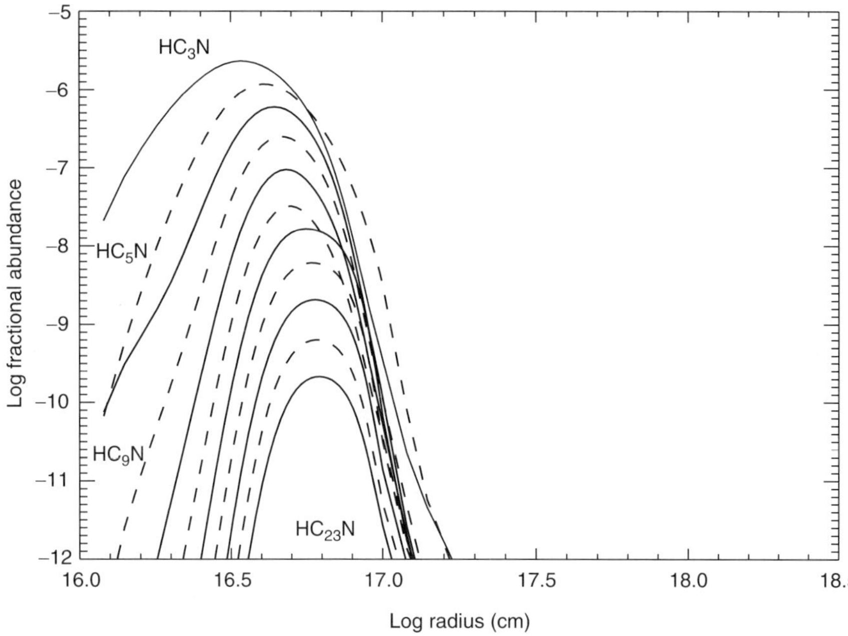

Figure 2.2. The radial distributions of the cyanopolyynes, up to $HC_{23}N$, in a model for IRC+10216 (after Millar et al. 2000).

times get longer as the gas cools and expands with the result that the photospheric values are generally frozen into the gas. There is a possibility that the gas interacts with the grains, and indeed some evidence from the IR line profiles of molecules such as ammonia indicates that there is an additional region of formation just outside the zone in which dust forms. Whether this implies a causal link between grain formation and molecular formation is unclear. Certainly, the grain temperature in this region is still fairly large, indicating that it will be difficult to bind gas phase atoms and molecules to the surface.

4 CHEMISTRY

In this section I review the basic chemistry which converts carbon atoms into organic molecules in dense interstellar clouds. Diffuse clouds have an inefficient chemistry due to their low densities, which imply long collision times, and their high UV fluxes, which imply short destruction times. The basic reaction which initiates and drives chemical complexity in dense, dark clouds is the proton transfer:

$$H_3^+ + C \longrightarrow CH^+ + H_2.$$

Subsequent reactions with H_2 lead to the formation of CH_5^+ which undergoes proton transfer with the abundant CO to form CH_4. Because the reaction between CH_3^+ and H_2 is a slow radiative association, the methyl ion is relatively abundant and plays an important role in forming species containing two carbon atoms, for example:

$$CH_3^+ + CH_4 \longrightarrow C_2H_5^+ + H_2$$

which, followed by dissociative recombination, produces ethylene, C_2H_4. Similarly reaction between C^+ and CH_4 produces protonated acetylene whose recombination with electrons produces the species C_2H and C_2H_2.

Further growth to large molecules is accomplished by reactions involving C atoms, the C_2H radical, and C^+ ions with hydrocarbons, such as:

$$C + C_mH_n \longrightarrow C_{m+1}H_{n-1} + H$$

$$C_2H + C_mH_n \longrightarrow C_{m+2}H_n + H$$

$$C^+ + C_mH_n \longrightarrow C_{m+1}H_{n-1}^+ + H$$

Such reactions are, when exothermic, relatively fast, occuring on almost every collision, and quickly build-up complexity when those reactions which break carbon bonds are inefficient. Reactions which reduce complexity generally involve O atoms, He^+ ions and UV photons. In diffuse interstellar clouds photons dominate the destruction on a time-scale of about 1000 years, typically much smaller than the cloud lifetime. Thus, the degree of complexity from *in situ* chemistry is small. In dense dark clouds, UV photons are extinguished efficiently by dust grains and binary reactions become more important. However, precisely because the clouds are dense, ionisation is provided by cosmic ray particles and leads to a low fractional abundance of He^+ ions, typically around $10^{-5}/n$, where n is the total density per cm^3. The O atom abundance is generally much higher but many reactions posses significant activation energy barriers, although more experimental work is needed in this area. In summary, formation dominates over destruction and complexity develops. Fig. 2.2 shows precisely this result for the case of IRC+10216, in which the parent molecules, HCN and C_2H_2, are injected at 10^{16} cm. At a radial distance of around $10^{16.6}$ cm, large molecules grow in a relatively dense, $n(H_2) \sim 10^5$ cm^{-3}, low UV environment. Beyond 10^{17} cm, the external UV field is large enough to destroy complex molecules unless the chemistry is able to produce photoresistant molecules. Fig. 2.2 shows also that the abundance of large molecules produced in the outer envelope decreases with increasing size, a reflection of the inefficiencies of the chemical synthesis in a flow expanding into a region of increasing UV flux. Thus it is highly unlikely that a sufficient abundance of photostable species can be produced in this region - by 'sufficient' I mean an abundance able to account for the DIB and UIR bands. We note that small ring molecules which are produced in the inner CSE are likely to be rapidly photodissociated if they are injected into the outer envelope. Peeters et al. (2003) have shown that the N-substituted rings adenine and uracil have large photodissociation cross-sections and would survive for only a few hundred years in the diffuse interstellar medium. Of course, conditions in the inner CSE are such that molecules with a large rather than small number of rings are produced (Cau 2002). If these large PAH molecules are photostable in the gas phase, or can be incorporated into the solid carbonaceous particles, then they may survive their journey as organic molecules into and through the interstellar medium.

The formation of very large hydrocarbons and carbon clusters, although not PAHs, in interstellar clouds has been considered by Bettens and Herbst (1996, 1997), for diffuse and dense clouds, respectively. In diffuse clouds, they find that although molecules as complex as C_{60} can form, the absolute abundance per species is much less than that required to produce an individual DIB. Indeed their model requires the existence of seed particles, containing around 10 carbon atoms, to fuel growth, particles which the authors note cannot be formed in the clouds.

In dense clouds, growth is more efficient but the detailed results are sensitive to the particular model adopted. Individual hydrocarbons still tend to have relatively low abundances, with fractional abundances less than 10^{-12}, when they contain more than 10 carbon atoms. Thus, although routes to complex molecule formation exist in interstellar clouds, it is unlikely that sufficient abundances to account for the DIBs or UIR bands can be produced in such environments.

5 FROM INTERSTELLAR TO PROTOSTELLAR ENVIRONMENTS

As noted previously, organic molecules are ubiquitous in the dense interstellar medium, with the additional possibility that photostable gas-phase species, or species trapped in the carbonaceous grains, might survive the journey from their origin in a carbon-rich circumstellar envelope to a molecular cloud. Star formation requires the formation of dense pre-stellar cores of gas which then collapse to high density. This collapse is isothermal in its initial stages and the gas remains cold, typically 10K. Gas-phase molecules which collide with dust grains will be accreted or frozen out forming an ice mantle on a time scale of $t_{ac}(yr) \sim 3 \times 10^9/n(cm^{-3})$. Thus, for $n > 10^6$ cm^{-3}, freeze out occurs very rapidly, much faster than the physical evolution of the gas. As the density of the collapsing core increases, conservation of angular momentum results in the formation of a thin accretion disk through which material from the molecular cloud is transferred to the central protostar. Material from the molecular cloud which falls on to the disk passes through an accretion shock, with a shock velocity which varies with radial distance from the protostar and with a median value given by (Cassen & Moosman 1981; Neufeld & Hollenbach 1994):

$$V_s = 17.2[M(t)/r_d(t)]^{1/2} \text{ km s}^{-1}$$

where $M(t)$ is the mass of the accreting protostar in solar masses and $r_d(t)$ is the centrifugal radius (in AU) and is proportional to t^3 where t measures the time since the start of the accretion phase. In the model for protostellar formation by Galli and Shu (1993a,b), material falls in on free-fall until it meets the edge of the centrifugally supported disk at a radius of perhaps 1–10 AU. Thus, in this model, the shock velocity is relatively low. The effect of the shock is to compress and heat the gas, with the immediate post-shock temperature $T_{ps} \propto V_s^2$. Low velocity shocks can lead to the evaporation of grain mantles. If shock velocities are much larger, $\sim 30 - 50$ km s^{-1}, say, then the dust grains themselves can be destroyed. Thus subsequent chemical evolution which occurs in the disk will have no memory of its interstellar history. For low shock velocities it is likely that a fraction of interstellar molecules survive intact and can act as a starting point for a subsequent protostellar disk chemistry.

One can search for evidence of an interstellar/circumstellar component to the chemistry by comparing the molecular composition of comets, the most primitive unprocessed bodies in the solar system, to that in hot molecular cores in regions of massive star formation. Like comets, the molecular composition of such cores appears to be dominated by the evaporation of molecular ices. A more detailed comparision can be achieved by comparing isotopic ratios, specifically those involving D/H and $^{13}C/^{12}C$. We note that the HDO/H_2O and DCN/HCN ratios measured in comets (Meier et al. 1998a; Meier et al. 1998b; Blake et al. 1999) are very similar to those measured in hot molecular cores (Hatchell et al. 1998a), although less than those measured in cold pre-stellar cores. Whether this means that D/H preserves an interstellar component is still unclear. A more clear cut test may come from the distribution of ^{13}C in HC_3N which, in molecular clouds, points to an *in situ* formation route involving the fast neutral-neutral reaction (Takano et al. 1998):

$$C_2H_2 + CN \longrightarrow HC_3N + H$$

with the absolute level of fractionation dominated by local chemical kinetics (Langer et al. 1984). Survival of interstellar HC_3N on to icy grain mantles and thence incorporation into comets would result in a similar isotopic distribution in cometary HC_3N. If the accretion shock is strong enough to destroy molecules, then re-formation of HC_3N in the pre-solar nebula would likely lead to a different isotopic distribution.

Finally, one should note that present models of chemistry in protoplanetary disks assume the survival of interstellar molecules on interstellar grains as the initial condition of the chemistry (Finocchi et al. 1997; Bauer et al. 1997; Willacy et al. 1998; Aikawa & Herbst 1999; Markwick et al. 2002; Millar et al. 2003). Transport of the grains in the accretion flow eventually leads to the evaporation of interstellar molecules in the inner disk, typically at radii less than 10 AU (see Fig. 2.3). In this figure, the dark band which is bounded in the mid-plane by the 5–8 AU region, corresponds to fractional abundances of $H_2CO \sim 10^{-7} - 10^{-8}$, and follows closely the isotherms between 100 and 300 K. The lower bound corresponds to the temperature at which H_2CO evaporates from the dust, and the higher bound to the temperature at which neutral reactions begin to destroy the gas-phase molecule. In this model, a significant evaporation of organic molecules occurs in the region of terrestrial planet

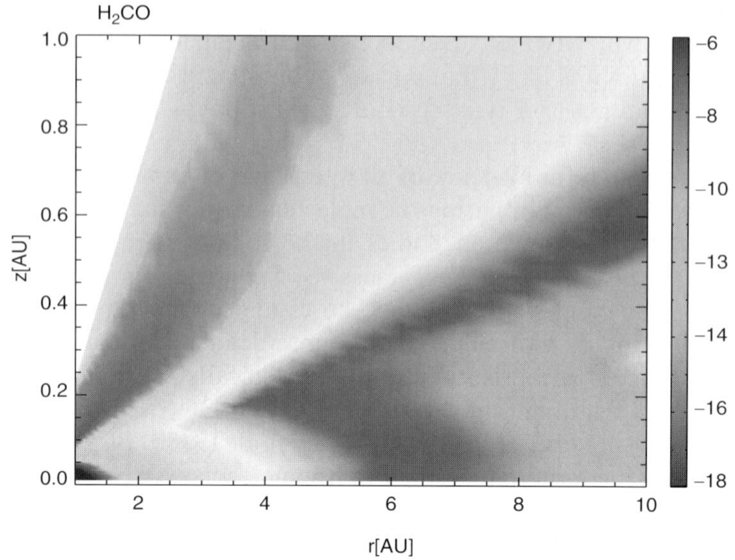

Figure 2.3. The distribution of formaldehyde, H_2CO, in the inner disk. The scale shows the log of the fractional abundance relative to total hydrogen nuclei (Millar et al. 2003).

formation, perhaps providing a significant initial chemical momentum to the formation of pre-biotic and biological molecules necessary for the formation of life.

Acknowledgments

Astrophysics at UMIST is supported by a grant from PPARC.

References

Aikawa, Y., Herbst, E., ApJ, 351, 233 (1999)

Allamandola, L.J., Tielens, A.G.G.M., Barker, J.R., in *Interstellar Processes*, eds. D.J. Hollenbach & H.A. Thronson, Jr., Dordrecht: D. Reidel, p. 471 (1987)

Bauer, I., Finocchi, F., Duschl, W., Gail, H.-P., Schlöder, J., A&A, 317, 273 (1997)

Bettens, R.P.A., Herbst, E., ApJ, 468, 686 (1996)

Bettens, R.P.A., Herbst, E., ApJ, 478, 585 (1997)

Blake, G.A., Qi, C., Hogerheijde, M., Gurwell, M.A., Muhleman, D.O., Nature, 398, 123 (1999)

Cassen, P., Moosman, A., Icarus, 48, 353 (1981)

Cau, P., A&A, 392, 203 (2002)

Charnley, S.B., Ehrenfreund, P., Kuan, Y-J., Spectochimica Acta, 57, 685 (2001)

Dickens, J. et al. ApJ, 542, 870 (2000)

Dickens, J. et al. Spectochimica Acta, 57, 643 (2001)

Finocchi, F., Gail, H.-P., Duschl, W., A&A, 327, 825 (1997)

Galli, D., Shu, F.H., ApJ, 417, 220 (1993a)

Galli, D., Shu, F.H., ApJ, 417, 243 (1993b)
Hatchell, J., Millar, T.J., Rodgers, S.D., A&A, 332, 695 (1998a)
Hatchell, J., Thompson, M.A., Millar, T.J., Macdonald, G.H., A&AS, 133, 29 (1998b)
Ikeda, M., Ohishi, M., Nummelin, A., Dickens, J.E., Bergman, P., Hjalmarson, Å., Irvine, W.M., ApJ, 560, 792 (2001)
Kuan, Y-J., Charnley, S.B., Huang, H-C., Tseng, W-L., Kisiel, Z., ApJ, 593, 848 (2003a)
Kuan, Y-J., Yan, C-H., Charnley, S.B., Kisiel, Z., Ehrenfreund, P., Huang, H-C., MNRAS, 345, 650 (2003b)
Langer, W.D., Graedel, T.E., Frerking, M.A., Armentrout, P.B., ApJ, 277, 581 (1984)
Liszt, H.S., Lucas, R., A&A, 370, 576 (2001)
Lucas, R., Liszt, H.S., in *Molecules in Astrophysics: Probes and Processes*, ed. E.F. van Dishoeck, Dordrecht: Kluwer, p. 421 (1997)
Lucas, R., Liszt, H. S., A&A, 384, 1054 (2002)
Maier, J.P., Lakin, N.M., Walker, G.A.H., Bohlender, D.A., ApJ, 553, 267 (2001)
Maier, J.P., Walker, G.A.H., Bohlender, D.A., ApJ, 566, 332 (2002)
Markwick, A.J., Ilgner, M., Millar, T.J., Henning, Th., A&A, 385, 632 (2002)
Meier, R., Owen, T.C., Matthews, H.E., Jewitt, D.C., Bockelée-Morvan, D.K., Biver, N., Crosvisier, J., Gautier, D., Science, 279, 842 (1998a)
Meier, R., Owen, T.C., Jewitt, D.C., Matthews, H.E., Senay, M., Bockelée-Morvan, D.K., Biver, N., Crosvisier, J., Gautier, D., Science, 279, 1707 (1998b)
Millar, T.J., in *Asymptotic Giant Branch Stars*, eds. H.J. Habing & H. Olofsson, New York: Springer, p.247 (2003)
Millar, T.J., Herbst, E., Bettens, R.P.A., MNRAS, 316, 195 (2000)
Millar, T.J., Macdonald, G.H., Gibb, A.G., A&A, 325, 1163 (1997)
Millar, T.J., Nomura, H., Markwick, A.J., Astrophys. Sp. Sci., 285, 761 (2003)
Neufeld, D., Hollenbach, D.J., ApJ, 428, 170 (1994)
Ohishi, M., Kaifu, N., Faraday Disc., 109, 205 (1998)
Peeters, Z., Botta, O., Charnley, S.B., Ruiterkamp, R., Ehrenfreund, P., ApJ, 593, L129 (2003)
Sarre, P.J., Miles, J.R., Kerr, T.H., Hibbins, R.E., Fossey, S.J., Somerville, W.B., MNRAS, 277, L41 (1995)
Sellgren, K., ApJ., 277, 623 (1984)
Takano, S., Masuda, A., Hirahara, Y., Suzuki, H., Ohishi, M., ApJ, 329, 1156 (1998)
Turner, B.E., ApJ, 542, 837 (2000)
Walker, G.A.H., Webster, A.S., Bohlender, D.A., Krelowski, J., ApJ, 561, 272 (2001)
Wiklind, T., Combes, F., A&A, 328, 48 (1997)
Wiklind, T., Combes, F., ApJ, 500, 129 (1998)
Willacy, K., Klahr, H., Millar, T.J., Henning, Th., A&A, 388, 995 (1998)

Chapter 3

CHEMISTRY OF PROTOPLANETARY DISKS

Relation to Primitive Solar System Material

A.J. Markwick* and S.B. Charnley
Space Science Division, MS 245-3
NASA Ames Research Center
Moffett Field, CA 94035-1000
USA
ajm@ajmarkwick.com
charnley@dusty.arc.nasa.gov

Abstract: We consider the chemistry occurring in protoplanetary disks and its possible contribution to the organic inventory of primitive solar system bodies. First, we outline the main physical and chemical processes associated with the formation of solar-type stars and their accretion disks. We then summarise the current observational status of protoplanetary disks and review chemical models of them. Finally, we discuss possible signatures of nebular chemistry in primitive organic material, paying particular attention to those related to isotope fractionation in comets and meteorites.

Keywords: astrobiology – solar system: formation – stars: circumstellar matter – stars: planetary systems: protoplanetary disks

1 INTRODUCTION

Delivery of extraterrestrial organic matter by comets and asteroids probably provided the molecules needed to initiate prebiotic chemistry on the early Earth (Oró 1961; Chyba et al. 1990; Ehrenfreund et al. 2002).

* NAS/NRC Resident Research Associate.

P. Ehrenfreund et al. (eds.), Astrobiology: Future Perspectives, 33–66.

Protoplanetary disks are therefore of fundamental importance for Astrobiology because all the starting materials available for the assembly of planetary systems, including many biomolecules, must be obtained from this reservoir (Ehrenfreund & Charnley 2000). Astronomical observations of molecular clouds, star-forming cores and protoplanetary disks can be employed to set constraints on the chemical composition and processing of the interstellar material that is first incorporated into planet-forming disks (Ehrenfreund & Charnley 2000). Another suite of chemical constraints lies within the measured composition of primitive solar system material. The meteoritic record contained in carbonaceous chondrites indicate that while interstellar organic material underwent a very significant degree of processing (Cronin & Chang 1993; Botta & Bada 2002), some volatile interstellar molecules may have entered the nebula in rather pristine condition (Messenger 2000; Irvine et al. 2000). Comets have a volatile ice composition that is broadly similar to that observed in molecular clouds, and many of their trace species are also well known in the interstellar medium (see the chapter by Crovisier in this volume). Thus, the ability to rewind the chemical clock for this evolution may, in principle, provide us with a window into the possible prevalence of similar organic inventories throughout the Universe. However, comets also display evidence for processed refractory dust grains. These crystalline silicates (Wooden 2002) are not observed in the dense interstellar medium and their presence in comets signifies modification and mixing within the protosolar nebula. So, however tantalizing it may be to make an astrobiological connection from interstellar organics to comets, IDPs and asteroids, and thence to prebiotic chemistry on the early Earth and extrasolar terrestrial planets, one must also consider chemical processing within the protosolar disk.

In this Chapter we review the possible relative contributions of interstellar and nebular chemistries to the organic composition of the young protosolar nebula. We begin with a brief summary of the state of molecular clouds and the theory of low-mass star formation (see also the Chapter by Millar in this book). We then describe the physical evolution of the protosolar nebula and relate these to the evolution of astronomical disks. It is now possible to partially observe the chemical composition of protoplanetary disks, and we give a summary of the existing observations. The basic physical processes which affect disk chemistry are then described and a detailed, comparative review of recent disk chemistry models is given. Finally, we discuss the implications of disk chemistry for understanding various isotopic and chemical characteristics of primitive solar system bodies.

2 STAR FORMATION

2.1 Molecular Clouds

Dense molecular clouds form from more diffuse interstellar material by processes probably related to the dissipation of interstellar turbulence (see Mac Low & Klessen 2003). In them, almost all the hydrogen is in molecular form due to efficient catalysis on dust grains surfaces (Hollenbach & Salpeter 1971). Molecular clouds are extremely cold. The relative balance of heating by cosmic rays and cooling by molecular rotational emission tends to maintain them at a temperature of around 10K (Goldsmith & Langer 1978). They display gas densities in the range $\sim 10^3 - 10^5\,\mathrm{cm}^{-3}$ for large-scale ($\sim$ few pc) dense clumps, which are in turn embedded in a lower-density ($\sim 10^2 - 10^3\,\mathrm{cm}^{-3}$) interclump medium (Evans 1999). Small-scale ($\sim$ 0.1 pc) dense cores have much higher densities ($> 10^5\,\mathrm{cm}^{-3}$) relative to their parent clumps. Low-mass stars like the Sun result from the gravitational collapse of dense molecular cores. Observations are suggestive of a low Galactic star formation rate (Zuckerman & Evans 1974); dense cores are not forming stars on a time-scale determined by that of free-fall gravitational collapse, but are somehow internally supported. This support can be by a combination of thermal, magnetic and turbulent pressures.

The observation that interstellar magnetic fields could be dynamically important led to the so-called 'standard model' of low-mass star formation (e.g. Shu et al. 1987). In this picture, dense cores are supported against gravity by a combination of magnetic field pressure, and the support afforded by internal MHD turbulence. Dense cores exhibit suprathermal linewidths, and tend to confirm the idea that the pressure exerted by the MHD waves, believed responsible, is substantial. In the standard model, formation of single and binary low-mass stars proceeds when dense core material loses magnetic support against gravity. Magnetic flux loss occurs through the relative drift (ambipolar diffusion) between ions and neutrals, on a time-scale of (e.g. McKee et al. 1993)

$$t_{\mathrm{AD}} = 7.3 \times 10^{13} x_{\mathrm{e}}\ \mathrm{yr} \tag{1}$$

where x_{e} is the fractional ionization and is $\sim 10^{-7}$ at these densities, assuming a standard cosmic ray ionization rate and depletion of the elements. As magnetic flux is lost, the density of the central regions increases until the unsupported core material undergoes an approximately isothermal, rotating, free-fall collapse, with a protostar and its accretion disk forming at its centre. The time-scale over which this occurs is the gravitational free-fall time

$$t_{\rm ff} = \frac{4.35 \times 10^7}{n_{\rm H}^{1/2}}\ {\rm yr} \qquad (2)$$

Clearly, $t_{\rm ff} << t_{\rm AD}$, and magnetised cores will evolve on a timescale of $\sim 7 \times 10^6$ years, or longer for lower x_e, before undergoing a rapid dynamic collapse. This picture then accounts for inefficient star formation as due to slow, quasistatic, magnetically-mediated cloud evolution and that dense molecular clouds a very long-lived entities $\gtrsim 10^7$ years.

However, over the past few years the standard model has been challenged by the proposition that, in fact, driven supersonic turbulence is responsible for the formation of molecular clouds, their destruction, and the formation of stars within them (Elmegreen 2000; Hartmann et al. 2001; Mac Low & Klessen 2003; Larson 2003). This picture is based on the observation that there is no evidence for T Tauri stars older than 1–4 Myr that are associated with molecular cloud gas (Hartmann et al. 2001). This implies that star formation proceeds rapidly once molecular clouds are formed, and that molecular cloud dissipation occurs soon after star formation is initiated. Also, numerical simulations of MHD turbulence show that, in fact, the turbulence decays rapidly, roughly on a timescale comparable to the free-fall time; turbulence cannot sustain a dense core against gravity (Mac Low & Klessen 2003). Hence, star formation occurs on approximately the free-fall time-scale, and not that determined by ambipolar diffusion (Elmegreen 2000). Star formation is inefficient because interstellar clouds spend a significant fraction of their lives in a low-density, primarily atomic, state. They are incapable of star formation until compressed to higher densities and molecular hydrogen forms. It has been noted that all the evidence from almost 20 years of modelling molecular cloud chemistry suggests chemical, and hence, dynamical youth (Elmegreen 2000; Hartmann et al. 2001). An enormous effort has been expended to construct chemical models which reproduce this chemical youth in the very long-lived clouds suggested by the standard model (e.g. Charnley et al. 1988).

2.2 Core Collapse

Irrespective of the preceding evolutionary scenario, protostars form from gravitational collapse of a dense molecular core (Evans 1999). The (isothermal) collapse physics is determined by the initial conditions and two semi-analytic prescriptions have been developed. Shu (1977) found self-similar collapse solutions starting from the 'singular isothermal sphere' (SIS) which initially has a r^{-2} density distribution throughout, and leads to an 'inside-out' collapse. The SIS was assumed to be the state to which dense cores would evolve as magnetic field support was lost by ambipolar diffusion; however MHD simulations show that the inner regions undergo a dynamical collapse before this structure can be established (e.g. Desch & Mouschovias 2001). Since it is intrinsically unstable, it is doubtful if the SIS could ever be realised in nature (Mac Low & Klessen 2003). Similarity solutions were also found for the collapse of an initially uniform density sphere by Larson (1969) and Penston (1969). Many candidates have now been identified for such pre-collapse, starless cores and detailed observations of some of them (e.g B68, Alves et al. 2001; Bergin et al. 2002) show their physical structure to be close to Bonnor-Ebert equilibrium spheres. Bonnor-Ebert spheres have flat radial density profiles and a steep drop in the outer regions. Thus, we might expect that the Larson-Penston (LP) solution would be more appropriate as a representative model of low-mass star formation. Radiative transfer models designed to identify infall for both the Shu and LP solutions (Zhou 1992) have been developed. Recent chemical models of both collapse scenarios have also been made (Rodgers & Charnley 2003; Aikawa et al. 2001).

2.3 Disk Formation

When the dense core is rotating, the angular momentum possessed by the original natal cloud means that most collapsing material does not fall directly onto the star, but rather onto a centrifugally-supported protostellar disk. As material falls inwards, conservation of angular momentum necessitates that angular velocity increases, and so at some point the trajectories diverge from spherically symmetric solutions. Since the kinetic energy gained by the gas as it falls into the gravitational potential well is much larger than the initial energy, the trajectories of the infalling material are effectively parabolas (e.g. Cassen & Moosman 1981; Tereby, Shu, and Cassen 1984). This disk evolution occurs on a similar timescale to the rapid accretion phase, $\sim 10^5$ yr, after which the collapse is halted by stellar winds or exhaustion of the available core material.

2.4 Observational Classification of Protostars

Observationally, forming protostars have been classified in a manner which broadly describes their evolutionary state and is based upon their spectral energy distributions (Lada & Shu 1990). In the prestellar phase a central object forms and, because of the high densities, the infall becomes optically thick, the core heats up, H_2 is dissociated, and collapse restarts and eventually forms the second hydrostatic object, the protostar. Submillimetre and FIR observations can locate Class 0 sources. These are sources where the deeply embedded protostar/disk system is still growing by accreting material from the dense core envelope; the mass of circumstellar material is greater than that of the protostar plus disk. Class I sources are optically invisible (depending upon viewing angle) IR protostars. Throughout the Class 0 and Class I phases powerful bipolar outflows develop and begin to clear away the surrounding envelope, to the point where the circumstellar mass is decreased to about one tenth of a solar mass in the Class I case. Class II objects are optically visible T Tauri stars and most of the circumstellar gas is found in the protostellar disk; it is at this stage that planets are believed to form, i.e. a *protoplanetary* disk is now present. Class III objects are mainly weak-line T Tauri stars with very tenuous disks. The chemistry at each phase, from pre-collapse (and prestellar) cores, Class 0 sources, through to Class III objects and their related 'debris disks' (e.g. β Pic), have been the subject of many detailed observational studies (van Dishoeck & Blake 1998) and chemical models (Millar, this book). Materials like that found in primitive Solar System matter are the probable survivors of material accreted during the Class II-Class III evolutionary phases.

2.5 Destruction and Survival of Interstellar Material

From the point of view of nebular chemistry starless cores are interesting because they are regions which are cold enough, for long enough, that a significant amount of the heavy molecular material (e.g. CO) can be depleted onto the dust grains. These 'depletion cores' can have very high isotopic fractionation, particularly in deuterium, and, since this material will ultimately appear in the protostellar disk, may be important for the high D/H ratios seen in, albeit processed, solar system material. Nevertheless, as cold interstellar gas and ice-mantled dust grains fall inwards,

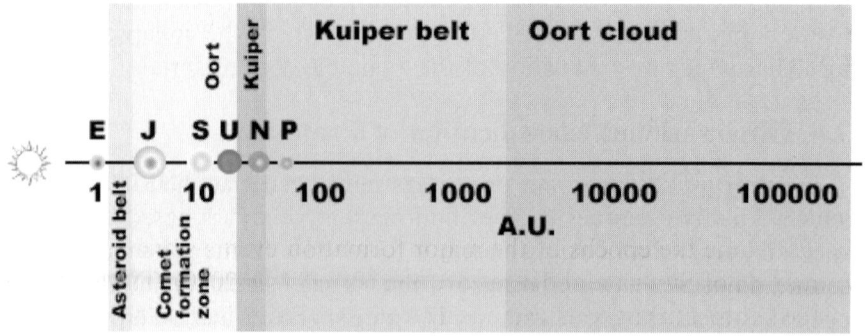

Figure 3.1. Schematic diagram of the protosolar nebula. The positions of the Kuiper belt and Oort cloud are shown, together with the formation zones of comets comprising these two families: Oort cloud (long period) comets are thought to form closer to the protostar than the Kuiper belt (short period) ones.

the condition of isothermality breaks down and interstellar material can begin to undergo physical and chemical change. For example, heating by thermal radiation, and the friction of gas-grain drag, can erode grain mantles as dust grains approach and enter the disk (e.g. Lunine et al. 1991). As material approaches the protoplanetary disk, many more endothermic chemical reactions come into play, driven by the increasing temperature. Radiation chemistry involving X-rays and UV from the accretion shock also play a role. Eventually, the accretion shock is encountered and shock processes come to dominate the chemistry of the material first entering the nebula (Lunine 1989; Neufeld & Hollenbach 1994). Specific regions of the disk favour survival of various interstellar materials (refractory metals, refractory and volatile organics, and ice) (see also Simonelli et al. 1997; Chick & Cassen 1997). The smallest shock speeds and preshock densities favour the survival of the most volatile material and occur in the outermost regions.

This is the first chemical processing distinct from interstellar chemistry that organic material experienced upon incorporation into the nebula. One important issue is to determine under what conditions largely unmodified interstellar matter could first arrive in the protosolar nebula. That is, in which regions of the nebula are ice mantles evaporated preshock, survive the shock, and simply recondense as ices of similar composition in the postshock flow.

3 STRUCTURE AND EVOLUTION OF THE PROTOSOLAR NEBULA

In this section we briefly summarise the physical processes which dominate the evolution of protoplanetary disks, paying particular attention to the protosolar nebula and the assembly of primitive bodies and planets.

3.1 Nebular Timescales

Astrophysical theory and observations, models of Solar System evolution and planet formation, and constraints from meteoritics, all allow us to determine the epochs of the major formation events starting from the formation of Sun's natal dense core (e.g. Benz et al. 2000). The evolution of the protosolar nebula and other protoplanetary disks can be roughly divided into four phases (Cameron 1995): (1) accretion of molecular cloud material (lasting a few hundred thousand years) most of which is consumed by the protosun; (2) disk dissipation (lasting about 50,000 years) where there is large scale mass and angular momentum transport;

(3) final solar accumulation (1–2 million years) where giant planet formation is almost complete but less so for the terrestrial planets; (4) removal of nebular gas by winds and stellar photoevaporation (3–30 million years). Comets started to be assembled towards the end of stage (2) when viscous effects dominated the nebular evolution. Thus, it should be noted that the chemistry of young *protostellar* disks (i.e. around 10^5 years old and with substantial gas masses) may have no relevance for Astrobiology, since all their accreted material is eventually consumed by the protostar. Later evolutionary phases are more relevant, when accreted matter has a higher chance of survival in the disk, and can subsequently be assembled into (proto)planets, asteroids and comets, i.e. when the transition has been made to a *protoplanetary* disk.

3.2 Physical Evolution

The protoplanetary disk undergoes differential rotation and shearing motions leading to frictional (or viscous) forces which heat the disk and cause radial motions. Possible sources of this dissipation are turbulence induced by thermal convection (e.g. Ruden & Pollack 1991); magnetorotational instabilities (e.g. Balbus & Hawley 2000); and various gravitational instabilities (Adams & Lin 1993). For the case of a turbulent viscosity, the disk temperature is determined by this viscous dissipation and by convection in the vertical plane. In all cases, radial motions involve the inward transport of most of the gas and dust and the outward transport of most of the angular momentum. All previous and current disk chemistry models (see Section 4) have assumed a turbulent viscosity (see Balbus & Hawley 2000 for an alternative view).

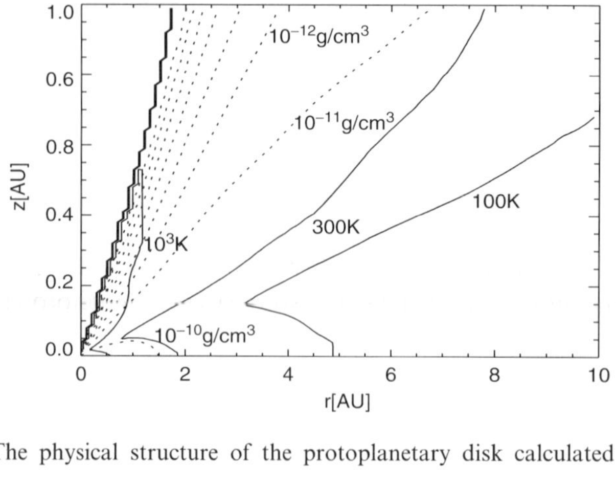

Figure 3.2. The physical structure of the protoplanetary disk calculated by Nomura (2002).

Dynamical models predict the existence of a hot, high-pressure inner region where dust can be vaporised. In this region (1 AU or less), chemical timescales are so rapid, relative to dynamical time-scales, that chemical equilibrium is an excellent approximation (Fegley & Prinn 1989). A radial temperature profile that decreases monotonically to around 10 K in the farthest outer disk (i.e. 100 AU or so) is also predicted by most dynamical models (e.g. Boss 1998). Beyond 5 AU, the time-scales of chemical processes eventually become much longer than all dynamical time-scales with increasing distance; at the farthest radii they are longer than the age of the disk. Intermediate to these regions is one where some dynamical and chemical time-scales can be comparable. This is the giant-planet formation zone and where the comets which bombarded the early Earth were produced.

4 CHEMISTRY IN PROTOPLANETARY DISKS

4.1 Observations of Molecules in Disks

Observations of molecules in the disks around several T Tauri stars indicate the possibility of directly studying the chemistry of protoplanetary accretion disks (see Dutrey et al. 2000). Many disks have been resolved in CO by interferometry (e.g. GG Tau, Dutrey et al. 1994). From these data, the disks are found to be large (300–800 AU) and in Keplerian rotation. The observation of molecules in disks is at the forefront of current receiver technology, and while new facilities like the Submillimetre Array (SMA) or the Atacama Large Millimetre Array (ALMA) will greatly increase our knowledge of these objects, many simple organic molecules, also found in the interstellar medium and in comets, have already been detected in such disks. Dutrey et al. (1997) reported the molecular content of the disks around the T Tauri stars DM Tau and GG Tau. They detected transitions in CO, HCN, CN, CS, H_2CO and HCO^+ in both disks. Since then, the list of known molecules in disks has increased, as has the number of objects studied (see Table 3.1). The derived abundances are lower than those in dense interstellar clouds, indicating that depletion onto dust grains is significant. The presence of ions and radicals confirms that, at least in the outer layers, an ionizing source drives the chemistry; either UV photons (Willacy & Langer 2000) or X-rays (Kastner et al. 1997; Glassgold et al. 1997). The large observed C_2H/HCO^+ and CN/HCN ratios are indicative of a photon-dominated chemistry. Mapping the distributions of molecules in disks is a task which pushes current observing facilities to the limit. Recently, Aikawa et al. (2003) mapped H_2CO in LkCa15 with the

Table 3.1. Molecular observations of protoplanetary disks.

Object	Detections	Upper limits	Reference
DM Tau	^{12}CO, ^{13}CO, $C^{18}O$ HCN, HNC, CN, CS $C^{34}S$, H_2CO, HCO^+, C_2H	$H^{13}CO^+$, N_2H^+, SiO SiS, H_2S, C_3H_2, HCO HC_3N, CH_3OH, CO^+ SO, SO_2, SiC_2, HNCS $HCOOCH_3$	Dutrey et al. (1997, 2000)
GG Tau	^{12}CO, ^{13}CO, $C^{18}O$, HCN CN, CS, H_2CO, HCO^+	HNC, $C^{34}S$, $H^{13}CO^+$ C_2H, N_2H^+	Dutrey et al. (1997)
L1157	CH_3OH		Goldsmith et al. (1999)
LkCa15	CO, HCN, CN, CS HCO^+, N_2H^+	OCS, HDO, DCN HNC, CH_3OH	Qi (2001)
TW Hya	CO, HCN, DCO^+ HCO^+, $H^{13}CO^+$		Kastner et al. (1997) van Dishoeck et al. (2003) van Zadelhoff et al. (2001)

Nobeyama Millimeter Array. The derived column density for this object is much higher than that derived for DM Tau, a trend seen in the column densities of other organic molecules between these two sources (Qi 2001). This is perhaps indicative of a difference either in the physical structure or evolutionary stage of these two disks (Aikawa et al. 2003). The only deuterated molecule unambiguously detected so far in a disk is DCO^+, detected in TW Hya by van Dishoeck et al. (2003). The derived fractionation ratio is 0.035, close to that found in pre-stellar cores, and to the DCN/HCN ratio observed in cometary comae. Upper limits for the DCN and HDO abundance in disks were reported by Qi (2001) and Kessler et al. (2002). For DCN, a fractionation ratio of >0.002 was reported; for HDO, no value is derived because of the lack of an observation of H_2O. The similarity of molecular D/H ratios between interstellar clouds, protoplanetary disks and cometary material suggests either that interstellar gas is incorporated into disks unprocessed (van Dishoeck et al. 2003), or that the ratios are chemically reset within the disk lifetime.

4.2 Physical Processes Affecting Chemistry

We now review the major physical processes which have to be accounted for in protoplanetary disks.

4.2.1 Ionisation

Sources of ionisation for driving chemistry in the disk include X-ray and UV radiation from the central object (Glassgold et al. 1997; Willacy & Langer 2000), UV from the interstellar radiation field and energetic particles formed by the decay of radioisotopes (see Figure 3.3). The latter process is principally from the radioactive decay of ^{26}Al (Finocchi &

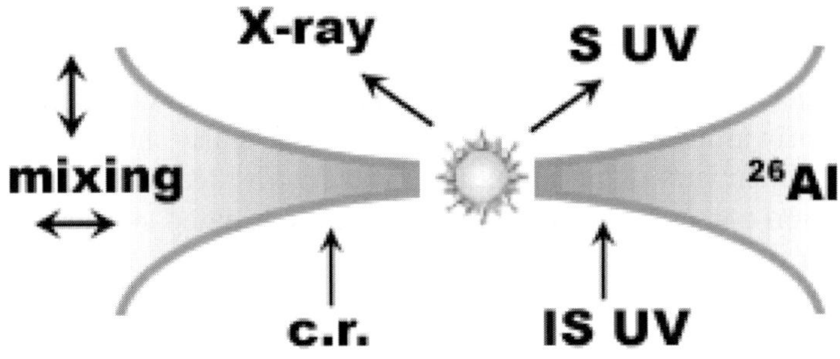

Figure 3.3. Schematic diagram of the protosolar nebula showing physical processes which affect chemistry. These include sources of ionisation (stellar X-rays, stellar UV, interstellar UV, cosmic rays, radionuclides), heating (viscous dissipation, stellar radiation) and transport (e.g. radial mixing, diffusion).

Gail 1997). Briefly, this isotope decays into excited ^{26}Mg by either positron decay or electron capture. In both cases, the excited magnesium isotope de-excites radiatively, releasing a photon of energy 1.809 MeV. This line has been observed in the Galactic Centre (Mahoney et al. 1984). In the case of positron decay, two such positrons can annihilate and form a pair of photons. The ionisation comes from two sources. Either the positron can collisionally ionise material itself, or the photons produced in the decay can Compton scatter their energy into recoil electrons, which can then collisionally ionise the molecules. The net ionisation rate due to this process was found to be $\zeta_R = 6.1 \times 10^{-19}$ s^{-1} by Umebayashi & Nakano (1981). An excess of ^{26}Mg was found in the Allende meteorite (Lee et al. 1977), proving that ^{26}Al was present at the time when the meteoritic chrondules formed. ^{26}Al can be produced in AGB stars and ejected during mass loss.

X-ray ionisation is only effective near the surface of the disk because the attenuation length of X-rays is small. Observations by X-ray satellites have found that the X-ray luminosity of T Tauri stars is in the range $10^{28.5} - 10^{31.5}$ erg s^{-1} (Glassgold et al. 1997). Igea & Glassgold (1999) presented a study of X-ray ionisation in protoplanetary disks. They found that the ionisation rate in the region $1 \leq R \leq 10$ AU is in the range $10^{-13} - 10^{-15}$ s^{-1}.

Aikawa & Herbst (1999a) calculated the variation of the total ionisation rate (X-rays plus cosmic rays) with height in a protoplanetary disk. Based on this calculation, the disk can be divided vertically into three regions; midplane, surface, and intermediate. In the midplane, radionuclides are the only major source of ionisation, while at the surface, the ionisation rate is dominated by radiation from the central object. Hence, the ionisation rate varies by orders of magnitude throughout the disk, with considerable implications for the chemistry. This is particularly important in the outer nebula, where the midplane temperature is so cold that all species are frozen onto grain surfaces. The observations of disks are currently limited to these outer regions, and clearly show evidence of a gas phase chemistry. This can be reconciled with theory by realising that the emission comes from the warm, photon-dominated surface regions of the disk, rather than the midplane.

4.2.2 Transport

Turbulent motions, whether convective or magnetohydrodynamic in origin, lead to a situation where there is an outward diffusion of material from the inner nebula, and an inwards advection of material from the outer nebula (Morfill & Volk 1984; Stevenson 1990). The key question is how much mixing actually occurred (see Lunine 1997; Fegley 1999). In the protosolar nebula, radial mixing was probably most pronounced in the region of giant planet formation. Hence, the spatial and temporal chemical evolution of this region was crucial in determining the composition of comets, especially the inventory of organics, that bombarded the terrestrial planets.

Theoretical models of radial mixing involve solving the related systems of advection-diffusion equations, usually for a stationary disk structure (Morfill & Volk 1984; Stevenson 1990). Recently, there have been several studies which investigated the effect of radial mixing upon cometary composition: water ice transport (Cyr et al. 1998); deuterium fractionation (Drouart et al. 1999; Hersant et al. 2001); and the oxidation of carbon dust (Gail 2002). Mixing models of disk chemistry involve incorporating chemical reactions networks into these models as source and sink terms. This leads to a system of reaction-advection-diffusion equations (Gail 2001). As the age of protoplanetary disks is longer than the diffusion time-scale in the outer disk, non-stationary disk models are required for a long-term treatment of the mixing process. In the case of a turbulently viscous disk, the disk evolution (Ruden & Pollack 1991) is also solved with the chemical evolution (Wehrstedt & Gail 2002).

Amongst the attractive features of radial mixing models is that they offer a means of transporting crystalline silicates into the comet-forming region (Gail 2001; Bockelée-Morvan et al. 2002). However, nebular shocks are another possible candidate for the origin of annealed silicates at around 10 AU (Harker & Desch 2002). Whichever explanation is correct, one might expect that the organic chemistries associated with each will be different. Also, Gail (2002) showed that while mixing of oxidised C-dust can explain the presence of some hydrocarbons in comets (CH_4 and C_2H_2), it cannot explain the presence of others (C_2H_6 and CH_3OH).

The chemistry of mass accretion involving inward radial transport onto the central protostar was treated by Bauer et al. (1997) and by Aikawa et al. (1999). In this case, they solved the continuity equation for species being advected into the inner disk. However, as one moves inwards, more physical and chemical processes come into play.

In addition to radial mixing, one must also consider whether there is any vertical transport of material in the disk. Ilgner et al. (2003) coupled the chemistry of Markwick et al. (2002) with a hydrodynamic model of the disk structure. The only vertical transport mechanism considered in this work was mass diffusion, and the chemical structure of the disk was calculated with the effects of diffusion switched on and off. The difference in distribution of some observable molecules (e.g. CS, SO) was pronounced, indicating that material can move vertically through the different layers of the disk, and that such transport processes should be accounted for in models. Whether or not these differences result in actual observable differences, however, is not yet apparent (see below).

4.3 Models of protoplanetary disks

There are now several models of the chemistry in protoplanetary disks. Table 3.2 compares the major characteristics of prominent chemical kinetic models from the literature.

Chemical modelling of the inner nebula has been based on chemistry in thermodynamic equilibrium, with the introduction of kinetic inhibition and quenching of chemical abundances in the low-temperature, comet-forming, regions where equilibrium is not valid (see Fegley and Prinn 1989). One qualitative difference concerns the gas-grain chemistry. In the inner nebula, catalysis can involve dissociative chemisorption onto metallic particles (Fegley 2000). This is fundamentally different from the cold surface catalysis, driven by physisorption, that occurs in interstellar clouds and, presumably, in the outer nebula. A key issue is how much these chemistries contributed to the composition of comets. A product distribution similar to that found in Fischer-Tropsch-Type (FTT) catalysis, for example high molecular weight aliphatic hydrocarbons, could be produced in the inner nebula (Fegley and Prinn 1989). If this is the origin of ethane in comets then it requires outwards radial mixing in the nebula. Alternatively, reduction of acetylene on cold (10K) grains could form ethane; this would require that it survive transport to thc comet forming region from its place of origin (natal cloud or 100 AU in the nebula).

The recognition that the outer protosolar nebula was probably a chemically active region has motivated several detailed studies. These began by simply applying simple interstellar gas-grain chemistry, either in a steady-state for an assumed disk structure (Charnley 1995), or time-

Table 3.2. Recent models of chemistry in protoplanetary disks. A comparison of the main features of some prominent models in the literature. The columns are as follows. (1) Physics: the underlying description of the disk structure; Semi-Analytical (S); Bell et al. (1997; B); Hayashi minimum mass solar nebula (Hayashi 1981; H); Steady Accretion Disk (A; Lynden-Bell & Pringle 1974); D'Alessio et al. (D; 1998); Chiang & Goldreich (1997, 1999; C); Nomura (N; 2002). (2) Chemistry: Prasad & Huntress (1980; PH); Mitchell (1984), Baulch et al. (1992) (MB); explicit grain surface reactions, not including gas-grain interaction, H_2 formation or recombination of ions (S); UMIST reaction kinetics (U; Le Teuff et al. 2000; Millar et al. 1997); Ohio State reaction kinetics (O; Terzieva & Herbst 1998); deuterium fractionation reactions (D). (3) Dimension: 1-D midplane model, or 1+1D vertical structure calculation. (4) Ionisation: cosmic ray (C), UV photon (U), X-ray photon (X), ionisation due to the decay of extinct radionuclides (R). (5) Gas-grain interaction: freeze-out and thermal/non-thermal desorption. (6) Dust destruction (carbon, troilite, silicate). (7) Observables reported; column densities (CD); D/H ratios; line profiles resulting from radiative transfer calculations. (8) Coupling: true coupling of chemistry and dynamics.

Model	P	Chem	D	I	G	D	Obs	C
Aikawa et al. (1996)	H	PH	1	U,C	Y	N	CD	N
Bauer et al. (1997)	S	MB	1	–	N	Y	No	N
Finocchi & Gail (1997)	S	MB	1	–	N	Y	No	N
Willacy et al. (1998)	B	U,S	1	C,R	Y	N	No	N
Aikawa & Herbst (1999a)	H	O	1+1	X,C,U	Y	N	CD	N
Aikawa et al. (1999)	A	U	1	C	Y	N	No	N
Aikawa & Herbst (1999b)	A	O,D	1	C	Y	N	D/H	N
Willacy & Langer (2000)	C	U,S	1+1	U,C,R	Y	N	CD	N
Aikawa & Herbst (2001)	H	O,D	1+1	X,C,U	Y	N	CD,D/H	N
Aikawa et al. (2002)	D	O	1+1	U,C	Y	N	CD,Line	N
Markwick et al. (2002)	B	U	1+1	X,C,U,R	Y	N	CD	N
van Zadelhoff et al. (2003)	D	O	1+1	U,C	Y	N	Line	N
Millar et al. (2003)	N	U	1+1	U,C,R	Y	N	CD	N
Ilgner et al. (2003)	B	U	1+1	U,C,R	Y	N	No	Y

dependent calculations at a fixed point under nebular conditions (Aikawa et al. 1996). Because the dynamical and chemical evolution can be decoupled in this region, there have been several time-dependent chemical models which assume a stationary density and temperature structure. The chemistry is very temperature-sensitive and is greatly influenced by the dynamical model adopted. For example, in deuterium fractionation by ion-molecule reactions it really matters whether the gas is at 10K or 20K or 30K (cf. Aikawa & Herbst 1999b; Aikawa et al 1999).

This decoupling also allows the vertical chemical structure to be computed straightforwardly (Aikawa & Herbst 1999b, 2001; Willacy & Langer 2000; Markwick et al. 2002). These models consistently show that ionisation is greater in the upper regions of the disk and can explain the existence of the molecules observed (cf. Dutrey et al. 2000). Depletion onto grains becomes increasingly important as one approaches the midplane. The CO depletions observed in disks have been attributed to this.

At disk densities the gas-grain interaction is a key process and either X-ray desorption (Najita et al. 2001) or photodesorption (Willacy & Langer 2000) could be responsible for those CO molecules that are detected.

A problem in comparing the results of chemical models of disks is that in general, they all are constructed on a different underlying physical prescription for the disk structure. From Table 1.2, we see that there are no less than 7 different descriptions in use in chemical models to date. Furthermore, the Table also shows that the models do not all include the same processes (e.g. gas-grain interaction, transport). For example, with respect to the chemistry in the inner nebula, a significant difference between the chemical kinetic models of Bauer et al. (1997) and Finocchi et al. (1997) and the rest of those outlined above, is that the former also considered the destruction of dust and volatiles as they are transported into the hot inner disk regions. In these models, the midplane temperature at 1 AU is 1500 K (hot enough for carbonaceous dust to oxidise), whereas in all other studies, the midplane temperature rarely exceeds 700 K, the temperature for troilite (FeS) oxidation, outside 1 AU. In the model of Markwick et al. (2002), for example, the midplane temperature at 1 AU is 650 K. Aikawa et al. (1996) find the midplane temperature to be 280 K at 1 AU. Also, in constrast to most, the models of Bauer et al. and Finocchi et al. do not include a detailed treatment of the chemical gas-grain interaction. Consequently, a major difference is that Finocchi et al. (1997) have an enormous injection of carbon (from dust oxidation) just outside 1 AU, which subsequently forms methane, right around 1 AU. This is in the vicinity of the terrestrial planets – a tantalising result, but totally dependent on their choice of physical model, which gives the temperature needed to oxidise carbon dust, and their omission of gas-grain chemistry, which, as shown in the other models, produces an appreciable abundance of methane well outside 1 AU, where it is desorbed from grains. In most models of protoplanetary disk chemistry, the inclusion of dust destruction was rendered unnecessary by the choice of physical model.

Some highlights from Table 3.2 are as follows. Aikawa & Herbst (1999b) presented the first model to include deuterium fractionation reactions, and therefore were able to draw conclusions between the fractionation in disks and comets. They showed that the fractionation ratios acheived in the outer nebula are more in line with cometary ratios, and that the results depended on the ionisation rate. The level of cosmic-ray ionisation used in the model to achieve agreement with cometary values is an order of magnitude lower than the canonical value. They also found that the DCN/HCN ratio depends on whether recombinations of ions on grain surfaces are dissociative or not. Willacy & Langer (2000) described a detailed photoprocessing calculation based on the underlying model of Chiang & Goldreich (1997) – a flared disk model. They included photodesorption, and found that significant quantities of molecules can be kept in the gas phase even at visual extinctions in excess of 4 magnitudes. This was based on experimental data for photodesorption by Westley et al. (1995). The main conclusion from the paper was that the observed emission comes from the upper layers of the disk, and not from the midplane where most of the mass resides. Aikawa et al. (2002) made a major advance by calculating for the first time the actual line profiles resulting from the calculated chemical distributions. These are obviously desirable (see below). This work was based on the flared disk model of D'Alessio et al. (1998), and the line profiles produced considering radiative transfer in two dimensions. van Zadelhoff et al. (2003) also used radiative transfer to model UV photoprocesses in the disk. Results were calculated for different stellar radiation fields, and found that the CN/HCN ratio in particular shows a strong dependence on the stellar spectrum, whereas HCO^+/CO for example, does not. They also concluded that the photodissociation of CO produces reactive carbon in the surface layer which enhances the abundances of molecules, especially the radicals CN and C_2H. Finally, Ilgner et al. (2003) succeeded in a true coupling of dynamical and chemical structure, including mass diffusion. A drawback is that the model is constructed atop the Bell et al. (1997) physical prescription, which does not include the effects of stellar radiation. Never-the-less, Ilgner et al. found that the inclusion of mass transport processes has a significant effect on the distribution of observable molecules, especially sulphur-bearing species like CS and SO. However, the lack of calculation of observables means that the true significance of these processes for observations is still unclear. Like the model of Markwick et al. (2002), Ilgner et al. (2003) differs from other models in Table 3.2 in that it only considers the inner regions of the protoplanetary disk ($R \leq 10$ AU).

Despite their differences, the models do all agree on some qualitative level. For example, they all produce significant quantities of molecules indicative of a photon-dominated chemistry (e.g. CN, C_2H) in the surface regions. In the midplane regions, they all show significant depletions of molecules. It is not clear, however, that a comparison of the detailed results is meaningful. One hopes that with the advent of state-of-the-art observing facilities like the SMA and ALMA, we will be able to use observations to discriminate between the models in Table 1.2. Unfortunately, because they are all based on different underlying physical descriptions, and include different physical processes affecting chemistry, this may prove impossible. A goal for chemical models should be to try to provide observables with this target in mind. One way to achieve this would be to fix the chemical processes and vary the underlying physical description, or vice versa. Subsequent application of radiative transfer codes (e.g. van Zadelhoff et al. 2003) to results from these models would then allow us to gauge the effects of changing either the physics or chemistry on the actual observations. At the time of writing, a comparison of the models and the observables they produce in this scientific manner has not been forthcoming. The models of Markwick et al. (2002) and Millar et al. (2003; see below) use exactly the same chemistry, but different physical descriptions. The former is based on a hydrostatic model where the only source of heating is from viscous dissipation. The latter adds a treatment of the radiation from the central object and a detailed calculation of the 2D radiative transfer throughout the disk (Nomura 2002). This makes a significant difference to the temperature and density structure, even at radii less than 10 AU (see Figure 3.2).

We now briefly present some results from the model of Millar et al. (2003). The physical model used therein differs from that of Nomura (2002) only in that it contains a 1D treatment of H_2 and CO self-shielding. The structure of the disk can be split into a hot surface layer where irradiation from the central object dominates, and a colder inner region. Closer to the central object, the midplane temperature rises, as does the density, which has implications for the gas-grain interaction. The output from this physical model is then used to provide the conditions for the chemical model, which is identical to that described in Markwick et al. (2002), and includes gas-grain interaction (freezeout, thermal desorption) and photoprocesses. The chemical model was run over the whole 2D region between 1 and 10 AU (Millar et al. 2003) and in the midplane out to 100 AU (Markwick & Charnley 2004). Contour plots in Figures 3.4 and 3.5 show the distributions of HCO^+ and C_2H, while Figure 3.3 in the Chapter by Millar (this book) shows the distribution of H_2CO. The latter, formaldehyde, and C_2H represent two classes of species in the model: products from the thermal desorption of grain ice mantles, and products from the photodissociation of other molecules. For example, C_2H is seen to be more abundant towards the surface of the disk, because it is a photodissociation product of acetylene. A similar result holds for CN, a photodissociation product of HCN. In contrast, the abundance of formaldehyde peaks in the interior of the disk, and shows a distribution reminiscent of the temperature profile (cf. Figure 3.2). This is because H_2CO is mainly produced from its thermal desorption from grain surfaces. HCO^+ (Figure 3.4) represents a third class of species – those which show both a peak due to a species desorbed from grains – in this case, CO which reacts with H_3^+, and which peak towards the surface due to the increased ionisation rate there. It is because of the significant effects on the gas phase chemistry that thermal desorption has that the underlying

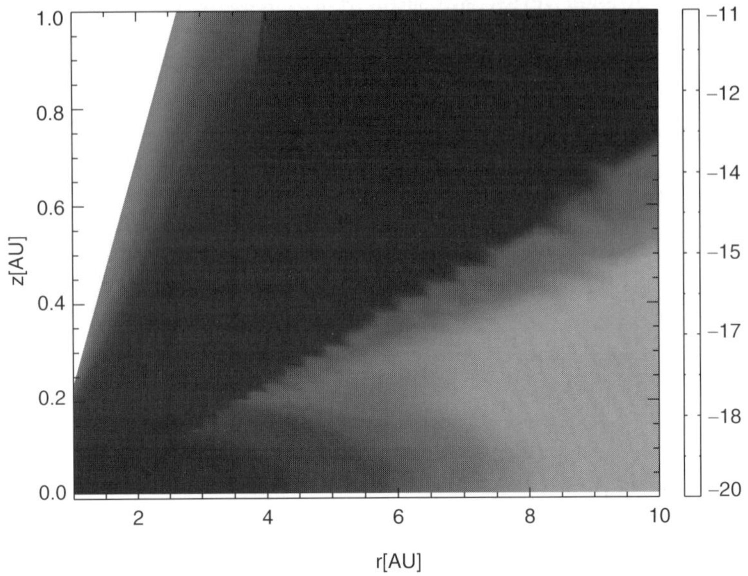

Figure 3.4. The abundance of HCO^+ in the model. Contours show the log abundance relative to total hydrogen, varying with radius and height in astronomical units.

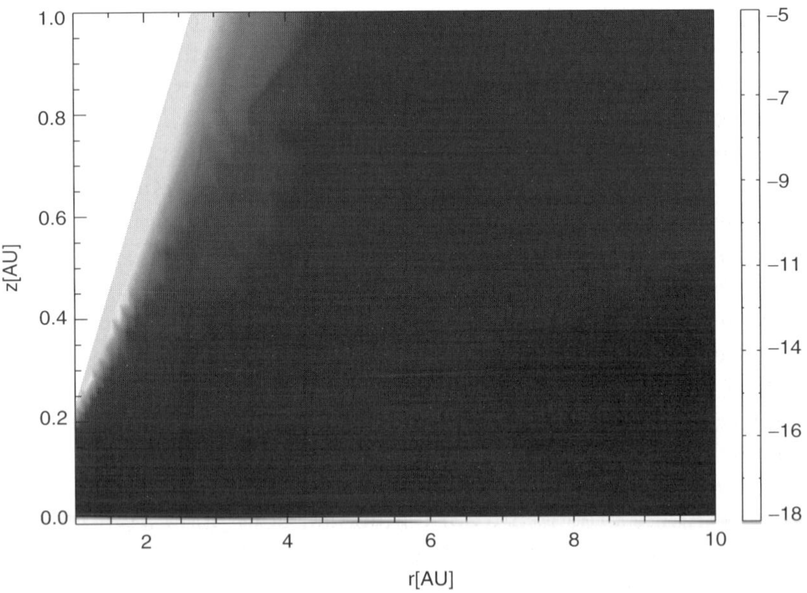

Figure 3.5. As Figure 1.4, for C_2H.

physical description (which provides the temperature and density) and the inclusion of gas-grain interaction is so important.

5 PRIMITIVE SOLAR SYSTEM MATERIAL

We now discuss the implications of current disk chemistry models for understanding the *volatile* organic composition of comets and meteorites (see the Chapters by Molster and Binet). Recent reviews of the composition and possible origin of more refractory meteoritic and cometary materials can be found in Zinner (1998), Benz et al. (2000) and Wooden (2002). Many existing disk chemistry models are based on thermodynamic equilibrium (e.g. Fegley & Prinn 1989; Drouart et al. 1999), here we focus on the more modern models described above, which are based on nonequilibrium chemistry. We also outline how ideas from interstellar chemistry could be implemented into future disk chemistry models to help shed further light on the origin of the isotopic signatures characteristic of primitive material.

5.1 Cometary Organic Molecules

For comets, the key issue is elucidating how much of their organic composition and chemical heterogeneity can be attributed to being either pristine ISM, or partially-processed ISM, or purely nebular, or a complex mix of of the three (Irvine et al. 2000). By comparing results from a model of disk chemistry such as one of those described above with observations of comets, one might hope to discriminate between these possibilities. For example, A'Hearn et al. (1995) showed that there is a variation in the abundance of C_2 and C_3 in the comae of comets, depending on whether the comets are of long- or short-period. Based on dynamical considerations, the short period comets formed at roughly the radius in the nebula at which they reside today (around 40 AU), while long period comets formed closer to the central object, at around 10 AU, but now have orbital radii approaching 50 000 AU. As C_2 and C_3 are coma photoproducts, differentiation between the comet families must be due to chemical differences in the nuclear ices of the two populations. Possible nuclear parents are C_2H_2 (acetylene) and CH_3CCH (methyl acetylene). The variation of carbon-chain species between the comet families is indicative of different chemical processing of the material in the two populations, either in the initial protoplanetary disk (nature), or in the photo-processing history of the comets (nurture). The model is that described in the previous section. Figure 3.6 shows the variation of some solid phase carbonaceous species in the midplane of the disk. The region between 10 and 40 AU is where comets are thought to form, and encouragingly we see that molecules in the ice are abundant there, due to the temperature structure of the midplane. If we are to take the 'nature' perspective described above, then we should see a difference in the carbon-chain abundances between 10 and 40 AU. Specifically, we would like to have more solid state C_3 or C_2 bearing species, molecules which would then photodissociate in the coma. However, Figure 3.6 shows that there is no variation in the abundances of these species in the region of interest. We have checked the abundances of all the species with carbon chains of lengths 2 and 3 atoms in the model, and find the same result for each. There is no variation between 10 and 40 AU. The reason for this is as follows: the gas-grain interaction in the model is described by 'freeze-out', where molecules collide with dust grains and stick, and thermal desorption, which depends on the temperature and the binding energy of the molecule to the dust grain – a material property. The thermal desorption rate involves a term in exp (-binding energy/ temperature), and the temperature structure in the midplane is such that this term effectively switches the solid state abundances between negligible and significant. The binding energies are simply not low enough to allow a variation across the comet-forming region, where the temperature varies by only about 10 K. Hence, using the results from this model, we conclude that any differences in carbon-chain composition between long-and short-period comets arises due to their different processing histories. The short-period comets will have had many more orbits than the long-period comets, and one can speculate that perhaps the ice has been processed into a more complex (stable?) form which precludes the release of suitable parents for C_2 and C_3 into the coma gas.

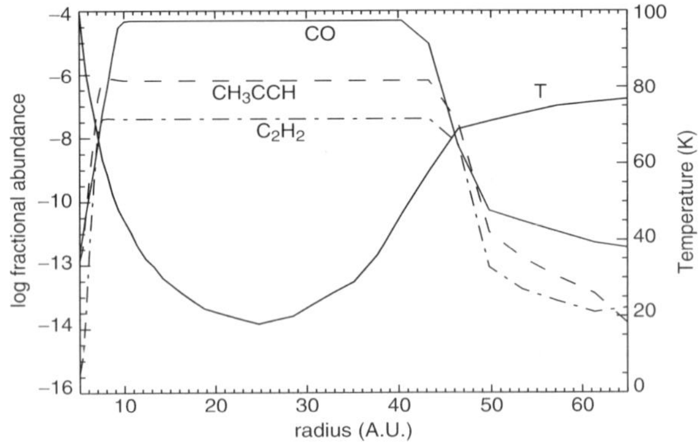

Figure 3.6. The midplane variation of the possible C_2 and C_3 photodissociation parents C_2H_2 and CH_3CCH (in the solid phase) in the model. Also shown is the variation of the midplane temperature in this model.

Also plotted in Figure 3.6 is the temperature distribution in the midplane of the Nomura disk, which shows a marked deviation from all previous studies in that the temperature is seen to increase as one moves outward radially in the midplane. This characteristic is produced because Nomura considered radiation transport in 2D throughout the disk, so heat produced either from viscous dissipation in the inner regions, or from UV radiation from the central protostar gets scattered and redistributed in the disk. The implications of this profile are significant, because, as Figure 3.6 shows, the midplane temperature is 25–35 K in the region where comets are believed to form (10–40 AU), and higher elsewhere. This temperature history will have affected the level of deuterium fractionation in comets. At 25–35 K, it is unlikely that the fractionation ratios will have had time to fully regain their interstellar values before being incorporated into the cometary material (see e.g. Markwick et al. 2001), and one might therefore expect the level of deuterium fractionation in comets to be lower than that observed in dense interstellar clouds.

Finally, the ortho/para ratios of ammonia and water have been determined in several comets (Kawakita et al. 2004). The ratios found are all similar, and are consistent with a spin temperature of 26–34 K, regardless of whether the comet observed is of long or short-period. This strongly implies that the ortho-para ratios in these molecules were equilibrated at the same place in the disk, and hence that long and short period comets formed in a similar region. Furthermore, the spin temperatures derived are consistent with the temperature of the comet formation zone in the model of Markwick & Charnley (2004).

Since the writing of this chapter, it has been brought to our attention that the calculation of the temperature profile discussed above is incorrect at radii greater than 30 AU (Nomura, private communication). The corrected calculation shows no temperature increase like that in Figure 3.6 - instead, the temperature levels out and remains constant at 15K, in qualitative agreement with other models. Inside 30AU, the temperature is as in the Figure. This does not greatly affect the discussion in section 3.5.1 however, because the solid-state abundances of carbon chain species will still be constant over the comet forming region of the disk, and the level of deuterium fractionation in comets will still provide a good test of the temperature profile in the Solar nebula.

5.2 Meteoritic Material

Carbonaceous meteoritic material represents a highly processed sample of interstellar matter. The challenge for astrochemical modelling is to determine which, if any, meteoritic organics are of true interstellar origin (Butterworth et al. 2004), and to identify possible interstellar precursors and reaction pathways leading to those molecules which are the result of parent-body processing, such as aqueous alteration.

The organic inventories of primitive meteorites generally display isotopic anomalies, including large enrichments in deuterium and ^{15}N (e.g. Messenger 2002; O'D Alexander et al. 1998); this has been taken as evidence of their retention of an interstellar heritage. Alternatively, the isotopic fractionation processes which are expected to occur in the low-temperature interstellar medium could occur in the nebular disk, prior to incorporation of the molecules into comets and asteroids. We now briefly review the potentially important processes.

5.2.1 Organics in Carbonaceous Chondrites

Organic compounds identified in carbonaceous C1 and C2 chondrites include amines and amides; alcohols, aldehydes, and ketones; aliphatic and aromatic hydrocarbons; sulfonic and phosphonic acids; amino, hydroxy-carboxylic, and carboxylic acids; purines and pyrimidines; and kerogen-type material (Sephton 2002). For a given number of carbon atoms there is complete structural diversity, all stable isomeric forms are present, and branched-chain isomers are the most abundant. For example, an extensive range of amino acids is found in the Murchison CM chondrite. Strecker-cyanohydrin synthesis during aqueous alteration is a viable source of meteoritic α-amino and α-hydroxy acids (Cronin & Chang 1993; Botta & Bada 2002). Using simpler interstellar aldehyde or ketone precursors, Strecker-cyanohydrin synthesis could account for the structural and isotopic characteristics of both amino and hydroxy acids, whereas Michael addition, with cyanoacetylene or vinyl cyanide as the precursors (Ehrenfreund et al. 2001), could produce other amino acids. The enhanced values of $\delta^{13}C$, δD and $\delta^{15}N$ found in meteoritic amino acids (Cronin 1989) support the view that they were derived from precursors formed in a low temperature environment.

Current theories of interstellar organic chemistry, mediated by gas-grain interactions, can provide all the molecular precursors required to start parent body chemistry, and may provide mechanisms for producing many more complex molecules (e.g. higher amino acids), found in meteorites but not yet detected in molecular clouds (Charnley 2001). Here several of the larger organic species form on or in icy grain mantles (e.g. simple alcohols, aldehydes and sugars), with many more (e.g. ketones, amino acids) probably formed through ion-molecule reactions in warm gas where the products of grain-surface chemistry have recently been deposited (Charnley et al. 1995; Blagojevic et al. 2003). Such an 'interstellar' organic chemistry occurring in a dynamically evolving protoplanetary disk (i.e. including transport of gas and dust) might produce many of the known interstellar organics in abundance. Such calculations have not yet been attempted.

5.2.2 Deuterium Fractionation

Ion-molecule chemistry at low temperatures, where zero-point energy differences become relevant, leads to enhanced D/H ratios in interstellar molecules. The reaction

$$H_3^+ + HD \longrightarrow H_2D^+ + H_2$$

can proceed efficiently (e.g. Millar et al. 2000) and initiates the deuterium chemistry which produces enhanced D/H ratios in interstellar molecules of between 10–10,000 times the cosmic value (D/H $\sim 3 \times 10^{-5}$). The gas phase atomic D/H ratio is also enhanced (~ 0.01) and so D and H atom reactions on cold grains will lead to highly fractionated molecules (Tielens 1983).

When significant molecular depletion onto dust grains occurs, although the gas phase abundances of heavy molecules such as CO are lowered overall, the abundances of the remaining D atoms and molecular ions can increase substantially (see Millar et al. 2000). At higher gas densities, D_3^+ and HD_2^+ ions become important in the depleted gas (Roberts et al. 2003) and deuterium fractionation by gas and grain-surface processes is then considerably more efficient, leading to enormous D/H ratios $\sim 0.1 - 1$). Depletion effects are believed to be responsible for the high D/H ratios and multiple deuteration recently observed in many interstellar molecules: ammonia, hydrogen sulphide, formaldeyde and methanol (van der Tak et al. 2002; Vastel et al. 2003; Loinard et al. 2000; Parise et al. 2002).

The high D/H ratios attainable by interstellar processes have led to the identification of the enhanced D/H ratios found in meteoritic material as the signature of it originating from an interstellar reservoir. Aikawa & Herbst (1999b) have shown that deuterium chemistry in a protoplantary disk could also provide this reservoir. They found that, under appropriate ionization and temperature conditions, D/H ratios similar to those obtainable from 'normal' dense cloud chemistry could be produced ($\sim 0.001 - 0.01$), with a maximum value of ~ 0.1 in some molecules. An important issue is the degree to which initial D/H ratios, produced either in interstellar clouds or in the nebula, would be reduced by parent-body processing. If this reduction is large then the D/H ratio in the initial reservoir must be correspondingly higher, perhaps closer to the highest ISM values. Some interplanetary dust particles (IDPs) show large, almost interstellar, D/H ratios ($\sim 0.001 - 0.01$, Messenger 2000, 2002). If substantially higher D/H ratios (> 0.1) are eventually discovered in primitive material, these would also put pressure on nebular models. In either case, it will eventually be necessary to re-examine deuterium chemistry in disks, employing a fractionation chemistry similar to that of Roberts et al. (2003).

5.2.3 ^{15}*N Enhancements*

Meteoriticists have generally appealed to ion-molecule reactions in cold gas as the origin of the measured ^{15}N enhancements and anomalies (e.g. Messenger 2000; Aléon et al. 2003; Sephton et al. 2003; Aléon & Robert 2004). In principle there are potentially several ways of accomplishing ^{15}N fractionation in interstellar molecules, Terzieva & Herbst (2000) identified and evaluated the key ion-molecule reactions. They found that atom-molecular ion reactions were most important in enriching molecules in heavy nitrogen:

$$^{15}N + ^{14}N_2H^+ \rightleftharpoons ^{14}N + ^{15}N^{14}NH^+$$
$$^{15}N + ^{14}N_2H^+ \rightleftharpoons ^{14}N + ^{14}N^{15}NH^+$$

Charnley & Rodgers (2002) considered a depletion-enhanced fractionation scenario where all C and O-bearing species were depleted, except for N_2 and its isotopomers. They found ^{15}N enhancements in ammonia ices as high as 80%, much higher than found by Terzieva & Herbst ($\sim$ 25%), and in the region needed by the meteoritic data. Charnley & Rodgers pointed out that this fractionation could occur either in a prestellar core, or in the cold regions of a protoplanetary disk. Recent experiments by Geppert et al. (2004) found, unexpectedly, that there is a destructive channel to electron recombination of N_2H^+. Updated calculations by Rodgers & Charnley (2004) show that inclusion of this process leads to a substantial reduction in the maximum ^{15}N enhancements achievable, but that the high $^{15}N/^{14}N$ ratios can be recovered if the chemistry takes place under conditions where N_2H^+ recombination of negatively-charged dust grains dominates its destruction. This result suggests the high-density regions of a disk may be the location of the ^{15}N fractionation.

A necessary feature of this theory is significant chemical destruction of N_2 at low temperatures, and implies that the high ^{15}N fractionation seen in primitive solar system material and the depletion of N_2 in cometary ices are related. Intriguingly, Arpigny et al. (2003) have found that the $^{15}N/^{14}N$ in CN ratio in several comets is significantly enhanced over that of its main photochemical parent, HCN. Arpigny et al. conclude that the additional source of ^{15}CN is photolysis of isotopically-enhanced organic macromolecules possessing $^{15}N/^{14}N$ ratios comparable, or in excess of, those found in IDPs and predicted by theory. It would therefore be of great interest to examine the spatial and temporal evolution of the specific conditions needed for maximal ^{15}N fractionation (cold gas, high density, selective depletion of CO and N_2, N_2 destruction) within the context of protoplanetary disk models of the comet-formation region.

5.2.4 ^{13}C Fractionation

The carbon in carbonaceous meteorites is distributed amongst aromatic and aliphatic hydrocarbons, various organic molecules (e.g. amino and hydroxy acids), and complex macromolecular networks. Whilst the ^{13}C fractionation found in meteoritic material is variable and enhancements, when found, are generally lower than found for D and ^{15}N (O'D Alexander et al. 1998; Sephton et al. 1998; Messenger 2002), recent analyses of $^{12}C/^{13}C$ ratios have identified several interesting trends.

Pizzarello et al. (2003) measured $^{13}C/^{12}C$ ratios in amino acids from Murchison and found a general ^{13}C enrichment. They also found that $\delta^{13}C$ was different between two subgroups of the α-amino acids and that it decreased with increasing carbon number. This $\delta^{13}C$ trend was reversed for the non-α-amino acids studied.

A key question concerns the $^{12}C/^{13}C$ ratios attainable in the larger hydrocarbon molecules which were probably the precursors of complex macromolecular networks (see Chapter by Cataldo). Polycyclic aromatic hydrocarbons (PAHs) are widespread in meteorites and IDPs (Clemett et al. 1998). Many of the PAHs found in meteorites probably originated in the cool envelopes of carbon AGB stars and planetary nebulae, where they are intermediates in carbon stardust formation (Frenklach and Feigelson 1989). Gilmour & Pillinger (1994) determined $^{12}C/^{13}C$ ratios for individual PAHs : phenanthrene, pyrene, fluoroanthene, chrysene, benzo[g,h,i]fluoroanthene, benzo[e]pyrene and benzo[j]fluoroanthene. They found that the larger (molecular weight) PAHs tended to show a greater enrichment in ^{12}C relative to the smaller ones, and concluded the data was consistent with PAH growth by an acetylenic mechanism such as the Frenklach-Feigelson one. Naraoka et al. (2000) analysed several meteoritic PAHs, showed that the condensed PAHs were more depleted in ^{13}C, and suggested that there may be two kinetically-controlled pathways to meteoritic PAHs: the 'pyrene series' and the 'fluoranthene series'. Messenger et al. (1998) also found evidence for ^{13}C-depletion in some PAHs. Sephton et al. (1998) found that macromolecular aromatic structures were more enriched in ^{13}C and there may a 'genetic' connection between these aromatic structures and free PAHs, perhaps indicative of radiation-induced interstellar chemistry (Sephton & Gilmour 2000).

Interstellar ^{13}C fractionation is controlled by the gas temperature and CO photochemistry. Ion-molecule reactions can proceed at low temperatures (Langer et al. 1984) through the the exchange reaction

$$^{13}C^{+} + ^{12}CO \longrightarrow ^{13}CO + ^{12}C^{+}$$

At around 10K, this fractionation chemistry acts to lock most of the available ^{13}C nuclei in CO, yielding a $^{13}CO/^{12}CO$ ratio that is enhanced relative to the other gas phase species (e.g. C_2H_2 and CH_4), which are depleted in ^{13}C with $^{12}C/^{13}C \approx 100 - 350$ possible. Of course, at higher gas temperatures the amount of free ^{13}C available in the gas will increase and the $^{13}C/^{12}C$ ratios of organics formed there will be larger, at the expense of CO. Another important process, largely independent of temperature, is isotope-selective photodestruction (ISP). As ^{12}CO is destroyed by line photons in dark clouds, it can self-shield whereas ^{13}CO does not. This leads to the release of more ^{13}C atoms into the gas and can affect the $^{12}C/^{13}C$ fractionation in other species. Hence, in protoplanetary disks, radial and vertical mixing and of material between regions of varying UV flux and temperature gradient should lead to a strongly-varying $^{12}C/^{13}C$ fractionation chemistry, based on the above processes.

To form more volatile organic material relatively enriched in ^{13}C at low temperatures, it must be derived from CO. High $^{13}CO/^{12}CO$ ratios would probably be preferred in the cooler midplane region of the outer disk. The ^{13}C fractionation in amino acids found by Pizzarello et al. (2003) could have originated in molecular precursors formed on icy dust particles, as suggested for interstellar grains (Charnley et al. 2004), followed by gas phase chemical reactions (Charnley 2001, Blagojevic et al. 2003) after transport into warmer regions of the disk.

For PAH fractionation, McEwan et al. (1999) recently showed that benzene could be formed from simple hydrocarbon ions in cold gas and so the (high) $^{12}C/^{13}C$ ratio expected in these simple precursors could be passed on to larger aromatics. Alternatively, in the upper layers of flared disks, more free ^{13}C would be available to the gas due to ISP; evidence for this effect has recently been found in the DM Tau disk by Dartois et al. (2003). Morgan et al. (1991) have extended models of PAH formation down to much lower pressures than found in circumstellar envelopes to study PAH production in the protosolar nebula. They computed PAH production in the pressure range $10^{-8} - 10^{-4}$ atm and found that efficient PAH formation was especially favourable at lower pressures. Morgan et al. 1991 et al. pointed out that the optimal sites for PAH growth in the nebula were in fact above and below the midplane. Thus, if PAHs can form and grow in a dynamic nebular environment, one might expect that they could have significantly different $^{12}C/^{13}C$ ratios within their distribution, reflecting the competing processes which determine ^{13}C fractionation. However, any purely nebular PAH chemistry should also provide a means for fractionating PAHs in deuterium, at least as efficiently as those proposed for molecular clouds (e.g. Sandford et al. 2001).

5.2.5 *Oxygen Isotope Fractionation*

Finally, Clayton (2002) has proposed that isotope-selective photodestruction of $^{12}C^{18}O$ and $^{12}C^{17}O$ in the *inner* disk played an important role in setting the oxygen isotope ratios in the early protosolar nebula (see Wiens et al. 1999; Thiemens 1999). The higher abundances of ^{17}O and ^{18}O available to the chemistry in this region, followed by ejection farther out into the nebula by an X-wind (O'D Alexander 2001), could then produce the $^{16}O/^{18}O$ and $^{16}O/^{17}O$ ratios measured in meteorites. However, preliminary calculations of the fractionation chemistry at the X-point (~ 0.1 AU), indicate that the desired oxygen isotope ratios may be difficult to achieve there (Lyons & Young 2003). In a complementary study, Young & Lyons (2003) demonstrated that the photochemically-active upper layers of the *outer* disk were in fact more conducive to producing material enriched in ^{17}O and ^{18}O. If correct, this explanation would have other chemical implications (e.g. for $^{15}N/^{14}N$ ratios) and should ultimately be testable by astronomical observations.

6 SUMMARY

The physics and chemistry of low-mass star formation provides the boundary conditions for the chemical evolution of both protostellar and protoplanetary disks (e.g. van Dishoeck & Blake 1998). It is now possible to observe in detail the organic composition of each phase prior to disk formation, a chemical sequence analogous to that which led to the protosolar nebula. The organic chemistry of protoplanetary accretion disks is of fundamental importance for understanding the composition and origin of comets and meteorities. Future theoretical models of disk fractionation chemistry will be important to highlight the differences and similarities between the molecular inventory of the natal cloud, the disk, and that detected in primitive solar system material.

ALMA will allow molecular distributions and the radial and vertical structures (temperature, density, turbulence) of disks to be measured at scales of the order of a few AU (Wootten 2001; Dartois, Dutrey & Guilloteau 2003). ALMA will also allow many large, low-abundance, organics to be detected for the first time. These astronomical observations will be used to constrain theoretical models of the chemistry occurring in each phase. A comprehensive and quantitative understanding of the chemistry of the major organic species currently detectable by remote sensing of protostellar cores and disks will allow us to predict with confidence which larger biomolecules should also be present. A future goal is to develop dynamical-chemical models that will allow the detailed radial and vertical chemical structure of protoplanetary disks to be routinely calculated. These models will represent the state-of-the-art and, in the near future, should allow direct comparisons between theory and ALMA observations of disks.

Acknowledgments

This work was supported by NASA's Origins of Solar System Program and Astrobiology Institute with funds allocated under Cooperative Agreements with the SETI Institute.

References

Adams F.C., Lin D.N.C., 1993, *Transport processes and the evolution of disks*, in Protostars and Planets III, eds. Levy E.H., Lunine J.I., University of Arizona Press. Tuscon. pp 721–748.

A'Hearn M.F., Millis R.L., Schleicher D.G., Osip D.J., Birch P.V., 1995, *The ensemble properties of comets: Results from narrowband photometry of 85 comets, 1976-1992*, Icarus **118**, 223–270.

Aikawa Y., Momose M., Thi W.-F., van Zadelhoff G.-H., Qi C., Blake G.A., van Dishoeck E.F., 2003, *Interferometric Observations of Formaldehyde in the Protoplanetary Disk around LkCa15*, PASJ, in press, astro-ph/0211440.

Aikawa Y., van Zadelhoff G.-H., van Dishoeck E.F., Herbst E., 2002, *Warm molecular layers in protoplanetary disks*, A&A **386**, 622–632.

Aikawa Y., Ohashi N., Inutsuka S., Herbst E., Takakuwa S., 2001, *Molecular evolution in collapsing prestellar cores*, ApJ **552**, 639–653.

Aikawa Y., Herbst E., 2001, *Two-dimensional distributions and column densities of gaseous molecules in protoplanetary disks. II Deuterated species and UV shielding by ambient clouds*, A&A **371**, 1107–1117.

Aikawa Y., Umebayashi T., Nakano T., Miyama S., 1999, *Evolution of Molecular Abundances in Protoplanetary Disks with Accretion Flow*, ApJ **519**, 705–725.

Aikawa Y., Herbst E., 1999a, *Molecular evolution in protoplanetary disks. Two-dimensional distributions and column densities of gaseous molecules*, A&A **351**, 233–246.

Aikawa Y., Herbst E., 1999b, *Deuterium Fractionation in Protoplanetary Disks*, ApJ **526**, 314–326.

Aikawa Y., Miyama S.M., Nakano T., Umebayashi T., 1996, *Evolution of Molecular Abundance in Gaseous Disks around Young Stars: Depletion of CO Molecules*, ApJ **467**, 684–697.

Aléon J., Robert F., 2004, *Interstellar chemistry recorded by nitrogen isotopes in Solar System organic matter*, Icarus **167**, 424–430.

Aléon J., Robert F., Chaussidon M., Marty B., 2003, *Nitrogen isotopic composition of macromolecular organic matter in interplanetary dust particles*, Geochimica et Cosmochimica Acta **67**, 3773–3783.

Alves J.F., Lada C.J., Lada E.A., 2001, *Internal structure of a cold dark molecular cloud inferred from the extinction of background starlight*, Nature **409**, 159–161.

Arpigny C., Jehin E., Manfroid J., Hutesmekers D., Zucconi J.-M., Schulz R., Stuewe J.A., 2003, *Anomalous Nitrogen Isotope Ratio in Comets.*, Science **301**, 1522–1524.

Balbus S.A., Hawley J.F., 2000, *Solar Nebula Magnetohydrodynamics*, Space Sci. Rev. **92**, 39–54.

Bauer I., Finocchi F., Duschl W.J., Gail H.-P., Schloeder J.P., 1997, *Simulation of chemical reactions and dust destruction in protoplanetary accretion disks*, A&A **317**, 273–289.

Baulch D.L., Cobos C.J., Cox R.A., Esser C., Franck P., Just Th., Ker J.A., Pilling M.J., Troe J., Walker R.W., Warnatz J., 1992, Journ. Phys. Chem. Ref. Data **21**, 411.

Bell K.R., Cassen P.M., Klahr H.H., Henning Th., 1997, *The Structure and Appearance of Protostellar Accretion Disks: Limits on Disk Flaring*, ApJ **486**, 372–387.

Benz W., Kallenbach R., Lugmair G.W. (Eds.), 2000, *From Dust to Terrestrial Planets*, 421pp, Kluwer Academic Publishers.

Bergin E.A., Alves J., Huard T., Lada C.J., 2002, *N_2H^+andC^{18}ODepletion in a Cold Dark Cloud*, ApJ **570**, L101–L104.

Blagojevic V., Petrie S., Bohme D.K., 2003, *Gas-phase syntheses for interstellar carboxylic and amino acids*, MNRAS, **339**, L7–L11.

Bockelée-Morvan D., Gautier D., Hersant F., Hure J.-M., Robert F., 2002, *Turbulent radial mixing in the solar nebula as the source of crystalline silicates in comets*, A&A **384**, 1107–1118.

Boss A.P., 1998, *Temperatures in Protoplanetary Disks*, Ann. Rev. Earth Planet. Sci. **26**, 53–80.

Botta O., Bada J., 2002. *Extraterrestrial Organic Compounds in Meteorites*, Surveys in Geophysics **23**, 411–467.

Butterworth A.L., Aballain O., Chappellaz J., Sephton M.A., 2004, *Combined element (H and C) stable isotope ratios of methane in carbonaceous chrondrites*, MNRAS **347**, 807–812.

Cameron A.G.W., 1995, *The first ten million years in the solar nebula*, Meteoritics **30**, 133–161.

Cassen P., Moosman A., 1981, *On the formation of protostellar disks*, Icarus **48**, 353–376.

Charnley S.B., Ehrenfreund P., Millar T.J., Boogert A.C.A., Markwick A.J, Butner H.M., Ruiterkamp R., Rodgers S.D., 2004, *Observational tests for grain chemistry: posterior isotopic labelling*, MNRAS **347**, 157–162.

Charnley S.B., Rodgers S.D., 2002, *The End of Interstellar Chemistry as the Origin of Nitrogen in Comets and Meteorites*, ApJ **569**, L133–L137.

Charnley S.B., 2001, *Interstellar Organic Chemistry*, in The Bridge Between the Big Bang and Biology, Ed. F. Giovannelli, Consiglio Nazionale delle Ricerche President Bureau, (Rome:Italy), Special Volume, 139–149.

Charnley S.B., Kress M.E., Tielens A.G.G.M., Millar T.J., 1995, *Interstellar Alcohols*, ApJ **448**, 232–239.

Charnley S.B., 1995, *The Interstellar Chemistry of Protostellar Disks*, Ap&SS **224**, 441–442.

Charnley S.B., Dyson J.E., Hartquist T.W., Williams D.A., 1988, *Chemical limit cycles for models of a region of low-mass star formation*, MNRAS **235**, 1257–1271.

Chiang E.I., Goldreich P., 1999, *Spectral Energy Distributions of T Tauri Disks: Inclination*, ApJ **519**, 279–284.

Chiang E.I., Goldreich P., 1997, *Spectral Energy Distributions of T Tauri Stars with Passive Circumstellar Disks*, ApJ **490**, 368–376.

Chick K.M., Cassen P., 1997, *Thermal Processing of Interstellar Dust Grains in the Primitive Solar Environment*, ApJ **477**, 398–409.

Clayton R.N., 2002, *Self-shielding in the solar nebula*, Nature **415**, 860–861.

Clemett S.J., Chillier D.F., Gillette S., Zare R.N., Maurette M., Engrand C., Kurat G., 1998, *Observation of indigenous polycyclic aromatic hydrocarbons in 'giant' carbonaceous Antarctic micrometeorites*, Orig. Life Evol. Bios. **28**, 424–448.

Cronin J.R., Chang S., 1993, *Organic matter in meteorites: molecular and isotopic analyses of the Murchison meteorite*, in The Chemistry of Life's Origins, ed. J.M. Greenberg et al. Kluwer Academic Publishers, pp. 209–258.

Cronin J.R., 1989, *Origin of organic compounds in carbonaceous chondrites*, Adv. Space. Res. **9**, 59–64.

Chyba C.F., Thomas P.J., Brookshaw L., Sagan C., 1990, *Cometary Delivery of Organic Molecules to the Early Earth*, Science

Cyr K.E., Sears W.D., Lunine J.I., 1998, *Distribution and Evolution of Water Ice in the Solar Nebula: Implications for Solar System Body Formation*, Icarus **135**, 537–548.

D'Alessio P., Canto J., Calvet N., Lizano S., 1998, *Accretion Disks around Young Objects I. The Detailed Vertical Structure*, ApJ **500**, 411–427.

Dartois E., Dutrey A., Guilloteau S., 2003, *Structure of the DM Tau Outer Disk: Probing the vertical kinetic temperature gradient*, A&A **399**, 773–787.

Desch S.J., Mouschovias T., 2001, *The Magnetic Decoupling Stage of Star Formation*, ApJ **550**, 314–333.

Drouart A., Dubrulle B., Gautier D., Robert F., 1999, *Structure and Transport in the Solar Nebula from Constraints on Deuterium Enrichment and Giant Planets Formation*, Icarus **40**, 129–155.

Dutrey A., Guilloteau S., Guelin M., 2000, *Observations of the Chemistry in Circumstellar Disks*, in Astrochemistry: From Molecular Clouds to Planetary Systems. (Y.C. Minh, E.F. van Dishoeck Eds.), pp. 415–423. Astronomical Society of the Pacific, Sogwipo.

Dutrey A., Guilloteau S., Guelin, M., 1997, *Chemistry of protosolar-like nebulae: The molecular content of the DM Tau and GG Tau disks*, A&A **317**, L55–L58.

Dutrey A., Guilloteau S., Simon M., 1994, *Images of the GG Tauri rotating ring*, A&A **286**, 149–159.

Ehrenfreund P., Irvine W.M., Becker L., Blank J., Brucato J.R. et al., 2002, *Astrophysical and astrochemical insights into the origin of life*, Rep. Prog. Phys. **65**, 1427–1487.

Ehrenfreund P., Glavin D.P., Botta O., Cooper G., Bada J.L., 2001, *Extraterrestrial amino acids in Orguil and Ivuna: tracing the parent body of CI type carbonaceous chondrites* Proc. Natl. Acad. Sci. **98**, 2138–2141.

Ehrenfreund P., Charnley S.B., 2000, *Organic Molecules in the Interstellar Medium, Comets, and Meteorites: A Voyage from Dark Clouds to the Early Earth*, Annu. Rev. Astron. Astrophys. **38**, 427–483.

Elmegreen B.G., 2000, *Star Formation in a Crossing Time*, ApJ **530**, 277–281.

Evans N., 1999, *Physical Conditions in Regions of Star Formation*, Annu. Rev. Astron. Astrophys. **37**, 311–362.

Fegley B., 2000, *Kinetics of Gas-Grain Reactions in the Solar Nebula*, Space Sci. Rev. **92**, 177–200.

Fegley B., 1999, *Chemical and Physical Processing of Presolar Materials in the Solar Nebula and the Implications for Preservation of Presolar Materials in Comets*, Space Sci. Rev. **90**, 239–252.

Fegley B., Prinn R.G., 1989, *Solar nebula chemistry – Implications for volatiles in the solar system*, in The formation and evolution of planetary systems (eds. H.A. Weaver, L. Danly), pp. 171–205. Cambridge University Press.

Finocchi F., Gail H.-P, 1997, *Chemical reactions in protoplanetary accretion disks. III. The role of ionisation processes*, A&A **327**, 825–844.

Frenklach M., Feigelson E.D., 1989, *Formation of polycyclic aromatic hydrocarbons in circumstellar envelopes*, ApJ **341**, 372–384.

Gail H.-P., 2001, *Radial mixing in protoplanetary accretion disks. I. Stationary disc models with annealing and carbon combustion*, A&A **378**, 192–213.

Gail H.-P., 2002, *Radial mixing in protoplanetary accretion disks. III. Carbon dust oxidation and abundance of hydrocarbons in comets*, A&A **390**, 253–265.

Geppert W. et al., 2004, *Branching ratios and cross sections of astrophysically important dissociative recombination reactions*, in Pfalzner S. et al. eds., Proc. 4th Cologne-Bonn-Zermatt Symposium, Springer-Verlag, Berlin, in press.

Gilmour I., Pillinger C.T., 1994, *Isotopic compositions of individual polycyclic aromatic hydrocarbons from the Murchinson meteorite*, MNRAS **269**, 235–240.

Glassgold A.E., Najita J., Igea J., 1997, *X-Ray Ionization of Protoplanetary Disks*, ApJ **480**, 344–350.

Goldsmith P.F., Langer W.D., Velusamy T., 1999, *Detection of Methanol in a Class 0 Protostellar Disk*, ApJ **519**, L173–L176.

Goldsmith P.F., Langer W.D., 1978, *Molecular cooling and thermal balance of dense interstellar clouds*, ApJ **222**, 881–895.

Harker D.E., Desch S.J., 2002, *Annealing of Silicate Dust by Nebular Shocks at 10 AU*, ApJ **565**, L109–L112.

Hartmann L., Ballesteros-Paredes J., Bergin E.A., 2001, *Rapid Formation of Molecular Clouds and Stars in the Solar Neighborhood*, ApJ **562**, 852–868.

Hayashi C., 1981, *Structure of the solar nebula, growth and decay of magnetic fields and effects of magnetic and turbulent viscosities on the nebula*, Prog. Theor. Phys. Suppl. **70**, 35–53.

Hersant F., Gautier D., Hure, J-M., 2001, *A Two-dimensional Model for the Primordial Nebula Constrained by D/H Measurements in the Solar System: Implications for the Formation of Giant Planets*, ApJ **554**, 391–407.

Hollenbach D., Salpeter E.E., 1971, *Surface Recombination of Hydrogen Molecules*, ApJ **163**, 155–164.

Igea J., Glassgold A.E., 1999, *X-Ray Ionization of the Disks of Young Stellar Objects*, ApJ **518**, 848–858.

Ilgner M., Henning Th., Markwick A.J., Millar T.J., 2003, *The influence of transport processes on the chemical evolution in steady accretion disk flows*, A&A, in press.

Irvine W.M., Crovisier J., Fegley B., Mumma M.J., 2000, *Comets: A Link Between Interstellar and Nebular Chemistry*, in Protostars and Planets IV (eds. Mannings V., Boss A.P., Russell S.S.), pp. 1159–1200. University of Arizona Press, Tucson.

Kastner J.H., Zuckerman B., Weintraub D.A., Forveille T., 1997, *X-ray and molecular emission from the nearest region of recent star formation*, Science **277**, 67–71.

Kawakita H., Watanabe J.-I., Furusho R., Fuse T., 2004, *Ortho-to-para ratios of cometary water and ammonia*, Formation of Cometary Material, 25th Meeting of the IAU, Joint Discussion 14, *Highlights of Astronomy*, in press.

Kessler J.E., Qi C., Blake G.A., 2002, *Millimeter observations of HDO and DCN in the circumstellar disks of protostars LkCa 15, MWC 480 and HD 163296*, in Chemistry as a Diagnostic of Star Formation, ed. C.L. Curry, M.I. Fich (NRC press), in press.

Lada C.J., Shu F.H., 1990, *The formation of sunlike stars*, Science **248**, 564–572.

Langer W.D., Graedel T.E., Frerking M.A., Armentrout P.B., 1984, *Carbon and oxygen isotope fractionation in dense interstellar clouds*, ApJ **277**, 581–604.

Larson R.B., 2003, *The Physics of Star Formation*, Rep. Prog. Phys. **66**, 1651–1701.

Larson R.B., 1969, *Numerical calculations of the dynamics of collapsing protostar*, MNRAS **145**, 271–295.

Lee T., Papanastassiou D.A., Wasserburg G.J., 1977, *Aluminium-26 in the early solar system – Fossil or fuel*, ApJ **211**, L107–L110.

Le Teuff Y.-H., Millar T.J., Markwick A.J., 2000, *The UMIST database for astrochemistry 1999*, A&AS **146**, 157–168.

Loinard L., Castets A., Ceccarelli C., Tielens A.G.G.M., Faur, A., Caux, E., Duvert G., 2000, *The enormous abundance of* D_2CO *in IRAS 16293–2422*, A&A **359**, 1169–1174.

Lunine J.I., 1997, *Physics and Chemistry of the Solar Nebula*, Orig. Life Evol. Biosph. **27**, 205–224.

Lunine J.I., Engel S., Rizk B., Horanyi M., 1991, *Sublimation and reformation of icy grains in the primitive solar nebula*, Icarus **94**, 333–344.

Lynden-Bell D., Pringle J., 1974, *The evolution of viscous discs and the origin of the nebula variables*, MNRAS **168**, 603–637.

Lyons J.R., Young E.D., 2003, *Towards an evaluation of self-shielding at the X-point as the origin of the oxygen isotope anomaly in CAIs*, Lunar & Planetary Science XXXIV, 1981.

Mac Low M.-M., Klessen R.S., 2003, *Control of star formation by supersonic turbulence*, Rev. Mod. Phys., in press.

Mahoney W.A., Ling J.C., Wheaton W.A., Jacobson A.S., 1984, *HEAO 3 discovery of Al-26 in the interstellar medium*, ApJ **286**, 578–585.

Markwick A.J., Charnley S.B., 2004, *Disk Chemistry and Cometary Composition*, Formation of Cometary Material, 25th Meeting of the IAU, Joint Discussion 14, *Highlights of Astronomy*, in press.

Markwick A.J., Ilgner M., Millar T.J., Henning Th., 2002, *Molecular distributions in the inner regions of protostellar disks*, A&A **385**, 632–646.

Markwick A.J., Charnley S.B., Millar T.J., 2001, *Deuterium fractionation along the TMC-1 ridge*, A&A **376**, 1054–1063.

McEwan M.J., Scott G.B.I., Adams N.G., Babcock L.M., Terzieva R, Herbst E., 1999, ApJ, *New H and* H_2 *reactions with small hydrocarbon ions and their roles in benzene synthesis in dense interstellar clouds*, ApJ, **513**, 287–293.

McKee C.F., Zweibel E.G., Goodman A.A., Heiles C., 1993, *Magnetic Fields in Star Forming Regions – Theory*, in Protostars and Planets III, eds. Levy E.H., Lunine J.I., University of Arizona Press. Tuscon. p 327.

Messenger S., 2002, *Deuterium enrichments in interplanetary dust*, Planetary & Space Science **50**, 1221–1225.

Messenger S., 2000, *Identification of molecular-cloud material in interplanetary dust particles*, Nature **404**, 968–971.

Messenger S., Amari S., Gao X., Walker R.M., Clemett S.J., Chillier X.D.F., Zare R.N., Lewis R.S., 1998, *Indigenous polycyclic aromatic hydrocarbons in circumstellar graphite grains from primitive meteorites*, ApJ **502**, 284–295.

Millar T.J., Nomura H., Markwick A.J., 2003, *Two-dimensional models of protoplanetary disk chemistry*, Ap&SS **285**, 761–768.

Millar T.J, Roberts H., Markwick A.J., Charnley S.B., 2000, *The role of* H_2D^+ *in the deuteration of interstellar molecules*, Phil. Trans. R. Soc. Lond. A **358**, 2535–2547.

Millar T.J., Farqhuar P.R.A., Willacy K., 1997, *The UMIST Database for Astrochemistry 1995*, A&AS **121**, 139–185.

Mitchell G.F., 1984, *Effects of shocks on the molecular composition of a dense interstellar cloud*, ApJS **54**, 81–101.

Morfill G.E., Volk H.J., 1984, *Transport of dust and vapor and chemical fractionation in the early protosolar cloud*, ApJ **287**, 371–395.

Morgan W.A., Jr., Feigelson E.D., Wang H., Frenklach M., 1991, *A new mechanism for the formation of meteoritic kerogen like material*, Science, **252**, 109–112.

Najita J., Bergin E.A., Ullom J.N., 2001, *X-Ray Desorption of Molecules from Grains in Protoplanetary Disks*, ApJ **561**, 880–889.

Naraoka H., Shimoyama A., Harada K., 2000, *Isotopic evidence from an Antarctic carbonaceous chrondrite for two reaction pathways of extraterrestrial PAH formation*, Earth Planet. Sci. Lett. **184**, 1–7.

Neufeld D., Hollenbach D.J., 1994, *Dense molecular shocks and accretion onto protostellar disks*, ApJ **428**, 170–185.

Nomura H., 2002, *Structure and Instabilities of an Irradiated Viscous Protoplanetary Disk*, ApJ **567**, 587–595.

O'D Alexander C.M., Boss A.P., Carlson R.W., 2001, *The Early Evolution of the Inner Solar System: A Meteoritic Perspective*, Science **293**, 64–68.

O'D Alexander C.M., Russell S.S., Arden W., Ash R.D., Grady M., Pillinger C.T., 1998, *The origin of chrondritic macromolecular organic matter: a carbon and nitrogen isotope study*, Meteoritics Planet Sci., **33**, 603.

Oró J., 1961, *Comets and the formation of biochemical compounds on the primitive Earth*, Nature **190**, 389–390.

Parise B., Ceccarelli C., Tielens A.G.G.M., Herbst E., Lefloch B., Caux E., Castets A., Mukhopadhyay I., Pagani L., Loinard L., 2002, *Detection of doubly-deuterated methanol in the solar-type protostar IRAS 16293–2422*, A&A **393**, L49–L53.

Penston M.V., 1969, *Dynamics of self-gravitating gaseous spheres III. Analytical results in the free-fall of isothermal cases.*, MNRAS **144**, 425–448.

Pizzarello S., Huang Y., Fuller M., 2003, *The carbon isotopic distribution of Murchinson amino acids*, Lunar & Planetary Science XXXIV, 1036.

Prasad S.S., Huntress W.T., 1980, *A model for gas phase chemistry in interstellar clouds. I &ndash-The basic model, library of chemical reactions, and chemistry among C, N, and O compounds*, ApJS **43**, 1–35.

Qi C., 2001, PhD Thesis, California Institute of Science and Technology, Pasadena, California.

Roberts H., Herbst E., Millar T.J., 2003, *Enhanced Deuterium Fractionation in Dense Interstellar Cores Resulting from Multiply Deuterated*H_3^+, ApJ **591**, L41–44.

Rodgers S.D., Charnley S.B., 2004, *Interstellar diazenylium recombination and nitrogen isotopic fractionation*, MNRAS., submitted.

Rodgers S.D., Charnley S.B., 2003, *Chemical Evolution in Protostellar Envelopes – Cocoon Chemistry*, ApJ **585**, 355–371.

Ruden S.P., Pollack J.B., 1991, *The dynamical evolution of the protosolar nebula*, ApJ **375**, 740–760.

Sandford S.A., Bernstein M.P., Dworkin J.P., 2001, *Assessment of the interstellar processes leading to deuterium enrichment in meteoritic organics*, Meteoritics Planet Sci., **36**, 1117–1133.

Sephton M.A., Verchovsky A.B., Bland P., Gilmour I., Grady M.M., Wright I.P., 2003, *Investigating the variations in carbon and nitrogen isotopes in carbonaceous chrondrites*, Geochimica et Cosmochimica Acta **67**, 2093–2108.

Sephton M.A., 2002, *Organic compounds in carbonaceous meteorites*, Nat. Prod. Rep. **19**, 292–311.

Sephton M.A., Gilmour I., 2000, *Aromatic moieties in meteorites: relics of interstellar grain processes?*, ApJ **540**, 588–591.

Sephton M.A., Pillinger C.T., Gilmour I., 1998, δ^{13}*C of free and macromolecular aromatic structures in the Murchison meteorite*, Geochimica et Cosmochimica Acta **62**, 1821–1828.

Shu F.H., Lizano S., Adams F.C., 1987, *Star formation in molecular clouds – Observation and theory*, Ann. Rev. Astron. Astrophys. **25**, 23–81.

Shu F.H., 1977, *Self-similar collapse of isothermal spheres and star formation*, ApJ **214**, 488–497.

Simonelli D.P., Pollack J.B., McKay C.P., 1997, *Radiative Heating of Interstellar Grains Falling toward the Solar Nebula: 1-D Diffusion Calculations*, Icarus **125**, 261–280.

Stevenson D.J., 1990, *Chemical heterogeneity and imperfect mixing in the solar nebula*, ApJ **348**, 730–737.

Terebey S., Shu F.H., Cassen P., 1984, *The collapse of the cores of slowly rotating isothermal clouds*, ApJ **286**, 529–551.

Terzieva R., Herbst E., 2000, *The possibility of nitrogen isotopic fractionation in interstellar clouds*, MNRAS **317**, 563–568.

Terzieva R., Herbst E., 1998, *The Sensitivity of Gas Phase Chemical Models of Interstellar Clouds to C and O Elemental Abundances and to a New Formation Mechanism for Ammonia*, ApJ **501**, 207-.

Thiemens M.H., 1999, *Mass-Independent Isotope Effects in Planetary Atmospheres and the Early Solar System*, Science **283**, 341–345.

Tielens A.G.G.M., 1983, *Surface chemistry of deuterated molecules*, A&A **119**, 177–184.

Umebayashi T., Nakano T., 1981, *Fluxes of Energetic Particles and the Ionization Rate in Very Dense Interstellar Clouds*, PASJ **33**, 617–636.

van der Tak F.F.S., Schilke P., Müller, H.S.P., Lis D.C., Phillips T.G., Gerin M., Roueff E., 2002, *Triply deuterated ammonia in NGC 1333*, A&A **388**, L53–L56.

van Dishoeck E.F., Thi W.-F., van Zadelhoff G.-H., 2003, *Detection of* DCO^+ *in a circumstellar disk*, A&A **400**, L1–L4.

van Dishoeck E.F., Blake G.A., 1998, *Chemical Evolution of Star-Forming Regions*, Ann. Rev. Astron. Astrophys. **36**, 317–368.

van Zadelhoff G.-J., Aikawa Y., Hogerheijde M.R., van Dishoeck E.F., 2003, *Axi-symmetric models of ultraviolet radiative transfer with applications to circumstellar disk chemistry*, A&A **397**, 789–802.

van Zadelhoff G.-J., van Dishoeck E.F., Thi W.-F., Blake G.A., 2001, *Submillimeter lines from circumstellar disks around pre-main sequence stars*, A&A **377**, 566–580.

Vastel C., Phillips T.G., Ceccarelli C., Pearson J., 2003, *First Detection of Doubly Deuterated Hydrogen Sulfide*, ApJ **593**, L97–100.

Wehrstedt M., Gail H.-P., 2002, *Radial mixing in protoplanetary accretion disks. II. Time dependent disk models with annealing and carbon combustion*, A&A **385**, 181–204.

Westley M.S., Baragiola R.A., Johnson R.E., Baratta G.A., 1995, *Photodesorption from Low-Temperature Water Ice in Interstellar and Circumsolar Grains*, Nature **373**, 405.

Wiens R.C., Huss G.R., Burnett D.S., 1999, *The solar oxygen-isotopic composition: Predictions and implications for solar nebula processes*, Meteoritics Planet Sci., **34**, 99–107.

Willacy K., Langer W.D., 2000, *The Importance of Photoprocessing in Protoplanetary Disks*, ApJ **544**, 903–920.

Willacy K., Klahr H.H., Millar, T.J., Henning, Th., 1998, *Gas and grain chemistry in a protoplanetary disk*, A&A **338**, 995–1005.

Wooden D.H., 2002, *Comet Grains: Their IR Emission and Their Relation to ISM Grains*, Earth, Moon & Planets **89**, 247–287.

Wootten H.A., ed., 2001, *Science with the Atacama Large Millimeter Array*, Astron. Soc. Pac. **365**, San Francisco.

Young E.D., Lyons J.R, 2003, *CO self shielding in the outer solar nebula: an astronomical explanation for the oxygen isotope slope-1 line*, Lunar & Planetary Science XXXIV, 1923.

Zhou S., 1992, *In search of evidence for protostellar collapse – A systematic study of line formation in low-mass dense cores*, ApJ **394**, 204–216.

Zinner E., 1998, *Stellar nucleosynthesis and the isotopic composition of presolar grains from primitive meteorites*, Annu. Rev. Earth Planet. Sci. **26**, 147–188.

Zuckerman B., Evans N.J., 1974, *Models of massive molecular clouds*, ApJ, **192**, L149–L152.

Chapter 4

PLANET FORMATION

Towards Estimating Galactic Habitability

G. Wuchterl
Max-Planck-Institute for extraterrestrial Physics
Postfach 1312, D–85741 Garching, Germany
wuchterl@mpe.mpg.de

Abstract: The present census of the galactic planetary population shows that planets exist outside the Solar System. But the empirical evidence is still insufficient to estimate the overall preconditions for the diversity and abundance of habitats that planets may provide in our galaxy. Detecting planets is still centered around giant planets in orbits with periods comparable and less than Jupiter's around nearby solar like stars. For a global picture, a general understanding of the planet formation process is needed to extrapolate what is known from the Solar System and about extrasolar planets to other stellar types and environments in the galaxy. A predictive theory of planet formation may contribute to this task.

1 INTRODUCTION: PLANET FORMATION AND STAR FORMATION

Planet formation is closely interrelated with star formation and early stellar evolution (e.g. Mannings et al. 2000). The star formation process leads to the formation of a central body, incorporating most of the mass, and of a circumstellar disk, which retains most of the initial angular momentum and will be the environment where planets will form. The circumstellar disks separate mass and angular momentum and allow the gas to accumulate in a central condensation—the young star. The gas density in the disk is enhanced by a factor of $\sim 10^{10}$ relative to the densest pre-stellar cloud, and reaches typically 10^{12} hydrogen molecules per cm^{-3}

P. Ehrenfreund et al. (eds.), Astrobiology: Future Perspectives, 67–96.

from an initial 10^4. Dust particles, which represent about 2% in mass, naturally grow by accretion in this denser environment from submicron sizes to much larger sizes. Most of the present theoretical work on planet formation relies on the planetesimal hypothesis which assumes that the dusty disks produce km-sized, mutually gravitationally interacting bodies called planetesimals, made of elements that are condensible under nebula conditions and constitute the planetary building blocks. Such planetesimals grow by mutual collisions up to terrestrial planet masses and beyond, if sufficient material is available. If sufficiently massive large planetary embryos form before the nebula or remnants of it are dissipated, they attract and permanently bind nebula gas.

Models of the interiors and evolution of giant planets in our Solar System point to a bulk enrichment of heavy elements by more than a factor of two above solar composition and imply heavy element cores ranging from greater than one Earth mass to a considerable fraction of the total mass. That is evidence for the importance of condensible elements in giant planet formation and hence in planet formation in general. The key role of planetesimal accretion is even more visible when looking at the cratered surfaces of terrestrial planets and moons as well as small objects in the Solar System. Indeed many of the comets and asteroids may be leftover planetesimals or are closely related to them.

Detailed models of planet formation explain the diversity of solar system and extrasolar giant planets by varying the assumptions about the protoplanetary nebulae. Variations of nebula properties result in variations of the core growth rates caused be a coupling of the dynamics of planetesimals and the contraction of the massive planetary envelopes they dive into, as well as changes in the hydrodynamical accretion behavior of the envelopes caused by differences in nebula density, temperature and orbital distance.

The key question is how great a diversity of protoplanetary nebulae will provide the preconditions needed for the planetesimal hypothesis to be true and which pathways of planetary growth are working and which are dominating. The diversity of galactic habitats will reflect the diversity of protoplanetary nebulae and the richness of the planet formation process. The question to be answered in the next years is to what extent planet formation is a well-defined self-organizing structure formation process or whether 'anything goes',[1] as might be concluded from the literature following the discovery in 1995 of 51 Peg b, the first extrasolar giant planet in orbit around an ordinary star.

[1] I.e. many qualitatively different pathways lead to planets with planet formation being due to a collection of mechanisms leading to a loosely common outcome with great variety.

The pioneering study of planet formation around a diversity of stars has been performed by Nakano (1987, 1988a,b). The Kyoto model of Solar System planet formation is applied to stars of 0.5 to 10 solar masses.[2] Planet forming regions are identified by requiring conditions for appropriate condensed material to be present and formation within the respective stellar life-time and the age of the universe. While being careful to note that given numbers are likely to change, Nakano concluded that there must be an upper limit for the masses of stars that can be accompanied by planets and one for stars accompanied by planets with surface temperatures nearly equal to that of Earth.

Presently the general theoretical understanding of planet formation is limited by (1) the fact that it is unclear if and how planetesimals must form, given the presently available assumptions, approximations and knowledge about the protoplanetary nebula conditions, (2) the fact that the nebula that forms the planetary building blocks and embryos might lead to their loss via various planet-nebula interactions and (3) uncertainties relating to the duration of final stages of planetary growth and the nebula lifetimes.

1.1 Solar System Formation

Our four giant planets contain 99.5% of the angular momentum of the Solar System, but only 0.13% of its mass. On the other hand, more than 99.5% of the mass of the planetary system is in those four largest bodies. The angular momentum distribution can be understood on the basis of the nebula hypothesis (Kant 1755). The nebula hypothesis assumes concurrent formation of a planetary system and a star from a centrifugally supported flattened disk of gas and dust with a pressure supported, central condensation (Laplace 1796, Safronov 1969, Lissauer 1993, Wuchterl et al. 2000, Markwick and Charnley 2004). Observationally inferred disk sizes and masses are overlapping theoretical expectations and confirm the nebula hypothesis. High resolution observations at millimeter-wavelengths are sensitive to disk conditions at orbital distances > 50 AU (see Markwick and Charnley 2004). However, observations thus far provide little information about the physical conditions in the respective nebulae on scales of 1 to 40 AU, where planet formation is expected to occur. Planet formation studies therefore obtain plausible values for disk conditions from nebulae that are reconstructed from the present planetary system and disk physics.

[2] That approximately corresponds to effective temperatures of 4000 to 30000 K and luminosities of 0.04 to 4000 solar luminosities on the stellar main sequence.

2 STELLAR PROPERTIES DURING PLANET FORMATION

The properties of a young star, that result from a given pre-stellar cloud can be evaluated assuming that the cloud collapse is not drastically delayed by the formation and evolution of the disk. Fluid-dynamical calculations can reach the ages of observed young stars—a few to a hundred Ma—from initial conditions that correspond closely to observed cloud fragments (Wuchterl and Klessen 2001, Wuchterl and Tscharnuter 2003). These calculations assume basic physical principles and a calibration of the convection prescription to the Sun. They provide self-consistent quantitative histories of protostars and predict global properties (ages, masses, luminosities, effective temperatures) for young stars during their early evolution, from embedded stages (Class 0) to the T-Tauri-star phenomena for Larson-Penston-type collapse (cf. Markwick and Charnley 2004). They presently theoretically constrain the effects of a forming star on the protoplanetary nebula during the early phases of planet formation (first few Ma). According to these collapse studies, the young Sun had a luminosity of 4 ± 0.4 times its present luminosity, $L_\odot$ with a temperature of 4950 ± 20 K, hence was twice as bright and 500 K hotter at an age of 1 Ma than according to the classical calculations of early stellar evolution. Classically stellar evolution studies start by assuming initially fully convective structures without considering the protostellar collapse in detail.

3 GIANT PLANET FORMATION

3.1 Basic Principles

Pre-planetary disks are rotating structures in (quasi)equilibrium. The gravitational force is balanced in the radial direction by the centrifugal force augmented by gas pressure, and in the vertical direction by gas pressure alone. This gravitational force is mostly that of the central star; the gravity due to the disk itself (self-gravity) remains weak in comparison.

The key problem in planet formation is that, in addition to this stable situation, any isolated, orbiting object below a critical density (the Roche density) is pulled apart by the stellar tides. Typical nebula densities are more than two orders of magnitude below the Roche density, so compression is needed to confine a condensation of mass M inside its tidal or Hill-radius at orbital distance a:

$$R_T = a\left(\frac{M}{3\mathrm{M}_\odot}\right)^{1/3}. \tag{3}$$

A local enhancement of self-gravity is needed to overcome the counteracting gas pressure.

Giant planet formation theories may be classified by how they provide this enhancement: (i) the *nucleated instability* model relies on the extra gravity field of a sufficiently large solid core (condensed material represents a gain of ten orders of magnitude in density and therefore self-gravity compared to the nebula gas), (ii) a *disk instability* may operate on length scales between short scale pressure support and long scale tidal support, or (iii) an *external perturber* could compress an otherwise stable disk on its local dynamical time-scales, e.g., by accretion of a clump onto the disk or rendezvous with a stellar companion.

If the gravity enhancement is provided by a dynamical process as in the latter two cases, the resulting nebula perturbation, (say of a Jupiter mass of material) is compressionally heated because it is 'optically thick' under nebula conditions and cannot radiate rapidly its internal heat. Giant planet formation would then involve a transient phase of tenuous giant gaseous protoplanets, that would be essentially fully convective and contract on a timescale of typically $\sim 10^6$ yr (Bodenheimer and Pollack 1986).

3.2 Nebula Stability

Preplanetary nebulae with solar composition and a mass that is just sufficient to accommodate the Solar System's inventory of condensible elements—a few percent of a solar mass—are stable. This probably excludes planet formation via disk instability. Substantially more massive disks resulting from the collapse of cloud cores are self-stabilizing due to transfer of disk mass to the stabilizing central protostar (see Bodenheimer et al. 1993). Nevertheless a moderate mass nebula disk might be found that can develop a disk instability leading to a strong density perturbation, especially when forced with a finite external perturbation. Giant gaseous protoplanets (GGPPs) might form when the instability has developed into a clump (DeCampli and Cameron 1979, see Bodenheimer 1985, 1986 for a discussion). These clumps, provided they are stable on a few cooling times, are candidates to become proto-giant planets via an intermediate state as tenuous giant gaseous protoplanet (GGPP).

A key issue, as in any theory involving an instability of the disk gas, is then the a posteriori formation of a core. Only condensates that are present initially would rain out to form a core, while material added later by impacts of small bodies after the GGPP had formed would be soluble in the envelope (Stevenson 1982a). To account for the bulk heavy element compositions of Saturn and certainly Jupiter planetesimal accretion has to occur anyway after the GGPPs have formed their cores. Boss (2003) applied the disk-instability model to the formation of outer giant planets and investigated whether clumps with a few Jupiter masses could result from a disk instability in the outer solar nebula. If the clumps he found indeed later develop into giant planets they would have to subsequently form a core and then loose $\sim$ 97% of their mass by photoevaporation driven by assumed massive stars neighboring the early Sun to finally become 'Neptunes'. See Wuchterl et al. (2000) for further discussion.

3.3 Nucleated Instability-Core-Envelope Accretion

Planet formation in the context of the core-nucleated instability hypothesis is the consequence of the formation and growth of the solid building blocks. According to the 'planetesimal hypothesis' planets grow within circumstellar disks via pairwise accretion of km-sized solid bodies, the so-called planetesimals. Sufficiently massive planetesimals embedded in a gravitationally stable preplanetary nebula can capture large amounts of gas and become the cores of giant planets (e.g. Fig. 4.1, Wuchterl 1995b, Tajima and Nakagawa 1997, Ikoma et al. 2001). The onset of formation of massive envelopes is characterized by the so-called critical and crossover masses. Typical values for the critical mass (largest static envelope for a growing core) are in the range of 7 to 15 earth masses, for standard (minimum mass) assumptions about the preplanetary nebula.

3.4 The Gaseous Envelopes

Planetesimals in the solar nebula are small bodies surrounded by gas. A rarefied equilibrium atmosphere forms around such objects. Detailed atmospheric models with radiative and convective energy transfer (Mizuno 1980) showed that cores of typically ten earth masses are needed to gravitationally bind a comparable amount of nebula gas. Such envelopes were found to continue mass-accretion up to a few tens of Earth masses if the contraction of their envelopes was accounted for in quasi-hydrostatic models (Bodenheimer and Pollack 1986). Static and quasi-hydrostatic models rely on the assumption that gas accretion from the nebula onto the core is very subsonic, and the inertia of the gas and dynamical effects as dissipation of kinetic energy do not play a role—in short the motion of gas is neglected. To check whether gas motion remains slow during accretion hydrodynamical investigations

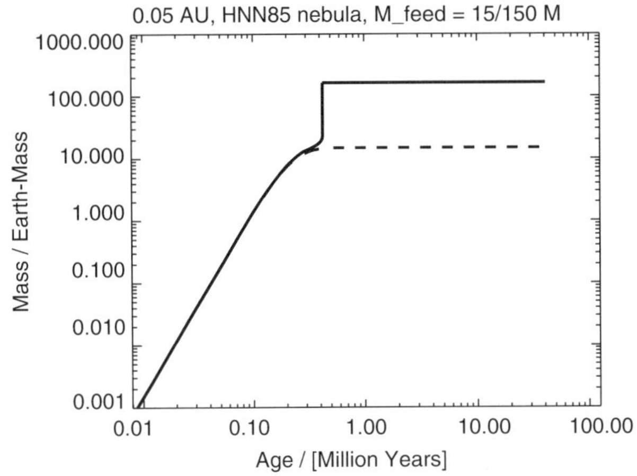

Figure 4.1. Total mass (full line) and core mass (dashed line) of a Pegasi-planet forming at 0.05 AU from a solar mass star. The accumulation of a gaseous envelope surrounding a condensible element core that grows by planetesimal accretion is shown for the first 100 million years. The structure of the envelope is calculated dynamically including time-dependent theory of convection that is calibrated to the Sun and including detailed equations of state and opacities. Note the gradual build-up of the core by planetesimal accretion and the step-like rapid capture of the gaseous envelope.

are necessary. A linear adiabatic dynamical stability analysis of envelopes evolving quasi-hydrostatically (Tajima and Nakagawa 1997) found that motions would remain slow and hence the gaseous envelopes could relatively slowly grow to giant planet masses. Nonlinear, convective radiation-hydrodynamical calculations of core-envelope proto-giant planets (see Wuchterl et al. 2000 for review) found two pathways for the further evolution. The first one, typical for lower nebula-densities—that lead to mostly radiative envelopes—produces Uranus/Neptune type planets with relatively small hydrogen-helium envelopes around a large core. The second one, for higher nebula-densities—that lead to mostly convective envelopes (Wuchterl 1993, Ikoma et al. 2001)—leads to massive giant planets as Jupiter and Saturn (Wuchterl 1995b).

The two pathways are separated by the transition of the outer protoplanetary envelopes from radiative to convective energy transfer. An approximate condition separates the two pathways depending on the midplane nebula density. The resulting criterion gives a minimum nebula density $\varrho_{\rm neb}^{\rm crit}$ that is necessary for a convective outer envelope (Wuchterl 1993, Ikoma et al. 2001):

$$\frac{\varrho_{\rm neb}^{\rm crit}}{[10^{-10}{\rm g\,cm^{-3}}]} = 2.2\left(\frac{T}{[100{\rm K}]}\right)^{3}\left(\frac{\nabla_{\rm s}}{[2/7]}\right)\left(\frac{\mu}{[2.2]}\right)\left(\frac{\kappa}{[{\rm cm^2\,g^{-1}}]}\right)^{-1}$$
$$\left(\frac{\dot{M}_{\rm core}}{[10^{-6}\,{\rm M_\oplus\,yr^{-1}}]}\right)^{-1}\left(\frac{M_{\rm core}}{[10\,{\rm M_\oplus}]}\right)^{\frac{1}{3}}\left(\frac{\varrho_{\rm core}}{[5.5]}\right)^{-\frac{1}{3}}, \qquad (1)$$

for midplane nebula temperature T, logarithmic adiabatic gradient (with respect to pressure), $\nabla_{\rm s}$, mean molecular mass μ, opacity κ, core-mass and -density, $M_{\rm core}$, $\varrho_{\rm core}$, resp. and planetesimal accretion-rate $\dot{M}_{\rm core}$. Protoplanets that grow under nebula conditions above that density have larger envelopes for a given core, a reduced critical mass and accrete envelopes that are more massive than the core.

3.5 Extrasolar Planets

More than a hundred extrasolar planet candidates have thus far been discovered to orbit main sequence stars other than the Sun. The extrasolar planets currently known all orbit nearer to their stars than Jupiter does to the Sun. This is primarily an observational bias because the dominating radial velocity techniques can only safely detect a planet when data are collected for a full orbital period of the planet—about 12 years in the case of Jupiter. Some of these planets orbit on highly eccentric paths, suggesting that after they formed they were subjected to close encounters with other giant planets (Weidenschilling and Marzari 1996, Lin and Ida 1997, Levison et al. 1998) or secular perturbations from companions. Some of the extrasolar planets are separated from their stars by less than one percent of the Jupiter-Sun distance. Guillot et al. (1996) showed that giant planets are stable over the lifetime of a 1 $M_\odot$ star even if they are as close as 0.05 AU. Models involving migration caused by disk-planet interactions are favored by many researchers for the formation of these objects (e.g., Lin et al. 1996; Trilling et al. 1998; Ward and Hahn 2000, and by Lin et al. 2000). However, simulations also show that it may be possible to form giant planets very close to stars.

3.6 Hydrodynamical Models of Pegasi-Planets

Pegasi-planets are massive planets that orbit their star with orbital periods of a few days and distances of a few hundreds of an AU. Applying the above density-criterion immediately leads to the conclusion that they may form by accretion induced by a core of a few Earth masses. Detailed convective radiation hydrodynamical calculations of core-envelope growth at 0.05 AU, indeed show gas accretion beyond 300 $M_\oplus$ for core masses around ten earth masses (Figs. 4.1, 4.2, Wuchterl 1996, 1997). Hence Pegasi-planets may form in-situ if sufficient mass of gas and dust is

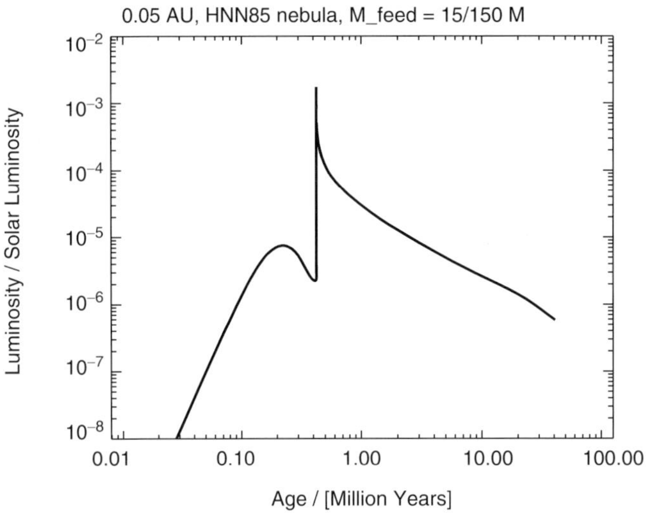

Figure 4.2. Luminosity of a Pegasi-planet during the first 100 million years according to a fluid-dynamical model. The two maxima correspond to peaks in the accretion of planetesimals and gas respectively (cf. Wuchterl et. al 2000).

available in their feeding zones. That requires surfaces densities that are enhanced above the minimum mass nebula but are plausible if a diversity of nebulae is considered.

3.7 Extrasolar Planets and Stellar Metallicity

To show the present state of the discussion and possible problems with biases in the sample of discovered extrasolar planets, I will briefly discuss the planet-metallicity relation (Santos et al. 2003, Fisher et al. 2004). The authors find the frequency of planets to increase with stellar metallicity (i.e. the [Fe/H] stellar metallicity indicator). Because their stellar samples are related to the star-lists that are used in planet searches, I will call them the *planet hunting samples* in the following. Fuhrmann (2002, 2003) studied a volume complete sample of nearby F,G and K stars. Because it correctly represents the galactic stellar population I will call it the *galactic sample* in the following. The complete galactic sample overlaps with the planet hunting samples. That allows to compare the galactic sample to hunting samples allowing a check whether the discovery procedures induce a bias into the hunting samples. We consider the Fisher et al. results—more planets for stars with a larger heavy element abundance, measured by [Fe/H] for their planet hunting sample. Assuming they use the Nidever et al. (2002) *volume limited* hunting sample the following is noticed (Fuhrmann 2003 pers. communication) when comparing their stars to those in the overlapping *volume complete* galactic sample: of the 166 stars in the galactic sample only 90 (54%) are in the hunting sample despite the smaller volume of the volume complete galactic sample. Hence the hunting sample does not contain the stars it should contain, given the overlap—a consequence of incompleteness. In particular it is interesting which stars are absent. Missing in the hunting sample are:

1. a few subgiants,
2. fast rotating stars with $v \sin i > 10\,\mathrm{km/s}$,
3. a few young and chromospherically active stars,
4. stars without a precise luminosity class,
5. binaries and multiple systems.

These are all properties that make planet hunting more difficult. From these findings of Fuhrmann, I can only conclude that stars that are (considered) unfavorable for planet hunting are under-represented. The average $[\mathrm{Fe/H}] = -0.02$, of the 90 stars in the hunting sample, that fall into the intersection with the galactic sample is only 0.01 dex higher than Fuhrmann's respective value for the galactic sample. However, for $[\mathrm{Fe/H}] \geq 0.2$ only 5 stars are missing but for $[\mathrm{Fe/H}] \leq -0.2$, 13 stars are missing. Hence there is a slight trend that metal-poor stars are preferentially missing in the hunting sample. Considering the still small numbers that are available for comparison, and the average metallicity of $[\mathrm{Fe/H}] = 0.00$ for thin disk stars derived from the galactic sample, there could be a metallicity effect of $+0.10$ to $+0.15$ (contrary to the $+0.25$ that are favored by Santos et al. 2003 on the basis of their hunting sample).

With the Sun at the thin disk average of [Fe/H] for our distance to the galactic center (Fuhrmann 2003) the role of metallicity may well be a slight increase overall due to a significant increase for the Pegasi-planet's host stars. That would put the Solar System into normality as far as metallicity and planet hosting are concerned. It is such a kind of bias contained in hunting samples, that I think might still be present in the planetary discoveries, and might render the first 100 discovered extrasolar planets somewhat atypical for planetary systems in the galaxy.

4 THE GROWTH OF SOLIDS IN CIRCUMSTELLAR DISKS

4.1 Grain Growth and Planetesimal Formation

Protoplanetary disks have the same elemental composition as the parent cloud and the growing star, that is primarily hydrogen and helium, with $\sim 1-2\%$ heavier elements. Sufficiently far from the central star, it is cool enough for some of this material to be in solid form, either remnant interstellar grains or condensates formed within the disk. Dust agglomerates via inelastic collisions, and gradually settles towards the disk midplane as it grows large enough to be able to drift relative to the surrounding gas (Weidenschilling and Cuzzi 1993). Either by continuing agglomeration and binary accretion or by a gravitational instability (Safronov 1960, Goldreich and Ward 1973) kilometer-sized planetesimals are thought to form. The details of the process, however, are incompletely known and stagnation of growth or loss to the central young star may be common.

Hueso and Guillot (2003) modelled planetesimal formation in a pragmatic approach that includes important processes in planetesimal formation in a balanced way. A simple evolving nebula model is combined with radial transport of gas, condensation and evaporation, and the drift and growth of dust and planetesimals. They show that km-sized planetesimals may form within their approximate but synoptic model and they find an increase in size and number of the planetesimals somewhat interior to the snow-line (cf. Fig. 4.3).

4.2 Terrestrial Planet Formation

The primary perturbations on the Keplerian orbits of kilometer-sized and larger bodies in protoplanetary disks are mutual gravitational interactions and physical collisions (Safronov 1969). These interactions lead to

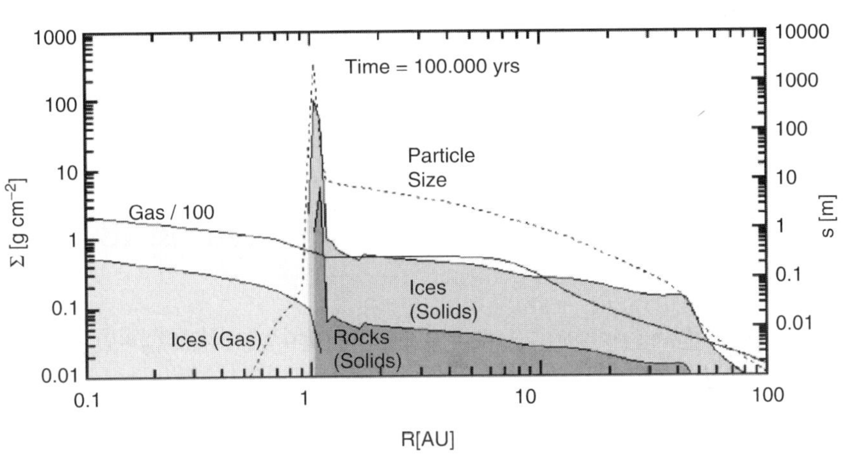

Figure 4.3. Planetesimal formation in a viscous α-disk model. The surface density of solid material is plotted for a model containing initially 99 % 'gas' and 1% 'ices' (a mixture of water, methane ammonia, etc.), and 0.1 % 'rocks' (silicates and refractory materials). Particle size is shown as a dotted line—right axis (Hueso and Guillot 2003). Note the increased surface-density and size of planetesimals near the snow-line.

accretion (and in some cases erosion and fragmentation) of planetesimals. The most massive planets have the largest gravitationally-enhanced collision cross-sections, and accrete almost everything with which they collide. If the random velocities of most planetesimals remain much smaller than the escape speed from the largest bodies, then these large 'planetary embryos' grow extremely rapidly (Safronov 1969). A few large bodies grow much faster than the rest of the swarm in a process known as runaway accretion (Wetherill and Stewart 1989, Kokubo and Ida 1996). Eventually, planetary embryos accrete most of the solids within their gravitational reach, and the runaway growth phase ends. Planetary embryos can continue to accumulate solids rapidly beyond this limit if they migrate radially relative to planetesimals as a result of interactions with the gaseous component of the disk (Tanaka and Ida 1999). A typical planetary embryo in the terrestrial planet zone of the Solar System would have the size and mass of Mars, i.e. about half the radius and a tenth of the mass of Earth.

The eccentricities of planetary embryos in the inner Solar System were subsequently pumped up by long-range mutual gravitational perturbations; collisions between these embryos eventually formed the terrestrial planets (Wetherill 1990, Chambers and Wetherill 1998). However, timescales for this type of growth in the outer Solar System (at least 10^8 years, Safronov 1969) are longer than the lifetime of the gaseous disk (cf., Lissauer et al. 1995). Moreover, unless the eccentricities of the growing embryos are damped substantially, embryos will eject one another from the star's orbit (Levison et al. 1998). Thus, runaway growth, possibly aided by migration (Tanaka and Ida 1999), appears to be the way by which solid planets can become sufficiently massive to accumulate substantial amounts of gas while the gaseous component of the protoplanetary disk is still present (Lissauer 1987).

5 PLANET FORMATION THEORY—CRITICAL SUMMARY

The previous discussion focused on detailed models of planet formation that describe the main accumulation processes for solids and gas and are able to cover the entire formation process until the final planetary masses are reached. Known planets, solar and extrasolar, are accounted for by studying a range of plausible nebula properties that allow a consistent overall picture. The detailed models combine two elements of planet formation theory that are most advanced and to some extend capable to obtain quantitative results for a wide range of possible nebula situations. These elements are

1. planetesimal accretion,
2. planetary structure and dynamics.

Modern planetesimal accretion theory includes, both, significant analytical detail and long-range numerical simulations that are able to cover large fractions of the planetary growth process in a single calculation and for a large number of planetesimals, planetary embryos and protoplanets. Using computer-hardware that is specially designed to calculate the gravitational force of a large ensemble of point masses the formation of planetary embryos, and the properties of the planetesimal swarm can be calculated directly. Furthermore the growth of planetary embryos can be followed to the final planetary masses by solving the n-body problem with collisions with symplectic integrators (Chambers and Wetherill 1998). Growth processes in the outer regions of gas-free disks of planetary embryos can be followed to quasi-stable, final configurations, when assumptions are made about the evolution of the gaseous envelopes of the outer planets. Long-term integrations of the solar system planets can explain the sculpturing of the outer Solar System by gravitational erosion of an initial swarm of small bodies and successfully predicted the location of trans-neptunian objects in the Solar System (Malhotra et al. 2000). Hence there are quantitative and predictive elements when the role of gas is minor and the effects of the potentially large number of small bodies (dust and planetesimals) can be neglected for simulations of the final stages of planet formation.

The second, key element, planetary structure and evolution theory, can be confronted with present solar system objects, most prominently Jupiter and Saturn (Guillot 1999). It has many elements in common with stellar structure and evolution theory and hence components, as the convection prescription, can be tested in stars and the Sun before being applied to the planet formation problem. The theory has proven to be of significant predictive power as the possibility of the existence of brown dwarfs and their properties have been quantitatively predicted theoretically (Kumar 1963) more than three decades before the observational capabilities allowed detection of such objects.

But numerous processes in planet formation have not reached a comparable maturity and have been neglected in the previous discussion. That does not mean that they are not needed to arrive at a better overall understanding, but just that they have not been included in models that have sufficient time-span or generality to exclude at least some planetary formation modes in general.

5.1 Problems of the Standard Planet Formation Models

The models of planet formation that have been worked out in detail are still closely related to models of Solar System formation (Wuchterl et al. 2000). That is mostly due to the fact that the Solar System provides by far most of the constraints that are needed to determine the many unknowns in theoretical planetary formation models. For the astrobiological application, i.e. the formulation of an expectation of what planetary systems may look like in general and what were the conditions in their early evolution, the predictive power will be dominated by the problems of planet formation theory. Hence the uncertainties and unsolved problems will determine how successfully we can extrapolate from the present solar system state to the planet formation time here and elsewhere. I will therefore discuss the weaknesses of our understanding to provide a guide for what we might encounter in other systems. Where we know the least, other systems are most likely to differ significantly.

There are three mayor problems in planet formation theory. The qualitative problem of planetesimal formation, the qualitative problem of migration that might be turned into a quantitative one when migration-rate estimates decrease further, and finally the purely quantitative formation time-scale issue.

5.2 Formation of the Protoplanetary Disk

There is presently no way, to infer even the most basic properties of protoplanetary disks with a precision that constrains planet formation theories (Markwick and Charnley 2004). Observationally the disks are very likely observed at relatively evolved stages, when critical processes like planetesimal formation and growth to the planetary embryos are already over. Furthermore the spatial resolution is presently insufficient to infer properties at orbital positions that are comparable to any known planet. Theoretically the formation of protoplanetary disks is technically elusive, because investigations that are able to model a fragmenting collapsing cloud with significant geometrical detail (in 3D) are based on explicit numerical fluid-dynamics schemes and hence unable to cover the disk-evolution time-scales to a steady or quasi-steady state, that could be considered a starting point for planet formation. Furthermore all non-spherical calculations of star formation in general and protostellar cloud collapse in particular, that might be able in principle to determine the mass and angular momentum of circumstellar disks have to make highly idealized assumptions about the thermal budget so that they rather prescribe temperatures one way or the other than to determine it. But it seems to be exactly the thermal evolution once clouds become optically thick and heat up that is responsible for the final outcome of the fragmentation process and hence how much mass ends up with what angular momentum in orbit around what kind of stellar system. Even the formation of binary stars, requiring 'only' the determination of the distributions of much larger fractions of cloud fragment mass then the few percent that go into protoplanetary disks is still far from a prediction of the masses and angular momentum of stellar companions. Hence it is essentially unknown how the protoplanetary disks look like in general. Information is available from reconstructions for the Solar System and it is known from observations that disks around young stars have masses that are consistent with the masses theoretically expected for gravitationally stable disks.

5.3 Evolution of the Protoplanetary Disks

It is clear that with an unknown initial state the evolution is difficult to access. Viscous accretion disks models provide important information about the mass and angular momentum evolution during stellar accretion. They also have the property that they forget about their initial state on their evolutionary time-scale of typically a few Ma. But planet formation, especially planetesimal formation occurs on the same time-scale and there is no consensus yet about what disk models are realistic in the sense that they are corroborated by observations. There is also no consensus about the mechanism that provides the assumed viscosity. There is even the possibility, that there is no such anomalous viscosity of a magnitude that is of relevance to the star formation process, and angular momentum redistribution was essentially completed by gravitational torques early in the stellar accretion history.

5.4 Structure of Protoplanetary Disks

Given a distribution of mass and angular momentum, momentary or initial, the structure and appearance of disks can be calculated in great detail and with some reliability. Assumptions about their overall properties and physical structure can be made that are consistent with observations of material in orbit around young stars and brown dwarfs. But the overall agreement can hardly be exploited for inversions that would serve quantitative planet formation theory. The result is that there are two basic categories of idealized disks used for planetary formation models: (1) 'passive', quiet, non-accreting disks, as the Kyoto nebula (Hayashi et al. 1985) with a thermal structure determined by the (proto-) stellar energy input or (2) 'active', viscous α-accretion discs with a net mass accretion rate onto the star and the thermal structure governed by the dissipative process that allows accretion and angular momentum redistribution.

5.5 Stability of Protoplanetary Disks

Classical protoplanetary disks models, similar to the minimum reconstitutive nebulae and α-accretion disks are gravitationally stable. But alternative protoplanetary nebulae can be imagined that might be gravitationally unstable (see section on disk instability, above). Actually in early stages of star formation it seems likely, for plausible assumptions about the rotation of the cloud fragment before it starts to finally collapse, that significant amounts of mass do not fall onto the central star directly but first onto the disk. Subsequently the mass is accreted through the disk onto the star (resulting in an active nebula as described above). Depending on the efficiency of the angular momentum transfer processes in the disk—that is essentially unknown and only weakly constrained—the mass flow either (1) proceeds in a quasi-steady manner, with the flow onto the star balancing the rate of material that drops onto the disk from the ambient cloud (standard accretion disk models) or, if the angular momentum transfer efficiency is insufficient (2) the gas piles up in the disk, increasing its mass. The third case with disk accretion being faster than disk mass loading is likely to act stabilizing in any case. The second case will likely result in a gravitational instability. The development of the instability is controversial, not only because it depends on the assumptions that are made about the initial disk and the mass-influx, from the cloud, but also because different investigators found different results for the same problem. In any case such stability calculations essentially answer the question whether there is an instability of the nebula and how it will develop. Present studies are essentially non-linear stability analysis and can only address the early phases of planet formation, both in terms of compression and evolutionary phase. All stability investigations presently published are performed with explicit hydrodynamical techniques, that are limited by basic numerical stability requirements, to a few dynamical time-scales. In our problem that is a few orbital time-scales of the respective disk, or to be more precise those of the innermost parts of the disk that are in the domain of calculation. The overall time covered is typically a few to a few hundred orbits, with planet formation requiring a few million to a few hundred million orbits. The limitations in nebula gas compression are even more severe. Due to resolution limitations the maximum compression typically reached in such calculations is less then 1000, while planet formation requires a robust factor of 10^{10}.

A synoptic interpretation of the present literature is that, mass loaded disks are driven towards an instability, that develops into restabilizing the situation by accretion of mass onto the central object and outward transfer of angular momentum. If that holds, the cloud collapse onto a disk would self-produce the very efficient angular momentum transfer process, that is needed, and disks would be limited to the gravitationally stable regime. The instabilities that might develop would stabilize the nebulae and ultimately lead to a relatively quiet protoplanetary nebula situation. I refer to Wuchterl et al. (2000), for a more detailed discussion relating disk-stability and planet formation.

5.6 Problems of Dust-Growth and -Dynamics

When considering dust growth in the nebula from a global perspective we encounter the uncertainties in the nebula structure again. For the two extreme regimes—quiet or viscous-turbulent—dust growth proceeds in different macroscopic gas-flows, and under different temperatures and vertical nebula structures, for given orbital radius, say. The midplane temperature, e.g. in turn depends on the opacities, that are governed by dust throughout the planet formation zone. The optical properties of dust depend on the chemistry, mineralogy, and size-distribution of the grains. Progress is made for assumed nebulae and consistent determinations of local nebula structure, chemistry, dust growth and mineralogy are presently developed. The enormous advantage for the dust growth problem considered locally, is that laboratory experiments have confirmed theoretical considerations and can be used for calibrating the models (Blum and Wurm 2000). Together with a consistent determination of the mineralogy that will provide the opportunity to predict the production of 'sand' an 'ice'-properties for given global nebula structure. So there are good perspectives to determine the general properties of planetesimal raw-material in the near future (cf. Markwick and Charnley 2004, Hueso and Guillot 2003).

Planetesimal formation obviously then results as the key unsolved problem because its precursor steps are incompletely understood. However, a theory of planetesimal formation is needed to predict the size distribution, chemical composition, mineralogy and orbital properties of this planetary building blocks everywhere in the protoplanetary disk.

5.7 Dynamical Planet-Disk Interaction

A key problem, starting already at the embryo size is the gravitational interaction of the growing planet with the ambient nebula gas. It determines if and how much nebula gas can flow from the nebula into the planet's gravitational sphere and whether the planet gains or looses mass and angular momentum to the nebula. Furthermore it determines how the planet modifies the disk in its neighborhood. Detailed structure and dynamical models of planetary growth cannot self-consistently account for that interaction because its geometry is necessarily non-spherical and hence beyond the capabilities of present planetary structure models based on implicit numerical techniques. Two alternative approximate approaches have been followed:

The first one is to approximate the growing planet by a smoothed gravitational potential of point mass that acts as a hole to account for accretion. Nebula gas that approaches an inner region that cannot be resolved by the numerical grid is assumed to proceed to the hydrostatic parts of the planet. These are not calculated as part of the flow-model. It is assumed that the mass is accreted by the planet and removed from the flow and added to the point-mass. With this procedure it is possible to calculate the protoplanets' effect on the quasi-keplerian nebula flow in 2D and 3D (Kley 1999, D'Angelo et al. 2002, 2003 for recent models). Calculations show the effect of approximate point-mass planets on initially unperturbed nebula disks. They are twofold: (1) the point-mass planet reduces the nebula density in a region of the order of a tidal radius around the planet, a situation, referred to as *gap-opening*. (2) outside the gap, the potential of the point mass induces density waves that are found to produce a net torque of the disk on the planet. The magnitude and even the sign of this torque somewhat depends on the physical assumptions and properties of the calculations (how the planetary potential is smoothed on small, subgrid scales, the spatial resolution and the method). Disk-planet interactions of growing planets are also attractive mechanisms to self-limit planetary masses. In particular gaps induced in the protoplanetary disk by grown-up giant planets have been proposed to cut off gas-accretion. High-resolution calculations have shown however, that the gas finds a way onto the planet even through well developed gaps and hence the process is unable to stop accretion. If the torques found in these 'disk-response to a point mass calculations' cumulate on secular time-scale, so called type II planetary migration will result—i.e. a protoplanet in a gap moving with the nebula towards the star. That may ultimately determine the planet's survival.

The second approach is to parameterize the results of planetesimal accretion, planetary structure and use simplified, usually analytical formulae for the planet-nebula interaction (Trilling et al. 1998, Bryden et al. 2000) and look at the combined outcome. It assumes that the results obtained for the component processes are not changed by the mutual interaction. That allows to look at the combined problem of protoplanetary mass evolution, interaction with the nebula and orbital evolution. Planets are introduced into an assumed nebula and their masses and orbits are evolved according to the parameterized prescriptions of the various processes, e.g. simple envelope structure and growth models for the protoplanet, subject to roche-lobe constraints for mass overflow and combined with a viscous α-accretion nebula disk and an analytical formula for the torque of the disk onto the planet due to disk planet interaction (e.g. Papaloizou and Terquem 1999). The results provide overall histories for simplified physics and a choice of parameters. Generally severe compromises have to be made to connect the different model parts and it is doubtful whether possible artifacts, due to the inconsistencies that are likely to be introduced, can be reliably excluded. The uniqueness of the choice of parameters is unlikely even if it is possible to sufficiently constrain all the parameters in the first place. With the obvious lack of calibrations, the parameterizations introduce new uncertainties into the calculations. That is in addition to the uncertainties about the nebula and hence the predictive power may be low, even if the procedures turn out to capture the most important physical processes in a qualitatively correct way. What is obtained is mostly a balance between the time-scales of the competing component processes.

Both approaches address two questions of disk-planet interaction:

1. Mass budget: During the final stages of gas-accumulation of massive giant planets, fluid-dynamical effects govern the accretion process. Control of the further evolution then passes from the ability of the inner hydrostatic parts of the planet to contract to the supply of nebula gas. With gas from the feeding zone already accumulated onto the planet, nebula gas accretion into the planetary feeding zone becomes the process limiting further growth. That relates to the key problem of how the final masses of gaseous envelopes and hence giant planets are determined. The detailed models of planetary growth (all with spherical symmetry) presently explicitly or implicitly assume a given reservoir of mass, from which the planet may accrete. In general it is unknown at what values the final masses of giant planets would arrive if that reservoir is not prescribed.

   Studies calculating the flow of gas from the ambient nebula into the feeding zone of the planet (Kley 1999, D'Angelo et al. 2002, 2003) generally find that gas supply from the nebula to the planet occurs on a time-scale that is very small compared to both the nebula life-time and the contraction and cooling times of the planets found in the detailed planetary structure models. That is due to the fact, that the nebula densities are relatively high and dynamical time-scales in protoplanetary envelopes and the transition regions into the nebula are short (10000a) compared to the overall planet formation time of Ma. Hence even a very inefficient dynamical process is capable of transferring mass from the nebula into the feeding zone from which the protogiant can capture gas. As a consequence stopping giant planet growth requires a very efficient cut-off mechanism or a very complete removal of nebula residues to shut off gas accretion and determine the planetary masses. To demonstrate a complete shut-off is a great challenge to present computational techniques.

2. Angular momentum budget and orbital evolution: Generally it is found that violent migration has to be expected, if the nebulae are viscous. Planets migrate significant distances after they are formed and a potentially very large number of planets moves into the host star. Additional processes that stop migration at orbital distances that are consistent with those observed for extrasolar planets are invoked. Since parameters are adjusted to account for what is known, the predictive power as far as the general population of planets is concerned is small, especially for planetary systems like the Solar System where migration-effects apparently were small and there seems to be no evidence for violent migration.

5.8 Nebula Dispersal

Many effects, as the fade-out of violent migration and mass-limitation of giant planets, in particular of Uranus/Neptune type relate to how the nebula gas dissipates during or after the final stages of planet formation. Hollenbach et al. (2000) discuss four candidate processes, (1) viscous accretion, (2) close encounters, (3) stellar winds (considered unlikely) and (4) photoevaporation. But it is still unclear how nebulae disappear after the end of the planet formation era especially close to the stars where the escape speeds are largest. An obvious possibility is that the nebulae disappear because they condense into planets. Photo-evaporation typically operates efficiently in the outer parts of proto-planetary disk, accretion of the residual material onto the star is most efficient for close-in parts of the disk. If violent migration dominates planet formation, the details of nebula dispersal will control the planetary outcome, in terms of survival and mass limitation. Planet formation may then be mostly a competition between planetary and nebula loss processes—of course assuming efficient formation in the first place—with an outcome that will be theoretically very hard to predict.

5.9 Late Accretion and Planet-Planet Interactions

This regime starts when the disk mass has been significantly reduced and the disk stops to be dynamically important for the planets' orbital evolution on the secular time-scales. What remains is the accretion of the residual gaseous nebula material, the interactions with other planets, the remaining planetary embryos and planetesimals (Levison et al. 1998, Thommes and Lissauer 2003, Levison and Agnor 2003). The relevant overall masses in all components may or may not be comparable to the mass of the largest planet. There is probably still a large and locally dominant number of bodies around, that in case of the Solar System are responsible, e.g. for the formation of the Moon and the late heavy bombardment. Late accretion affects are apparently important in the asteroidal region of the Solar System where it is possible that Jupiter's perturbations precluded the accretion of embryos into a planet. The important remaining dynamical process is then the orbital evolution of planets and embryos due to secular mutual perturbations. Eccentricity excitation is possibly damped by a remaining small-size planetesimal population via dynamical friction. The time-scale is larger than 10^5 orbital periods leading to the completion of the planetary system formation process on time-scales of tens to hundreds of Ma.

5.10 Gravitational Erosion

Secular orbit-dynamical instabilities seem to shape orbital structures including the planets after the early formation period with its strong chaos. The distribution and evolution of orbital elements in the Solar System can presently be understood if the major masses, i.e. the major planets are set up at the backward integrated orbits, that they occupied at the time of Solar System formation. Although *the Solar System is not stable, just old*, the stabilization process of *gravitational erosion* leads to well separated orbits that satisfy radial exclusion principles (Lecar et al. 2001). They are based on non-overlapping of the gravitational spheres of influence around the orbits. Their half widths are estimated to be of the order of 10 Hill-radii for stability over the age of the Solar System. These structuring processes are expected to operate in general planetary systems and provide relatively quiet conditions in terms of impact history as an 'aging effect'. But presently it is unknown for how large a class of planetary systems they are important. The key challenge here is to transform the understanding that has been developed by very long range time integrations of the Solar System to planetary systems in general. A task that is just beginning.

This enumeration of incompletely studied processes that are relevant for the overall outcome of planet formation reflects the piecewise approach that is necessary. Solving the overall problem in particular for arbitrary nebulae is presently beyond reach. Only after the various stages in a step-wise approach are heuristically joined together, overall statements about the final outcome, the planetary yield, can be made. An alternative approach is to statistically analyze the final outcome, i.e. the possible and probable end-states of the process, the planetary equilibria, as the basis of the structure of planets (Pečnik and Wuchterl 2004).

5.11 Problems of Planetesimal Formation

Planetesimal formation by coagulation and agglomeration may stall at dm to m size where loss processes by radial drift may halt the planet formation process. Planetesimal formation by a gravitational instability of a dust-subdisk may require special nebula conditions that are incompletely explored to decide under how wide a range the instability will operate and whether the non-linear outcome are the consolidated condensible element bodies that are envisaged and assumed in the planetesimal hypothesis. The related key question is how wide a diversity of nebulae will lead to instabilities that produce appropriate planetesimals. Being appropriate mostly means a size large enough to decouple from the head-wind of the nebula gas, an event that typically occurs at km size. Other, yet unknown planetesimal production processes do not automatically imply that planet formation will not proceed as presently imagined but new pathways in a theoretically essentially unexplored regime have to be worked out in that case.

5.12 Planetary Migration

This second problem becomes severe at larger sizes because it is proportional to the planetesimal mass. It is expected by many investigators to become important in the mass range resulting from\the early fast *runaway* mode of planetesimal growth. The runaway phase ends when all planetesimals that are within the gravitational range of the locally largest body have been accreted and hence its feeding zone has been emptied. Grown-up planetesimals in that mass range of order\0.1 to 1 earth masses are referred to as *planetary embryos*. Planetary embryos gravitationally\interact with the ambient gas disk, planetesimal disk and other planetary embryos or planets. As a result *planetary migration* can come about (see Thommes and Lissauer (2004), for a review).

Planetary migration had been noticed by Fernandez and Ip (1984) and has been considered by Hayashi et al. (1985) as a process that shortens the growth times of Uranus and Neptune. In the late stages of outer Solar System formation these planets would move outward as a consequence of angular momentum exchange when passing comets down to Jupiter for subsequent ejection to the Oort cloud. After the discovery of 51 Peg b it has become a custom in planet formation theory to denote many kinds of change in the planetary semi-major axis or orbital distance as planetary migration. That is usually independent of the physical process underlying the orbital change. With many proposed processes and a suggested terminology of types I to III, that only covers certain regimes in particular described by Ward (1997), we distinguish between migration processes that modify the orbit by less than a factor of e^2 (or 10) and those that may lead to larger changes up to orders of magnitude in the orbital radius, and may ultimately result in the loss of the planet. The latter processes we will call *violent migration* in the following. They may dominate the planet formation processes if they operate in many and diverse nebulae.

Violent migration is a back-reaction of the planetary embryo's 'bow wave' in the nebula onto the embryo itself. As the embryo orbits the star, its gravitational potential adds a bump to the stellar one. At the embryo's orbit—at the corotation resonance, in the linear terminology of migration theory—the embryo and its potential move almost at the same velocity as the nebula gas. That is co-orbital motion as in the case of Jupiter and the Trojan asteroids. Inside the embryo's orbit, the gas in a quasi-keplerian disk orbits faster and hence the embryos potential and gravitational acceleration travel at a different speed. This accelerates the disk gas and excites a pressure-and density-wave that travels with the embryo. Because matter deeper in the primaries potential must orbit faster, the waves are dragged forward inside, and backwards relative to the embryo, outside the embryos orbit. These rather particular protoplanetary bow-waves include density enhancements that gravitationally back-react onto the planet. Due to the inherently asymmetric nature of the situation (Keplerian orbital velocities changing $\propto r^{-1/2}$) and the particular wave-pattern the forces (and in particular the torques) on the planet may not cancel out, leading to a net exchange of angular momentum between the planetary embryo and the gaseous disk if the waves dissipate or break in the disk neighboring the planet's gravitational sphere of influence. The result is the familiar reaction of orbiting matter if angular momentum exchange is allowed: most matter (the embryo) moves in and a small amount (some gas) moves out carrying away the angular momentum. Hence, the very growing of the embryo would lead to orbital decay and gradual movement towards the star on time-scales of disk evolution or much smaller. Many studies are presently devoted to determine the strength of the effect and evaluate the rates of orbital decay and hence the possible survival times for planets of given mass in a given disk. If migration dominates it has the potential to wipe out any and many generations of planets. In that case and because it originates in a delicate torque balance, it seems unlikely that we will be able to reliably predict much about planet formation in the galaxy any time soon.

5.13 Planet Formation Times

The standard model is centered around the planetesimal hypothesis that has been successful to understand a wide range of solar system bodies, to a large extent in a quantitative way. But observational results obtained for nearby star-forming and young star regions quantitatively challenge the standard model because indicators of the presence of circumstellar disks (Haisch et al. 2001) suggest disk depletion time-scales that are comparable or shorter than some calculated formation times for solar system giant planets. Modern theories of planetesimal accretion (Kokubo and Ida 2002) could account for rapid growth if increased solid-surface densities are allowed. In addition, the theoretical estimates for planet formation times have been known to be idealized because of the many approximations that are necessary to reliable calculate at least an idealized situation to allow a quantitative discussion and theoretical progress. Recently, the collisional cross-sections of planetesimals with gaseous envelopes have been studied (Inaba and Ikoma 2003, Inaba et al. 2003). Accretion-rates were found to increase significantly, reducing the planet-growth times considerably. This is especially important for the giant planet regime where the envelopes may become comparable to the condensible element cores during the runaway phase. Hence growth times decrease when the nebula gas is not neglected in determining the collision cross sections.

6 ASTROBIOLOGICAL GOALS OF PLANET FORMATION THEORY

Theoretically characterizing the galactic habitat can be coined into two questions: What can be deduced in terms of astrobiological relevance from planet formation theory as sketched above? What is the expectation value of potentially habitable planets per star or stellar system? To address the questions we may distinguish three items:

1. the number of planet forming star systems,
2. critical events that may potentially quench the planet formation process,
3. efficiency issues, that determine how many planets form per star system.

The second issue relates to the question of whether there is only one star with terrestrial planets, i.e. the question how unique the Solar System is.

Theoretical error bars are worth to be mentioned here. They are a result of technical limitations in approximations and computational techniques as well as lack of knowledge about contingencies (as nebula conditions) and entire processes (as planetesimal formation).

It is very hard to estimate the accuracy of planet formation theories because there is not yet a test to assess the real errors. Detailed theories in planet formation have an intrinsic precision, obtained by inter-comparison of results between different investigations of about 10% for global planetary properties as mass and luminosity for standard cases as Jupiter-formation and in regimes where the assumptions overlap (Wuchterl 1995b). The differences are much larger for non-standard issues like the formation of Pegasi-planets (see Wuchterl et al. 2000). If we take the optimistic view that the precision above is representing contemporary theoretical accuracy, we have a ten percent domain of uncertainty that essentially makes it impossible to make a relevant statement about the galactic issue of quenching planet formation. A prediction of $0 \pm 10\%$ of star-systems that contain terrestrial planets would allow a lot of planets in the galaxy overall. It may, however lead to significant limitations of observationally accessible planets in the galactic neighborhood, and in particular concerning the about few hundred stars of similar type as the Sun within a distance of 25 pc, say. If the empirical estimates extrapolated from planet searches of about 20% of stars hosting detectable planets hold in general, a theoretical result that excludes terrestrial planets in 90% of the detected systems could easily leave us with a zero to a handful for the expectation value for the number of neighboring extrasolar terrestrial planets.

6.1 The Number of Appropriate Star Systems in Time

Planet formation theories require evidence for star formation with some heavy elements to allow planetesimal formation as a side effect. A large fraction of stars in the milky way formed early in the history of the galaxy, with Arcturus being the most prominent member of this so called thick disk population. Together with the presently more conspicuous thin disk population, a galactic population of about 10^{12} star systems has been around for almost 14 Ga, with little indication of an age-metallicity relation, and hence an early enrichment in heavy elements (50% in the first Ga), at least in some parts of the galaxy. There is evidence for a star formation gap (12-8 Ga) after the 'starburst' of the milky way (Fuhrmann 2004). Hence there are 10^{12} potentially planet hosting systems possibly producing planets in the galaxy for 10^{10} years with a 4 Ga gap.

6.2 Quenching Planet Formation

There are two processes that might inhibit planet formation:

1. the failure to form planetesimals,
2. violent migration of planetary embryos.

Planetesimals may fail to form because dust growth stalls and/or gravitational instability of the dust subdisk does not occur. That is halting in a size range (between cm and giant planet size) where all presently available detection techniques are blind for extrasolar systems and hence widespread quenching would be consistent with observations for sufficiently small planets, less then 50 earth masses, say. The indirect evidence of dust-size growth and disappearance of the observational signatures of smaller solids is consistent with the standard view of planet formation but also with an unexpected stalling below the planetesimal size. Replenished disks, like the famous example of β-Pictoris, offer additional evidence for larger bodies via the requirement to resupply the smaller particles, that are observationally accessible, by larger, invisible parent bodies. But their sizes are not sufficiently constrained to conclude about the planetesimal nature of such sources. Possible *falling evaporating bodies* that appear as additional spectral signals of condensible elements and from time to time in high resolution spectra of β-Pictoris and a few other stars provide evidence for extrasolar comets and hence for possible presence of residual extrasolar planetesimals (Lagrange et al. 2000).

At the other end of the size range the detected extrasolar planets and planetary candidates provide evidence for the build up of planetary masses but not (yet) for a growth via planetesimals and condensible element core build-up. Investigations of candidate alternative formation mechanism for those objects, as the disk-instability theories, may succeed to show that they are capable of forming the observed bodies *without* planetesimals. However, detailed models of giant planet formation show that extrasolar planet candidates can be formed in a way consistent with the planetesimal hypothesis, if a wide range of nebulae is assumed to exist (Wuchterl et al. 2000).

The Solar System provides some evidence that the planetesimal formation process operates successfully under a wide range of stellar tides, nebula temperatures and densities, and microphysical and chemical environments. There are icy cometary bodies in the outer Solar System and rocky asteroidal bodies of the sizes assumed in the hypothesis. Hence the process successfully operated widespread, both in heliocentric distance as well as in chemical/mineralogical domains. We may conclude that despite the theoretical uncertainties the pathway to planetesimals does have a finite width but it is still possible that it is very narrow on the scale of the overall galactic stellar population.

Violent migration of *planetary embryos*, if indeed an important process, is more an efficiency issue because repeated planetary embryo generations can be envisaged to balance ongoing losses. If careful timing of nebula removal and planetary migration histories occurs, and that seems plausible at least in a statistical view, at least some embryos—and as a consequence the planets they grow into—will stop migrating with the nebula removed just in time to leave them in a safe orbit (Lin et al. 2000). Violent migration might make terrestrial planets rare but it will not be able to quench the process in the entire galaxy.

7 TOWARDS ESTIMATING GALACTIC HABITABILITY

With the possibility of quenching planet formation due to the failure to produce planetesimals we finally consider the question of what processes may limit the planetary yield. Hence we assume the planetesimal hypothesis to hold for a large fraction of the galactic stellar population in the following.

Given a large fraction of condensible elements in planetesimals there are two possibilities to arrive at an estimate for the number of potentially habitable planets in the galaxy:

1. calculate the planet-forming and planet-degrading processes and look at the outcome,
2. calculate all possible planetary systems, i.e. all combinations of stable planets in stable orbital configurations.

The first approach is analogous to determining the result of a process from reaction kinetics, the second is analogous to thermodynamics.

In the first approach planetesimals grow into embryos rapidly and continue to grow into terrestrial protoplanets. The processes seem to be limited by the amount of available planetesimals and the fact that massive planets start clearing out the remaining building blocks. Kokubo and Ida (2002) have given an estimate of the diversity of expected planetary systems based on a variation of the assumed surface density of planetesimals in the protoplanetary nebula. Giant planets are then an automatic consequence of a terrestrial planet of a few earth masses and availability of nebula gas.

The second approach seems more appropriate to giant planets because their envelopes rapidly reach equilibrium conditions due to their short dynamical time-scales. A variety of proto-giant planets are constructed to arrive at a general overview over masses and properties of such planets (Wuchterl 1993, Ikoma et al. 2001 and Pečnik and Wuchterl 2004). Giant planet formation is found to occur under very general conditions with some indications that massive, Jupiter like planets and close-in Pegasi planets do require more massive nebulae, or at least higher gas- and solid-surface-densities in their feeding zones.

8 CONCLUSIONS

The nucleated instability hypothesis provides a viable model for the formation of the giant planets observed in our Solar System and beyond if the planetesimal hypothesis holds in general. Presently known extra-solar planets may have accreted *in situ* if their preplanetary nebula provided sufficient amounts of gas and solids. Alternatively, according to studies of disk-induced migration, planets may have moved considerably during and after their formation. In that case the orbits of most, if not all, planets known to be bound to main sequence stars *other than the Sun* suffered substantial orbital evolution.

The key question determining how widespread planet formation is in the galaxy, is planetesimal formation. Presently an answer cannot be given, but if it is positive then planet formation should occur in any gravitationally stable circumstellar disk that lives long enough to produce planetary embryos, 10^5a, say.

Planet formation studies are starting to develop predictive elements to determine the properties of planetary systems in general. But the transition from theories to account for the formation of the Solar System, to a general theory of planet formation is far from complete.

Because of that, the completion of the copernican revolution, in the sense that we can safely say that the Earth is not the center of habitability in the galaxy and the Sun not the center of the galactic planet formation process, is still ahead of us.

References

Blum, J. and G. Wurm (2000) Icarus **143**, 138

Bodenheimer, P. (1985) In Protostars and Planets II, eds. D.C. Black and M.S. Matthews, Tucson: Univ. of Arizona Press, 493

Bodenheimer, P. (1986) In Cosmogonical processes, eds. W.D. Arnett, C. Hansen, J. Truran, S. Tsuruta, Utrecht, VNU Sci. Press, 249

Bodenheimer, P., and J.B. Pollack (1986) Icarus, **67**, 391

Bodenheimer, P., Ruzmaikina, T., and Mathieu, R.D. (1993) In Protostars & Planets III, eds. E.H. Levy and J.I. Lunine (Tucson: Univ. of Arizona Press), 367

Boss., A.P. (2003) ApJ **599**, 577

Bryden, G., D.N.C. Lin, and S. Ida (2000), Astrophys. Journ. **544**, 481

Chambers, J. and G.W. Wetherill, (1998) **136**, 304

D'Angelo, G., T. Henning and W. Kley (2002) Astron. Astroph. **385**, 647

D'Angelo, G., W. Kley and T. Henning (2003) Astroph. Journ. **586**, 540

DeCampli, W.M., and Cameron, A.G.W. (1979) Icarus, **38**,367

Fernandez, J.A. and W.-H-Ip (1984) Icarus **58**, 109
Fisher, D., Valenti, J.A., and Marcy, G. 2004 in IAUS 219 Stars as Suns: Activity Evolution and Planets, in press.
Fuhrmann, K. (2000) http://www.xray.mpe.mpg.de/~Fuhrmann
Fuhrmann, K. (2004) Astron. Nachr. **325**, 3
Goldreich, P., and W.R. Ward (1973) Astroph. J. **183**, 1051
Guillot, T. (1999), **286**, 72
Guillot, T., A. Burrows, W.B. Hubbard, J.I. Lunine and D. Saumon (1996) Astrophys. J **459**, L35
Haisch, K.E., E.A. Lada and C.J. Lada (2001) Astrophys. Journ. **533** L153
Hayashi, C., K. Nakazawa and Y. Nakagawa (1985) In Protostars and Planets II, eds. D.C. Black and M.S. Matthews, Univ. of Arizona Press, Tucson, 1100
Hollenbach, D.J., H.W. Yorke and D. Johnstone (2000) In Protostars and Planets IV, ed. V. Mannings, A.P. Boss & S.S. Russell, Tucson, University of Arizona Press, 401
Hueso, R. and Guillot, T. 2003. Space Science Rev. **106** 105
Ikoma, M., H. Emori, K. Nakazawa (2001) Astrophys. Journ. **553**, 999
Inaba, S., and M. Ikoma (2003) Astron. Astrophys. **410**, 711
Inaba, S., G.W. Wetherill, and M. Ikoma (2003) Icarus **166**, 46
Kant, I. (1755) Allgemeine Naturgeschichte und Theorie des Himmels. Johann Friederich Petersen, Königsberg und Leipzig. English translation: W. Hastie "Universal Natural History and Theories of the Heavens" In Kant's Cosmology, Greenwood Publishing, New York 1968.
Kley, W. 1999, MNRAS **303**, 696
Kokubo, E., and S. Ida, (1996) Icarus **123**, 80
Kokubo, E., and S. Ida, (2002) Astrophys. Journ. **581**, 666
Kumar, S.S. (1963) Astrophys. Journ. **137**, 1121
Lagrange, A.-M., D.E. Backmann, and P. Artymovicz (2000) In Protostars and Planets IV, ed. V. Mannings, A.P. Boss & S.S. Russell, Tucson, University of Arizona Press,639
Laplace, P.S. (1796) Exposition du Système du Monde. Circle-Sociale, Paris. English translation, Pond, J. 1809, The System of the World, Richard Phillips, London
Lecar, M., F.A. Franklin, M.J. Holman and N.W. Murray (2001) Ann. Rev. in Astron. and Astrophys. **39**, 581
Levison, H.F., and C. Agnor (2003) Astron. Journ. **125**, 2692
Levison, H.F., J.J. Lissauer, and M.J. Duncan (1998) Astron. J, **116**, 1998
Lin, D.N.C., P. Bodenheimer, and D. Richardson (1996) Nature **380**, 606
Lin, D.N.C., and S. Ida (1997) Astrophys. J. **477**, 781
Lin, D.N.C., J.C.B. Papaloizou, C. Terquem, G. Bryden and S. Ida (2000) In Protostars and Planets IV, ed. V. Mannings, A.P. Boss & S.S. Russell, Tucson, University of Arizona Press, 1111
Lissauer, J.J. (1987) Icarus **69**, 249
Lissauer, J.J. (1993) Planet formation. Ann. Rev. Astron. Astrophys. **31**, 129
Lissauer, J.J., J.B. Pollack, G.W. Wetherill, and D.J. Stevenson (1995) In Neptune and Triton, ed. D.P. Cruikshank, Tucson: Univ. of Arizona Press, 37
Mannings, V., Boss, A.P., and Russell, S.S. (eds.) (2000) Protostars and Planets IV Tucson: University of Arizona Press
Malhotra, R., M.J. Duncan and H.F. Levison (2000) In Protostars and Planets IV, ed. V. Mannings, A.P. Boss & S.S. Russell, Tucson: University of Arizona Press, 1231
Markwick, A. and Charnley, S. (2004) in Astrobiology: Future Perspectives (Proceedings of the ISSI Workshop April 1–4, 2003 Bern), Ehrenfreund et al. (eds).

Mizuno, H. (1980) Prog. Theor. Phys. **64**, 544
Nakano, T. (1987) MNRAS **224**, 107
Nakano, T. (1988a) MNRAS **230**, 551
Nakano, T. (1988b) MNRAS **235**, 193
Nidever et al. 2002 ApJS 141, 503
Papaloizou, J.C.B. and C. Terquem (1999) Astrophys. Journ. **521**, 823
Pečnik, B. and Wuchterl, G. 2004, Astron. Astrophys. **subm.**
Safronov, V.S. (1960) Ann. d'Astrophys. **23**, 901
Safronov, V.S. (1969) Evolution of the Protoplanetary Cloud and Formation of the Earth and Planets. Nauka Press, Moscow (also NASA–TT–F–677, 1972)
Santos, N.C., Israelian, G., Mayor, M., Rebolo, R., and Udry, S. (2003) A&A **398**, 363
Stevenson, D.J. (1982a) Planet. Space Sci. **30**, 755
Tajima, N., and Nakagawa, Y. (1997) Icarus **126**, 282
Tanaka, H., and S. Ida (1999) Icarus, **139**, 350
Thommes, E.W., and J.J. Lissauer (2003) Astrophys. Journ. **597**, 566
Thommes, E.W., and J.J. Lissauer (2004) astro-ph/020941, to appear in proceedings of STScI Astroph. of Life Symposium (May 6–9, 2002)
Trilling, D., W. Benz, T. Guillot, J.I. Lunine, W.B. Hubbard and A. Burrows (1998) **500**, 428
Ward, W.R. (1997) Astrophys. Journ. **482**, L211
Ward, W.R., J.M. Hahn (2000) In Protostars and Planets IV, ed. V. Mannings, A.P. Boss & S.S. Russell, Tucson, University of Arizona Press, 1135
Weidenschilling, S.J., and J.N. Cuzzi (1993) In Protostars and Planets III, eds. E.H. Levy and J.I. Lunine, Tucson, Univ. of Arizona Press, 1031
Weidenschilling, S.J., and F. Marzani (1996) Nature **384**, 619
Wetherill, G.W. (1990) Ann. Rev. Earth Planet. Sci. **18**, 205
Wetherill, G.W., and G.R. Stewart (1989) Icarus **77**, 330
Wuchterl, G. (1993) Icarus **106**, 323
Wuchterl, G. (1995b): In Comparative Planetology, eds. M.T. Chahine, M.F. A'Hearn, J.H. Rahe, Kluwer Acad. Publ., Dordrecht, and Earth, Moon, and Planets **67**, 51
Wuchterl, G. (1996) Bull. Amer. Astron. Soc. **28**, 1108
Wuchterl, G. (1997) In Science with the VLT Intereferometer ed. F. Paresce, Springer, Berlin, 64
Wuchterl, G., T. Guillot and J.J. Lissauer, (2000) In Protostars and Planets IV, ed. V. Mannings, A.P. Boss & S.S. Russell, Tucson: University of Arizona Press, 1081
Wuchterl, G., Klessen R.S, (2001) Astroph. Journal Letters, **560**, L185
Wuchterl, G. and Tscharnuter, W.M. (2003) Astron. Astrophys., **398**, 1081

Chapter 5

FROM ELEMENTAL CARBON TO COMPLEX MACROMOLECULAR NETWORKS IN SPACE

Franco Cataldo
Soc. LUPI Chemical Research Institute, Via Casilina 1626/A, 00133 Rome, Italy

Abstract: Elemental carbon is formed in stars at the end of their life cycle and is ejected in the interstellar medium. Through a review of the elemental forms of carbon, it is possible to have a clearer picture of the formation and the evolution of this element in space, its ability to build complex molecular and macromolecular species, and its basis for the subsequent formation, under prebiotic conditions, of the molecules that are the precursors of life. The aim of the present chapter is to combine, in a tutorial way, the chemical knowledge of elemental carbon forms with the knowledge and hypotheses concerning elemental carbon present in the interstellar medium, from micron-sized grains to nanosized particles and gas-phase molecules.

Key words: Elemental carbon; allotropy; carbon dust; carbon black; diamond; graphite; polyynes; fullerenes; nanotubes; interstellar light extinction curve; UIBs; DIBs.

1. INTRODUCTION

Carbon is a fundamental element in the Universe. It plays a crucial role in stellar nucleosynthesis, and it is the fourth most abundant element in the cosmos (Fig.1). Carbon has a rich chemistry. It can form an enormous number of molecules and macromolecules, and it is the basis of all living things. The versatility of this element lies in its electronic structure which allows it to form stable chemical bonds both with other carbon atoms and with other elements. The foundation of organic chemistry is the ability of elemental carbon to combine with itself, and with hydrogen, oxygen, nitrogen and sulphur, leading to the formation of molecules and macromolecules of extreme complexity which possess chiral properties (Cataldo and Keheyan, 2003a) and which are able to exert extremely complex synthetic and regulatory functions in living organisms and in biological matter.

P. Ehrenfreund et al. (eds.), Astrobiology: Future Perspectives, 97–126.

Organic chemistry occurs in the interstellar medium, in molecular clouds and comets and meteorites, as well as in the atmospheres, surfaces, and interiors of planets. One of the key objectives of astrobiology concerns the question of whether the higher expression of organic chemistry, the biochemical and biological systems and the living organisms are really a peculiarity of the Earth or if also these extremely complex systems are widespread in the Universe.

Because of the limited knowledge we still have of the molecules comprising living organisms, the phenomenon of life is still a mystery for us. A great challenge is now in front of all scientists: to contribute to the solution of the mentioned mystery in an interdisciplinary way. In this context, astrobiology may provide decisive contributions by demonstrating that life, even in its elementary form, is a common feature of the Universe.

Of primary importance is the understanding of how complex organic molecules are formed in space or in certain bodies of the solar systems and the role they may play as a reservoir and as precursors of biochemical molecules. It is also important to know how elemental carbon is formed in stars, in which form it is ejected into the interstellar medium, and its chemical and physical transformations in space. Another important topic concerns how complex organic molecules are generated from the elemental carbon dust that represents from 1-3% of the mass of our Galaxy (Gribbin & Gribbin 1996).

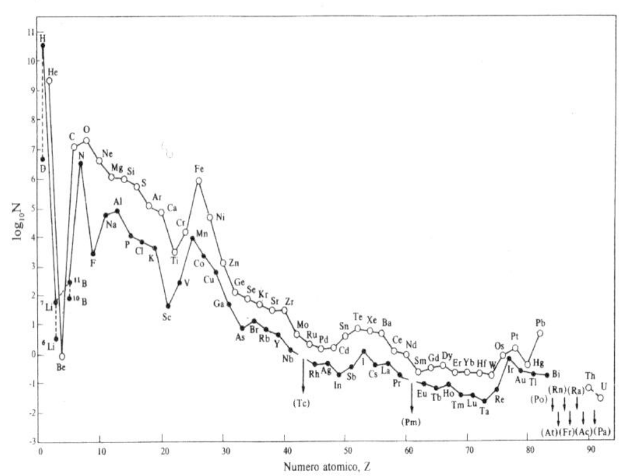

Figure 1. The cosmic abundance of the elements (after Cameron, 1973)

The objective of this chapter is to give a general tutorial overview of elemental carbon in the Universe and the macromolecular structures it forms. We would like to give to the reader a general picture of elemental carbon from its synthesis in stars to the formation of complex carbonaceous macromolecules which may have become the organic matter necessary for life. We will start by explaining nucleosynthesis in stars which lead to the formation of carbon. The allotropy of the elemental carbon used to interpret the chemical structure of the carbon dust present in various astronomical environments is also discussed.

2. CARBON NUCLEOSYNTHESIS: NUCLEAR CHEMICAL ASPECTS

The nucleosynthesis path of elemental carbon in stars has been well established since the second half of the last century (Burbidge et al. 1957; Taylor 1966; Kaler 1992; Gribbin and Gribbin 1996; Unsold and Baschek 2002). In young stars, the so-called proton-proton (PP) processes produce energy through nuclear fusion reactions in which hydrogen is converted to helium. This PP process can be distinguished from the P_1 process, which occurs when low amount of ^4He is available:

$$2\,^1H + ^1H \rightarrow 2\,^2D + 2e^+ + 2\nu$$
$$2\,^2D + 2\,^1H \rightarrow 2\,^3He + 2\gamma$$
$$^3He + ^3He \rightarrow ^4He + 2\,^1H$$

$$4\,^1H \rightarrow ^4He + 2e^+ + 2\nu + 2\gamma \qquad [R.1]$$

The P_1 process involves the fusion of 4 protons into a ^4He atom in a series of reactions involving deuterium ^2D and ^3He as intermediates and with liberation of positrons (e^+), neutrons (ν), and gamma quanta (γ). The total energy produced by the P_1 process is 26.20 MeV. When ^4He is available in sufficient quantity, the P_1 process may be replaced by the P_2 :

$$^1H + ^1H \rightarrow e^+ + \nu + ^2D$$
$$^2D + ^1H \rightarrow ^3He + \gamma$$
$$^3He + ^4He \rightarrow ^7Be + \gamma$$
$$^7Be + e- \rightarrow ^7Li + \nu$$
$$^7Li + ^1H \rightarrow ^4He + ^4He$$

$$4\,^1H \rightarrow ^4He + e^+ + 2\nu + \gamma \qquad [R.2]$$

and P_3 processes:

$$^1H + ^1H \rightarrow e^+ + \nu + ^2D$$
$$^2D + ^1H \rightarrow ^3He + \gamma$$
$$^3He + ^4He \rightarrow ^7Be + \gamma$$
$$^7Be + ^1H \rightarrow ^8B + \gamma$$
$$^8B \rightarrow ^8Be + e^+ + \nu$$
$$^8Be \rightarrow ^4He + ^4He$$

$$4\,^1H \rightarrow ^4He + 2e^+ + 2\nu + 3\gamma \qquad [R.3]$$

All the PP processes involve the transformation of four protons into one ^{4}He atom; these processes cannot happen in a four body collision but involve several steps and some heavier intermediates such as ^{2}D, ^{3}He, ^{7}Li, ^{8}Be and ^{8}B. The energy liberated with the P_2 process (25.67 MeV) is comparable to that liberated with the P_1 process (26.20 MeV) while the P_3 process produces much less energy (19.2 MeV). The PP nuclear reactions occur in relatively cold stars. In warmer stars, PP-processes may be out-competed by CNO nuclear reactions (the "Bethe cycle"):

$$^{12}C + {}^{1}H \rightarrow {}^{13}N + \gamma$$
$$^{13}N \rightarrow {}^{13}C + e^{+} + \nu$$
$$^{13}C + {}^{1}H \rightarrow {}^{14}N + \gamma$$
$$^{14}N + {}^{1}H \rightarrow {}^{15}O + \gamma$$
$$^{15}O \rightarrow {}^{15}N + e^{+} + \nu$$
$$^{15}N + {}^{1}H \rightarrow {}^{12}C + {}^{4}He$$

$$4{}^{1}H \rightarrow {}^{4}He + 2e^{+} + 2\nu + 3\gamma \qquad [R.4]$$

Notice that carbon, together with oxygen and nitrogen, plays a crucial role in the CNO cycle by 'catalyzing' the nuclear reactions which lead again to production of a ^{4}He atom. Notice also that the ^{12}C employed at the beginning is recovered intact at the end of the cycle. For this reason the CNO process is also called the "carbon cycle" as the carbon present in the star is preserved and acts as a catalyst, thereby playing a crucial role. The CNO cycle produces 25.02 MeV of energy, a value very close to that of the P_2 process. All steps of the CNO cycle involve the tunneling of a proton into the given nucleus followed by the decay of the unstable derivative.

With respect to the temperature required for these reactions, in general PP processes occur from $4x10^6$ K to $3x10^7$ K. The CNO nuclear cycle occurs from $1.2x10^7$ K to $3x10^8$ K and hence governs the nuclear chemistry of hot giant stars belonging to the O, B and A spectral classes.

As a giant star consumes more hydrogen, the CNO cycle becomes less effective. Radiation pressure, which counteracts gravitational collapse, diminishes, and the star's nucleus contracts. When the density in the star's nucleus reaches a value of $\sim 10^{4.5}$ g/cm^3, the temperature in the nucleus exceeds 10^8 K. At this point, the CNO cycle is replaced by the so-called He flash or 3α process which involves the expansion of the star. At the end of the CNO cycle, the star abandons the main sequence trajectory in the HR diagram, undergoes further expansion, and migrates toward the M and K spectral classes. The 3α process can be described by the following nuclear equations

$$2\,{}^{4}He + 99keV \rightarrow {}^{8}Be$$
$${}^{8}Be + {}^{4}He \rightarrow {}^{12}C + 2\gamma$$

$$3\,{}^{4}He \rightarrow {}^{12}C + 2\gamma \qquad [R.5]$$

and is called "He flash" because He is consumed on a time scale much shorter than H in the CNO cycle. Moreover, the nuclear reactions involved in the 3α process are not as exothermal as the reactions of the PP and CNO cycles. In fact the energy released in the 3α process is about 7.3 MeV per ${}^{12}C$ nucleus and the first stage of ${}^{4}He$ fusion into ${}^{8}Be$ is even slightly endothermic.

The 3α process is of paramount importance for biochemistry and the existence of life. Without this series of nuclear reactions no elemental carbon can be accumulated in the Universe (Barrow & Tipler, 1986). The importance of the 3α process can be further evidenced if we consider that the formation of elemental carbon is the fundamental step which leads to heavier elements such as ${}^{16}O$:

$${}^{12}C + {}^{4}He \rightarrow {}^{16}O \qquad [R.6]$$
$${}^{16}O + {}^{4}He \rightarrow {}^{20}Ne \qquad [R.7]$$
$${}^{20}Ne + {}^{4}He \rightarrow {}^{16}Mg \qquad [R.8]$$

The fore-mentioned "He flash" is replaced by the "C flash" at $T > 4x10^{8}$ K, whereby part of the C formed in the "He flash" is consumed in the synthesis of other elements:

$${}^{12}C + {}^{12}C \rightarrow {}^{23}Na + {}^{1}H \qquad [R.9]$$
$${}^{12}C + {}^{12}C \rightarrow {}^{20}Ne + {}^{4}He \qquad [R.10]$$

The "C flash" occurs over very short timescales, even shorter than the "He flash" and involves the production of 6.85 MeV of energy.

3. THE NUCLEAR SYNTHESIS OF CARBON AND THE ANTHROPIC PRINCIPLE

Fred Hoyle first mused that life in our Universe exists due to a series of coincidences in nucleosynthesis reactions that permit elemental carbon to accumulate and to become one of the most abundant elements (Hoyle, Wickramasinghe, 1977). In fact, the carbon abundance in the Sun involves from 350-470 carbon atoms per million of H atoms while in the local interstellar medium the average abundance is 225 carbon atoms per million of H atoms. Red giant stars are the most effective contributors to carbon stardust ejection in the ISM. Each red giant may eject 0.002 solar masses of carbon dust per year.

The nuclear reaction [R.5] may occur through the formation of the unstable ^{8}Be only because the average lifetime of ^{8}Be (10^{-17} s) is much longer than the collision time between 2 ^{4}He (10^{-21} s). Thus ^{8}Be lives long enough to collide with ^{4}He and produce a carbon atom, assuming the density of these atoms is sufficient. Another important aspect of this reaction is the ability of the ^{8}Be to form a ^{12}C atom rather than decomposing; this occurs because certain resonance energies of these two sides of the reaction (7.3667 MeV for ^{8}Be + ^{4}He and 7.6549 MeV for ^{12}C) are very similar (Barrow & Tipler, 1986). In addition, when ^{12}C reacts with a ^{4}He nucleus to form ^{16}O, the energy of that system ^{12}C + ^{4}He (7.1616 MeV) is just *above* the level of ^{16}O (7.1187 MeV), and thus the transformation of carbon into oxygen is *not* resonant. If instead the level of the system ^{12}C + ^{4}He had been just below the level of ^{16}O (as it happens for ^{8}Be + ^{4}He in the formation of ^{12}C), the nuclear reaction would have been resonant and all carbon formed would have been transformed into oxygen. If this had been the case, carbon would have been absent and no life would have originated our Universe (Barrow and Tipler, 1986).

4. ELEMENTAL CARBON IN THE INTERSTELLAR MEDIUM: CARBON DUST

Asymptotic giant branch (AGB) stars play a key role in the chemical evolution of the interstellar medium (ISM). The elemental carbon and oxygen formed in the core of these stars, as outlined in the previous sections, are released into the circumstellar environment and ISM. O-rich and C-rich molecules and refractory solids may form in these stellar envelopes. In the case of carbon-rich giants, the elemental ratio, C/O, in the mass outflow is larger than 1 and all oxygen is present as CO. From the CO, carbonaceous grains, carbides and sulfides are formed via chemistry analogous to that known for hydrogen-rich atmospheres. As discussed by Henning (1998) and by Cataldo and Pontier-Johnson (2002), the formation of the carbonaceous materials in the outflows of evolved C-rich stars appears to be analogous to known mechanisms of soot chemistry in hydrocarbon flames. In fact, pyrolysis in certain flame zones produces simple molecules such as acetylene; polyyne chains are derived from this precursor and may be precursors of polycyclic aromatic hydrocarbons (PAHs), which, in turn, are considered to be the 'nuclei' from which larger graphene sheets grow in the gas-solid interface of dust. The graphene sheets, once they are sufficiently large in size, cluster together to form soot.

Carbon grain formation also takes place in hydrogen-deficient atmospheres of carbon-rich stars (e.g., R Coronae Borealis and Wolf-Rayet). Here carbon dust should form by chemistry analogous to synthesis involving pure carbon vapor (cf., Kraetschmer and Huffmann, 1992). These reactions involve both linear carbon chains and carbon ring structures comprised of alternating single- and triple-bonded carbon atoms (acetylenic rings). When carbon vapor is quenched in a low-pressure inert gas, larger mono- and polycyclic rings presumably derived from the condensation and crosslinking of acetylenic rings. Extensive crosslinking leads to fullerenes, principally fullerene black or fullerite (Henning, 1998; McCarthy and Thaddeus, 2001). Acetylene (or its chains) is considered to be the universal precursor of soot and carbon dust, because acetylene is able to form both polyyne chains and, by cyclization, also PAHs.

Chemical growth of dust takes place between 1100-800 K (Henning, 1998; Henning and Salama, 1998). Below 1100 K, the addition of acetylene at the edges of a PAH or graphene sheet is an irreversible process. Below 800 K, further growth is hindered by the low rate of hydrogen abstraction from the PAH edges. When present, metal carbides may also form at 1100K and contribute to the soot nucleation process.

In the case of O-rich stars (where $C/O < 1$), the available carbon occurs as CO. No carbon dust is observed and silicate and other metal oxides are predominant.

As reported by Henning and Salama (1998), the red giant stars are the largest contributors to the stardust ejection in the ISM of the Galaxy. Supernovae may also contribute in a significant way but their efficiency in dust production as well as the chemical composition of the related dust is not well known. The mass loss rate for carbon-rich stars has been calculated to be in the range of 10^{-5}–10^{-7} solar masses per year and the total production rate of solid C from carbon-rich stars is about 0.002 solar masses per year (Kaler, 1992 and 1997).

Snow and Witt (1995) pointed out that the average carbon abundance in the sun (from 350 to 470 carbon atoms per 10^6 H atoms), its nearby galactic environment, and in general in our solar system is higher than the average abundance of carbon in the interstellar medium (about 225 carbon atoms per 10^6 H atoms), which suggests that a supernova event was closely associated with the formation of the solar system.

5. SOME GENERAL ASPECTS OF THE INTERSTELLAR CARBON DUST

The observation of a reddening and dimming of stellar light, referred to as an interstellar extinction, was first observed in the 1930's and interpreted to be a consequence of both absorption and scattering of light by small solid particles.

These solid particles, or carbon dust, include species ranging from gas-phase molecules to micro-sized grains. It is generally thought that the absorption feature observed at ~217 nm of the average normalized interstellar extinction curve of our Galaxy is due to carbon dust in general (Ehrenfreund et al. 2002; Ehrenfreund and Charnley, 2000; Savage and

Mathis, 1979) and until recently many researchers held that this dust is essentially made of graphitic particles. To verify this hypothesis, many attempts were made to reproduce in the laboratory both the interstellar conditions of dust formation and the formation of the carbon dust itself. These efforts were successful, and it is now widely accepted that the feature at ~217.5 nm (Fig.2) of the interstellar light extinction curve is due to carbon materials, dominantly PAHs. The peak position of this UV "bump" in the interstellar extinction curve is remarkably constant, with variation of 1% and a half-width of about 40 nm (Savage and Mathis, 1979).

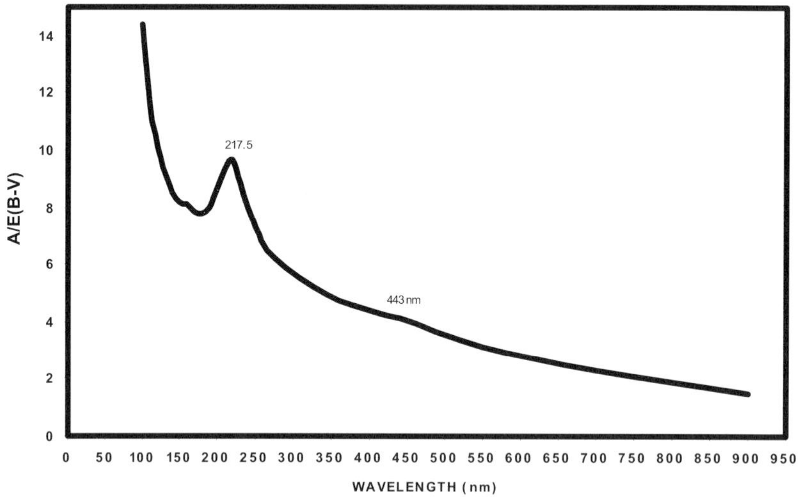

Figure 5.2. Interstellar light extinction curve. Constructed with the data of Savage and Mathis, 1979.

Theoretical calculations have shown that the band profile of the UV peak can be explained by a population of small spherical graphite particles about 6 nm in diameter (Jager et al. 1999). However, such a uniform particle size is improbable, and many other forms of carbon have been proposed to explain the feature (e.g., Ehrenfreund and Charnley 2000). These include graphitic onions (onion-like carbon), amorphous carbon (AC), hydrogenated amorphous carbon (HAC) (eventually nano-sized and UV-processed), silicate or graphite core particles with a mantle of absorbed PAH, coal-like material (anthracite-like), mixtures of PAHs, fullerenes, quenched carbon condensates (QCC), small carbon particles with sp^2/sp^3 hybridization bonding (nanodiamonds in sp^2 amorphous carbon coating). Among the various proposed structures, the most convincing are those involving carbon black particles made by a mixed graphitic and amorphous phase (Jaeger et al. 1999; Michel et al. 1999) and hydrogenated and radiation processed nano-sized carbon grains (Mennella et al. 1995; 1996; 1997; 2001).

Experimentally, carbon vaporization in an electric arc discharge usually produces carbon black particles with an absorption maximum in the UV at 230 nm (Kraetschmer and Huffmann 1992). Adequate quench conditions provided by an inert gas such as He produce a fullerene-rich carbon soot with characteristic fullerene absorption features at 215, 265 and at 340 nm (Kraetschmer and Huffmann 1992). We now know that C_{60} fullerene, present in very low amounts in interstellar dust and molecular clouds (Ehrenfreund et al. 2002; Ehrenfreund and Charnley 2000), cannot account for the UV interstellar feature at 217.5 nm. Fulleranes, the hydrogenated fullerenes, appear to be a more promising candidate for the bump at 217 nm (Webster 1995;Cataldo 2003a) and would form readily in space due to the presence of excess hydrogen.

It has been proposed that the carbon dust surface catalyzes the formation of molecular hydrogen from atomic hydrogen (Pirronello et al. 1997). Carbon dust composed of curved graphene sheets and fullerene-like structures (Cataldo 2002a), probably provides an even better model than common carbon dust or idealized graphene planar sheets for this reaction (Cataldo, 2003b).

Cataldo (2002b) has shown that the carbon black produced by industrial processes cannot create an adsorption at 217 nm but instead show absorptions above 250 nm, usually at 265 nm. Similarly, fullerite (the carbon soot formed by arc discharge between two graphite rods) does not absorb at this energy either. In fact the two carbon materials have analogous absorption spectra. It must be said, for the sake of completeness, that no feature has been detected at 217 nm in the circumstellar shells of late type stars but rather at higher wavelengths (e.g., 240-250 nm; Jager et al. 1999); this may be a consequence of thermal processing or "graphitization" of the dust (Cataldo 2002b).

6. CARBON ALLOTROPY AND THE STRUCTURE OF CARBON DUST

Broadly speaking, the term allotropy refers to the phenomenon that an element can assume different chemical and/or crystalline structures. However, the concept of allotropy sometimes appears too limiting when applied to carbon due to the diversity of its chemical structures (Diederich and Rubin 1992). For this reason it is convenient to classify carbon allotropes based on the valence orbital hybridization (Heimann et al. 1997). According to the proposed classification (Fig. 3), there are three fundamental carbon allotropes: diamond, graphite and carbyne. In each of these elemental forms the carbon atoms are have hybridized sp^3, sp^2 and sp orbitals, respectively. Fullerenes, carbon nanotubes and related structures (Fig.3) are intermediate forms of carbon because of the deviation from planarity imposed by the curved surfaces in these molecules or materials; the carbon hybridization is no longer pure sp^2 but an intermediate value between sp^2 and sp^3 (Cataldo, 2002a). An acetylenic or polyyne carbon chain is also called carbyne; it is an almost linear carbon chain composed of a sequence of sp hybridized carbon atoms. However, when a long polyyne chain forms a ring, we obtain a cyclopolyyne which again is no longer bonded by purely sp hybridized carbon. The distortion from linearity introduces a higher degree of hybridization and hence the carbon atoms are at an intermediate level of hybridization between sp and sp^2.

There are also mixed forms of carbon combining carbon atoms with different types of hybridization in the same solid. Examples of mixed forms of carbon are carbon black, glassy carbon, diamond-like carbon, and carbynoid structures (Fig. 3).

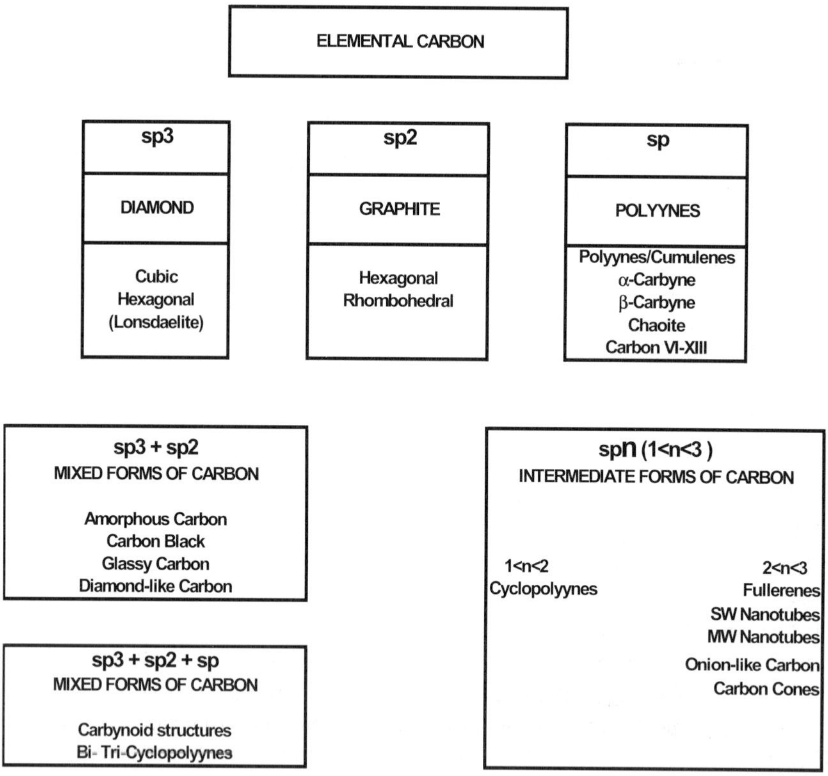

Figure 5.3. Carbon allotropes and other forms of carbon. Elaborated and updated by the author; original reference: Heimann, Evsyukov and Koga, 1997.

7. CARBON ALLOTROPY: A SHORT SURVEY

7.1 Diamond

Diamond is essentially a macromolecular crystal made by sp^3 hybridized carbon in which each atom is bonded tetrahedrally to four other carbon atoms and the whole crystal possesses cubic symmetry with a_o = 0.356679 nm . The C-C bond is given by $a_o\sqrt{3/4}$ = 0.15545 nm which is the normal bond length for alkanes (Holliday et al. 1973; Greenwood and Earnshaw 1984). The strong bonding throughout the crystal makes diamond one of the hardest known substances. Its high boiling point and low coefficient of cubical expansion are due to this bonding. In the absence of impurities, there is a large energy gap (6 eV) between the highest occupied and lowest unoccupied molecular orbitals resulting in a colourless material that is an excellent electrical insulator. The presence of impurities (e.g., nitrogen or boron) or defect-induced by ionizing radiation may produce colour. Defects can be annealed from a diamond to a point, depending on the degree of radiation damage. Diamonds have two main classifications: Type I and Type II. Both have a cubic unit cell but Type I is almost ideal in its structure and Type II shows fine parallel laminations. Furthermore the infrared spectrum of Type I diamonds have four main absorption bands at ~3333, 2439, 2083 and 1280 cm^{-1}. Type II diamonds lack the absorption at the lowest wavelength, which has been attributed to the presence of N impurities in Type I diamonds. In the UV region, the absorption edge of a Type I diamond lies below 300 nm but occurs at 225 nm for Type II (Holliday et al. 1973).

Lonsdaelite is a special type of diamond formed by shock impact (Pillinger, 1993). It was first identified in 1967 in samples derived from Canon Diablo iron meteorite in Arizona. Lonsdaelite is characterized by a hexagonal wurzite-type crystal structure. Lonsdaelite can be produced in the laboratory using common diamonds through shock loading or static high pressures and temperatures (>1000°C).

Nanodiamonds and diamond nanograins, both having cubic structures, are interesting because of their occurrence in meteorites (Pillinger, 1993, Heymann, 2001), in which they appear to predate impact delivery. Belonging to the family of presolar grains, the nanodiamonds found in meteorites can be used as a fossil record of nuclear and chemical processes in stars and interstellar medium (Heymann, 2001). The nanodiamonds apparently formed by carbon condensation in cool stellar atmospheres with Xe, from supernova explosion (Heymann, 2001).

Individual nanodiamonds from meteorites have diameters >3 nm which corresponds to about 2000 carbon atoms. Recently, it has been shown (Koscheev et al. 2001) that 4 nm synthetic nanodiamonds (UDD) impregnated with ionic noble gases, were able to release the gases when heated. The release pattern of trapped noble gases and their isotopic fractionations have been found to be analogous to those observed in presolar nanodiamonds isolated from meteorites. In a recent work (Cataldo and Koscheev, 2003) it has been shown that pure nanodiamond grains in water have a maximum of absorption at 203 nm and a shoulder at 250 nm.

7.2 Graphite

Graphite is made by sp^2 hybridized carbon atoms. Graphite is a weak and lubricating material consisting of parallel planar layers of carbon atoms arranged in sheets of condensed hexagonal rings. These layers are also called graphene sheets. In normal α (or hexagonal) graphite, the graphene sheets are arranged in the sequence -ABAB-, whereas in β (or rhombohedral) graphite, the stacking sequence is -ABCABC-. In both forms, the carbon-carbon bond length within the basal plane is 0.1421 nm and the interplanar spacing is 0.33539 nm. This distance increases slightly in poorly crystalline carbons. The dimensions of the unit cell of hexagonal graphite are a_o= 0.24611 nm and c_o=0.67078 nm. The C-C bond length in graphite suggests a bond order of 1.33. Within each graphene sheet, each carbon atom is bonded to three other carbon atoms. The non-bonded p_z atomic orbitals of each carbon atom overlap to give a completely delocalized system of molecular orbitals, allowing delocalization of π electrons. Graphite is a moderate electrical conductor although it shows anisotropy in this property. The black colour of graphite is due to the excitation of the delocalized π electrons. Although some naturally occurring graphite contains traces of the β form, all synthetic graphite belongs to the α-form. The β-form may be obtained by grinding of the α-form but can be converted back to the α-form by thermal annealing at 1000°C (Holliday et al. 1973).

It is important to mention here that synthetic graphite can be obtained through a relatively complicated industrial process. Graphitization efficiency depends on the precursor materials. For instance, coke (a thermal residue produced from coal) heated in an electric furnace in the presence of silica at 2600°-3000°C for one-two days produces synthetic graphite and is considered a good precursor. Other materials such as carbon black and glassy carbon do not yield graphite under any conditions (Oberlin et al. 1973).

Another aspect of graphite is its high sensitivity to radiation damage. It has been known for a long time that corpuscular radiation causes atom displacements and C-C bond breakage resulting in curling and distortions in the graphene sheets. Similarly, prolonged, high-energy radiation such as γ-rays introduce defects in graphite (Galvan et al. 1998; Cataldo 2000a). Such defects can be partially repaired only by a prolonged annealing at 1000°C. These experimental results suggest that micrometer to nanometer size graphite grains immersed in a field of cosmic rays, solar winds and high energy ionizing electromagnetic radiation cannot survive, but must undergo a transformation into carbon black.

Graphite, disordered graphite and especially carbon black are paramagnetic solids. The paramagnetism arises from the presence of dangling bonds at the edges of each graphene sheet. The smaller the graphene sheets, the higher the disorder in these solids and the larger the concentration of the carbon radical sites. Obviously, not all the carbon atoms at the end of the graphene sheets have dangling bonds; some are saturated by impurities. Surfaces of diamonds and especially nanodiamonds are terminated by hydrogen atoms chemisorbed on the surface or by other elements.

7.3 Carbon Vapour and Polyynes/Carbyne

At atmospheric pressure, carbon sublimes above 3367°C. The complex nature of carbon vapour has been known for fifty years. Mass analysis of carbon vapour formed in an electrical discharge between graphite electrodes has revealed the presence of all ions from C_1^+ up to C_{33}^+ and C_{35}^- (Holliday et al 1973). By contrast, laser ablation of graphite by ruby laser produces a series of species ranging from C_1 to C_{10}, the most abundant being C_2 and C_3 (Holliday et al. 1973). Tanuma and Pal'nichemko (1995) and Lagow et al. (1995) have demonstrated polyyne character of the carbon vapour through laboratory investigation. Molecular orbital theory has been applied to the study of carbon vapour, predicting the growth of linear polyyne molecules. Extrapolation of the calculations to liquid carbon predicts an infinite carbyne chain (Holliday et al 1973). Laser vaporization of graphite (Rohlfing et al, 1984) confirmed the earlier observations that odd and even clusters were obtained up to C_{38}, but only even clusters above that value. It was shown later by Kroto and Walton (1993) that the carbon clusters larger than C_{38}, having only even number are consistent with fullerenes dominated by C_{60} and C_{70}. Conversely, the clusters having less than 38 carbon atoms have been assigned to polyynes and/or cyclopolyynes which are considered to be the precursors of fullerenes (Alekseev and Dyuzhev 2002; Kroto and Walton, 1993).

The presence of non-carbon end groups in the polyyne chains precludes these chains as pure carbon allotropes. However, carbyne can be described as a polyyne because it is a almost infinite carbon chain made by sp hybridized carbon atoms (Heimann et al. 1999):

$(-C{\equiv}C-C{\equiv}C-C{\equiv}C-C{\equiv}C-C{\equiv}C-)_n$

As such, it occurs as a solid. Polyyne chains may be partially crosslinked and embedded into other forms of carbon, thereby becoming more stable structures. The allenic isomer of acetylenic carbyne may be a viable structure as well (Heimann et al. 1999):

$=(C=C=C=C=C=C=C=C=C)_n=$

When acetylenic and allenic isomers are compared, the former is more stable (Cataldo 1997a). Acetylenic carbyne is relatively stable especially if its chain is end-capped for instance by copper atoms (Heimann et al 1999). In an excellent work of Lagow et al. (1995), it was clearly shown that end-capping of acetylenic chains with relatively bulky groups engender some kinetic stability to a thermodynamically unstable material.

Low molecular weight end-capped oligomers of carbyne (cyanopolyynes) have been identified in the interstellar clouds especially near carbon-rich stars (Heimann et al 1999). Thermodynamically, a carbyne is considerably less stable than graphite and diamond. Another interesting result on carbyne is that it shows a tendency to crosslink and the resulting derivatives are diamond-like carbon, carbynoid structures and graphite (Cataldo and Capitani 1999). Recently, a real breakthrough occurred in our knowledge of carbyne, when Ravagnan and co-workers (2002) demonstrated that it is possible to obtain pure carbyne chains by soft deposition of carbon vapour on reactive surfaces, demonstrating once again that the real nature of the carbon vapour is under the form of acetylenic carbon chains.

When carbyne is mixed with other carbon atoms in different degree of hybridization, the materials are classified as carbynoids. The presence of carbynoid structures in carbonaceous matter has been confirmed by Raman, FT-IR and solid state ^{13}C-NMR spectroscopy (Cataldo 1999a-c; Cataldo and Capitani 1999). In these solids, carbynoid structures occur in combination with diamond-like carbon and disordered graphite (Cataldo 1999a; Cataldo and Capitani 1999).

Polyynes and fullerenes are observed in the electric arc discharge between graphite electrodes when the quenching gas is He. When hydrogen at high pressure is introduced in excess, PAHs, fullerenes, and polyynes are formed in the arc (Hirsch, 1994). When HCN or C_2N_2 are present as quenching medium for carbon vapour, cyanopolyynes are formed (Lagow et al. 1995, Hirsch, 1994). All the mentioned species are known to exist in the galactic medium. Essentially the same intermediates, i.e., PAHs, polyyne chains and fullerenes have been detected by *in situ* mass spectrometry of laminar acetylene (or other hydrocarbon) flames (Richter et al. 2000; Hall-Roberts et al. 2000; Fialkov et al. 2001). Fullerene formation involves an isomerization of polyyne chains to polyyne rings, which then rearrange to fullerenes. Other mechanisms of fullerene formation for instance from coronene and other PAHs have been reviewed (Richter et al. 2000).

7.4 Polyyne Formation in Bulk Quantity with the Submerged Electric Arc Technique

Another breakthrough in the field of polyyne formation is represented by the simple synthesis of these carbon chains by the electric arc discharge between graphite electrodes submerged in a solvent (Cataldo, 2003c-f). This is referred to as "submerged electric arc technique". The electric arc between carbon electrodes is known to generate polyynes under vacuum or in presence of a quenching gas. This classic method is useful to detect the polyyne chains by mass spectrometry but it is not useful for bulk synthesis. The same applies for laser ablation of graphite targets. When an electric arc is struck between two graphite electrodes submerged in a solvent such as methanol or n-hexane, the polyynes which are formed from the carbon vaporization are quenched in the solvent and remain in solution. They can accumulate in solution by prolonging the arc treatment. The polyynes can be readily detected in solution using electronic absorption spectroscopy because they have extremely characteristic absorption spectra. Furthermore, the polyyne series $H\text{-}(C{\equiv}C)_m\text{-}H$ with m = 3,4,5,6,7,8,9 obtained in solution with the submerged electric arc technique can be easily separated by liquid chromatography (HPLC) and detected with a diode-array detector (Cataldo, 2003c-f). The longest polyyne carbon chain detected in solution until now has consisted of 18 carbon atoms (m=9) and was obtained by arcing graphite electrodes submerged in acetonitrile cooled to -40°C. The polyynes obtained in solution were essentially hydrogen-capped. In solution the polyynes are stable for weeks even if the solutions are exposed continuously to air. This stability is quite amazing because previously it was thought that it was a peculiarity only of polyynes end-capped by bulky groups. However, the polyyne solutions are easily photolyzed in UV light although the photolysis rate appears rather slow (Cataldo, 2003c-f). Moreover, distilling off the solvent under reduced pressure causes the decomposition of polyynes at room temperatures.

The submerged electric arc synthesis of polyynes has demonstrated that polyynes (and hence carbyne) can be easily synthesized and are quite stable. While the main products of the submerged electric arc discharge are polyynes, it is worth mentioning here that also PAHs have been detected as secondary products. This result is of extreme importance because demonstrates once again how these two classes of molecules, polyynes and PAHs, are linked. Knowledge of polyyne production may thus throw new light on our knowledge of the chemistry of carbon vapour and help us to better understand the mechanism of formation of PAHs, soot, fullerenes and nanotubes, among other compounds (Cataldo, 2003c-f).

7.5 Fullerenes, Carbon Nanotubes and Onion-like Carbon: The Concept of "Curved Graphene Sheets"

Our knowledge of carbon allotropy has changed in recent years due to a series of important discoveries involving the identification of fullerenes (Kroto et. al. 1985) and the subsequent discovery of carbon nanostructures such as multiple-walled (Ijima, 1991) and single-walled carbon nanotubes (Ijima et al. 1993). These new molecules were initially produced by laser ablation of graphite targets (Kroto et. al. 1985) and later created by a simple resistive heating of graphite or graphite vaporization via electric arc discharge under Helium atmosphere (Kraetschmer et al. 1990). Fullerene formation was detected in low pressure premixed flames (Howard et al. 1991).

Fullerenes are carbon molecules having a closed shell or cage-like structure. The most studied fullerene has 60 carbon atoms and possesses a highly symmetric structure of truncated icosahedron. The 3D structure of C_{60} fullerene consists of 12 pentagonal rings of carbon atoms which are fully annealed by hexagonal rings. The building principle of fullerene series is governed by the Euler's rule, in which 12 pentagons and X hexagons are needed to built a fullerene cage according to the following relationship: $2(10 + X) = N°$ carbon atoms. The smallest fullerene possible is C_{20} when $X = 0$. Starting from C_{20} any even-membered carbon cluster, except C_{22} can form at least one fullerene structure. With increasing X the number of possible fullerene isomers rises dramatically, from only 1 for X=0 to 20,000 for X=29 and so on. The reason why the C_{60} fullerene is the smallest stable and most studied fullerene is elucidated by the isolated pentagons rule, which requires that each pentagonal site must be fully surrounded by five hexagonal sites. This rule eliminates out all fullerene homologues having adjacent pentagonal sites, for instance, all isomers of C_{60} that have adjacent or fused pentagons. Hence C_{60} fullerene is a unique molecule that does not have stable isomers.

Another feature of C_{60} fullerene (and its homologues) is that the double bonds present in these molecules are localized and weakly conjugated. This is due to the strong deviation from planarity of the curved surfaces of these molecules. The double bonds occur at the junction of two hexagons and are shorter (1.38 A) than the single bonds which exist at the junction of a hexagon and a pentagon (1.45 A).

The topic of fullerenes in nature and in the cosmos has been discussed in a recent paper by Heymann et al. 2003.

Carbon nanotubes (Saito et al. 1998) and fullerene onions (or nested fullerenes) have different morphologies and involve a variable number of carbon atoms, but all share a cage-like polyhedral geometry. Regardless of the number of carbon atoms in a closed-shell fullerenic structure, the cage is always comprised of exactly 12 pentagons distributed over the surface of the cage and the remaining carbon atoms are arranged in hexagons that complete the cage. The structure of carbon nanotubes is derived by rolling up a graphene sheet, which is then sutured to form a cylinder. Each end of the nanotube cylinder is capped with a fullerene half- cage. It is possible to obtain an open nanotube by oxidizing and removing the fullerene caps. The open nanotubes can be filled with a variety of molecules including C_{60} fullerene (fullerene peapods).

The presence of the pentagonal rings in a graphene sheet causes strain that can be relieved by the introduction of curvature at each pentagon (Cataldo 2002a). The pentagons provide sufficient curvature for a sheet of otherwise hexagonally bound carbon atoms to form a closed shell, the shape of which is determined by the arrangement of the pentagons around the shell. If the sheet contains fewer than 12 pentagons the layer is curved in an open shell, the depth of which can range from shallow if only a few pentagons are present to deep or even approaching closure as the number of pentagons increases or approaches 12. The variety of morphologies of curved layers is very large, given the many different possible arrangements of pentagons (Cataldo 2002a). The presence of a pentagon introduces a positive local curvature in a graphene sheet, while a heptagon introduces a negative local curvature. The presence of pentagons may induce the formation of even more exotic nanostructures including carbon nanocones.

8. CARBON BLACK AND SOOT

It is possible to introduce different degrees of disorder in graphite and also to reduce the dimension of crystallite sheets of graphite. Amorphous carbon is composed of fragments of carbon sheets or graphene layers arranged without long-range order while carbon black shows a turbostratic distribution of the mentioned layers. Usually X-ray diffraction of these materials reveals that they are practically amorphous although a short range order can be observed in certain cases.

8.1 Carbon Black Preparation and General Aspects

Carbon black is of great interest to the astrophysicists because it was discovered that the interstellar carbon dust may have a comparable structure (Mennella et al. 2001, 1997,1996,1995; Jager et al. 1999) and that the mechanism of formation of carbon black in the circumstellar medium may mimick that in terrestrial sooting flames (Henning, 1998, Cherchneff 1998). Carbon black is a well-known material in the Earth and is widely used as filler in rubber and plastics.

Below is an outline of the basic aspects of terrestrial carbon black, which is industrially produced by four main processes (Donnet et al. 1993):

i. The furnace process involves the combustion of heavy oils in stoichiometric deficiency of oxygen. Carbon black is therefore produced by the partial pyrolysis of the oil during the combustion.
ii. The thermal process involves the decomposition at high temperature and in the absence of air of a heavy hydrocarbon feedstock similar in its components to that used in the furnace process. Methane or other hydrocarbons may be used in this process.
iii. The acetylene process, involves thermal decomposition in the absence of air of acetylene with formation of carbon black via the liberation of hydrogen.
iv. The plasma process involves thermal decomposition of methane or other hydrocarbons in an electric arc discharge producing acetylene and carbon black.

It is also possible to produce carbon black through the amorphization of graphite, fullerenes and nanodiamonds using high energy radiation.

Carbon black is amorphous carbon made up of essentially a turbostratic aggregation of very small graphene sheets (Donnet et al. 1993). The graphene sheets may contain fullerene-like defects (Cataldo 2002a). Additionally, carbon black may contain variable nanocrystalline domains, depending on the conditions in the production processes and of subsequent processing.

Another classification of carbon blacks and soot may be undertaken based on morphology. Figs. 4 and 5 illustrate this concept with high-resolution transmission electron microphotographs (HRTEM) of a furnace-produced carbon black (process i) and an acetylene carbon black (process iii), respectively. It is clear that the microstructure of acetylene black exhibits a larger degree of order, both with respect to length and organization, when compared to carbon black generated in the furnace process. This is due to the higher temperatures reached in the acetylene reactor and the long residence time and slower formation process of the acetylene black in comparison to very fast formation associated with the furnace process. In Fig. 4, the turbostratic arrangement of the graphene sheets in a furnace carbon black can be easily appreciated together with the presence of onion-like carbon nanostructures and other curved surfaces. The curved surfaces suggest the presence of fullerene-like structures. The fact that the graphene sheets are not always flat can be easily detected in the HRTEM image (Fig. 5). In addition, ovoidal features can be detected on the carbon surface. These structures may be interpreted again as fullerene-like structures and as giant fullerenes, which are not extractable with solvents.

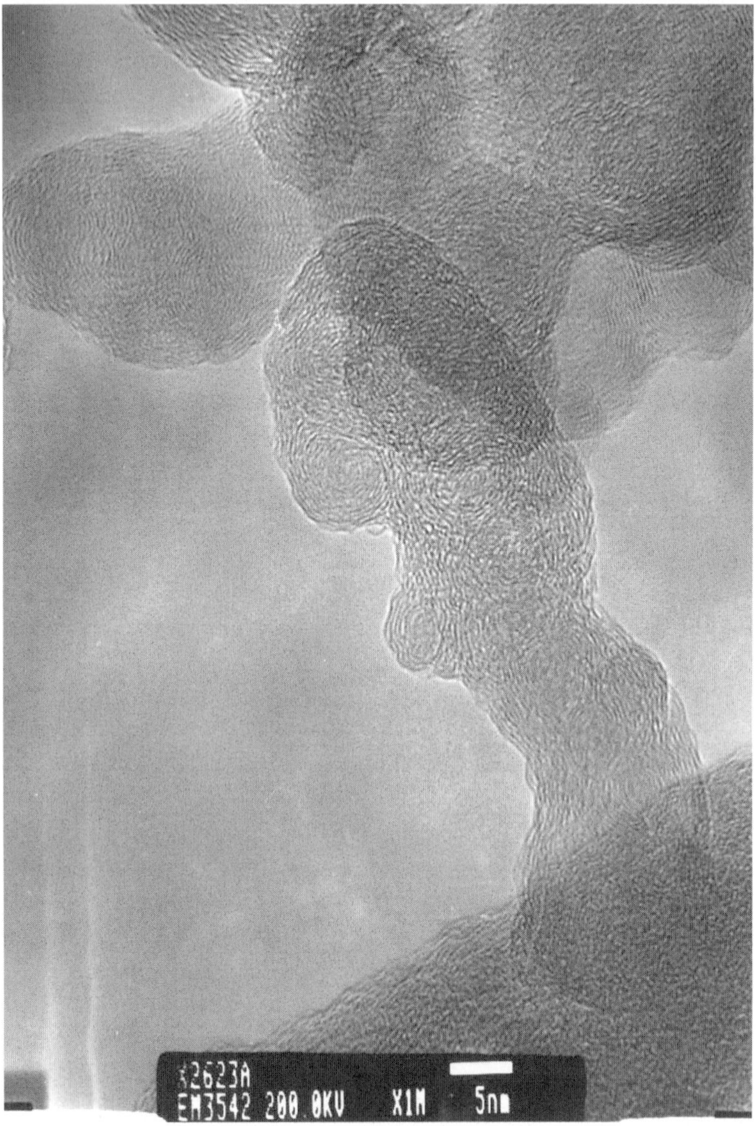

Figure 5.4. HRTEM of furnace carbon black. Note the turbostatic arrangement of the graphene sheets together with the presence of onion-like carbon nanostructures and other curved surfaces.

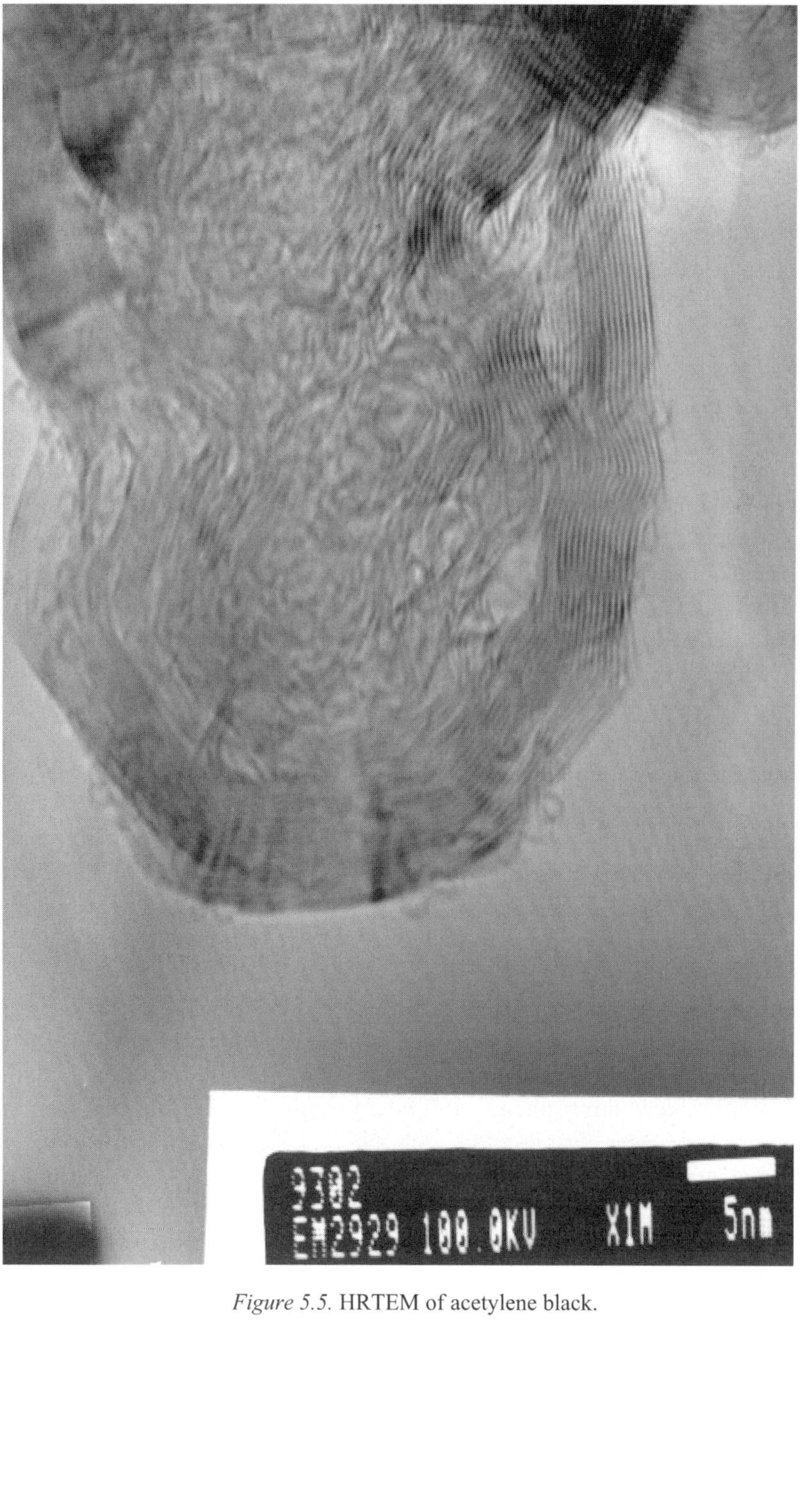

Figure 5.5. HRTEM of acetylene black.

Carbon black processing at high temperature (2000°-3000°C) avoiding contact with air always causes some, but incomplete graphitization. This is of capital importance because it implies that perfect graphite grain formation from carbon black is impossible in interstellar media. Instead, the so-called glassy carbon is formed, showing anyway an increased degree of order in comparison to the pristine carbon black.

Today, it appears more likely that the carbon dust in the universe may have structure closer to that of carbon black than graphite. However, graphite is still used as a model for interstellar dust.

It is also worthwhile to mention here that Donnet and co-workers (1993) reported an elemental composition of furnace carbon black which appears to be closer to that reported for interstellar carbon dust. It is also quite similar to the composition of the carbon black produced in minute amounts experimentally by resistive heating of graphite electrodes under vacuum at temperatures using quench gases such as He or H_2, designed to mimic conditions simulating circumstellar outflow regions (Mennella et al. 2001; 1997; 1996; 1995; Jaeger et al. 1999).

Particles of carbon black generated in a furnace range in size from 10-80 nm with a modal maximum of ~33 nm. In comparison, particles generated via the thermal process are larger and range in size from 120-500 nm. Acetylene carbon black particles more closely resemble those of furnace carbon black and range from 32-42 nm. The particle size of carbon black obtained by resistive evaporation under conditions mimicking the circumstellar medium ranges from 10-45 nm is thus closer to acetylene carbon black and near the peak size of furnace carbon black.

Interstellar carbon dust and furnace carbon black or flame soot have more in common than their structures (Jaeger et al. 1999) since they also share common formation mechanisms and precursors. The most abundant precursor molecule for formation of the carbon grains in the carbon star outflow is acetylene (Unsold and Baschek, 2002), which may undergo a thermal decomposition reaction to form polyynes and PAHs. These two molecular classes are recognised precursors of soot (Richter et al. 2000; Hall-Roberts et al. 2000; Fialkov et al. 2001; Krestinin 2000). In the hydrogen-deficient atmospheres of carbon-rich stars, we believe that the formation of carbon dust begins with polyyne chains and progresses to fullerenes and finally to dust (Henning 1998, Cherchneff 1998). Based on the above discussion, the closest terrestrial analog of interstellar carbon dust is therefore acetylene carbon black, which is produced in conditions of low free oxygen and high temperatures.

8.2 Soot

In flame combustion, soot is formed initially by pyrolysis and fragmentation of the hydrocarbon used as feedstock. These fragments may be ions or neutral radicals. For instance, a key building unit of soot is the molecular species C_2H; the most accepted mechanism in soot formation involves the repeated multiple addition of such units, leading to the formation of polyynes and PAHs (Richter et al. 2000; Hall-Roberts et al. 2000; Fialkov et al. 2001; Krestinin 2000). According to Krestinin (2000), polyynes yield soot directly. However, others conclude that soot is formed by a complex process of addition reactions among acetylenic molecular species and oligomeric polyyne chains followed by cyclization reactions. Larger PAHs form by an accretion process involving the addition of acetylenic species in the gas phase over the smaller PAHs resulting in graphene sheets (Richter et al. 2000). The graphene sheets then agglomerate to form crystallites which, in turn, are the building blocks of the spherical particles which constitute the soot aggregates. The following steps have been proposed for soot formation (Richter et al. 2000):

TIME	EVENT	Molecular Mass (in atomic mass units)
0.5 ms	primary radicals and/or ions formation	25 Da
2 ms	polycyclic ions	10^2 - 10^3 Da
3 ms	incipient particles (size of fullerene cage)	10^4 Da
5 ms	crystallite formation	10^5 Da
8-30 ms	spherical particles	max 10^7 Da
>40 ms	soot aggregate	up to 10^{10} Da

On the basis of *in situ* mass spectrometry of laminar flames (such as those derived from benzene or acetylene), the formation of both polyyne chains and fullerenes has been confirmed (Richter et al. 2000; Hall-Roberts et al. 2000; Fialkov et al. 2001; Krestinin 2000). The concentrations of these species are at a maximum in the hottest part of the flame, > 2000°C, although the main reaction zone of PAHs is 1400°-1850°C (Fialkov et al. 2001). The nascent soot is highly hydrogenated, as circumstellar carbon dust is assumed to be (Jaeger et al. 1999), and is most likely derived from agglomeration of PAHs. Among the PAHs formed in a flame are a certain number containing pentagonal rings fused with hexagonal rings. These have been extracted from the carbon black surface together with minute amounts of C_{60} fullerene (Donnet 1999; Pontier-Johnson et al. 2002). It seems to be logical to think that these molecules are the precursors of fullerenes (Richter et al. 2000) and may be also the seeds for the introduction of fullerene-like or pentagonal sites in a graphene sheet.

These "defective" sites in the graphene sheets may also form during late-stage growth of the soot particle. However, they may also be the products of oxidation reactions.

It is interesting to notice that both polyynes and PAHs which have been detected in flames together with fullerenes, are present in the interstellar space and in molecular clouds. Thus, the formation processes of carbon dust in circumstellar space may be similar to those occurring in a flame or in a plasma.

Fullerene-like structures form during soot formation and produce the observed curvatures in the graphene sheets of carbon blacks. Additionally, half-cage shaped fullerenic moieties have been detected on the surface of furnace carbon black (Grieco et al. 2000). Highly curved graphene sheets were observed in soot, especially among samples collected relatively far from the burner (Grieco et al.2000). Curved surfaces or graphene sheets have been called fullerene-like structures and have been detected also in furnace carbon blacks (Donnet et al. 1999; Pontier-Johnson et al. 2002).

8.3 Effect of Radiation on Elemental Carbon

Ugarte (1995) reported a mechanism through which fullerene black (morphologically comparable to carbon black), upon heating, transforms into a polyhedral particle with a closed-shell nanotube structure. Alternatively, electron or ion bombardment results in onion-like graphitic particles (Ugarte, 1995). With HRTEM it was shown that arc-discharge-produced soot has a tendency to form curved surfaces after both thermal processing and electron irradiation and ion bombardment (Miki-Yoshida et al. 1998; Takeuchi et al., 1997). Evidence for the formation of fullerene onions and curved sheets was obtained by electron irradiation of soot, carbon black and char (Burden and Hutchinson, 1998). Curved layers with pentagonal sites such as those in fullerenes were also observed in diesel engine soot (Ishiguro et al. 1997). Ion bombardment and irradiation also produced nanodiamonds (Daulton et al., 1999).

Using Raman spectroscopy, it has been shown that γ radiation of graphite can induce the formation of diamond microdomains, glassy carbon and onion-like carbon (Cataldo, 2000a and 2000b). Previously, Galvan and his colleagues (1998) demonstrated that γ radiation induces the rotation of graphite planes. Banhart and Ayan (1996) have also shown that electron irradiation and annealing cause carbon onions to collapse into ultradisperse diamond. This transformation is reversible with repeated annealing.

All these studies support the theory that graphite transforms into disordered carbonaceous material (such as carbon black) in the interstellar medium when exposed to radiation. A similar destiny awaits C_{60} and C_{70} fullerenes bombarded with He^+ ions (Cataldo, Baratta, Strazzulla 2002; Cataldo, Baratta, Ferini, Strazzulla 2003). However, when C_{60} is irradiated with UV light or compressed at very high pressure, it becomes a new material in which all the fullerene cages remain intact but are linked together by cyclobutane rings. It is therefore reasonable to propose that C_{60}, if adsorbed on dust grains, may undergo a photopolymerization reaction.

9. COMPLEX ORGANIC MACROMOLECULAR NETWORK IN SPACE – SUMMARY

In the previous sections we reviewed how carbon is formed in late-type, carbon rich stars by nuclear reactions and then ejected in the circumstellar and interstellar media. We discussed the way in which carbon vapor tends initially to form acetylenic chains known as polyynes. These chains can be end-capped by H and CN (nitrile) groups and achieve modest stability. Under this form the polyyne chains are known as cyanopolyynes and have been detected by radioastronomers (Kroto and Walton 1993). However, we know from laboratory studies of carbon black and soot formation that the polyyne chains are just early intermediates which lead to the formation of PAHs. This can happen as a consequence of a crosslinking reaction of the polyynes in the presence of sufficient hydrogen. Alternatively, when hydrogen is absent, the polyynes are precursors leading, via cross-linking reactions, to fullerenes (which are a type of three-dimensional PAHs) or to carbon black. In a simplified view, carbon black (or soot, or carbon dust) can be formed when a PAH becomes large enough to constitute a graphene sheet. When graphene sheets agglomerate in a turbulent flow, carbon black results. Accretion of PAHs is due to gas-solid interactions followed by cross-linking reactions. Because of its mechanism of formation, interstellar carbon dust is supposed to be characterized by paramagnetism, with different degrees of hydrogenation resulting from reaction with H present in the interstellar medium. Fullerene-like structures are anticipated in the interstellar carbon dust just as they are present in "terrestrial" soot. Carbon black displays an interesting surface catalytic activity to yield both molecular hydrogen and also more complex molecules (Cataldo 2003b).

We have firm yet indirect evidence that carbon dust exists in the interstellar medium, namely, the compelling absorption feature at 217 nm. Other, diffuse interstellar bands (DIBs) observed as spectral features have been attributed to polyyne chains and/or to PAHs (Thielens and Snow 1995).

Likewise, unidentified infrared bands (UIBs) are generally accepted to be caused by the presence of aromatic organic matter and are observed in the emission spectra from a large variety of astronomical objects; for this reason they are considered ubiquitous in the Universe. IR spectroscopic investigations of mixtures of PAHs (Cataldo, Keheyan and Heymann 2002; Cataldo and Keheyan, 2003b) have yielded emission band patterns matching those of certain protoplanetary nebulae. Similar results were obtained by Reynaud and others (2001), using coal and pyrolyzed aromatic substrates as reference material for modeling the UIBs.

10. CONCLUSIONS

Having a basic knowledge of the elemental carbon forms, their stability and transformations, it is possible to better understand the chemical and physical nature of both elemental carbon dust and complex macromolecular networks present in space as well as the chemical and physical phenomena occurring on the surfaces of the carbon dust.

The aim of the present tutorial review was to give to the reader a chemical basis for interpretation of the structure and the role played by the interstellar carbon dust.

Atomic carbon is synthesized in late-type stars by the 3α process and plays a role also in warm stars because it catalyzes the CNO cycle. In stars at the end of their lifecycle, it is active in the synthesis of heavier elements such as oxygen and metals.

Carbon dust particles are comprised of nano- to micron-sized grains. The UV absorption feature at 217 nm in the interstellar background extinction curve is mainly attributed to carbon grains, which are best described as radiation-processed and partially-hydrogenated carbon black. Carbon black can be adopted as a generic word to describe amorphous carbon. It is made by small graphene sheets arranged in a turbostratic fashion with a low range order. Macroscopically it results amorphous but under a certain scale it shows a certain degree of order. It is commonly accepted that the carbon black formation in space follows a similar process to the soot formation in flames thus it is possible to apply our knowledge to soot formation in flame to the formation of carbon dust in the ISM. This model gives us a strong predictive power: the interstellar carbon dust must be thought as carbon black and not as graphite or graphite-like matter. Carbon black cannot be transformed into graphite in the sense that thermal annealing causes a partial reordering of the structure (called graphitization) but it never yield graphite. Another rather new concept in the carbon black model is the fact that the graphene sheets may be either flat and planar or bent curled. In the past the graphene sheets were thought completely and perfectly planar being made exclusively by condensed hexagonal carbon rings. With the discovery of fullerenes it has been realized that the mentioned model was too restrictive. The graphene sheets can be thought with "defects" made by pentagonal, heptagonal or other kinds of rings other than hexagonal. The mentioned defects, collectively called "fullerene-like structures" induce open or closed curvatures in the graphene sheets and represent a better model of carbon black explaining in a better way its reactivity. Radiation processing of graphite causes graphene sheet curling and the transition of graphite to carbon black. Thus, graphite formed in the space cannot survive for a long time: being submerged in a cosmic and high energy radiation field it must be converted into carbon black. The fullerene-like structures in the interstellar carbon black may explain in a better way both the optical properties of the carbon dust and also its role in molecular hydrogen formation in space starting from atomic hydrogen.

Using the concept of carbon allotropy, we have shown that three fundamental carbon allotropes can be thought: diamond, graphite and polyynes or carbyne respectively made by sp^3, sp^2 and sp hybridized carbon atoms. All the other forms of carbon have intermediate degree of hybridization (e.g. fullerenes) or are mixed forms in the sense that in these solids are simultaneously present carbon atoms having variable degree of hybridization. Concerning the sp-hybridized carbon, the polyyne chains have been detected and identified now in many different conditions so that their existence is beyond any doubt. They constitute also the carbon vapor and are the key intermediates which lead to cyanopolyynes (detected by radioastronomy in molecular clouds), to PAH if hydrogen is present or even to carbon black. It is remarkable that the DIBs are thought to be caused by PAH and polyynes occurring as large molecules or ions. The same is believed to be the case for the UIBs. Thus, very complex molecules may be present in the interstellar medium and serve as a source for even more complex chemical derivatives. These in turn may be incorporated in the carbonaceous part of meteorites and deliver precious organic matter to more fruitful environments such as the early Earth or earth-like planets of other solar systems.

ACKNOWLEDGEMENTS

Our sincere thanks and gratitude to Dr. Vito Mennella and Dr. John Robert Brucato from INAF, Osservatorio Astrofisico di Capodimonte, Via Moiariello 16, 80131, Naples, Italy, who have read and corrected the present manuscript giving precious advice to the author. Our thanks also to Prof. Pascale Ehrenfreund, Dr. Jen Blank, Dr. Sylvie Derenne and Prof. Dieter Heymann who have helped the author in editing the manuscript and in making it more readable.

Many thanks also to Dr. Marie-Antoniette Pointier-Johnson from Rice Univesity, Houston, TX, USA, for the HRTEM images.

We wish to thank also ASI, the Italian Space Agency for the financial support of the present work with the grant I/R/070/02.

REFERENCES

Alekseev, N.I., Dyuzhev, G.A., 2002. *Technical Physics*, **47**, 634 and 643.

Banhart, F., Ajayan, P.M. 1996. *Nature*, **382**, 433.

Barrow, J.D., Tipler, F.J. 1986. *The Anthropic Cosmological Principle*. Chapter 4. (Italian translation by Adelphi Edizioni, Milan, 2002).

Burbidge, E.M., Burbidge, G.R., Fowler, W.A., Hoyle, F. 1957. *Rev. Mod. Phys.* **29**, 547.

Burden, A.P., Hutchinson, J.L. 1998. *Carbon*, **36**, 1167.

Cameron, A.G.W. 1973. *Space Sci. Rev.* **15**, 121.

Cataldo, F. 2003a. *Fullerenes, Nanotubes and Carbon Nanostructures*, **11**, 295.

Cataldo, F. 2003b. *Fullerenes, Nanotubes and Carbon Nanostructures*, **11**, 317.

Cataldo, F. 2003c. *Carbon* **41**, 2671.

Cataldo, F. 2003d. *Fullerenes, Nanotubes and Carbon Nanostructures*. In press.

Cataldo, F. 2003e. *Carbon* **42**, 129.

Cataldo, F. 2003f. *Tetrahedron Lett,* **45,** 141.

Cataldo, F., Baratta, G.A., Ferini, G., Strazzulla, G. 2003. *Fullerenes, Nanotubes and Carbon Nanostructures*, **11**, 191.

Cataldo, F., Keheyan, Y. 2003a. An Essay on Asymmetric Polymerization and the Origin of Chirality in the Biologically Active Macromolecules; in *International Perspectives on Chemistry and Biochemistry Research,* Chapter 3. Zaikov G.E., Lobo V.M.M. and Guarrotxena editors, Nova Science Publishers, New York.

Cataldo, F., Keheyan, Y. 2003b. *Int. J. Astrobiology*, **2**, 41.

Cataldo, F., Koscheev, A.P., 2003. *Fullerenes, Nanotubes and Carbon Nanostructures*, **11**, 201.

Cataldo, F. 2002a. *Carbon* **40**, 157.

Cataldo, F. 2002b. *Fullerenes, Nanotubes and Carbon Nanostructures*, **10**, 155.

Cataldo, F., Baratta, G.A., Strazzulla, G. 2002. *Fullerenes, Nanotubes and Carbon Nanostructures*, **10**, 197.

Cataldo, F., Keheyan, Y., Heymann, D. 2002. *Int. J. Astrobiology*, **1**, 79 (see also: *Origin of Life and Evolution of Biosphere* 2004, **34**,13).

Cataldo, F., Pontier-Johnson, M.A. 2002. *Fullerenes, Nanotubes and Carbon Nanostructures*, **10**, 1.

Cataldo, F. 2000a. *Carbon*, **38**, 634.

Cataldo, F. 2000b. *Fullerene Science & Technology*, **8**, 577.

Cataldo,F. 1999a. *Carbon*, **37**, 161.

Cataldo, F. 1999b. *Polymer International*, **48,** 15.

Cataldo, F. 1999c. *Die Angewandte Makromol. Chem.*, **264**, 65.

Cataldo, F., Capitani, D. 1999. *Materials Chemistry and Physics*, **59**, 225.

Cataldo, F. 1997a. *Fullerene Science & Technology*, **5**, 1615.
Cherchneff, I. 1998. *The Molecular Astrophysics of Stars and Galaxies.* Edited by Harquist, T.V., Williams, D.A., Oxford University Press, Oxford.
Daulton, T.L., Lewis, R.S., Lehn, L.E., Kirk, M.A. 1999. *Mat. Res. Symp. Proc.*, **540**, 189.
Diederich, F., Rubin, Y. 1992. *Angew. Chem. Int. Ed. Engl.* **31**, 1101.
Donnet, J.B., Wang, T.K., Wang, C.C., Monthioux, M., Johnson, M.P., Norman, D.T., Wansborough, R.W., Bertrand, P. 1999. *Kautschuck Gummi Kunststoffe*, **52**, 340.
Donnet, J.B., Bansal, R.C., Wang, M.J. 1993. *Carbon Black.* M. Dekker editor, New York.
Eherenfreund, P., Irvine, W., Becker, L., Blank, J., Brucato, J.R., Colangeli, L., Derenne, S., Despois, D., Dutrey, A., Fraaije, H., Lazcano, A., Owen, T., Robert, F. 2002. *Rep. Prog. Phys.* **65**, 1427.
Ehrenfreund, P., Charnley, S.B. 2000. *Annu. Rev. Astron. Astrophys.* **38**, 427.
Fialkov, A.B., Dennebaum, J., Homann, K.H. 2001, *Combust. Flame*, **125**, 763.
Galvan, D.H., Garzon, I.L., Santiago, P., Josè-Yacaman, M. 1998. *Fullerene Science & Technology*, **6**, 867.
Greenwood, N.N., Earnshaw, A. 1984. *Chemistry of the Elements.* Chapter 1. Pergamon Press, London.
Gribbin, J., Gribbin, M. 1996. *Companion to the Cosmos.* The Orion Publishing House, London.
Grieco, W.J., Howard,J.B., Rainey, L.C., Vander Sande, J.B. 2000. *Carbon*, **38**, 597.
Hall-Roberts, V.J., Hayhurst, A.N., Knight, D.E., Taylor, S.G. 2000. *Combust. Flame*, **120**, 578.
Heimann, R.B., Evsyukov, S.E., Kavan, L. 1999. *Carbyne and Carbynoid Structures*, Kluwer Academic Publishers, Dordrecht, NL.
Heimann, R.B., Evsyukov, S.E., Koga, Y. 1997. *Carbon.* **35**, 1654-1658.
Heymann, D., Jenneskens, L.W., Jehlicka, J., Koper, C., Vliestra, E. 2003. *Fullerenes, Nanotubes and Carbon Nanostructures*, **11**, 333.
Heymann, D. 2001. *Astrophys. Space Sci.* **274**, 415.
Henning, Th., Salama F. 1998. *Science.* **282**, 2204.
Henning, Th. 1998. *Chem. Soc. Rev.* **27**, 315.
Hirsch, A. 1994. *The Chemistry of Fullerenes.* Thieme Verlag, Stuttgart.
Holliday, A.K., Hughes, G., Walkers, S.M. 1973. In: *Comprehensive Inorganic Chemistry* Edited by Bailar, J.C., Emeleus, H.J., Nyholm, R., Trotman-Dickenson, A.F. Pergamon Press, London.
Howard, J.B., McKinnon, J.T., Makarovsky, Y., Lafleur, A.L., Johnson, E.M. 1991. *Nature*, **352**, 139.
Hoyle, F., Wickramasinghe, N.C. 1977. *Nature*, **270**, 701.
Iijima, S. 1991. *Nature*, **354**, 56.
Iijima, S., Ichihashi, T. 1993. *Nature*, **363**, 603.
Ishiguro, T., Takatori, Y., Akihama, K. 1997. *Combust Flame*, **108**, 231.
Jager, C.; Henning, Th.; Schlogl, R.; Spillecke O. 1999. *J. Noncryst. Solids*, **258**, 161.
Kaler, J.B. 1992. *Stars.* Scientific American library, an imprint of W.H. Freeman and Co., New York.
Kaler, J.B. 1997. *Cosmic Clouds. Birth, Death and Recycling in the Galaxy.* Scientific American library, an imprint of W.H. Freeman and Co., New York.
Koscheev, A.P., Gromov, M.D., Mohapatra, R.K., Ott, U. 2001. *Nature* **412**, 615.
Kratschmer, W., Huffmann, D.R. 1992. *Carbon* **30**, 1143.
Kratschmer, W., Lamb, L.D., Fostiropoulos, K., Huffman, D.R. 1990. *Nature*, **347**, 354.
Krestinin, A.V. 2000. *Combust. Flame*, **121**, 513.

Kroto, H.W., Walton, D.R.M. 1993. *Phil. Trans. Roy. Soc. Lond.* **A343,** 103.

Kroto, H.W., Heath, J.R., O'Brien, S.C., Curl, R.F., Smalley, R.E. 1985. *Nature*, **318**, 162.

Lagow, R.J., Kampa, J.J., Wei, H.C., Battle, S.L., Genge, J.W., Laude, D.A., Harper, C.J., Bau, R., Stevens, R.C., Haw, J.F., Munson, E. 1995. *Science* **267**, 362.

McCarthy, M.C., Thaddeus, P. 2001. *Chem. Soc. Rev.* **30**, 177.

Mennella V.; Brucato J.R.; Colangeli L. 2001. *Spectrochim. Acta* **A57**, 787.

Mennella, V., Baratta, G.A., Colangeli, L., Palumbo, P., Rotundi, A., Bussoletti, E., Strazzulla, G. 1997. *Astrophys. J.* **481**, 545.

Mennella,V., Colangeli, L., Palumbo, P., Rotundi, A., Schutte, W., Bussoletti, E. 1996. *Astrophys. J.* **464**, L191.

Mennella,V., Colangeli, L., Blanco, A., Bussoletti, E., Fonti, S., Palumbo, P., Mertins, H.C. 1995. *Astrophys. J.* **444**, 288.

Michel, B.; Henning, Th.; Jager, C.; Kreibig, U. 1999. *Carbon* **37**, 391.

Miki-Yoshida, M., Castillo, R., Ramos, S. 1994. *Carbon* **32,** 231.

Oberlin, A., Boulmier, J.L., Villey, M. 1973 in *Kerogen: Insoluble Organic Matter from Sedimentary Rocks.* Chapter 7. Edited by B. Durand. Editions Technip, Paris.

Pillinger, C.T. 1993. *Phil. Trans. Roy. Soc. Lond.* **A343**, 73.

Pirronello, V.; Liu, C.; Shen, L.; Vidali, G. 1997. *Astrophys. J. Lett.* **475**, L69.

Pontier-Johnson, M; Donnet, JB; Wang, TK; Wang, CC; Locke, RW; Brinson, BE; Marriott, T. 2002. *Carbon* **40**, 189.

Ravagnan, L.; Siviero, F.; Lenardi,C.; Piseri, P.; Barborini, E. Milani, P. 2002. *Phys. Rev. Lett.* **89**, 285506.

Reynaud, C. Guillois, O., Herlin-Boime, N., Rouzaud, J.N., Clinard, C., Balanzat, E., Ramillon, J.M. 2001. *Spectrochim. Acta* **A57**, 797.

Richter, H., Howard, J.B. 2000. *Prog. Energy Comb. Sci.* **26**, 565.

Rohlfing, E.A.; Cox, D.M.; Kaldor, A.1984. *J. Chem. Phys.* **81**, 3322.

Savage, B.D., Mathis, J.S. 1979. *Annu. Rev. Astron. Astrophys.* **17**, 73.

Saito, R., Dresselhaus, G., Dresselhaus, M.S. 1998. *Physical Properties of Carbon Nanotubes*. Imperial College Press, London.

Snow, Th.P., Witt, A.N. 1995. *Science* **270**, 1455.

Takeuchi, M., Muto, S., Tanabe, T., Kurata, H., Hojou, K. 1999. *J. Nucl. Materials*, **271& 272**, 280 and 285.

Tanuma, S., Pal'nichenko, A. 1995. *J. Mater. Res.* **10**, 1120.

Taylor, R.J. 1966. *Rept. Prog. Phys.* **29**, 489.

Thielens, A.G.G.M. Snow, T.P. editors. 1995. *The Diffuse Insterstellar Bands.* Kluwer Academic Publishers, Dordrecht.

Ugarte, D. 1995. *Carbon*, **33**, 989.

Unsold, A., Baschek, B. 2002. *The New Cosmos*. 5th Edition. Springer-Verlag, Berlin.

Webster, A. 1995. in: *The Diffuse Insterstellar Bands.* Thielens, A.G.G.M. and Snow, T.P. editors. Kluwer Academic Publishers, Dordrecht; p. 349-358.

Chapter 6

ORGANIC MOLECULES IN PLANETARY ATMOSPHERES

Maarten Roos-Serote
Lisbon Astronomical Observatory
Tapada da Ajuda, 1349-018 Lisbon, Portugal
roos@oal.ul.pt

Abstract: In the Solar System three types of atmospheric environments can be found: the highly oxidized Earth-like atmospheres, the mildy reduced atmospheres of Titan, Pluto and Triton, and the highly reduced atmospheres of the giant planets. In the terrestrial atmospheres carbon, the most abundant element in the Universe after hydrogen, helium and oxygen, is mostly combined with oxygen to form carbon dioxide and carbon monoxide. On Earth methane exists in some significant quantity also, and this is due to the presence and action of life. In the non-terrestrial atmospheres, methane is very abundant and it is the carrier of most of the carbon. In these atmospheres other organics are formed from photochemical processes with methane at the origin. Titan has the most complex atmosphere when it comes to organic molecules. Titan's atmosphere may represent a low-temperature version of the conditions similar to those existant on Earth before the appearance of life.

The goal of this paper is to provide a short overview of the current knowledge on organic molecules in planetary atmospheres.

Keywords: Planetary Atmopheres, Organic Molecules

1 Some General Remarks

The Solar System contains a total of eight major planets, four of the terrestrial type and four of the jovian type. The jovian planets are orbited by tens of satellites. Some of these have sizes that would easily classify them as planets were they to orbit the Sun directly. Apart from these large

P. Ehrenfreund et al. (eds.), Astrobiology: Future Perspectives, 127–147.

bodies the Solar System contains tens of thousands of minor bodies with Pluto being the largest one, presently known.

Substantial atmospheres are found on all of the major planets except Mercury, and on Titan, Saturn's largest satellite. Tenuous atmospheres are found on Mercury, Pluto and Triton, Neptune's largest satellite.

Table 6.1 lists the three most abundant species in each atmosphere. In terms of composition, none of the known atmospheres resembles the interstellar medium or comets (see chapters by Crovisier and Millar). An important observation can be made about the composition and evolution of the atmospheres and is related to the gravity fields of the planets. The atmospheres of the most massive (jovian) planets have preserved their initial composition, as it is virtually impossible for even the lightest atom (H) to escape in any significant quantities. Evolution only occurs by means of irreversible chemical processes. The atmospheres of the smaller planets, *e.g.* the terrestrial planets plus Titan, Pluto and Triton, are subject to the escape of part of their lightest gases. The escape process has important consequences for the time evolution of the composition of their atmospheres. For example, it is thought that on early Venus liquid water was evaporated and transported to high altitudes, where it was photodissociated. Most of the H atoms escaped, but the oxygen remained in the atmosphere, reacting with the other atmospheric species and the surface. The decrease of the amount of liquid water reduced the capability to store carbon-dioxide (on Earth most carbon-dioxide is stored in the lithosphere as carbonates). An increasing carbon-dioxide abundance in the atmosphere raised the surface temperature, as

Table 6.1. Three most abundant species of each atmosphere

	species	species	species
Venus	CO_2 (0.96)	$N_2(3.5 \times 10^{-2})$	$SO_2(1.5 \times 10^{-4})$
Earth	N_2 (0.78)	$O_2(0.21)$	$Ar(9.3 \times 10^{-3})$
Mars	CO_2 (0.95)	$N_2(2.7 \times 10^{-2})$	$Ar(1.6 \times 10^{-2})$
Titan	N_2 (0.95)	$Ar(\leq 6-7 \times 10^{-2})$†	$CH_4(4 \times 10^{-2})$
Pluto	N_2 (0.98)	$CH_4(10^{-3} - 10^{-2})$	$CO(10^{-3})$
Jupiter	H_2 (0.864)	He (0.136)	$CH_4(1.81 \times 10^{-3})$†
Saturn	H_2 (0.885)	He (0.115)	$CH_4(5 \times 10^{-3})$‡
Uranus	H_2 (0.85)	He (0.15)	$CH_4(2 \times 10^{-2})$‡
Neptune	H_2 (0.85)	He (0.15)	$CH_4(2 \times 10^{-2})$‡

Mixing ratios are given in parenthesis.
† Inferred, not measured.
‡ H_2O is probably more abundant than CH_4 in the atmosphere below the visible clouds.

well as the rate of evaporation. This process entered a runaway state, resulting in the observed dry CO_2 rich atmosphere, with a high D/H ratio in the residual water vapor (D escapes less easily than H).

In Table 6.1 the atmospheres are classified in three groups, relative to their states of oxidation. Group 1 contains the terrestrial planets, whose atmospheres are dominated by carbon dioxide or molecular oxygen: highly oxidized atmospheres. Group 2 contains Titan, Pluto and Triton, with a CO/CH_4 ratio of order $10^{-1} - 10^{-2}$: mildy reducing atmospheres. Group 3 are the giant planets, where CO/CH_4 is around 10^{-6}: highly reducing atmospheres. Indeed, the atmospheres of the inner planets of Earth size or less will inevitably move to an oxidized state as a result of the escape of H and the mobilization of water due to the higher temperatures in the inner Solar System. It is the absence of water that keeps the atmosphere of Titan, for example, mildy reducing. Due to the low temperatures the water stays out of the atmosphere.

In the following sections I review the most important aspects of the current state of knowledge on the presence and fate of organic molecules in the Solar System atmospheres. I consider an organic molecule to be a molecule with a C-H bond. Methane, CH_4, is the simplest of all organic molecules, and it is found to be the most abundant among all detected organic molecules in the Solar System.

2 Venus

Venus has a very thick atmosphere, which has resulted from a run away greenhouse. The pressure and temperature at the surface are of the order of 90 bars and 730 K respectively.

The atmosphere of Venus seems to be void of any organic molecules. Three carbon containing molecules have been detected: OCS (Bézard et al. 1990), CO and of course CO_2, which dominates the atmosphere (Table 6.1). The chemical composition of the atmosphere is determined by different processes: equilibrium chemistry in the low atmosphere and at the surface (0–60 km), photochemistry in the upper atmosphere (60–110 km).

An early study of the chemical composition and processes by Florensky et al. (1978) assumed the atmosphere to be divided into three layers: a thermochemical equilibrium layer (0-33 km), a photochemical layer (> 50 km) and an intermediate layer. The model produced negligible quantities of some organic molecules (HCN, CH_2N_2), but these were not studied any further. Florensky et al. (1978) considered the abundance of methane, and predicted a very large variation from 10^{-13} near the surface to 10^{-6} at 30 km altitude. Thermochemical equilibrium was assumed up to about 50 km altitude. Krasnopolsky and Parshev (1979) argued that thermochemical equilibrium may indeed exist near the surface for most species due to the catalytic action of surface rocks, and that constant mixing ratios better approximate the vertical density profiles at higher altitudes. Donahue and Hodges (1992) reported on methane measurements by the Pioneer Venus Large Probe Mass Spectrometer in the lowest 20 km of the atmosphere, but cautioned that this may be an instrumental effect. Donahue et al. (1997) confirmed this to be the case.

From ground-based high spectral resolution spectra (Bézard et al. 1990) the OCS abundance at 33 km altitude has been determined to be 4.4 ± 1.0 ppm, increasing towards to the surface (Pollack et al. 1993). The same authors determined an abundance of 23 ± 5 ppm for CO at 36 km, decreasing towards the surface! Krasnopolsky and Pollack (1994) presented a detailed model to explain these features, and showed how they are related to the slow atmospheric cycle of sulfur. It is out of the scope of this review to discuss this model in detail and the interested reader is refered to Krasnopolsky and Pollack (1994) and references therein.

Venus is covered by a thick and global cloud deck. Several cloud layers occupy the vertical range between about 50 and 75 km altitude. The cloud particles are partly composed of concentrated sulfuric acid, but other components remain to be identified. A complex and still poorly understood chemical cycle takes place in the clouds involving water, SO_3, CO_2, etc. No organic molecules have been observed, nor predicted to exist in any significant quantities in the clouds.

3 Detecting Life on Earth and other Planets

Venus and Earth are very similar in terms of size and density. In fact, the total carbon dioxide content is also similar for both planets. The difference is that on Venus the carbon dioxide resides in the atmosphere, whereas on the Earth most of it is in the lithosphere (carbonate rocks). The same happens with other species, like SO_2 and HCl (NaCl in the Earth's oceans).

The atmosphere of the Earth is unique in the Solar System. Earth is the only planet we know that has liquid water on the surface. The interaction between the oceans and the atmosphere has important influence on the composition of the atmosphere. For some species, this composition is very far from thermodynamic equilibrium.

The departure from thermodynamic equilibrium of a planetary atmosphere is a necessary, but not sufficient condition for the existence of life as we know it (Lederberg 1965). Lovelock (1965) suggested that a strong signature for life on Earth is the presence of methane and nitrous oxide

(N_2O), both highly reduced species, along with the oxiding gas molecular oxygen.

Sagan et al. (1993) did a most interesting experiment. They attempted to find evidence for life on Earth, just by analysis of the observations made by the Galileo spacecraft when it flew close to its home planet in December of 1990. Galileo's instruments were not designed for Earth-observing purposes, but rather for the jovian environment.

Sagan et al. (1993) obtained the following results:

- Large quantities of ozone are observed in the stratosphere at UV wavelengths. Ozone is in thermodynamic disequilibrium, *i.e.* it is about 20 orders of magnitude more abundant than expected. This can be explained by the UV photochemical processes in which it is involved. It also indicates the presence of a significant amount of molecular oxygen.

  Oxygen can be formed from the photolysis of stratospheric water, with subsequent escape of the hydrogen atom to space. Estimates of the current production of oxygen by this process yield a value of 4×10^7 molecules cm^{-2} s^{-1}. A sink of oxygen is the Earth's crust, as well as reactions with volcanic gases such as CO and H_2. The volcanic sink rate could be as much as 60 times higher than the photolysis induced production (Schindler and Kasting 2000 and references therein). So, photolysis alone cannot account for the large amount of oxygen in the atmosphere.

  Sagan et al. (1993) also note that significant levels of ozone in the stratosphere will protect the surface from high UV irradiation. This means that molecular bonds that are important for organic chemistry (C-C, C-H, C-O, C-N) will suffer little photo-destruction, so that this chemistry can take place at the surface.
- Methane was observed by the Galileo spacecraft as well. Methane has absorption bands in the near-infrared, which are seen in the spectra of the Near Infrared Mapping Spectrometer (NIMS). The measured and known abundance of methane is about 140 orders of magnitude more than expected from equilibrium calculations, given the atmospheric levels of oxygen. Methane does not survive in an oxygen-dominated atmosphere, unless there is a source. The same is true for Venus and Mars.

  Methane in the Earth's atmosphere comes mainly from natural biological sources (methane bacteria) and from anthropogenic sources (rice fields, biomass burning, etc.). Abiotic sources (outgassing from the surface through different processes) contribute negligibly. Methane is homogeneously mixed in the troposphere, with a mixing ration of 1.6 ppmv. The global concentration of methane increases by 1 -2 % per year.

- The presence of nitrous oxide (N_2O) at a level of about 1 ppm is also derived from the Galileo/NIMS data. This is about 15 orders of magnitudes more than predicted for equilibrium. In the case of the Earth, N_2O is produced by bacteria and algae that convert soil and oceanic NO_3^- into N_2 and N_2O.
- For the Galileo case, the most convincing and direct evidence for intelligent activity on Earth is the detection of narrow band modulated radio transmissions.
- Except for the last one, all these observations could have been made as long as 2×10^9 years ago.

Galileo remote sensing observations have clearly identified strong indications for the presence and action of life on Earth.

Schindler and Kasting (2000) analysed the possibility of the identification of Earth-like atmospheres on planets around other stars, by measuring the infrared spectra of these planets. This is an objective set for the proposed NASA Terrestrial Planet Finder (TPF), a space based interferometer. They conclude that the presence of significant amounts of methane (more than 100 times that on Earth), as well as ozone, can be detected with the TPF. This amount of methane could be representative of an early Earth scenario.

Fridlund (2000) presented a report on the DARWIN mission, currently a mission idea under assessment in the European Space Agency's (ESA) Science Programme. DARWIN is an infrared space interferometer. Its goals are to detect Earth-like planets around nearby stars (several tens of parsecs) and to provide images and spectra of these panets in the 5 to 28 micron range. In this range, spectral signatures of water vapor, ozone and carbon dioxide can be found (see Fig. 6.1).

Both DARWIN and TPF could be operational in the next decade.

4 Mars

Mars has a very thin atmosphere (about 6 mbar pressure at the surface), and is transparent to all ranges of the solar radiation. To date, there is no confirmed detection of any organic species in the atmosphere of Mars.

Methane is not expected to be present in large amounts in the martian atmosphere, due to a lack of both H atoms and a (photochemical) source. If methane is present in the atmosphere, it either originates from outgassing from the interior, from biological sources, or both. Methane has a lifetime of several years in the martian atmosphere, so it can be expected to be

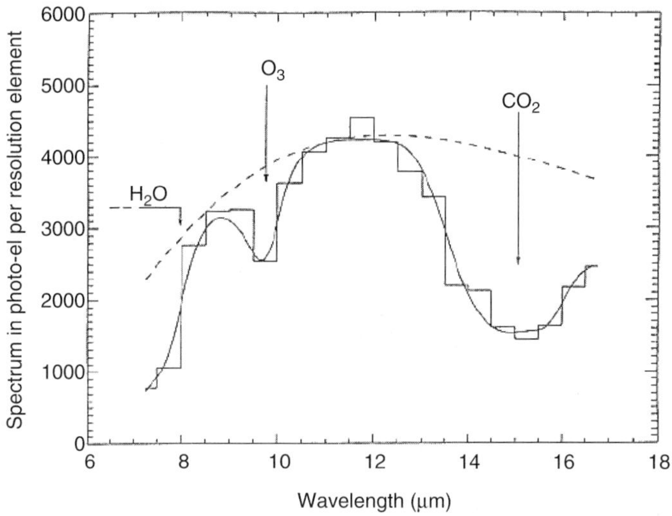

Figure 6.1. Spectrum from the simulation of an Earth-like planet observation with a SNR of 20, spectral resolution of 0.5 μm. The planet is located around a star at 10 pc. The solid line is the Earth spectrum at this resolution, the dashed line the emission from a blackbody at 300K and the stepped line the recovered spectrum. Absorption features of H_2O ($\lambda <\sim 8\mu$m), O_3 ($\lambda = 9.6\mu$m) and CO_2 ($\lambda = 15\mu$m) are visible. They would indicate that the exo-planet has an atmosphere with a feature common to all solar telluric ones, (CO_2), is habitable (H_2O) and is the place of a photosynthetic activity (H_2O and O_3 simultaneously). The latter observation would be a major discovery. Figure and caption from the Darwin report ESA-SCI(2000)12, 2000 (Fridlund 2000).

distributed homogeneously, if there is a sufficient supply. An upper limit for methane has been determined to be of order 20 ppb from measurements with the infrared spectrometer on the Mariner 9 spacecraft taken in 1971 (Maguire et al. 1977). Earth based measuments performed in mid 1988, using a high spectral-resolution Fourier spectrometer mounted on a 4-meter class telescope, are consistent with this result. Krasnopolsky et al. (1997) find an upper limit of 70 $\pm$ 50 ppb, derived from methane lines in the 3.6 and 8 μm range. Yet, both results must be considered with some caution. They refer to large areas on Mars. If the sources of methane are very localised, then a local enhancement of the methane abundance can easily be missed in these observations.

Formaldehyde (CH_2O has been tentatively detected in the atmosphere of Mars. Solar occultation data from the Auguste infrared spectrometer flown on the Russian Phobos mission show spectral features near 3.66 and 3.69 μm. Korablev et al. (1993) suggest that these features could be attributed to formaldehyde. The data refer to the equatorial region of the planet, and were taken during Mars Northern spring in March 1989.

The estimated average CH_2O volume mixing ratio is $0.5^{-0.8}_{-0.3}$ ppm above 8 km altitude, which is an order of magnitude higher than the upper limit for methane, mentioned before. The data are not sensitive to lower altitudes. Korablev (2002) rediscussed the results and noted the difficulty of the detection in the spectra. In fact, the identification of the spectral features refers to a 2 % change in depth of a band at 3.7 μm, in spectra that show distortions of up to a level of 5 %. Much lower upper limits of 0.11 ppb (Marten et al. 1995) – 3 ppb (Krasnopolsky et al. 1997) have been derived from ground-based observations.

Two other problems exist with high quantities of formaldehyde in the atmosphere. The first is that it implies high abundances of methane as well. In fact, the abundance of formaldehyde should be similar or lower than that of methane by up to several orders of magnitude. The second, and more constraining, is that the photochemical lifetime of formaldehyde is shorter than that of methane by six orders of magnitude (Wong et al. 2003).

Korablev (2002) refered to studies that produce simple organic compounds from UV irradiation of mixtures of CO_2, H_2O, CO and O_2 in the presence of a silicate and/or ferrous catalyst, like dust in the martian atmosphere. This may produce organics which exhibit spectral features near 3.7 μm, a typical wavelength where CH bonds are seen, and that are more stable than formaldehyde.

The idea of local outgassing still occuring on Mars has recently been proposed to explain several surface features seen in Mars Global Surveyor images. In particular, some of the features could be interpreted by invoking processes such as surface runoff and ground water seepage (Malin et al. 2000). If this is correct, then outgassing may still occur on a local scale. Gases like methane could be among the species released in outgassing events and could trigger a whole series of chemical reactions. Wong et al. (2003) model the consequences of outgassing events, predicting the species that would result from the photochemistry after the outgassing. They find that, in terms of organic molecules, methane should be the most important indicator of outgassing, followed possibly by formaldehyde and methanol (Fig. 6.2).

Future space missions and observations will search for organic molecules on Mars, and try to map their variation on a small spatial scale. The ESA Mars Express / Beagle 2 spacecraft carry instrumentation that will address this question from orbit and *in situ*.

5 Titan

Titan is by far and away the most interesting and complex object in the Solar System when it comes to organic molecules in the atmosphere.

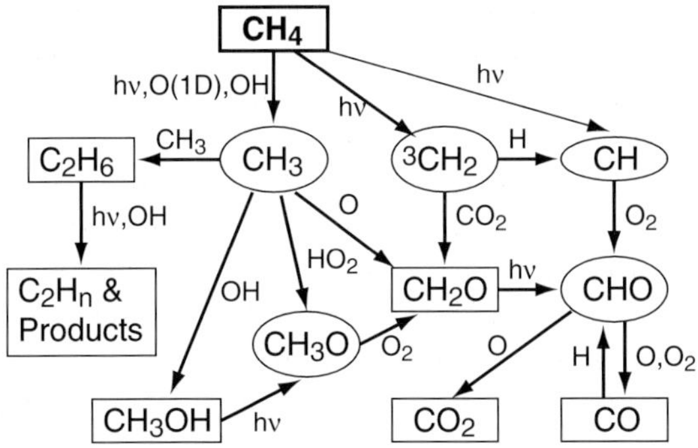

Figure 6.2. Important pathways of hydrocarbon chemistry in the Martian atmosphere. Stable species are in rectangles, radicals are in ovals, and outgassing species are in bold. Figure from Wong et al. (2003). Permission for reproduction granted by the American Geophysical Union.

The atmosphere of Titan has some similarities to the Earth's: both atmospheres have a total mass of about 5×10^{18} kg, the pressure at the surface are similar (1 bar for the Earth and 1.5 bar for Titan), as well as the main atmospheric constituent, which is molecular nitrogen in both cases (78 % for the Earth, and some 94 % for Titan). The second most important species identified in Titan's atmosphere is methane. However, as much as 6 % of the atmosphere could be primordial Argon (^{36}Ar, ^{38}Ar), which is not detectable by remote sensing (Table 6.1). It is thought that the composition of the current Titan atmosphere is similar to that of the Earth's atmosphere before the appearance of life, with the crucial difference that the low temperature of Titan's surface (94 K) keeps water vapor out of the atmosphere.

A very important difference is the temperature. Because Titan is about 10 times further from the Sun than the Earth, the surface temperature is much lower, *i.e.* 94 K. Also, Titan's solid body is a factor of 2.3 times smaller than the Earth's and less massive by a factor of about 44. The atmosphere of Titan is therefore less bound to the planet, and it is easier for atoms and molecules to escape.

The atmosphere is exposed to UV sunlight and bombarded by electrons from the Saturnian magnetosphere. At high stratospheric altitudes (> 500 km) photolysis of methane and subsequent chemical reactions lead to the formation of more complex molecules. The low temperatures in the lower stratosphere (between 70 and 160 K at altitudes between 50 and 200 km) and the presence of a global atmospheric haze throughout this part of the stratosphere and upper troposphere result in the condensation of organics onto the haze particles. The haze particles, probably composed of organic polymers, are expected to be coated with several layers of organics, as these condense onto the particles while they fall through the atmosphere. If these particles do indeed reach the surface, they could be an effective transport mechanism of organics formed high in the atmosphere down to the surface (Raulin 1997). Furthermore, the presence of areas (lakes, ponds) of liquid methane/ethane is possible.

Research activities have been increasing with the Cassini/Huygens mission due to arrive at Saturn in mid 2004. Remote sensing observations have unraveled many aspects of the processes in Titan's atmosphere. To date, the most detailed observations remain those made by the InfraRed Interferometer Spectrometer and Radiometer (IRIS) experiments on the Voyager spacecraft, in the early 1980's. Voyager 1 flew by Titan about nine months before Voyager 2, and some 200 times closer. As Titan's year is about 30 Earth years long, a nine month period corresponds to about one week in Titan's year. At the time of flyby, Titan had just passed its northern spring equinox, so that the north pole was passing into sunlight after years of darkness. This may be one of the reasons for the observed latitudinal distribution of several compounds (see discussion below). Recently new remote sensing observations became available, both from ground and Earth-orbiting observatories, particularly the ESA's Infrared Space Observatory (ISO). These new data added to the list of known species, and confirmed and/or refined Voyager results. Even though the Earth-based observations have much better spectral resolution than the Voyager data, they mostly refer to disk-averages.

Table 6.2 lists the current known species and their abundances (Table adapted from Coustenis and Taylor 1999). As can be seen most of the remote sensing information refers to the stratosphere, which extends between about 45 and 300 km altitude. This is inherent to the remote sensing technique, which uses the absorption lines and bands of the different molecules to determine their abundances. In Titan's case, the lines and bands probe stratospheric levels. Other species are expected to be present, but in too small quantities to be detectable with current remote sensing techniques.

Figure 6.3 shows the principal reaction pathways of the chemical processes in the atmosphere of Titan. Methane is being destroyed at the top of the chain. This process effectively eliminates methane from the atmosphere. The present abundance of methane and rate of destruction would destroy all methane in about 10^7 year, if there was no source (Yung et al. 1984). The identification of this source, probably at the surface, is one of the main problems that remains to be resolved. Possible sources have been proposed: evaporation from hydrocarbon (ethane/methane) lakes (McKay et al. 1993), diffusion from a sub-surface ocean (Kossacki and Lorenz 1996, Fortes 2000), cryovolcanism (Lorenz 1996), or a layer of methane hydrate below the surface (Loveday 2001). Methane is an

Table 6.2. Composition of Titan's Atmosphere

Constituent	Mole Fraction (atm. level)
N_2	0.98 – 0.85
Ar	$< 7 \times 10^{-2}$ (surface)
CH_4	$4.5 - 8.5 \times 10^{-2}$ (surface)
	$1.7 - 3.0 \times 10^{-2}$ (strat.)
H_2	1.1×10^{-3}
Ne	$< \times 10^{-2}$
C_2H_6	1.3×10^{-5} (130 km)
C_3H_8	5×10^{-7} (110 km)
C_2H_2	3×10^{-7} (130 km)
C_2H_4	1.5×10^{-7} (185 km)
CH_3C_2H	5×10^{-9} (80 km)
C_4H_2	1.4×10^{-9} (100 km)
C_2H_2	5.5×10^{-9} (170 km)
HCN	1.7×10^{-7} (110 km)
	4×10^{-9} (170 km)
HC_3N	$< 1.5 \times 10^{-9}$ (110 km)
	8.4×10^{-8} (170 km)
CH_3CN	1×10^{-8} (320 km)
	1.5×10^{-9} (180 km)
H_2O	8×10^{-9} (400 km)
CO	$5 \times 10^{-6} - 5 \times 10^{-5}$ (strato.)
	$1 - 6 \times 10^{-5}$
CO_2	1.4×10^{-8} (110 km)

Adapted from Coustenis and Taylor (1999).

important greenhouse gas on Titan, and if its supply stopped, the atmosphere could collapse. There could be a cyclic process, where the atmosphere is in a low pressure / low temperature state alternated by high pressure / high(er) temperature situation (Lorenz et al. 1997, 1999). Cassini/Huygens observations will tell us more about the present total volatile content on Titan, as well as the atmospheric evolution.

What exactly happens to methane in the troposphere is not quite clear. Analysis of the Voyager data seem to indicate that supersaturation of up to 200 % can exist and that the methane mole fraction near the surface changes from about 0.06 to 0.02 between the equatorial region and the pole (Courtin et al. 1995, Samuelson et al. 1997). McKay et al. (1997) analysed the changes in temperature lapse rate as a function of altitude between 0 and 40 km. They deduced that if methane condensation takes place, the methane humidity is lower than 60 %. However, if supersaturation is allowed for, then the methane humidity may have any value. Supersaturation can exist for example if eddy transport of methane from the surface is efficient against condensation, or if condensation is kinetically impossible

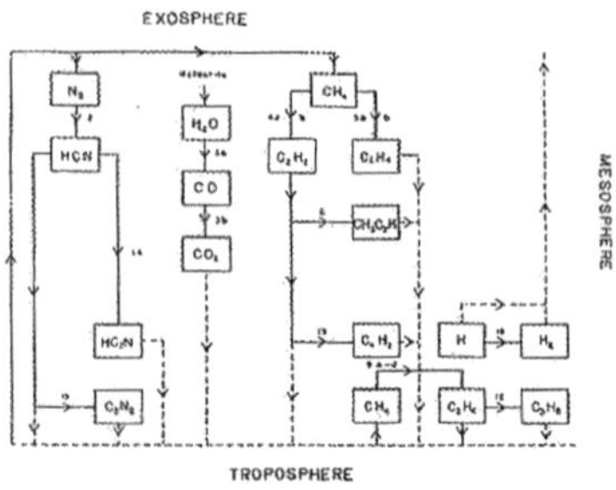

Figure 6.3. Principal photochemistry paths in Titan's atmosphere. Figure from Yung et al (1984). Permission for reproduction granted by the American Astronomical Society

by the insolubility of the possible condensation nuclei. Griffith et al. (1998) reported for the first time the observation of possible transient tropospheric clouds located at about 15 km altitude. Tokano et al. (2001) presented a 3-D study of the tropospheric methane cycle, simulated by a General Circulation Model. They concluded that supersaturation of about 150 % is more compatible with the Voyager data, and that clouds form if the supersaturation exceeds this value. They also concluded that the entire methane cycle can take place mostly in the atmosphere, as no methane rain drop should reach the surface.

From analysis of Voyager observations it was shown that the most abundant hydrocarbons (methane, ethane, acetylene and propane) are rather homogeneously mixed as a function of latitude. Some other species however show significant latitudinal variations, *i.e.* C_4H_2, C_3H_4, C_2H_4, HCN, HC_3N and C_2N_2 increase by a factor of about 10-20 in the north polar region, relative to the equator and south polar region (Coustenis and Bézard 1995). The variation of UV sunlight in the polar region as a function of the season, and the subsequent change in photochemistry was thought to be the main reason for this feature. However, Lebonnois et al. (2001) showed that this is not the case. They combined photochemistry and atmospheric dynamics in a 2-D (latitude-altitude) model, and were able to reproduce a large part of the observations. The global circulation of Titan's atmosphere can be described by one or two Hadley cells during solstice or equinox periods respectively (Hourdin et al. 1995). Downward transport in the polar regions brings upper stratospheric air (500 km), enriched in hydrocarbons, to lower levels (200 km) where the variations are seen.

Titan's atmosphere has a global haze layer, located thoughout the lower stratosphere. The haze particles are composed of organic molecules, but the exact composition is not known. In the first famous laboratory experiment by Khare et al. (1984) particles, called tholins, were produced through electric discharges and UV irradiation of a Titan-like gas mixture. Their optical properties are similar to the optical properties of Titan's haze, and can explain some of the observed spectral features. From later experiments similar results were found (see Table 6.1 in Lebonnois et al. 2002). Khare et al. (1984) found that tholins have more than 75 compounds (hydrocarbons and nitriles) and C-H-N polymers. Titan's haze particles were found to be of fractal nature (West and Smith 1991, Rannou et al. 1997). Lebonnois et al. (2002) presented a first theoretical study of how aerosols can be formed from gaseous species in Titan's atmosphere, linking photochemical models to microphysical models. The Huygens descent mission in the atmosphere of Titan, planned for January 2005, will give the first direct measurements of the detailed composition of the haze particles.

6 Pluto and Triton

Pluto and Triton are very similar in terms of mass and composition, and are located in the outer, colder parts of the Solar System.

In 1975, Hart published a paper on the possibility of the existence of an atmosphere at Pluto. Pluto's atmosphere was discovered in the early 1980's, from stellar occultation measurements (Fink et al. 1980). The surface pressure is of the order of tens to hundreds of μbar, and the surface temperature around 36 K. The observed large quantities of frozen nitrogen on the surface suggest that the main component of the atmosphere is molecular nitrogen (Owen et al. 1993), followed by methane and carbon monoxide (Table 6.1). Photochemical models of the atmosphere (Lara et al. 1997, Krasnopolsky and Cruikshank 1999) predicted the formation of more complex hydrocarbons and nitriles. Lara et al. (1997) showed that the methane and ethane can be destroyed between 50 and 300 km altitude, forming stable species like C_2H_2, C_2H_4, C_4H_2, and CH_3C_2H. No observational detection of these species other than methane exists for the moment.

Triton's atmosphere is very similar in composition and structure to Pluto's. It was observed from nearby with the Voyager 2 spacecraft in 1989. Methane was detected in the atmosphere (Herbert and Sandal 1991), and the background gas is molecular nitrogen, at similar pressures to those on Pluto. Also, Voyager 2 observed photochemical hazes throughout the atmosphere (Rages and Pollack 1991, Herbert and Sandal 1991). Kranopolsky and Cruikshank (1995) presented a model for the photochemistry of Triton's atmosphere. The same type of organic species are predicted as for Pluto, yet all without observational detection, except for methane and nitrogen (Cruikshank et al. 1984, Cruikshank and Apt 1984).

The presence of methane on Titan, Triton and Pluto suggests that there must be a significant amount of ethane as well, the most important photochemical product. A still open question is, where is this ethane?

7 The Giant Planets

Organic molecules are found in the observable parts of atmospheres of Jupiter, Saturn, Uranus and Neptune. The composition of the atmosphere of Jupiter is the best known at this moment, thanks to the successful Galileo Orbiter and Galileo Entry Probe mission. Even though the mission has ended, the Galileo data set has not yet been fully exploited. Important advances can still be made for example on the Galileo Entry Probe Mass Spectrometer results, as shown by Wong (2001).

The fact that there is a lot of methane in the atmosphere of the giant planets is related to the temperature dependence of the following reaction:

$$CH_4 + H_2O = CO + H_2 \tag{1.1}$$

At high temperatures, higher than about 1000 K, the equilibrium lies on the right hand side, at low temperatures on the left hand side. In the observable parts of the giant planets' atmospheres the temperatures are much lower than 1000 K.

As on Titan and Pluto, the chemistry that leads to organic molecules is triggered by UV photolysis of methane at high stratospheric levels. An important difference is the dominant background gas. This is molecular hydrogen for the giant planets, instead of molecular nitrogen for Titan and Pluto.

Methane is abundant in the atmospheres of the giant planets. The deep troposphere is the source for constant supply of methane to the stratosphere. Possible pathways of methane photochemistry in giant planets' atmospheres are illustrated for Saturn in Fig. 6.4. UV photodissociation occurs in the stratosphere at sub-millibar levels. From the products other species are formed at millibar and higher pressure levels. The main and stable products are ethane, acetelyne and polyacetelynes ($C_{2n}H_2$). Vertical motions in the atmosphere transport the molecules to different levels. Peaks of abundances are observed at levels between 1 μbar and 0.1 mbar. Eddy coefficients are derived by comparing the observations and the photochemical models. The more stable products can eventually diffuse downward into the troposphere. There, at higher temperatures and pressures, they are pyrolysed back into methane.

Table 6.3 lists the known organic molecules of the jovian atmospheres. The two smaller giants, Uranus and Neptune, have the highest methane

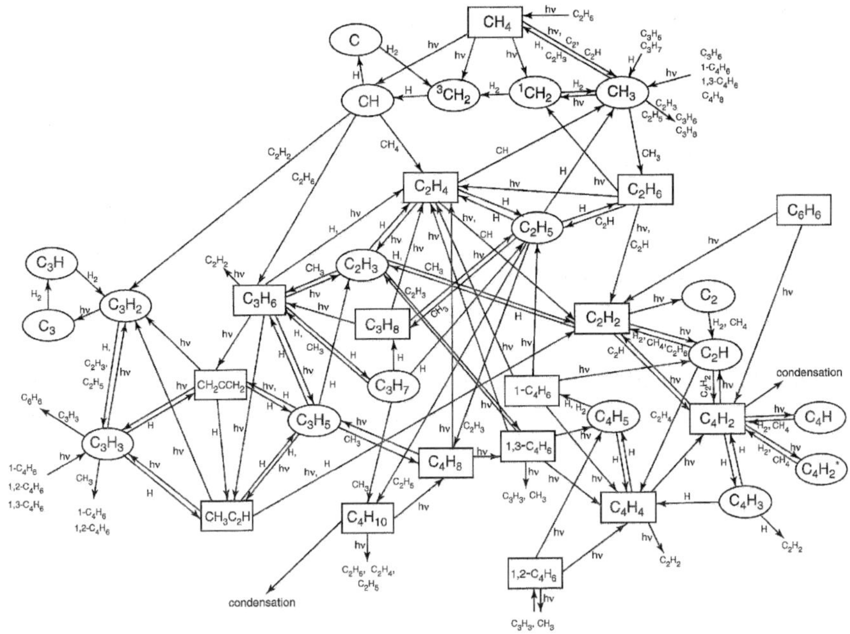

Figure 6.4. Important photochemical reaction pathways in Saturn's stratosphere. Methane photolysis initiates all the hydrocarbon reactions. Radicals are in ovals, stable molecules in rectangles. Figure reprinted from Moses et al. (2000). Copyright (2000), with permission from Elsevier.

abundances. These planets have accreted much less of the solar composition gas, after the formation of their ice-rich nuclei. The enrichments in the heavy elements (everything heavier than He) is therefore much higher than for Jupiter and Saturn.

Clouds and hazes composed of hydrocarbon particles are also formed in all giant planets. On Jupiter and Saturn, the stratospheric hazes are most probably composed of organic molecules.

Other clouds formed by condensation of volatiles are predicted to be composed of ammonia ice, ammonium hydrosulfide and water, with base pressures for Jupiter at about 1, 2 and 5 bar. The exact base pressure levels depend on the detailed composition of the atmosphere (Atreya 1986). Only ammonia ice has been detected (Baines et al. 2002). The

Table 6.3. Organic Molecules in Giant Planet Atmospheres

Constituent	Jupiter[a]	Saturn[a]
CH_4	$(1.82 \pm 0.35) \times 10^{-3}$	$4.5^{+2.4}_{-1.9} \times 10^{-3}$
CH_3	detection polar region	$(1.5 - 7.5) \times 10^{13}$ cm^2
C_2H_6	$(0.9 - 4.3) \times 10^{-6}$ (strat.)	$(3 \pm 1) \times 10^{-6}$ (strat.)
C_2H_2	$(2.6 - 8.7) \times 10^{-8}$ (strato.)	$(2.1 \pm 1.4) \times 10^{-7}$ (strato.)
	$< 2.5 \times 10^{-6}(1 - 10\,\mu$bar)	$(5 \pm 1) \times 10^{-8}$ (southern hemisph.)
C_2H_4	$6.1 \pm = 2.6 \times 10^{-9}$ (pole, auroral)	$(2 - 3) \times 10^{15}$ cm^2 (non auroral)
	$(2 - 3) \times 10^{15}$ cm^2 (non auroral)	
C_3H_4	$2.2^{+1.7}_{-0.9} \times 10^{-9}$ (pole)	$6 \times 10^{-10}(< 10$ mbar)
C_3H_8	Detection	–
C_4H_2	Detection	9×10^{-11} (10 mbar)
C_6H_6	$9^{+4.5}_{-7.5} \times 10^{14}$ cm^2(midlatp < 50 mbar)	$4.7^{+2.1}_{-1.1} \times 10^{13}$ cm^2(globalp < 10 mbar)
	$1.7^{+1.7}_{-0.9} \times 10^{-9}$ (pole, strat.)	–
HCN	$(0.9 - 3.6) \times 10^{-9}$ (b)	–
CH_3C_2H	2×10^{-10} (b)	7×10^{-10} (b)

Constituent	Uranus[b]	Neptune[b]
CH_4	$(1.6 - 2.3) \times 10^{-2}(p > 1.5$ bar)	$(1 - 2.2) \times 10^{-2}$ ($p > 1.5$ bar)
	$(0.3 - 1) \times 10^{-4}$ (tropopause)	
	3.5×10^{-4} (strat.)	
	$< 0.3 \times 10^{-6}$ (0.1 mbar)	
CH_3	–	–
C_2H_6	3×10^{-6} (0.1 mbar)	1.5×10^{-6} (strat.)
C_2H_2	4×10^{-7} (0.1 mbar)	6×10^{-8} (stratoph.)
C_2H_4	–	–
C_3H_4	–	–
C_3H_8	–	–
C_4H_2	–	–
C_6H_6	–	–
HCN	–	$1 \times 10^{-10} - 3 \times 10^{-9}$
CH_3C_2H	–	–

Abundances are in mole fractions, (a) Atreya et al (2003), (b) Irwin (2003)

composition of the other two clouds remains to be confirmed from future observations, although very strong indications for the existence of a water cloud on Jupiter have been found (Roos-Serote et al. 2000 and references therein).

Uranus and Neptune are much colder planets, with much higher enrichments of heavy elements. Consequently, condensation clouds are predicted to form at quite different levels, as compared to Jupiter and Saturn. For example, where on Jupiter a water cloud is predicted around 5 bar, on Uranus this figure lies somewhere between 100 and 1000 bar.

Due to the lower temperatures, methane is predicted to condense and form a cloud both on Uranus and Neptune around the 1-2 bar level. Clouds were indeed detected at these levels, even though with much lower optical depths (< 1) than predicted by models. A strong indication that these are methane clouds is the sharp drop in methane abundance at pressures lower than 1-2 bar. Beneath the optically-thin methane clouds the top of another, opaque, cloud is seen, perhaps the predicted H_2S cloud. No spectral identification has been possible yet. If it is the same cloud on both planets, as expected, then the different coloration (Uranus green, Neptune blue) has to be explained (see Irwin 2003 and references therein).

7.1 The impact of Comet D/Shoemaker-Levy 9 in Jupiter

During about 10 days in the month of July 1994, all 20 fragments of comet D/Shoemaker-Levy 9, crashed into the atmosphere of Jupiter at $-44°$ latitude. The impact phenomena and their consequences were extensively monitored and studied, both from ground and space based observatories. All fragements exploded in the atmosphere at pressure levels on the order of several bars, *i.e.* below the main cloud deck at 1-2 bar. These explosions caused plumes of heated and shocked jovian air mixed with cometary material to be ejected up to 3000 km above the main clouds, where the pressure is negligible. The plumes fell back and started to interact again with the atmosphere in high the stratosphere, at pressure levels around 0.1 mbar. Many chemical reactions took place, producing all sorts of new molecules. After suffering heating, the stratosphere returned to its normal temperature state in a matter of several weeks, due to cooling by dust particles.

Lellouch (1996) reviewed the observations. Hydrocyanide (HCN) was detected by several experiments. An enhancement of the mixing ratio by a factor of 100 to 1000 relative to the nominal value reported in Table 6.3 was clearly seen over the impact sites in the hours and days following each impact. Moreno et al. (2003) present a longterm study (1995–1998) of the evolution of the HCN, CS and CO abundance in the stratosphere after the impacts. They monitored the evolution (spatial distribution and abundance) using spectral signatures of these species in the mm/sub-mm range. The HCN abundance at the 0.1-0.2 mbar level at the impacts' latitude remained higer than the nominal value by a factor of 10 to 100 in the period of 1995 through 1998.

Variations in the emission spectra of CH_4 and C_2H_2 were observed above the sites of impact, for hours and days after the events. This can be attributed to an enhancement of the abundances of these species, to temperature effects, or to a combination of both.

Emission from C_2H_4 was also detected and a derived enhancement of its abudance by a factor of 1000-5000 over the impact sites means that it was probably formed by shock chemistry.

8 Conclusion

In this chapter, I have presented an overview of our current knowledge on the subject of organic molecules in planetary atmospheres. In the terrestrial planets' atmospheres organic molecules are scarce, except for some species in the atmosphere of the Earth, where there are biological and anthropogenic sources. The effects of life on Earth can be detected from remote sensing observations. Future experiments will address this issue for extrasolar Earth like planets.

In the atmospheres of the giant planets most of the organic molecules are found in the stratosphere and are products from the UV induced photochemistry chain with methane at the origin.

The huge variety of organic molecules already known to exist on Titan is the result of photochemistry in an atmosphere dominated by molecular nitrogen and methane. Other more complex molecules are predicted to exist, but could not yet be detected. Organic molecules should be abundantly present at the surface too. We hope to learn much more in the near future from the Cassini/Huygens mission. In terms of astrobiology and apart from the Earth, Titan appears the most obvious and interesting object, but surprises will surely come as we explore the individual objects of the Solar System more and more thoroughly (Mars and Europa in particular).

The impact of comets onto the main bodies of the Solar System transfers (organic) material to these objects, and induces the creation of new species, as was shown by the impact of comet D/Shoemaker-Levy 9 in the atmosphere of Jupiter.

Acknowledgements

MRS thanks Sushil K. Atreya, Ah San Wong, Mike H. Wong, Eric Wilson, Luisa Lara, André Martin, Tobias Owen and Margarida Serote Roos for fruitful comments, corrections and discussions, and the anonymous referee for useful input and corrections. MRS thanks Malcolm Fridlund, Ah San Wong, Julie Moses and Yuk Yung for kindly permitting the use of graphs from their published work. MRS thanks Pascale Ehrenfreund and the ISSI team for the invitation to participate in the Workshop *Future Perspectives and Strategies in Astrobiology*.

References

Atreya, S.K. (1986). Atmospheres and Ionospheres of the Outer Planets and Their Satellites. Springer, Berlin.

Atreya, S.K., P.R. Mahaffy, H.B. Niemann, M.H. Wong, and T.C. Owen (2003). Composition and Origin of the Atmosphere of Jupiter -an Update, and Implications For the Extrasolar Giant Planets. Plan. Space Sci. 21, 105–112.

Baines, K.H., R.W. Carlson, and L.W. Kamp (2002). Fresh ammonia ice clouds in Jupiter: Spectroscopic Identification, Spatial Distribution and Dynamical Implications.Icarus 159, 74–94.

Bézard, B. C. de Bergh, D. Crisp, J-P. Maillard (1990). The Deep Atmosphere of Venus Revealed by High-Resolution Nightside Spectra. Nature 345, June 7, 1990, 508–511.

Cabane, M., E. Chassefière and G. Israel (1992). Formation and Growth of Photochemical Aerosols in Titan's Atmosphere. Icarus 96, 176–189.

Cruikshank, D.P., and J. Apt (1984). Methane on Triton-Physical state and distribution. Icarus 58, 306–311.

Cruikshank, D.P., R.H. Brown, R.N. Clark (1987). Nitrogen on Triton. Icarus 58, 293–305.

Cruikshank, D.P. (1987). Pluto, Charon, and Triton: A Review of Their Physical Parameters, Atmospheres, and Surfaces. Bulletin of the American Astronomical Society, Vol. 19, 858.

Courtin, R., D. Gautier, and C.P. McKay (1995). Titan's Thermal Emission Spectrum: Reanalysis of the Voyager Infrared Measurements. Icarus 114, 144–162.

Coustenis, A. and B. Bézard (1995). Titan's Atmosphere from Voyager Infrared Observations: IV. Latitudinal Variations of Temperature and Composition. Icarus 115, 126–140.

Coustenis, A, and F. Taylor (1999). Titan -the Earth like Moon. Series on Atmospheric, Oceanic and Planetary Physics, vol.1, World Scientific.

Donahue, T.M., and R.R. Hodges, Jr. (1992) Methane measurement by the Pioneer Venus large probe neutral mass spectrometer. In Lunar and Planetary Inst., Papers Presented to the International Colloquium on Venus, p.29.

Donahue, T.M., D.H. Grinspoon, R.E. Hartle, and R.R. Hodges Jr. (1997). Ion/Neutral Escape of Hydrogen and Deuterium: Evolution of Water. In Venus II: Geology, Geophysics, Atmosphere and Solar Wind Environment (S.W. Bougher, D.M. Hunten and R.J. Phillips, Eds.), The University of Arizona Press, 385–414.

Fink, U., B.A. Smith, D.C. Benner, J.R. Johnson, and H.J. Reitsema (1980). Detection of a CH4 Atmosphere on Pluto. Icarus 44, 62–71.

Florensky, C.P., V.P. Volkov, and O.V. Nikolaeva (1978). A Geochemical Model of the Venus Troposphere. Icarus 33, 537–553.

Fortes, A.D. 2000. Exobiological Implications of a Possible Ammonia-Water Ocean Inside Titan. Icarus 146, 444–452.

Fridlund, C.V.M. (2000). Darwin -The Infrared Space Interferometry Mission. ESA Bulletin 103, 20–25. Link at esapub.esrin.esa.it/bulletin/bullet103/fridlund103.pdf

Griffith, C.A., T. Owen, G.A. Miller, and T. Geballe (1998). Transient Clouds in Titan's Lower Atmosphere. Nature 395, 575–578.

Hart, M.H. (1975). A Possible Atmosphere for Pluto. Icarus, vol. 21, 242–247.

Herbert, F. and B.R. Sandel (1991). CH4 and haze in Triton's lower atmosphere. J. Geoph. Res. 96, 19,241–19,252.

Hourdin, F., O. Talagrand, R. Sadourny, R. Courtin, D. Gautier, and C. McKay (1995). Numerical Simulation of the General Circulation of the Atmosphere of Titan. Icarus 117, 358–374.

Irwin, P. (2003) Giant Planets of Our Solar System. Springer -PRAXIS.

Khare, B.N., C. Sagan, E.T. Arakawa, F. Suits, T.A. Callott and M.W. Williams (1984). Optical Constants of Organic Tholins Produced in a Simulated Titatian Atmosphere: From Soft X-Rays to Microwave Frequencies. Icarus 60, 127–137.

Korablev, O.I. (2002). Solar Occultation Measurements of the Martian Atmosphere on the *Phobos* Spacecraft: Water Vapor Profile, Aerosol Parameters, and Other Results. Solar System Research 36, 12–34.

Korablev, O.I., M. Ackerman, V.A. Krasnopolsky, V.I. Moroz, C. Muller, A.V. Rodin, and S.K. Atreya (1993). Planet. Space Sci. 41, 441–451.

Kossacki, K.J., and R.D. Lorenz (1996). Hiding Titan's Ocean: Densification and Hydrocarbon Storage in an Icy Regolith. Planet. Space Sci. 44, 1029–1037.

Krasnopolsky , V.A. and V.A. Parshev (1979). On the Chemical Composition of the Troposphere and Cloud Layer of Venus Based on the Venera 11 and 12 and Pioneer Venus Measurements. Cosmic. Res. 17, 763–770.

Krasnopolsky, V.A., and J.B. Pollack (1994). H_2) -H_2SO_4 Systems in Venus' Clouds and OCS, CO and H_2SO_4 Profiles in Venus' Troposphere. Icarus 109, 58–78.

Krasnopolsky, V.A., and D.P. Cruikshank (1995). Photochemistry of Triton's atmosphere and ionosphere. J. Geoph. Res. 100, 21271–21286.

Krasnopolsky, V.A., G.L. Bjoraker, M.J. Mumma, and D.E. Jennings (1997). High-Resolution Spectroscopy of Mars at 3.7 and 8 μm: a Sensitive Search for H_2O, H_2CO, HCl, and CH_4, and Detection of HDO. J. Geoph. Res., 102, 6526–6534.

Krasnopolsky, V.A., and D.P. Cruikshank (1999). Photochemistry of Pluto's atmosphere and ionosphere near perihelion. J. Geoph. Res. 104, 21979–21996.

Kuiper, G.P. (1944). Titan: A Satellite with an Atmosphere. Astrophys. J. 100, 378–383.

Lara, L.M., W.H. Ip, and R. Rodrigo (1997). Photochemical Models of Pluto's Atmosphere. Icarus 130, 16–35.

Lara, L.M., E. Lellouch, J.J. Lopez-Moreno, and R. Rodrigo (1996). Vertical Distribution of Titan's Atmospheric Neutral Constituents. J. Geoph. Res. 101, 23261–23283.

Lebonnois, S., E.L.O. Bakes, and C.P. McKay (2002). Transition from Gaseous Compounds to Aerosols in Titan's Atmosphere. Icarus 159, 505–517.

Lebonnois, S, D. Toublanc, F. Hourdin and P. Rannou (2001). Seasonal Variations of Titan's Atmospheric Composition. Icarus 152, 384–406.

Lederberg, J. (1965). Signs of Life: Criterion-System of Exobiology. Nature 207, 9–13.

Lellouch, E. (1996). Chemistry induced by the impacts: Observations. In: The Collision of Comet Shoemaker-Levy 9 and Jupiter (Noll, K.S., Waever, H.A., Feldman P.D., Eds.), Science Institute Symposium Series 9, Cambridge University Press, Cambridge, 213–242.

Lorenz, R.D. (1996). Pillow Lava on Titan: Expectations and Constraints on Cryo-Volcanic Processes. Planet. Space Sci. 44, 1021–1028.

Lorenz, R.D., J.I. Lunine, P.G. Withers, and C.P. McKay (1997). Photochemically Driven Collapse of Titan's Atmosphere. Science 275, 642–644.

Lorenz, R.D., C.P. McKay, and J.I. Lunine (1999). Analytic Investigation of Climate Stability on Titan: Sensitivity to Volatile Inventory. Plan. Space Sci. 47, 1503–1515.

Loveday, J.S., R.J. Nelmes, M. Guthrie, S.A. Belmonte, D.R. Allan, D.D. Klug, J.S. Tse, and Y.P. Handa (2001). Stable Methane Hydrate Above 2 GPa and the Source of Titan's Atmospheric Methane. Nature 410, 661–663.

Lovelock, J.E. (1965). A Physical Basis for Life Detection Experiments. Nature 207, 568–570.

Maguir, W.S. (1977). Martian isotopic ratios and upper limits for possible minor constituents as derived from Mariner 9 infrared spectrometer data. Icarus 32, 85–97.

Malin, M.C, and K.S. Edgett (2000). Evidence for Recent Groundwater Seepage and Surface Runoff on Mars. Science 288, 2330–2335.

Marten, A., J. Rosenqvist, D. Moreau, and R. Moreno (1995). New Photochemical Assessments on Mars Derived from Millimeter Observations. Unpublished, private communication.

McKay, C.P., J.B. Pollack, J.I. Lunine and R. Courtin (1993). Coupled Atmosphere-Ocean Models of Titan's Past. Icarus 102, 88–98.

McKay, C.P., S.C. Martin, C.A. gri;th, R.M. Keller (1997). Temperature Lapse Rate and Methane in Titan's Troposphere. Icarus 129, 498–505.

Moreno, R., A Marten, H.E. Matthews, and Y. Biraud (2003). Long-term evolution of CO, CS and HCN in Jupiter after the impacts of comet Shoemaker-Levy 9. Plan. Space Sci. 51, 591–611

Moses, J.I., B. Bézard, E. Lellouch, G.R. Gladstone, H. Feuchtgruber, and M. Allen (2000). Photochemistry of Saturn's Atmosphere: I. Hydrocarbon Chemistry and Comparisons with ISO Observations. Icarus 143, 244–298.

Owen, T.C., T.L. Roush, D.P. Cruikshank, J.L. Elliot, L.A. Young, C. de Bergh, B. Schmitt, T.R. Geballe, R.H. Brown, and M.J. Bartholomew (1993). Surface ices and the atmospheric composition of Pluto. Science 261, 745–748.

Pollack, J.B., J.B. Dalton, D. Grinspoon, R.B. Wattson, R. Freedman, D. Crisp, D.A. Allen, B. Bézard, C. de Bergh, L.P. Giver, Q. Ma, and R. Tipping (1993). Near Infrared Light from Venus' Nightside: A Spectroscopic Analysis. Icarus 103, 1–42.

Rages, K.A., and J.B. Pollack (1991). Hazes and Clouds in Triton's Atmosphere. Bulletin of the American Astronomical Society, Vol. 23, 1207.

Rannou, P. M. Cabane, R. Botet, and E. Chassefière (1997). A New Interpretation of Scattered Light Measurements at Titan's Limb. J. Geoph. Res. 102, 10997–11013.

Raulin, F. (1997). Titan's Organic Chemistry and Exobiology. in Huygens -Science, Payload and Mission, ESA SP 117, 219–229.

Roos-Serote, M. A.R. Vasavada, L. Kamp, P. Drossart, P. Irwin, C. Nixon, and R.W. Carlson (2000). Proximate humid and dry regions in Jupiter's atmosphere indicate complexe local meteorology. Nature 406 (May 11, 2000), 158–160.

Sagan, C., W.R. Thompson, R. Carlson, D. Curnett, and C. Hord (1993). A Search for Life on Earth From the Galileo Spacecraft.

Samuelson, R.E., N.R. Nath, and A. Borysow (1997). Gaseous Abundances and Methane Supersaturation in Titan's Troposphere. Plan. Space Sci. 45, 959–980.

Schindler, T.L. and J.F. Kasting (2000). Synthetic Spectra of Simulated Terrstrial Atmospheres Containing Possible Biomarker Gases. Icarus 145, 262–271.

Tokano, T., F.M. Neubauer, and M. Laube (2001). Three-Dimensional MOdeling of the Tropospheric Methane Cycle on Titan. Icarus 153, 130–147.

West, R.A., and P.H. Smith (1991). Evidence for Aggregate Particles in the Atmospheres of Titan and Jupiter. Icarus 90, 330–333.

Wong, A.S, S.K. Atreya, and T. Encrenaz (2003). Chemical Markers of Possible Hot Spots on Mars. J. Geoph. Res. 108, 5026–5036.

Wong, M.H. (2001). Hydrocarbons and consensible volatiles in Jupiter's Galileo Entry Probe Site. Ph.D. Thesis, University of Michigan, USA.

Yung, Y.L, M. Allen, and J.P. Pinto (1984). Photochcmistry of the atmosphere of Titan: Comparison between model and observations. Astroph. J. 55, 465–506.

Chapter 7

OBSERVATIONS AND LABORATORY DATA OF PLANETARY ORGANICS

Ted L. Roush and Dale P. Cruikshank
NASA Ames Research Center

Abstract: Interpretations of telescopic observations show that H_2O ice is ubiquitous on surfaces throughout many regions of the outer Solar System. Additionally, carbon-bearing molecular material is emerging as a major component in the outer Solar System, where it appears entrained in H_2O ice on comets and many planetary satellites, and in the more volatile N_2 ice on Triton and Pluto. Complex macromolecular carbon, long known in carbonaceous meteorites, appears to be a notable component of comets, planetary satellites, and small trans-Neptunian bodies. Some of this material may be retained from the solar nebula, but some of it originates in the surface ices (and in a few cases, in the tenuous atmospheres) through energetic processing by uv radiation, cosmic rays, and magnetospheric particles. Laboratory studies of the reflectance properties and chemical reactions associated with C-, H-, and N-bearing precursor gases and ices show that the stable residues created exhibit a range of coloration at visual and near-infrared wavelengths. Modeling of the telescopic observations using these residues suggest they are capable of giving rise to the observed variable red color of many surfaces in the outer Solar System.

Key words: Solar System Surfaces, Composition, Organics, Ices

1. INTRODUCTION

Here we use the term organics to describe materials that contain chiefly C, H, N, and O that are stable at temperatures and pressures commonly associated with the surfaces of outer Solar System bodies. Throughout the outer Solar System organic chemistry in both the gaseous and solid states is currently in progress. In the atmospheres of Titan and all the gas giant planets photochemical reactions produce organic molecules and solid aerosol

P. Ehrenfreund et al. (eds.), Astrobiology: Future Perspectives, 149–177.

particles that constitute a significant feature of their atmospheres [e.g., Bruston et al. 1994]. Even on Triton and Pluto photochemical reactions produce hydrocarbons and nitrites in the cold, tenuous atmospheres that precipitate to the surfaces [Krasnopolsky and Cruikshank 1995, 1999] and organics may be forming directly in the surface ices [Bohn et al. 1994; Gerakines et al. 1996, 2001; Strazzulla 1998; Hudson and Moore 2001, Moore et al. 2003]. At the same time, irradiation of airless icy bodies, specifically the satellites of Jupiter, Saturn, Uranus, and Neptune, by the solar wind and ultraviolet, galactic cosmic rays, and precipitating magnetospheric particles, produces complex organic molecules in the solid state [Khare et al. 1993; McDonald et al. 1994, 1996; Salama 1998; Delitsky and Lane 1997, 1998, 2002].

In addition to the current production of organic molecules, there is evidence of the existence of primitive organic material in meteorites, in interplanetary dust particles derived from comets, and in the dust directly sampled from Comet P/Halley. The best current evidence favors the formation of much of this primitive organic inventory found in the outer parts of the Solar System in the outflows of carbon-rich stars and in the icy grains of the interstellar medium, rather than genesis in the solar nebula [Cronin and Chang 1993 Lunine 1997].

Insofar as the genesis of life on Earth and possibly other planets is linked to the abiotic formation and chemical history of the compounds of carbon and the elements H, N, and O we provide a summary of the distribution of these materials in the present-day Solar System. Here we focus on the outer Solar System, beyond Mars. Comets are discussed in more detail in the chapter by Crovisier contained within this volume. This manuscript builds upon the previous discussions of Cruikshank [1997], Roush et al. [1995], and Roush [2001].

1.1 Observational Information

Most of the information we have on the compositions of solid bodies in the Solar System comes from spectroscopic observations with ground-based telescopes and spacecraft in the visible and near-infrared wavelength regions. For most of the icy objects in the outer Solar System the detected radiation is entirely reflected sunlight because these objects are very cold. Because the incident sunlight penetrates the surface and interacts with the materials present there, the measured reflected sunlight contains information regarding the surface materials, and the ratio of the reflected to incident sunlight provides a mechanism of identifying those materials.

An initial identification of the species present on Solar System objects can be provided by a comparison of the planetary surface spectrum to the spectra of materials measured in the laboratory. For example, one can use laboratory spectra of pure ices to identify an ice on the basis of the position, number, and characteristics of the spectral features present. However, more detailed information, such as relative abundances of more than one species or estimates of the grain sizes of surface constituents, requires more detailed models that describe the interaction of sunlight with particulate surfaces.

1.2 Laboratory Studies

1.2.1 Ices

Organic ice studies include the spectroscopy of solid CH_4, the C_2-hydrocarbons, CH_3OH, and many more molecules [Fink and Sill 1983; Schmitt et al. 1998; Quirico et al. 1996; Quirico and Schmitt 1997a,b] both in transmission and in reflectance. The pure ices of a particular molecule typically have absorption bands at nearly the same wavelengths as the absorption bands in the gas phase, but with the rotational structure suppressed and with moderate wavelength shifts in the band centers. The combination and overtone bands are represented, as are the fundamental vibrational transitions. Laboratory spectra of ices are useful in identifying specific materials on a planetary surface, but models of a surface seen by diffusely reflected sunlight requires the complex refractive indices of ices derived from spectra and certain parameters of the laboratory specimen. Spectroscopy of ices in their different crystalline phases, such as the α and β phases of N_2 ice and amorphous phases of H_2O [Schmitt et al. 1992], are also critical to understanding planetary surfaces and environments. An important factor in analyzing multi-component icy surfaces is the matrix effect; for example, CH_4 frozen in a matrix of N_2 shows frequency shifts of $\sim$10 cm^{-1} in the central frequencies of many of the methane bands. The matrix shift of CH_4 in N_2 is seen in spectra of Triton and Pluto [Cruikshank et al. 1993; Quirico et al. 1999; Douté et al. 1999].

1.2.2 Organics

The low reflectance and red or black colors of a number of asteroids, planetary satellites, and comet surfaces have led investigators to study naturally occurring complex organic materials, such as terrestrial hydrocarbons (coal, tar sands, asphaltite, anthraxolite, kerite, etc.), which have similar properties of color and albedo. Cloutis [1989], Cloutis et al. [1994], and Moroz et al. [1998] have published reflectance spectra of these and related terrestrial materials, which show absorption bands of aliphatic and aromatic hydrocarbons in complex combinations, in addition to low albedo and strong red color. These terrestrial organic materials, the subjects of intensive investigations as fuels, typically contain a soluble, volatile component (largely aliphatic hydrocarbons) and an insoluble component having a complex three-dimensional structure of large sheets of aromatic rings linked loosely by short aliphatic bridging chains and having side chains (kerogen). The literature on kerogens is extensive [Duran 1980], and because of their structures they are relevant to planetary organics, even though terrestrial kerogens originate largely, or perhaps entirely from the decay of organisms. Organic solid materials in carbonaceous meteorites are dominantly an insoluble macromolecular component resembling terrestrial kerogen. The soluble organic fraction contains many different classes of compounds, some of which, such as carboxylic and amino acids, reveal great structural diversity and a predominance of branched-chain molecules, whereas others,e.g., aliphatic hydrocarbons, do not [Cronin and Chang 1993].

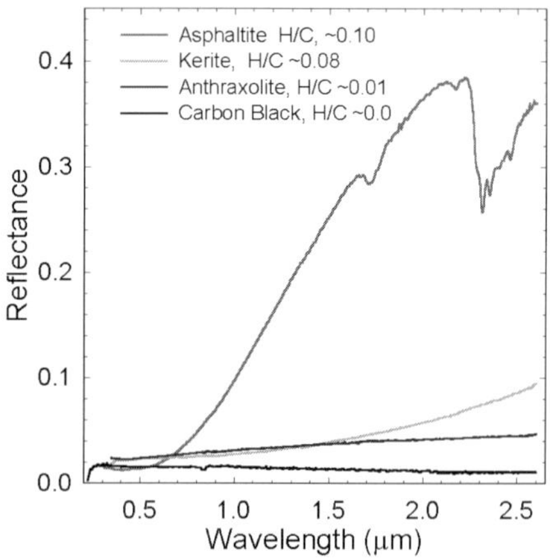

Figure 1. Reflectance spectra of various organics. Data from Moroz et al. [1998] and Cruikshank et al. [1991] show the change in color as the H/C ratio decreases.

Interplanetary dust particles (IDPs) collected in the stratosphere are microscopic fragments of comets and asteroids that enter the Earth's atmosphere without melting. Some of them contain significant amounts of organic macromolecular material with strong deuterium and ^{15}N enhancements indicating an origin in the interstellar medium. The organics in meteorites and IDPs are studied not only with spectroscopy [e.g., Flynn et al. 2002] and other standard chemical techniques [Cronin and Pizzarello 1983], but with increasingly sensitive instrumentation, such as nuclear magnetic resonance (Cody et al. 2002a], laser desorption mass spectroscopy [Clemett et al. 1993], and ion microprobe. This work is leading to a rapidly improving understanding of the composition and structure of the organic components of meteorites and IDPs, and is strengthening the links with their sites and conditions of origin [e.g., Aléon et al. 2003].

1.2.3 Processing of original organic materials

Materials on the surfaces of airless Solar System bodies are exposed to uv photons from the Sun and to cosmic rays, which are primarily H^+, He^+, and O^+ particles with energies ranging from MeV to GeV. Planets with magnetic fields trap solar wind charged particles, providing an environment in which orbiting natural satellites are exposed to a flux of protons and electrons. These kinds of radiation alter the chemical structure of minerals, ices, and organic materials in many ways. Charged particles and uv flux incident on a planetary or satellite atmosphere also causes chemical changes that result in the formation of photochemical haze particles, some of which remains aloft, and some of which precipitates to the surface. The photochemical smog in Titan's atmosphere has been simulated by the irradiation of mixtures of N_2 and CH_4 in the atmospheric ratio (9:1), the result of which is the formation of a fairly refractory complex solid of yellow or dark red color; this material is widely called tholin.

The irradiation by low-energy charged particles and uv (in a plasma discharge) of mixtures of hydrocarbon ices in H_2O ice produces tholins [Khare et al. 1993, McDonald et al. 1996]. The detailed composition of the tholin depends upon the ice composition and other factors, but for ices containing C, H, O, and N- bearing molecules, a range of materials including polyalcohols, ethers, esters, carboxylic acids, and hydrocarbons are produced. Overall, ice tholins can be termed organic heteropolymers.

Other experiments on H_2O- and N_2-rich ices containing CO and CH_4 by Moore et al. [2003] used 0.8 MeV protons from a Van de Graaff accelerator, with a focus on the new volatile molecules formed in the reactions. For example, the irradiation of H_2O + CO produces H_2CO and CH_3OH. The proton irradiation of mixtures of CH_4 and CO in N_2, which occurs on the surfaces of Triton and Pluto, produces a number of N-bearing molecules and radicals, including HNC, HCN, NH_3, HN_3, OCN- and CH_2N_2. Baratta et al. [2002] compared ion irradiation and uv photolysis of CH_4 and CH_3OH ices, finding that the effects of 30 KeV ions are comparable to those of uv on fresh ices (that is, at low doses), but are increasingly different as the dose increases. While the interaction of uv with the ice depends strongly on the optical properties of the ice (and the products of the irradiation), the interaction of ions is independent of the optical properties.

Several summaries are readily available that discuss the effects of uv photolysis [Gerakines et al. 1996, 2001; Salama, 1998; Hudson and Moore 2001] and radiolysis [Johnson and Quickenden 1997; Delitsky and Lane, 1997, 1998, 2002; Strazzulla, 1998; Johnson, 1998; Cottin et al. 1999; Moore and Hudson, 2000; Hudson and Moore 2001; Bahr et al. 2001; Schou and Pedrys 2001; Satorre et al. 2001; Gerakines et al. 2001] on ices germane to solid surfaces in the outer system. These generally focus on the materials produced by the energetic processing. Many studies germane to the investigation of the effects of photolysis and radiolysis of ices located in the ISM and diffuse clouds have been undertaken. Summaries are given by

Gerakines et al. [1996, 2001], Salama [1998], Strazzulla [1998], Cottin et al. [1999], and Ehrenfreund et al. [1997, 1999, 2001]. These studies show the ready production of daughter materials from the initial starting ices. However, these latter studies typically involve only thin films of ices, and measurements are performed at infrared wavelengths (>2.5 μm), making their application to Solar System observations, discussed below, limited. In a number of these studies the optical constants of parent and daughter products are derived.

Energetic processing of gases and solids containing hydrocarbons has produced a variety of refractory organic residues [Thompson et al. 1987; Khare et al. 1987, 1993; McDonald et al. 1996; Imanaka et al. 2004] whose optical constants have been reported [Khare et al. 1987, 1993; McDonald et al. 1996; Ramirez et al. 2002].

1.3 Impact Formation of Organics

The impacts of asteroids and comets on planetary surfaces and their disruption in planetary atmospheres have played a role in both the formation and destruction of organic materials. For the early Earth, Chyba and Sagan [1992] evaluated the synthesis of organic molecules by shocks generated by meteoritic air bursts and their vapor plumes, as well as by lightning and uv sunlight, finding that the yields are strongly dependent on the composition of the atmosphere. Impact shocks in a strongly reducing atmosphere are very effective, but in the more likely case of an atmosphere of intermediate oxidation state, impact shock is a weak source of organics.

Meteoroid impact on a solid, ice-rich planetary surface may have some effect on the formation of organic chemicals from the native target materials plus the projectile components. Borucki et al. [2002] observed that the impact of a 5 km s^{-1} projectile on an ice target produces an electrical discharge within the ice. Borucki et al. [2002] suggested the discharge arises from charge separation along fracture boundaries, drawing energy not only from the kinetic energy of the impact, but the stress energy stored in the ice target. They propose that this electrical energy might contribute to the formation of organic molecules within the ice, especially in a planetary (or satellite) surface that is highly (and repeatedly) stressed, as in the case of Europa.

1.4 Color in Complex Organic Molecules

Color in complex organic molecules is caused by absorption in the ultraviolet, violet, and visible spectral regions arising from electronic transitions connected primarily with carbon-carbon and carbon-oxygen bonding. These bonds consist of shared electrons, either singly or in pairs. Electronic transitions in simple molecules with C-H bonds (e.g., CH_4 methane) or single C-C bonds (e.g., C_2H_6 ethane) cannot be easily excited without disruption of the molecule.

The single bonds (denoted σ) can only be excited by the absorption of radiation in the far ultraviolet, hence the colorless nature of these molecules.

Hydrocarbon molecules with double bonds (π bonds), or two pairs of electrons shared by adjacent carbon atoms, have sufficient electronic freedom to permit excitation by longer wavelength radiation without disruption of the molecule. The simplest of these molecules is $H_2C{=}CH_2$, ethylene (C_2H_4) and the two electrons involved are π electrons. With only one π bond, ethylene absorbs in the deep ultraviolet, but when a large molecule has many π bonds, particularly when they alternate with single bonds, the absorption edge moves into regions of longer wavelength, and eventually into the visible spectral region. In molecules with conjugated bonds (alternating single and double or triple bonds), the π electron can be shared with more than one adjacent atom, yielding a complex delocalized π-electron distribution. Delocalization allows energy levels of the π electrons of the entire molecule to be more closely spaced than those in molecules with only one double bond, as in ethylene.

Delocalization of π electrons results in the colors of molecules that have very long conjugated chains or rings of double bonds alternating with single bonds. A compound with nine alternating double and single bonds (lutein) gives egg yolk its color, while carotene, with eleven double bonds in conjugation, has its maximum absorption at 0.451 μm, giving carrots their characteristic color. Much larger molecules in solid form with 20, 30, or more carbon atoms arranged in joined aromatic rings (polycyclic aromatic hydrocarbons, or PAHs) are often brightly colored; some are distinctly red and with low absolute reflectivity.

The electronic states in polyatomic molecules are extremely numerous. A molecule can undergo transitions from any number of ground states to any number of excited levels. Consequently, ultraviolet absorption bands are broad, particularly in liquids and solids, in contrast to the well-defined nature of infrared vibrational transitions.

The general behavior of large molecules is such that the absorption edge progresses toward longer wavelengths as the sequence of conjugated bonds lengthens. The net effect is that macromolecular carbon compounds are typically dark red, brown or black to the eye, and "red" in the sense that is that their spectral reflectance increases (often steeply) toward longer wavelengths in the near-infrared (0.8-2 μm).

1.5 Modeling the Observations for Compositional Interpretation

Various models are used to represent the interaction of electromagnetic energy (solar flux) with materials in the outer Solar System in order to interpret the telescopic observational data discussed below. These mathematical models of surface reflectance are useful for characterizing the chemical and physical nature of the surfaces under investigation. They can also be used to extend the results of laboratory studies to different mixtures and particle sizes that are not directly investigated. They can span the range of spatial mixing of components (scales of dm to km), intimate mixing of components (mixtures of grains on scales of μm to cm), and molecular mixtures of components (scales of Å to nm).

Spatial mixtures describe the situation where the photons reflected from the surface are dominated by interactions with individual surface species. The total reflectance of the surface (R_T) is expressed as a linear combination of the reflectances of each component (R_i) weighted by its areal extent (X_i), that is, $R_T = X_1{\times}R_1 + X_2{\times}R_2 + \cdots + X_n{\times}R_n$, where n is the total number of different components, and it is required that $\Sigma X_i = 1$. Examples of these types of surfaces are individual rocks or outcrops.

In intimate, or "granular", mixtures, the surface reflectance is a nonlinear function of the reflectance of each component because multiple scattering of the incident photon means it can encounter several grains before exiting the surface. Examples are individual minerals that are contained in rocks and planetary soils and regoliths.

Hapke [1993] and Shkuratov et al. [1999] developed approximations to the intimate mixture case that allows the reflectance of these mixtures to be calculated from the optical constants of the individual components [see Roush 1994]. The application of Hapke's approach to interpretation of satellite surface photometry is presented by Verbiscer and Helfenstein [1998] and to compositional interpretation by Cruikshank et al. [1998a]. Application of the Shkuratov approach to interpretation of surface spectra is presented in Poulet et al. [2002].

2. OUTER SOLAR SYSTEM

2.1 Meteorites-Asteroids

2.1.1 Meteorites

The recognition that organic matter occurs in certain classes of meteorites provides evidence that organics can be found in locations in the Solar System other than Earth. Important reviews of meteoritic organics have been given by Cronin et al. [1988] and Cronin and Chang [1993]; the discussion here is derived in part from those reviews and from Cruikshank and Kerridge [1992].

The great majority of meteoritic organic matter is found in the carbonaceous meteorites. These meteorites have the additional property that their elemental compositions closely resemble those of the Sun. The two properties are related at least to the extent that preservation of organic molecules would have been aided by a history that led to minimal chemical fractionation. Still, even carbonaceous chondrites have undergone considerable processing during their evolution, and episodes of localized heating, impact brecciation, irradiation by solar flares and galactic cosmic rays, and aqueous mineralization recorded in them have altered their organic components. The observed patterns of molecular and isotopic composition of meteoritic organic matter are difficult to reconcile within a single model of synthesis and suggest instead multiple origins for the organic species found in meteorites. The heterogeneous chemical nature of the solar nebula no doubt played a role in the incorporation, and possibly the synthesis (on a minor scale), of some of the materials.

The salient properties of meteoritic organic matter can be succinctly stated. The dominant organic fraction is an insoluble macromolecular component, resembling terrestrial kerogen discussed above. The unusual organic composition of the Tagish Lake meteorite is discussed below in section 2.1.2.4.

The soluble organic fraction contains many different classes of compound, some of which, e.g., carboxylic and amino acids, reveal considerable, perhaps complete, structural diversity and a predominance of branched-chain molecules, whereas others, e.g., aliphatic hydrocarbons, do not.

Mineralogical evidence in the most volatile-rich carbonaceous meteorites clearly shows that liquid water, probably resulting from ice melting within the meteorite parent bodies, has produced phyllosilicates, other hydrated minerals, and deposits of sulfates, carbonates, sulfides, and oxides [Zolensky and McSween 1988]. Water has also hydrolized the native organic material in these meteorites, resulting in amino acids, carboxylic acids, and numerous other complex products. In fact, only those meteorites in which originally anhydrous minerals have been converted by the activity of liquid water to hydrous minerals, notably the clays, show abundant organic compounds.

2.1.2 Low-Albedo Asteroids, Small Primitive Bodies in the Solar System

2.1.2.1 Taxonomy

A taxonomy of low-albedo asteroids was originally developed by Tholen [1984, 1989], based on photometry in the region 0.3-1.05 μm. In this scheme, the C asteroids are neutral in spectral reflectance (0.4-1.0 μm), but with a downturn in reflectance at the violet end of the spectrum, and are neutral to slightly reddish in the near-IR (1-2.5 μm), with no discrete spectral absorption bands. D-type asteroids are different from the Cs in that they have red colors (reflectance increases toward longer wavelengths) in both the photovisual and near-IR spectral regions. P-type asteroids are intermediate in "redness" between the neutral Cs and the very red Ds [see Table 2.1.3 of Wetherill and Chapman 1988].

In general terms, meteorites are primarily fragments of asteroids, and the organic-rich meteorites probably come from fragmentation of low-albedo asteroids (C-, P-, and D-types with geometric albedos of 0.02-0.07). The distribution of the very low albedo asteroids has been established by taxonomic studies of large samples. The observations show that the Cs are concentrated between 2.8 and 3.2 AU, the Ps between 3.2 and 4.2 AU, and Ds at >4.5 AU [Bell et al. 1989].

The presumption that organic material might occur on some of the asteroids arises from the fact that organic-bearing meteorites tend to be very low in albedo and have relatively featureless spectra when seen in diffuse reflectance. Detailed comparisons of asteroid and meteorite spectra have been in progress for 30 years. Convincing comparisons of certain meteorites and asteroids have been drawn chiefly for differentiated objects rich in silicate minerals. Less progress has been made in the detection of specific spectroscopic diagnostics linking the dark asteroids with the carbonaceous meteorites.

2.1.2.2 Hydrous Silicates and Water Ice on Asteroids

Some C-type asteroids show a distinct spectral absorption band at 3 μm attributed to H_2O and OH in hydrated silicates in the surface layer minerals. One-third of the Cs have no evident water band, and it appears that different degrees of hydration exist among them; the presence of hydrous silicates declines gradually among the Cs from 2.5-3.5 AU [Jones et al. 1990]. Feierberg et al. [1985] found a correlation of the presence of the 3- μm band with the strong decrease in albedo of the violet portion of the reflectance spectrum; the "wet" C-types are defined as the G subclass in Tholen's taxonomy. Additional evidence of former hydrothermal activity in low-albedo asteroids comes from the detection of iron alteration minerals consistent with phyllosilicates [Vilas et al. 1994; Vilas and Sykes 1996].

Jones et al. [1990] suggest that the apparently anhydrous Ps and Ds (and outermost Cs) point to a mixture of anhydrous silicates, water ice, and organic solids in the outermost part of the asteroid system, while the inner regions of the main belt asteroids were warmed to the point that hydration of the silicates occurred as water from the included ice was mobilized. Consistent with this view is Howell's [1995] possible detection of the hydrous silicate band in 336 Lacadiera, a main belt D asteroid. In this view, the P and D type asteroids probably contain water ice in their subsurface regions and an insulating mantle prevents the sublimation of the ice at a detectable rate.

A report of water ice on asteroid I Ceres, based on the interpretation of a spectral feature at 3.07 µm [Lebofsky 1978], has been reinterpreted as being due to an ammoniated clay mineral [King et al. 1992].

In summary, there is spectroscopic evidence for the aqueous alteration of low-albedo asteroids that is commensurate with the petrological evidence for such processes in the carbonaceous meteorites that show the most abundant organic matter [Cronin and Chang 1993].

2.1.2.3 Organic Solids on Asteroids

The low-albedo material on the surfaces of the C-, P-, and D-types of asteroids is generally thought to consist of complex organic solids of a class related to kerogen, but without the connotation of biogenesis.

Seen in diffuse spectral reflectance, as are the asteroids studied by remote sensing techniques, the aromatic moieties in kerogen contribute little to the spectrum beyond lowering the albedo at the shortest wavelengths and giving the red overall color in the region 0.3-4 µm. The aliphatic bridging units and side chains contain -CH_2 and -CH_3 functional groups, both exhibiting distinct spectral features attributed to C-H fundamental stretching and bending along with their overtones and combinations. The C-H stretching fundamentals near 3.4 µm are the strongest of these features. In carbonaceous meteorites, the 3.4 µm band complex arises partly from the kerogen side units, but mostly from the soluble cyclic aliphatic molecular material.

The results of spectroscopic searches for the 3.4- µm band complex in asteroids generally remains negative. The band may be intrinsically weak, and it is also somewhat quenched by the presence of low-albedo opaque materials. In addition, the spectral region that it occurs in is difficult to observe because of numerous problems associated with thermal background radiation of the sky and the telescope in the 2.5-5 µm spectral region, reduced transparency of the Earth's atmosphere, and the strongly wavelength-dependent thermal emission of the asteroid itself (in contrast to the situation with cold, outer Solar System bodies, as noted in Section 1.1). The singular report of the presence of the 3.4-µm band complex in the spectrum of the wet, low-albedo G asteroid 130 Elektra [Cruikshank and Brown 1987] is not confirmed with more recent higher quality data [Cruikshank et al. 2002]. Spectra of Electra and six other low-albedo asteroids do not show the 3.4-µm feature, which if present would indicate the presence of hydrocarbons and provide a stronger link between these asteroids and certain classes of carbonaceous meteorites [Cruikshank et al. 2002]. New high-quality spectra of a number of Trojan asteroids exhibit a distinct red slope, one possible interpretation would suggest an association of the red color with hydrocarbons, but no evidence is reported for the 3.4-µm band [Emery and Brown, 2003]. In any case, macromolecular hydrocarbons are so optically opaque that the subtle spectral absorption bands arising from CxHy and other functional groups may be entirely invisible in the diffusely reflected sunlight from asteroids.

Thus, there is no direct and convincing detection of organic solids on asteroids, although the circumstantial evidence that they are there is very strong and is discussed in the following section.

2.1.2.4 Carbonaceous Meteorites and Low-Albedo Asteroids

Correlations of asteroid types in the Tholen taxonomy with the carbonaceous meteorite classes are neither direct nor compelling. Still, collisions among asteroids in the Main Belt are the most likely sources of the fragments that fall to earth as meteorites, so the search for connections continues. The best information we can get on asteroid compositions comes from spectroscopy with telescopes and spacecraft, while meteorites can be studied in the laboratory both by spectroscopy and the many other analytical techniques.

In early 2000 an extremely bright fireball was observed over Canada [Brown et al. 2000] and one resident collected several samples of the surviving meteorite from the frozen surface of Tagish Lake [Zolensky et al. 2002, and references therein]. The Tagish Lake meteorite (TLM) has been suggested as a spectral analog for D-type asteroids based upon comparisons of spectral color and slope [Hiroi et al. 2001]. To the extent that this correspondence of color (and albedo) is valid, then TLM can be considered a sample of one of the types of materials encountered in the outer Solar System that may impart a characteristic red coloration to many of these surfaces [e.g. Wilson et al. 1994, Wilson and Sagan, 1995, Cruikshank et al. 1998b, Owen et al. 2001]. To this end, Roush [2003] derived the optical constants of TLM for future modeling of outer Solar System surfaces. Furthermore, careful laboratory analyses of TLM can provide significant insight into the specific materials found on objects in the outer Solar System.

The TLM, with abundant phyllosilicates, chondrules, and olivine grains in the matrix, require it's classification as a type 2 carbonaceous chondrite. While similar to CI1- and CM-meteorites, the bulk density of TLM (~1.67 g/cc) is significantly lower than the range observed for CI (2.2-2.3 g/cc) or CM (2.6-2.9 g/cc) chondrites and the porosity of TLM (~11%) is much higher than any other meteorite [Zolensky et al. 2002]. Two lithologies are reported for TLM; carbonate-poor and carbonate-rich [Zolensky et al. 2002]. The total carbon content of TLM is ~5.8 wt. % with slightly less than half of this amount as a soluble fraction (2.5 wt. %) [Grady et al. 2002]. The insoluble C-bearing component contains nanodiamonds, graphitic carbon, fullerenes, and SiC, all suggesting a presolar origin of these materials on the parent body, and a variety of carbonates, suggesting several aqueous events or temperature change on the parent body [Grady et al. 2002]. The TLM organic carbon contains a variety of complex organic components, including carboxylic and dicarboxylic acids [Pizarello and Huang 2002], minor amounts of amino acids [Kminek et al. 2002], and a strong aromatic component (mostly the polyaromatic hydrocarbons, fluoranthene, and pyrene) with a distribution of other functional groups [Cody et al. 2002b]. Morphologies of some organic components are consistent with materials produced by uv photolysis of interstellar ices having experienced subsequent aqueous alteration, suggesting preservation of primitive organic materials [Nakamura et al. 2002].

2.2 Jovian System

The icy satellites of Jupiter embedded in the energy-rich Jovian radiation belt complex [e.g., Cooper et al. 2001] are potential reservoirs of organic material synthesized from native materials and/or deposited by meteoroid impacts [Delitsky and Lane 1998]. The spectral evidence for H_2O ice on Europa, Ganymede, and Callisto has been well established (Roush et al. 1990; Calvin et al. 1995 and references in both). SO_2 frost is widespread on Io [e.g., Douté et al. 2001], and is also found on Europa and Callisto [Lane et al. 1981; Noll et al. 1995; 1997a]. O_2 and O_3 have been detected on Ganymede [Spencer et al. 1995, Noll et al. 1996, and references in all]. Recent summaries describe the ices observed on the satellites of Jupiter [Nash and Betts, 1998; Cruikshank et al. 1998c; Douté et al. 2001].

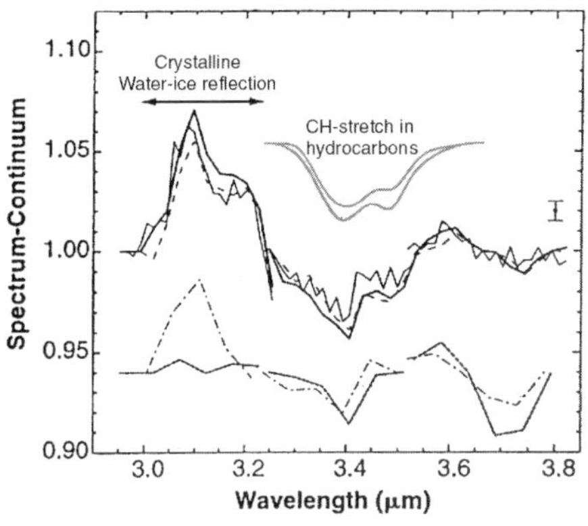

Figure 7.2. Galileo NIMS spectra of Callisto showing the 3.4 μm C-H features. Reprinted with permission from TB McCord et al. Science 278:271 ©[1997] AAAS.

The Galileo mission, with its Near Infrared Mapping Spectrometer (NIMS), has revealed additional ices and other materials on the Galilean satellites [Carlson et al. 1996, McCord et al. 1998]. Figure 2, from McCord et al. [1997] shows spectra in the 3.0-3.8 μm region for Callisto and compares these to absorption spectra of two hydrocarbons showing the C-H stretching band complex of the methylene ($-CH_2$) and methyl ($-CH_3$) functional groups at 3.4 μm. The coincidence of the hydrocarbon bands and an absorption in the Callisto spectrum is suggestive of hydrocarbon compounds in certain regions of Callisto's surface. Galileo NIMS spectra of both Ganymede and Callisto show spectral features that may be attributed to X-C≡N [McCord et al. 1998].

Hibbitts et al. [2000] discuss the genetic implications of the spatial distribution of CO_2 and SO_2 on Callisto and suggest that the distribution of CO_2 on Callisto is consistent with an exogenic effect related to Jupiter's co-rotating magnetic field while the SO_2 distribution is less well defined. Hibbitts et al. [2003] also find CO_2 in regions of Ganymede's surface that do not show H_2O ice; CO_2 may occur together with H_2O, but its detection limits are affected by the strength of the H_2O bands. Table 7.1 lists the ices and other surface materials recognized on the Galilean satellites.

Table 7.1. Surface Components Observed in the Outer Solar System

Jovian Satellites	Io: SO_2, SO_3 H_2S?, H_2O?
	Europa: H_2O, SO_2, CO_2, Hydrous sulfate salts, H_2O_2, H_2SO_4
	Ganymede: H_2O, O_2, CO_2, X-C≡N, CH, SO_2, Hydrated silicates, SH, O_3
	Callisto: H_2O, Hydrated silicates, SO_2, SH, CO_2, X-C≡N, CH
Saturn's Satellites	Mimas, Enceladus, Tethys, Hyperion, Phoebe: H_2O
	Dione: H_2O, C, HC, O_3
	Rhea: H_2O, HC?, O_3
	Iapetus: H_2O, C, HC, NH
	Titan: H_2O, HC, C≡N
	Rings: H_2O, CH
Uranus's Satellites	Ariel: H_2O, CO_2, OH?
	Umbriel: H_2O
	Titania: H_2O, C, HC, OH?
	Oberon: H_2O, C, HC, OH?
	Miranda: H_2O, NH_3 hydrate?
Neptune's Satellites	Triton: N_2, CH_4, CO, CO_2, H_2O
Pluto	N_2, CH_4, CO
	Charon: H_2O, NH_3, NH_3 hydrate
Centaurs and Trans-Neptune Objects	H_2O, HC-ices (e.g. CH_4, CH_3OH), HC, silicates

HC= miscellaneous hydrocarbons

2.3 Saturnian System

Cruikshank et al [1998c] provided a summary of the ices observed on the satellites of Saturn and these are summarized in Table 7.1 along with other surface components. Again these are dominated by H_2O ice, while O_3 is identified on Dione [Noll et al. 1997b]. Bell et al. [1985] invoked hydrocarbon materials as the darkening and coloring agents on the leading hemisphere of Iapetus, while Wilson and Sagan [1995] favored HCN polymer mixed with meteoritic kerogen. In a study using expanded wavelength coverage (0.4-3.9 μm), Owen et al. [2001] derived a simple and self-consistent model consisting of N-rich Triton tholin, H_2O ice, and elemental C (Fig. 3). In their model, the tholin provides not only the strong red coloration, but contributes significantly to the profile of the prominent 3-μm absorption band.

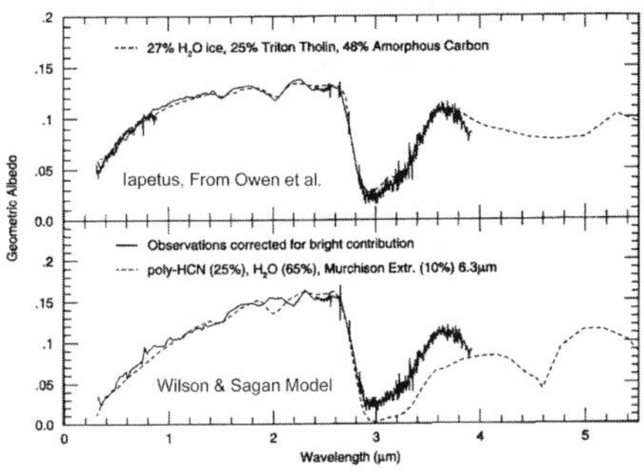

Figure 3. Comparison of the Owen et al. [2001] model for the leading hemisphere of Iapetus and that of Wilson and Sagan [1995]. Reprinted from Owen et al. 2001, figure 13, ©2001, with permission from Elsevier.

Spectra of the Iapetus' leading hemisphere in the 0.4-0.8 μm region by Jarvis et al. [2000] exhibit spectral features near 0.4-0.6, 0.67, and 0.73 μm that Vilas et al. [1996] attribute to charge transfer transitions in materials such as hematite and goethite, although the study by Owen et al. [2001] does not support this interpretation. Gudipati et al. [2003] illustrated a marked similarity between the broad luminescence from macromolecular organic residues created by uv irradiation of cosmic ices and the leading hemisphere spectrum of Jarvis et al. [2000]; providing an alternative explanation for the mineralogical interpretation of the observed spectral structure by Vilas et al. [1996] and Jarvis et al. [2000].

Cuzzi and Estrada [1998] suggested that water ice contaminated with minor amounts of organic macromolecules can explain the visual and near-infrared colors of Saturn's rings measured by the Voyager spacecraft. Poulet et al. [1999] qualitatively confirmed these color properties using Hubble Space Telescope imaging. Poulet et al [2003] report new observations of Saturn's rings that exhibit two absorption features near 1.73- and 3.4-μm providing direct evidence of C-H-bearing materials associated with Saturn's C-ring. The Poulet et al. [2003] spectrum of Saturn's rings are compared to other measurements in Figure 4.

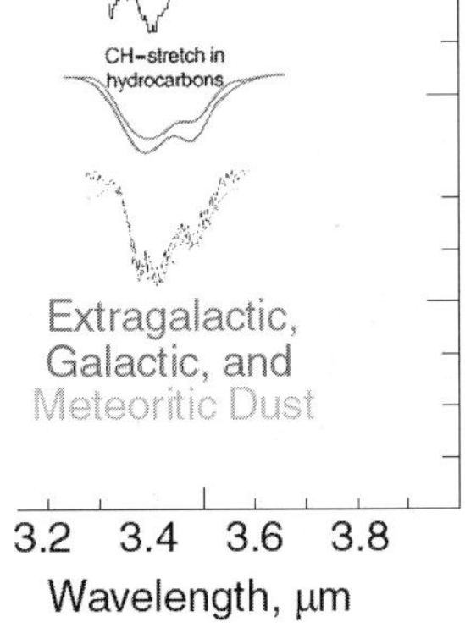

Figure 4. Ratio of the C- to B- ring spectra [solid black line, from Poulet et al. 2003] compared to CH-stretching fundamentals (solid red lines), and extragalactic (blue points) galactic (red points) and meteoritic (green dashed line) dust

[after Pendleton 1997].

If confirmed, these bands will be direct evidence of C-H-bearing materials associated with Saturn's C-ring.

Titan's atmospheric haze presents a challenge for assessing the materials present the surface. Recently, Griffith et al. [2003] report that the spectral signature of the surface is consistent with the presence of H_2O ice.

With its Visible-Infrared Mapping Spectrometer (VIMS) and Composite Infrared Spectrometer (CIRS), the Cassini spacecraft will conduct investigations of the Saturn system for 4 to 6 years beginning in 2004. High spatial resolution spectra of the icy satellites should result in the detection of new species and the resolution of long-standing questions about the compositions of the rings and most of the satellites. For example, because the VIMS instrument will measure to 5.2 μm, the data can be used to readily select between the two models of Iapetus' leading hemisphere shown in Figure 3, by evaluating the data near 4.6 μm.

2.4 Uranian System

Cruikshank et al. [1998c] also provide a summary of the ices and other components observed on the satellites of Uranus and these are listed in Table 7.1. Grundy et al. [2002] identified the presence of CO_2 ice on Ariel. Bauer et al. [2002] identified the presence of crystalline H_2O ice on Miranda and also noted a weak spectral feature near 2.21 μm. The theoretical models containing crystalline H_2O ice and ammonium hydrate support the suggestion that ammonium hydrate ice can produce a feature consistent with the Miranda data [Bauer et al. 2002]. More recent optical constants of ammonia hydrate suggest that it might also be able to explain the weaker feature near 1.99 μm [Schmitt et al. 1998]. If future observations of Miranda, with higher signal precision, confirm the assignment of the 2.21 μm feature to ammonia hydrate, then this will add to growing evidence of ammoniated species associated with icy objects in the outer Solar System (see discussion of Charon below). With no missions scheduled to visit the Uranian system, most future compositional information will likely come from ground-based observations.

2.5 Triton

Cruikshank et al. [1998a] provide a summary of the ices observed on Triton and these are listed in Table 7.1. In contrast to H_2O ice, inspection of Table 7.1 shows this to be the first icy surface discussed so far to contain volatile ices such as CH_4, N_2, and CO. As discussed by Cruikshank et al. [1993] the surface is dominated by coarse-grained N_2 ice (~1 cm, ~99.75% in regions where it occurs) with only minor amounts of CO (~ 1 mm, 0.1%)

and CH_4 (~0.2 mm, 0.05%) ices. The CO_2 ice abundance can vary from 0.10% to 10% depending upon whether it is mixed on a granular scale with, or spatially segregated from, the other ice species. The wavelength positions of the CH_4 features are consistent with this molecule being dissolved in the N_2 ice, and not present as pure CH_4 patches on the surface. Using more recent data [Cruikshank et al. 2000], Quirico et al. [1999] performed both direct spectral analysis, via comparison to laboratory spectra of various condensed species and their mixtures, and via theoretical modeling of Triton's spectrum. Quirico et al. [1999] confirmed the assignment of the four volatile ices, the dilution of CH_4 (0.11%) in N_2, suggested the CO is also diluted (0.05%) in N_2 and suggested that C_2H_6 isolated in N_2 may explain some additional weak spectral features in Triton's spectrum. The theoretical modeling of Quirico et al. [1999] also confirmed the large effective grain size of the N_2 ice dominated regions (~ 10 cm, 55% of the surface). In addition to exhibiting the CO_2 ice features the newer data also provided evidence for the presence of H_2O ice [Cruikshank et al. 2000]. In the models of Quirico et al. [1999] they included a spatially segregated region (45% of the surface) consisting of a granular mixture of H_2O (~0.1 mm, 13%) and CO_2 (~0.5 mm, 87%) ices. Additionally, Quirico et al. [1999] estimated that a maximum of 10% of the surface area could be pure CH_4. Cruikshank et al. [2000] analyzed the newer telescopic observations of Triton in an attempt to identify if the H_2O ice present on the surface was in the crystalline or amorphous phase. Figure 5 shows two theoretical models of Triton's geometric albedo based upon the results of Quirico et al. [1999]. One incorporates amorphous and the other crystalline water ice.

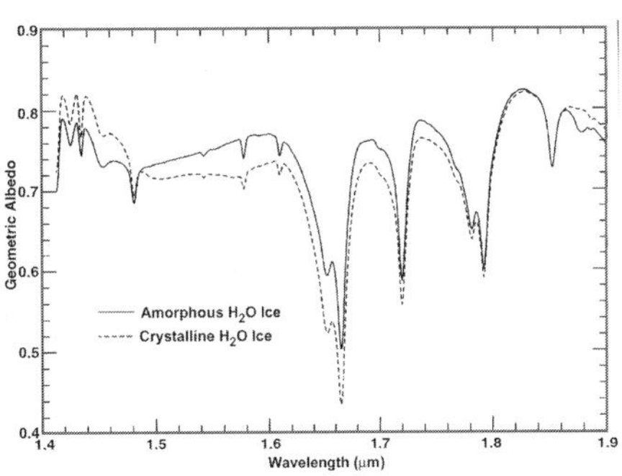

Figure 5. Two models of Triton's near-infrared spectrum, one contains amorphous (solid line) and the other crystalline (dashed line) ice. Reprinted from Cruikshank et al. 2000, Figure 2, ©2000, with permission from Elsevier.

Cruikshank et al. [2000] used the ratio of the equivalent widths of the 1.65 μm feature to the 1.72- and 1.79 μm features in the model spectrum for this determinaton. They compared these values to the same ratios calculated from the Triton spectrum and concluded that they were more consistent with the amorphous H_2O ice, but within the formal uncertainties of the measurements and hence could not unambiguously determine the phase of the H_2O ice.

Grundy et al. [2002] present the ~2.85-4.05 μm spectrum of Triton. Because H_2O ice has two fundamental vibrational stretching modes that occur near 3 μm these observations provide evidence regarding what phase is present on Triton's surface. Crystalline H_2O ice exhibits a Fresnel reflectivity peak near 3.12 μm [see figure 2 above and Owen et at 2001] whereas, amorphous H_2O ice does not exhibit such a peak. The Triton spectrum exhibits a distinctive spectral peak near 3.2 μm reminiscent of crystalline H_2O ice, but occurs at too long a wavelength (~0.08 μm). Grundy et al. [2002] suggests that this peak may be spurious and due to incomplete telluric corrections. Because of the lack of a distinctive crystalline H_2O ice Fresnel peak in the data of Grundy et al. [2002], it appears the data are more consistent with the presence of amorphous H_2O ice on the surface of Triton.

2.6 Pluto-Charon

2.6.1 Pluto

In the past decade our knowledge regarding the surface compositions of Pluto and Charon has grown considerably [Owen et al. 1993; Cruikshank et al. 1998a,d; Grundy and Fink 1996; Douté et al. 1999; Grundy and Buie, 2001, 2002; Grundy et al. 2002]. The spectrum of Pluto (0.3-4 μm) provides evidence for the clear evidence for the presence of solid N_2, CH_4, CO, and some coloring agent on the surface while the evidence for H_2O ice is weaker. Owen et al. [1993] modeled the Triton spectrum as a granular mixture of N_2 (~1 cm, 98%), CH_4, (~0.8 mm, 1.5%), and CO (~0.5 mm, 0.5%) ices. Although, the relative abundance of both CH_4 and CO are higher on Pluto compared to Triton, Owen et al. [1993] concluded that, like Triton, some CH_4 was dissolved in the N_2 ice on Pluto. Grundy and Fink [1996] modeled 0.5-1.0 μm spectra of Pluto obtained over a period of 15 years. Based upon models of the spectral data Grundy and Fink [1996] proposed three terrain types on the surface. The dominant terrain was a high albedo region located at a range of latitudes and composed chiefly of N_2 ice containing modest amounts (1-5%) of dissolved CH_4. The next largest terrain was a pure CH_4 region and located at northern latitudes and near the equator. The remaining terrain was spatially restricted to the equatorial region and consisted of an equal mixture of tholins and H_2O ice. Douté et al. [1999] examined a variety of complex models representing the surface of Pluto. The various models are shown schematically in Figure 6. In all models there are three terrain types. In the simplest model (middle of Fig. 6) fine-grained N_2, is one terrain, pure CH_4 is another, and N_2 in a molecular mixture with CH_4 and CO, is the third. In the intermediate complexity model N_2 and CH_4 ices are separate terrains, and the third terrain is a layered with CH_4 frost overlying a molecular mixture (bottom of Fig. 6). In the most complex model (top of Fig. 6) there are two layered terrains and the third terrain is pure N_2 ice. All these models suggest that like Triton, on Pluto, at least some CH_4 is dissolved in solid N_2, and there is evidence for a pure CH_4 component as well. Grundy and Buie [2001] investigated the distribution of volatiles ices (CH_4, N_2, and CO) on Pluto. They found that the spectral features of CO, N_2, and the weak features of CH_4 ice exhibited a longitudinal behavior different from both the observed visual light curve, and the strong CH_4 ice spectral features. However, they found no evidence for any changes in the distribution of the surfaces during 1995-1998. Grundy and Buie [2002] found a red near-infrared continuum slope associated with the CH_4-rich regions and included both Titan and Triton tholins in their models of this region. Unlike Triton, Grundy and Buie [2002] found only weak evidence for the existence of H_2O ice on the surface of Pluto. Using more recent telescopic data of Pluto spanning the 3-4 μm region Grundy et al. [2002] did not identify any spectral features associated with H_2O ice.

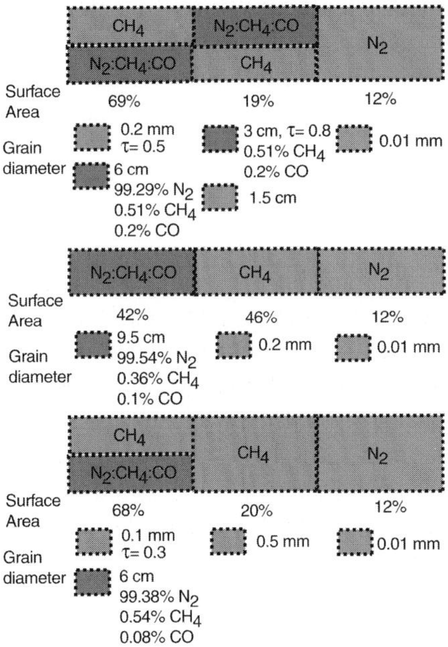

Figure 6. Schematic representation of the models of Pluto's surface of Douté et al. [1999].

Telescopic observations during the late 1980's [Buie et al. 1992; Young et al. 2001] and more recent Hubble Space Telescope images of Pluto and Charon [Buie et al. 1997; Stern et al. 1997] show evidence for the presence of large areas of surface material having a low albedo. In both instances, it is plausible that low-albedo material on the surface of a volatile-rich body could be complex organic solids, elemental carbon, or anything in between. It could also be silicate rocky material that would appear dark against the high albedo of the predominantly icy background surface.

In addition to the CH_4 ice, the visual coloration suggests organic matter on the Pluto's surface [McEwen 1990; Thompson and Sagan, 1990; Grundy and Fink, 1996]. Energy deposition in the surface ice can be expected to produce organic molecules much more complex than the observed CH_4. Indeed, a tholin produced by energy deposition in a mixture of C_2H_6 and H_2O has been studied in some detail (Khare et al. 1993), and there is a large literature of laboratory studies of photochemistry in interstellar ice analogs [e.g., review by Strazzulla 1998; Gerakines et al. 1996, 2001; Ehrenfreund et al. 1997, 2001].

Photochemical models (e.g. Krasnopolsky and Cruikshank 1999) of Pluto's atmosphere predict the formation and precipitation of significant quantities of the C_2 hydrocarbons, C_4H_2, HC_3N, and HCN, although none of these has been reliably detected in the infrared spectrum of the planet. The dominance of simple molecules on the surfaces of Pluto and Triton, rather than the abundant presence of photochemically produced complex hydrocarbons, is probably best understood in terms of the dynamical surface-atmosphere interactions over annual time-scales, in which the evaporation and recondensation of N_2, CO, and CH_4, buries precipitated heavier molecules to the degree that they are not, with current technology, detectable from Earth.

2.6.2 Charon

Marcialis et al. [1987] and Buie et al. [1987] identified H_2O ice on the surface of Pluto's moon Charon. Roush [1994] used theoretical models to investigate the possible presence of other volatile species (CH_4 and CO_2) on Charon's surface by comparison to the existing data. Roush concluded that up to 50% CO_2 ice might be present and 30% CH_4 ice might be present and remain undetected.

Using much higher spectral resolution data Brown and Calvin [2000] clearly identified the H_2O ice on Charon as being crystalline and noted an additional spectral feature near 2.21 μm that they attributed to ammonia or ammonia hydrate ice. In their models of the data they included a dark spectral neutral material.

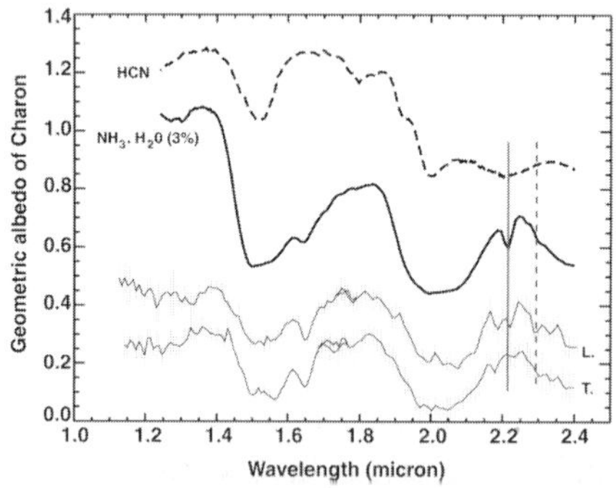

Figure 7. Spectra of Charon's leading (L.) and trailing (T.) hemispheres compared to HCN (dashed), and ammoniated hydrate (solid black line). Reprinted with permission from Dumas et al. [2001], Astron. J., 121, 1163, figure 6, ©2001 Am. Astron. Soc.

Dumas et al. [2001] presented data for both the leading and trailing hemispheres of Charon. Their data confirm the crystalline water ice identification and corroborates the presence of the 2.21μm feature on the leading hemisphere of Charon (see Figure 7). The 2.21 μm feature is much less apparent on the trailing hemisphere. Dumas et al [2001] suggested that HCN could give rise to a feature near this wavelength, but in their analysis favored an assignment to ammonia hydrate, in agreement with Brown and Calvin [2000]. Buie and Grundy [2000] report spectra of several different longitudes of Charon's surface. All data confirm the presence of crystalline H_2O ice. While there is a suggestion of the 2.21 μm feature, Buie and Grundy [2000] conclude the signal precision is insufficient to draw any definitive conclusions.

2.7 Centaurs

A few asteroid-like bodies have been found in orbits that cross one or more of the orbits of the giant planets. After the discovery of the first one in this dynamical class, 2060 Chiron in 1977 [Kowal 1979], they were designated Centaur objects. As discussed in more detail in Chapter 9, the Centaurs appear to be diverse in physical characteristics. Their spectral reflectances in the 0.3-1.0 μm region range from the very flat spectrum of Chiron to the extraordinarily steep (upward toward longer wavelength) spectra of Pholus and 1993 HA2 (7066 Nessus). This extreme degree of redness shown by Pholus caught the immediate attention of photometricists just after the object was discovered in 1991 [e.g., Mueller et al. 1992], and it was at once clear that no natural simple ice, mineral, or rock could match this color.

A plausible material showing the extreme red color of Pholus is a family of organic solids called tholins (Sagan and Khare 1979). As noted above in section 1.2.3, tholins are stable, refractory solids produced by energy deposition in gaseous mixtures (mainly strongly reducing). They consist of nitrogen heterocycles, with amino acids (dominantly glycine) released upon acid hydrolysis, and pyrimidines and purines identified in pyrolysis products of some tholins. Reflectance spectra of four different tholins are shown by Cruikshank et al. (1991) and chemical analyses of tholins produced in gaseous mixtures of N_2 and CH_4 are given by McDonald et al. (1994). Tholins are also produced by energy deposition in ice mixtures of condensed hydrocarbons and other volatiles (Khare et al. 1993, McDonald et al. 1996).

Only a few Centaurs are sufficiently bright to permit observations in the near-infrared where important diagnostic spectral features occur. For example, the near-infrared spectrum of Pholus exhibits distinct absorption bands, particularly at 2.27 μm (Davies et al. 1993), and has a higher overall reflectance than Chiron by a factor of four in the near-infrared. Luu et al. (1994) confirmed the presence of the 2.27 μm absorption band.

Wilson et al. (1994) used Hapke scattering theory to model the spectrum of Pholus with a mixture of tholin, HCN polymer, and ammonia ice. While the tholin accounts for the red color of Pholus at shorter wavelengths, the 2.27 μm band is not matched by the favored model of Wilson et al. More detailed modeling of the spectrum of Pholus by

Cruikshank et al. [1998b] suggest the presence of carbon covering about 62% of the surface and a complex mixture covering 38% of the surface. The complex mixture consists of olivine (55%), Titan tholin (15%), H_2O ice (15%), and methanol ice (15%). The methanol ice has a band complex at 2.27 μm, but the authors indicate that it is a representative of most relatively light hydrocarbons.

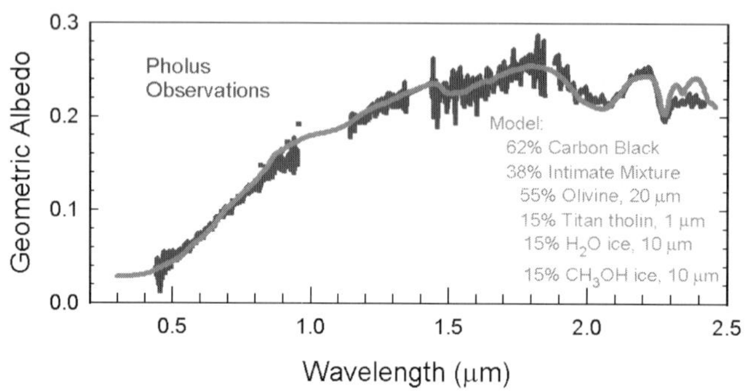

Figure 8. Comparison of Pholus observations (solid black line and data points) to a model of the surface (solid gray line), after Cruikshank et al. [1998b]. Reprinted from Cruikshank et al., figure 3, ©1998, with permission from Elsevier.

3. SUMMARY

Interpretation of the telescopic observations show that H- and O-bearing materials are ubiquitous throughout the outer Solar System as H_2O ice. Energetic processing of simple H_2O ice can result in the surface materials and radicals whose presence are beginning to be identified on a limited number of satellites. At large heliocentric distances other H-, N-, C-bearing volatile surface ices are identified. Their presence certainly provides the potential for creation of more complex organic components due to energetic processing of the pure ices. Laboratory studies of the reflectance properties and chemical reactions associated with C-, H-, and N-bearing precursor gases and ices show that the stable residues created exhibit a range of coloration at visual and near-infrared wavelengths. Modeling of the telescopic observations using these residues suggest they are capable of giving rise to the observed variable red color of many surfaces in the outer Solar System.

4. ACKNOWLEDGEMENTS

We acknowledge research support from NASA's Planetary Geology and Geophysics Program via RTOP 344-30-01-01 and NASA's Planetary Astronomy Program via RTOP 344-09-09-09 that allowed us to contribute this chapter.

5. REFERENCES

Aléon, J. and 12 coauthors 2003, Cometary organic macromolecules in interplanetary dust particles? Lunar Planet. Sci. Conf. 34, Abstract 1308.

Bahr, D.A. M. Famá, R.A. Vidal, and R.A. Baragiola 2001, Radiolysis of water ice in the outer Solar System: Sputtering and trapping of radiation products, J. Geophys. Res. 106, 33285-33290.

Baratta, G. A., Leto, G., and Palumbo, M. E. 2002, A comparison of ion irradiation and UV photolysis of CH_4 and CH_3OH. Astron. Astrophys 384, 343-349.

Bauer, J.M., Roush, T.L. T.R. Geballe, T.C. Owen, K.J. Meech, W.D. Vacca, and J.T. Rayner 2002, The near infrared spectrum of Miranda: Evidence for crystalline water ice, Icarus, 158, 178-190.

Bell, J.F., D.P. Cruikshank, and M.J. Gaffey 1985, The composition and origin of Iapetus dark material, Icarus, 61, 192-207.

Bell, J.F., D.R. Davis, W.K. Hartmann, and M.J. Gaffey 1989, Asteroids: The Big Picture, In *Asteroids II*, R.P. Binzel, T. Gehrels, and M.S. Matthews, eds. Univ. Arizona Press, 921-945.

Bohn, R.B. S.A. Sandford, L.J. Allamandola, and D.P. Cruikshank 1994, Infrared spectroscopy of Triton and Pluto ice analogs: The case for saturated hydrocarbons, Icarus, 111, 151-173.

Borucki, J. G., Khare, B. N., and Cruikshank, D., P. 2002, A new energy source for organic synthesis on Europa. J. Geophys. Res. 107, 24:1-24:5.

Brown, M.E. and W.M. Calvin 2000, Evidence for crystalline water and ammonia ices on Pluto's satellite Charon, Science, 287, 107-109.

Brown, P. and 21 authors 2000, The fall, recovery, orbit, and composition of Tagish Lake meteorite: A new type of carbonaceous chondrite, Science, 290, 320-325.

Bruston, P., M. Khlifi, Y. Benilan, and F. Raulin 1994, Laboratory studies of organic chemistry in planetary atmospheres: From simulation experiments to spectroscopic determinations, J. Geophys. Res. 99, 19047-19061.

Buie, M.W. and W.M. Grundy 2000, The distribution and physical state of H_2O on Charon, Icarus, 148, 324-339.

Buie, M.W., D.P. Cruikshank, L.A. Lebofsky, and E.F. Tedesco 1987, Water frost on Charon, Nature, 329, 522-523.

Buie, M.W., D.J. Tholen, and K. Horne 1992, Albedo maps of Pluto and Charon: Initial mutual event results, Icarus, 97, 211-227

Buie, M.W., D.J. Tholen, and L.H. Wasserman 1997 Separate lightcurves of Pluto and Charon, Icarus 125, 233-244

Calvin, W.M., R.N. Clark, R.H. Brown, and J.R. Spencer 1995, Spectra of the icy Galilean satellites from 0.2 to 5.0 μm: A compilation, new observations, and a recent summary, J. Geophys. Res. 100, 19041-19048.

Carlson, R., et al. 1996, Near-infrared spectroscopy and spectral mapping of Jupiter and the Galilean satellites: Results from Galileo's initial orbit, Science, 274, 385-388.

Chyba, C., and Sagan, C. 1992, Endogenous production, exogenous delivery and impact-shock synthesis of organic molecules: an inventory for the origins of life, Nature 355, 125-132.

Clemett, S. J., Maechling, C. R., Zare, R. N., Swan, P. D., and Walker, R. M. (1993) Identification of complex aromatic molecules in individual interplanetary dust particles. Science 262, 721-724.

Cloutis, E.A. 1989, Spectral reflectance properties of hydrocarbons: Remote-sensing implications, Science 245, 165-168.

Cloutis, E. A., M. J. Gaffey, and T. F. Moslow 1994, Spectral reflectance properties of carbon-bearing materials. Icarus 107, 276-287.

Cody, G. D., C. M. O'D. Alexander, and F. Tera 2002a, Solid-state (^{1}H and ^{13}C) nuclear magnetic resonance spectroscopy of insoluble organic residue in the Murchison meteorite: A self-consistent quantitative analysis. Geochim. Cosmochim. Acta 66, 1851-1865.

Cody, G.D., C.D. O'D Alexander, and F. Tera 2002b, Comparison of the complex organic functionality of the Tagish Lake and Murchison meteorites, Lunar. Planet. Sci. Conf., Abstract 1806, Lunar & Planetary Institute, CDROM.

Cooper, J. F., R. E. Johnson, B. H. Mauk, H. B. Garrett, and N. Gehrels 2001, Energetic ion and electron irradiation of the icy Galilean satellites, Icarus 149, 133-159.

Cottin, H., M.C. Gazeau, and F. Raulin 1999, Cometary organic chemistry: a review from observations, numerical and experimental simulations, Planet. Space Sci., 47, 1141-1162.

Cronin, J. R., and Chang, S. 1993, Organic Matter in Meteorites: Molecular and Isotopic Analyses of the Murchison Meteorite, In The Chemistry of Life's Origins, J. M. Greenberg et al., Eds. Kluwer Acad. Pub., pp. 209-258.

Cronin, J.R., S. Pizzarello, and D.P. Cruikshank 1988, Organic Matter in Carbonaceous Chondrites, Planetary Satellites, Asteroids, and Comets, In *Meteorites and the Early Solar System*, J. F. Kerridge and M.S. Matthews, eds., Univ. Arizona Press, 819-857.

Cronin, J.R. and S. Pizzarello 1983, Amino acids in meteorites, Adv. Space Res. 3, 5-18.

Cruikshank, D.P. 1997, Organic Matter in the Outer Solar System: From the Meteorites to the Kuiper Belt, in *From Stardust to Planetesimals*, Y.J. Pendleton and A.G.G.M. Tielens, eds, Astron. Soc. Pac. Conf. Ser. Vol. 122, 315-333.

Cruikshank, D.P., and R.H. Brown 1987, Organic matter on asteroid 134 Elektra, Science 238, 183-184.

Cruikshank, D.P., and J.F. Kerridge 1992, Organic Material: Asteroids, Meteorites, and Planetary Satellites, In *Exobiology in Solar System Exploration*, G. Carle, D. Schwartz, and J. Huntington, eds., NASA SP-512, 159-176.

Cruikshank, D.P., L.J. Allamandola, W.K. Hartmann, D.J. Tholen, R.H. Brown, C.N. Matthews, and J.F. Bell 1991, Solid C≡N bearing material on outer Solar System bodies, Icarus 94, 345-353.

Cruikshank, D. P., T.L. Roush, T.C. Owen, T.R. Geballe, C. de Bergh, B. Schmitt, R.H. Brown, and M.J. Bartholomew 1993, Ices on the surface of Triton, Science, 261, 742-745.

Cruikshank, D.P., T.L. Roush, T.C. Owen, E. Quirico, and C. de Bergh 1998a, The Surface Composition of Triton, Pluto, and Charon, in *Solar System Ices*, edited by B. Schmitt C. deBergh, and M. Festou, pp. 655-684, Kluwer Acad. Sci., Norwell, Mass..

Cruikshank, D.P., T.L. Roush, M.J. Bartholomew, T. R. Geballe, Y.J. Pendleton, S.M. White, J.F. Bell III, J. K. Davies, T.C. Owen, C. de Bergh, D.J. Tholen, M.P. Bernstein, R.H. Brown, K.A. Tryka, C.M. Dalle Ore 1998b, The composition of Centaur 5145 Pholus, Icarus, 135, 389-407.

Cruikshank, D.P., R.H. Brown, W.M. Calvin, T.L. Roush, and M.J. Bartholomew 1998c, Ices on the satellites of Jupiter, Saturn, and Uranus, in *Solar System Ices*, edited by B. Schmitt C. deBergh, and M. Festou, pp. 579-606, Kluwer Acad. Sci., Norwell, Mass.

Cruikshank, D.P., T.L. Roush, J.M. Moore, M.V. Sykes, T.C. Owen, M.J. Bartholomew, R.H. Brown, and K.A. Tryka 1998d, The surfaces of Pluto and Charon, in *Pluto and Charon*, edited by S.A. Stern and D.J. Tholen, pp. 221-267, Univ. of Ariz. Press, Tucson.

Cruikshank, D.P., B. Schmitt, T.L. Roush, T.C. Owen, T.R. Geballe, C. de Bergh, B. Schmitt, R.H. Brown, and M.J. Bartholomew 2000, Water ice on Triton, Icarus, 147, 309-316.

Cruikshank, D.P., T.R. Geballe, T.C. Owen, C.M. Dalle Ore, T.L. Roush, R.H. Brown, and J.H. Lewis 2002, Search for the 3.4-μm C-H spectral bands on low albedo asteroids, Icarus, 156, 434-441.

Cuzzi, J.N., and P.R. Estrada 1998, Compositional evolution of Saturn's rings due to micrometeroid bombardment, Icarus 132, 1-35.

Davies, J.K., M.V. Sykes, and D.P. Cruikshank 1993, Near-infrared photometry and spectroscopy of the unusual minor planet 5145 Pholus (1992AD), Icarus, 102, 166-169.

Delitsky, M.L., and A.L. Lane 1997, Chemical schemes for surface modification of icy satellites: A road map, J. Geophys. Res., 102, 16,385-16,390.

Delitsky, M.L., and A.L. Lane 1998, Ice chemistry on the Galilean satellites, J. Geophys. Res., 103, 31,391-31,403.

Delitsky M.L., and A.L. Lane 2002, Saturn's inner satellites: Ice chemistry and magnetosphere effects, J. Geophys. Res. 107, 3-1-3-17.

Douté, S., B. Schmitt, E. Quirico, T.C. Owen, D.P. Cruikshank, C. de Bergh, T.R. Geballe, and T.L. Roush 1999, Evidence for methane segregation at the surface of Pluto, Icarus, 142, 421-444.

Douté, S., B. Schmitt, R. Lopes-Gautier, R. Carlson, L. Soderblom, J. Shirley, and the Galileo NIMS Team 2001, Mapping SO_2 frost on Io by modeling of NIMS hyperspectral images. Icarus 149, 107-132.

Dumas, C., R.J. Terrile, R.H. Brown, G. Schneider, and B.A. Smith 2001, Hubble space telescope NICMOS spectroscopy of Charon's leading and trailing hemispheres, Astron. J., 121, 1163-1170.

Duran, B. (Ed.) (1980) *Kerogen.* Éditions Technip, Paris. 550 pp.

Ehrenfreund, P., A.C.A. Boogert, P.A. Gerakines, A.G.G.M. Tielens, and E.F. van Dishoeck 1997, Infrared spectroscopy of interstellar apolar ice analogs, Astron. Astrophys., 328, 649-669.

Ehrenfreund, P., O. Kerkhof, W.A. Schutte, A.C.A. Boogert, P.A. Gerakines, E. Dartois, L. d'Hendecourt, A.G.G.M. Tielens, E.F. van Dishoeck, and D.C.B. Whittet 1999, Laboratory studies of thermally processed H_2O-CH_3OH-CO_2 ice mixtures and their astrophysical implications, Astron. Astrophys., 350, 240-253.

Ehrenfreund, P., L. d'Hendecourt, and S. Charnley, and R. Ruiterkamp 2001, Energetic and thermal processing of interstellar ices, J. Geophys. Res. 106, 33291-33301.

Emery, J.P., and R.H. Brown 2003, Constraints on the surface composition of Trojan asteroids from near-infrared (0.8-4.0 μm) spectroscopy, Icarus 164, 104-121.

Feierberg, M.A., L.A. Lebofsky, and D.J. Tholen 1985, The nature of C-class asteroids from 3-μm spectrophotometry, Icarus 63, 183-191.

Fink, U., and G.T. Sill 1983, The infrared spectral properties of frozen volatiles, in *Comets*, edited by L.L. Wilkening, pp. 164-202, Univ. of Ariz. Press, Tucson.

Flynn, G., L.P. Keller, D. Joswiak, and D.E. Brownlee 2002, Infrared analysis of organic carbon in anhydrous and hydrated interplanetary dust particles: FTIR identification of carbonyl (C=O) in IDPs, Lunar. Planet. Sci. Conf. XXXIII, Lunar Planetary Inst., Houston, TX, CD-ROM, abstract 1320.

Gerakines, P., W.A. Schutte, and P. Erenfreund 1996, Ultraviolet processing of interstellar ice analogs, Astron. Astrophys., 312, 289-305.

Gerakines, P., M.H. Moore, and R.L. Hudson 2001, Energetic processing of laboratory ice analogs: Photolysis versus ion bombardment, J. Geophys. Res.106, 33381-33385.

Grady, M.M., A.B. Verchovsky, I.A. Franchi, I.P. Wright, and C.T. Pillinger 2002, Light element geochemistry of the Tagish Lake C2 chrondrite: Comparison with CI1 and CM2 meteorites, Meteoritics and Planetary Science 37, 713-735.

Griffith, C.A., T. Owen, T.R. Geballe, J. Rayner, and P. Rannou 2003, Evidence for the exposure of water ice on Titan's surface, Science 300, 628-630.

Grundy, W.M. and M.W. Buie 2001, Distribution and evolution of CH_4, N_2, and CO ices on Pluto's surface: 1995 to 1998, Icarus 153, 248-263.

Grundy, W.M. and M.W. Buie 2002, Spatial and compositional constraints on non-ice components and H_2O on Pluto's Surface, Icarus, 157, 128-138.

Grundy, W.M. and U. Fink 1996, Synoptic CCD spectrophotometry of Pluto over the past 15 years, Icarus, 124, 329-343.

Grundy, W.M., M.W. Buie, and J.R. Spencer 2002, Spectroscopy of Pluto and Triton at 3-4 microns: Possible evidence for wide distribution of nonvolatile solids, Astron. J. 124, 2273-3378.

Gudipati, M.S., J.P. Dworkin, X.D.F. Chillier, and L.J. Allamandola 2003, Luminescence from vacuum-ultraviolet-irradiated cosmic ice analogs and residues, Astrophys. J. 583, 514-523.

Hapke, B. 1993, *Theory of Reflectance and Emittance Spectroscopy*, Cambridge Univ. Press, New York.

Hibbits, C.A., T.B. McCord, and G.B. Hansen 2000, Distributions of CO_2 and SO_2 on the surface of Callisto, J. Geophys. Res. 105, 22541-22557.

Hibbitts, C. A., R. T. Pappalardo, G. B. Hansen, and T. B. McCord 2003, Carbon dioxide on Ganymede. J. Geophys. Res. 108 (E5), 2-1 to 2-22.

Hiroi, T., M.E. Zolensky, and C. Pieters 2001, Discovery of a sample of the D-class asteroids: The Tagish Lake meteorite, Science 293, 2234-2236.

Howell, E.S. 1995, Probing Asteroid Composition Using Visible and Near-Infrared Spectroscopy, PhD Thesis, Univ. Arizona, 186 pp.

Hudson, R.H. and M.H. Moore 2001, Radiation chemical alterations in Solar System ices: An overview J. Geophys. Res. 106, 33275-33284.

Imanaka, H., B. N. Khare, J. E. Elsila, E. L. O. Bakes, C. P. McKay, D. P. Cruikshank, S. Sugitta, T. Matsui, and R. N. Zare 2004, Laboratory experiments on Titan tholin formed in cold plasma at various pressures: Implications for nitrogen-containing fused heteroaromatics in Titan haze. Icarus, in press.

Jarvis, K.S., F. Vilas, S.M. Larson, and M.J. Gaffey 2000, Are Hyperion and Phoebe linked to Iapetus? Icarus 146, 125-132.

Johnson, R.E., Sputtering and desorption from icy surfaces 1998, in *Solar System Ices*, edited by B. Schmitt, C. de Bergh, and M. Festou, pp. 303-334, Kluwer Acad. Sci., Norwell, Mass.

Johnson, R.E., and T.I. Quickenden 1997, Photolysis and radiolysis of water ice on outer Solar System bodies, J. Geophys. Res., 102, 10,985-10,996.

Jones, T.D., L.A. Lebofsky, J.S. Lewis, and M..S. Marley 1990, The composition and origin of the C, P, and D asteroids: Water as a tracer of thermal evolution of the outer Solar System, Icarus 88, 172-192

Khare, B.N., C. Sagan, W.R. Thompson, E.T. Arakawa, and P. Votaw 1987, Solid hydrocarbon aerosols produced in simulated Uranian and Neptunian stratospheres, J. Geophys. Res. 92, 15,067-15,082.

Khare, B. N., Thompson, W. R., Cheng, L., Chyba, C., Sagan, C., Arakawa, E. T., Meisse, C., and Tuminello, P. 1993, Production and optical constants of ice tholin from charged particle irradiation of (1:6) C_2H_6/H_2O at 77K, Icarus 103, 290-300.

King, T.V.V., R.N. Clark, W.M. Calvin, D.M. Sherman, and R.H. Brown 1992, Evidence for ammonium-bearing minerals on Ceres, Science 255, 1551-1553.

Kminek, G., O. Botta, D.P. Glavin, and J.L. Bada 2002, Amino acids in the Tagish Lake meteorite, Meteoritics and Planetary Science 37, 697-701.

Kowal, C.T. 1979, Chiron, In *Asteroids*, T. Gehrels, ed., Univ. Arizona, 436-439.

Krasnopolsky, V.A. and D.P. Cruikshank 1995, Photochemistry of Triton's atmosphere and ionosphere, J. Geophys. Res. 100, 21271-21286.

Krasnopolsky, V.A. and D.P. Cruikshank 1999, Photochemistry of Pluto's atmosphere and ionosphere near perihelion, J. Geophys. Res. 104, 21979-21996.

Lane, A.L., R.M. Nelson, and D. Matson 1981, Evidence for sulphur implantation in Europa's UV absorption band, Nature, 292, 38-39.

Lebofsky, L.A. 1978, Asteroid 1 Ceres: evidence for water of hydration, Mon.Not. Royal. Astron. Soc. 128, 17-21.

Lunine, J.I. 1997, Processing of Material in the Solar, In *From Stardust to Planetesimals*, Y.J. Pendleton and A.G.G.M. Tielens, eds, Astron. Soc. Pac. Conf. Ser. Vol. 122, 271-279.

Luu, J., D. Jewitt, and E. Cloutis 1994, Near-infrared spectroscopy of primitive Solar System objects, Icarus 109, 133-144.

Marcialis, R.L., G.H. Rieke, and L.A. Lebofksy 1987, The surface composition of Charon: Tentative identification of water ice, Science 237, 1349-1351.

McCord, T.B., and 11 authors 1997, Organics and other molecules in the surfaces of Callisto and Ganymede, Science, 278, 271-275.

McCord, T.B., and 13 authors 1998, Non-water-ice constituents in the surface material of the icy Galilean satellites from the Galileo near infrared mapping spectrometer investigation, J. Geophys. Res., 103, 8603-8626.

McDonald, G.D., W.R. Thompson, M. Heinrich, B.N. Khare, and C. Sagan 1994, Chemical investigation of Titan and Triton Tholins, Icarus 108, 137-145.

McDonald, G.D., L.J. Whited, C. DeRuiter, B.N. Khare, A. Patnaik, and C. Sagan 1996, Production and chemical analysis of cometary ice tholins, Icarus, 122, 107-117.

McEwen, A.S. 1990, Global color and albedo variations on Triton, Geophys. Res. Lett. 17, 1765-1768.

Moore, M.H. and R.L.Hudson 2000, Far-infrared spectral studies of phase changes in water ice induced by proton irradiation, Astrophys. J., 401, 353-360.

Moore, M. H., Hudson, R. L., and Ferrange, R. F. 2004, Radiation products in processed ices relevant to Edgeworth-Kuiper-Belt objects. Earth, Moon and Planets, in press.

Moroz, L. V., G. Arnold, A. V. Korochantsev,and R. Wäsch 1998, Natural solid bitumens as possible analogs for cometary and asteroid organics, Icarus 134, 253-268.

Mueller, B.E.A.M., D.J. Tholen, W.K. Hartmann, and D.P. Cruikshank 1992, Extraordinary colors of asteroidal object 1992AD, Icarus 97, 150-154.

Nakamura, K., M.E. Zolensky, S. Tomita, S. Nakashima, and K. Tomeoka 2002, Hollow organic globules in the Tagish Lake meteorite as possible products of primitive organic reactions, Intl. J. Astrobiology, 1, 179-189.

Nash, D.B., and B.H. Betts 1998, Ices on Io - Composition and texture, in *Solar System Ices*, edited by B. Schmitt, C. de Bergh, and M. Festou, pp. 607-637, Kluwer Acad. Sci., Norwell, Mass.

Noll, K.S., H.A. Weaver, and A.M. Gonnella 1995, The albedo spectrum of Europa from 2200 Å to 3300 Å, J. Geophys. Res. 100, 19,057-19,059.

Noll, K.S., R.E. Johnson, A.L. Lane, D. Dominque, and H.A. Weaver 1996, Detection of ozone on Ganymede, Science, 273, 341-343.

Noll, K.S., R.E. Johnson, M. McGrath, and J.J. Caldwell 1997a, Detection of SO_2 on Callisto with the Hubble Space Telescope, Geophys. Res. Lett., 24, 1139-1142.

Noll, K.S., T.L. Roush, D.P. Cruikshank, R.E. Johnson, and Y. Pendleton 1997b, Detection of ozone on Saturn's satellites Rhea and Dione, Nature, 388, 45-47.

Owen, T.C., T.L. Roush, D.P. Cruikshank, J.L. Elliot, L.A. Young, C. de Bergh, B. Schmitt, T.R. Geballe, R.H. Brown, and M.J. Bartholomew 1993, Surface ices and atmospheric composition of Pluto, Science, 261, 745-748.

Owen, T.C., D.P. Cruikshank, C.M. Dalle Ore, T.R. Geballe, T.L. Roush, C. de Bergh, R. Meier, Y.J. Pendleton, and B.N. Khare 2001, Decoding the domino: The dark side of Iapetus, Icarus, 149, 160-172.

Pendleton, Y.J. 1997, Detection of organic matter in interstellar grains, Origins Life Evol. Biosphere 27, 53-78.

Pizzarello, S., and Y. Huang 2002, Molecular and isotopic analysis of Tagish Lake alkyl dicarboxylic acids, Meteoritics and Planetary Science 37, 687-696.

Poulet, F., E. Karkoschka, and B. Sicardy 1999, Spectrophotometry of Saturn's small satellites and rings from Hubble Space Telescope images, J. Geophys, Res. 104, 24095-24110.

Poulet, F., J.N. Cuzzi, D.P. Cruikshank, T. Roush, and C.M. Dalle Ore 2002, Comparison between Shkuratov and Hapke scattering theories for planetary surfaces: Application to the surface composition of two Centaurs, Icarus 160, 313-324.

Poulet, F., J.N. Cuzzi, D.P. Cruikshank, T.L. Roush, and R.G. French 2003, Saturn's ring composition from near-infrared spectroscopy: Implications for the origin and evolution of the rings, Astron. Astrophys, 412, 305-316.

Quirico, E., and Schmitt, B. 1997a, Near infrared spectroscopy of simple hydrocarbons and carbon oxides diluted in solid N_2 and as pure ices: Implication for Triton and Pluto. Icarus 127, 354-378.

Quirico, E., and Schmitt, B. 1997b, A spectroscopy study of CO diluted in N_2 ice: Applications for Triton and Pluto, Icarus 128, 181-188.

Quirico, E., Schmitt, B., Bini, R., and Salvi, P. R. 1996, Spectroscopy of some ices of astrophysical interest: SO_2, N_2, and N_2:CH_4 mixtures. Planet. Space Sci. 44, 973-986.

Quirico, E., S. Douté, B. Schmitt, C. de Bergh, D.P. Cruikshank, T.C. Owen, T.R. Geballe, and T.L. Roush 1999, Composition, physical state, and distribution of ices at the surface of Triton, Icarus, 139, 159-178.

Ramirez, S.I., P. Coll, A. da Silva, R. Navarro-González, J. Lafait, and F. Raulin 2002, Icarus 152, 515-529.

Roush, T.L. 1994, Charon: More Than Water Ice?, Icarus, 108, 243-254.

Roush, T.L. 2001, Physical State of Ices in the Outer Solar System, J. Geophys. Res.-Planets, 106, 33315-33323.

Roush, T.L. 2003, Estimated optical constants of Tagish Lake meteorite, Meteorit. Planet. Sci., 38, 419-426.

Roush, T.L., J.B. Pollack, F.C. Witteborn, J.D. Bregman, and J.P. Simpson 1990, Ice and minerals on Callisto: A reassessment of the reflectance spectra, Icarus, 86, 355-382.

Roush, T.L., D.P. Cruikshank, and T.C. Owen 1995, Surface ices in the outer Solar System, in *Volatiles in the Earth and Solar System, Am. Inst. Phys. Conf. Proc.*, vol. 341, pp. 143-152, AIP Press, New York.

Sagan, C. and B.N. Khare 1979, Tholins: Organic chemistry of interstellar grains and gas, Nature 277, 102-107.

Salama, F. 1998, UV photochemistry of ices, in *Solar System Ices*, edited by B. Schmitt, C. de Bergh, and M. Festou, pp. 259-279, Kluwer Acad. Sci., Norwell, Mass.

Satorre, M.A., M.E. Palumbo, and G. Strazzulla 2001, Infrared spectra of N_2-rich ice mixtures, J. Geophys. Res. 106, 33363-33370.

Schmitt, B., R., J. A. Grim, J. M. Greenberg, and J. Klinger 1992, Crystallization of water rich amorphous mixtures. In Physics and Chemistry of Ice (N., Maeno and T. Hondoh, Eds.). Hokkaido Univ. Press, Sapporo, pp. 344-348.

Schmitt, B., E. Quirico, F. Trotta, and W.M. Grundy 1998, Optical properties of ices from UV to infrared, in *Solar System Ices*, edited by B. Schmitt, C. de Bergh, and M. Festou, pp. 199-204, Kluwer Acad. Sci., Norwell, Mass.

Schou, J. and R. Pedrys 2001, Sputtering of carbon monoxide ice by hydrogen ions, J. Geophys. Res. 106, 33309-33314.

Shkuratov, Y., L. Starukhina, H. Hoffman, and G. Arnold 1999, A model of spectral albedo of particulate surfaces: Implications for optical properties of the moon, Icarus 137, 235-246.

Spencer, J.R., W.M. Calvin, and M.J. Person 1995, Charge-coupled device spectra of the Galilean satellites: Molecular oxygen on Ganymede, J. Geophys. Res., 100, 19,049-19,056.

Stern, S.A., M.W. Buie, and L.M. Trafton 1997, High resolution images and maps of Pluto, Astron. J. 113, 827-843.

Strazzulla, G. 1998, Chemistry of ice induced by bombardment with energetic charged particles, in *Solar System Ices*, edited by B. Schmitt, C. de Bergh, and M. Festou, pp. 281-301, Kluwer Acad. Sci., Norwell, Mass.

Tholen, D.J. 1984, Asteroid taxonomy from cluster analysis of Photometry, PhD Thesis, Univ. Arizona, 150 pp.

Tholen , D.J. 1989, Asteroid Taxonomic Classifications, In *Asteroids II*, R.P. Binzel, T. Gehrels, and M.S. Matthews, eds. Univ. Arizona Press, 1139-1150.

Thompson, W.R. and C. Sagan 1990, Color and chemistry on Triton, Science 250, 415-418.

Thompson, W.R., B.G.J.P.T. Murray, B.N. Khare, and C. Sagan 1987, Coloration and darkening of methane clathrate and other ices by charged particle irradiation: Application to the outer Solar System, J. Geophys. Res., 92, 14,933-14,947.

Verbiscer, A., and P. Helfenstein 1998, Reflectance spectroscopy of icy surfaces, in *Solar System Ices*, edited by B. Schmitt, C. de Bergh, and M. Festou, pp. 157-197, Kluwer Acad. Sci., Norwell, Mass.

Vilas, F. and M.V. Sykes 1996, Are low-albedo asteroids thermally metamorphosed?, Icarus 124, 483-489

Vilas, F., K.S. Jarvis, and M.J. Gaffey 1994, Iron alteration minerals in the visible and near-infrared spectra of low-albedo asteroids, Icarus 109, 274-283.

Vilas, F., S.M. Larson, K.R. Stockstill, and M.J. Gaffey 1996, Unraveling the zebra: Clues to Iapetus dark material composition, Icarus 124, 262-267.

Wetherill, G.W. and Chapman, C.R. 1988, Asteroids and Meteorites, In *Meteorites and the Early Solar System*, J.F. Kerridge and M.S. Matthews, eds., Univ. Arizona Press, 35-67.

Wilson P.D., C. Sagan, and W.R. Thompson 1994, The organic surface of 5145 Pholus: Constraints set by scattering theory, Icarus, 107 288-303.

Wilson, P.D. and C. Sagan 1995, Spectrophotometry and organic matter on Iapetus. I. Composition models, J. Geophys. Res. 100, 7531-7537.

Young, E.F., R.P. Binzel, and K. Crane 2001, A two-color map of Pluto's sub-Charon hemisphere, Astron. J. 121, 552-561.

Zolensky, M.E. and H.Y. McSween, Jr. 1988, Aqueous Alteration, In *Meteorites and the Early Solar System*, J.F. Kerridge and M.S. Matthews, eds., Univ. Arizona Press, 114-143.

Zolensky, M.E., K. Nakamura, M. Gounelle, T. Mikouchi, T. Kasama, O. Tachikawa, and E. Tonui 2002, Mineralogy of Tagish Lake: An ungrouped type 2 carbonaceous chondrite, Meteoritics and Planetary Science 37, 737-761.

Chapter 8

THE MOLECULAR COMPLEXITY OF COMETS

Jacques Crovisier
Observatoire de Paris
F-92195 Meudon, France
jacques.crovisier@obspm.fr

Abstract By delivering prebiotic molecules to the Earth, comets could have played a role in the early phases of the development of life on our planet. In order to explore this possibility, we present here an assessment of the molecular content of comets. The current methods of investigations — by both in situ analysis and remote sensing — are reviewed. The present status of our knowledge of the composition of cometary ices is reviewed. We probably now know most of the main components, but we still have a very partial view of the minor ones. A large diversity of composition from comet to comet is observed, so that no "typical comet" can be defined. No clear correlation between the composition and the region of formation of the comets and their subsequent dynamical history can yet be established. A crucial cometary component, both for cometary coma chemistry and for the possible delivery of organics to the Earth, is the (semi-)refractory high molecular-mass organic material present in grains. From the advent of new instrumentation, we can expect the detection of many new molecular species in future bright comets. However, the identification of really complex molecules will need in situ analysis or the return to Earth of nucleus samples.

Keywords: Astrobiology, comet, molecule, space mission, spectroscopy

1. Introduction

To understand how life could have appeared on Earth is a formidable project. We are only beginning to gather the pieces of the puzzle. One of them is the assessment of the chemical conditions in the early Earth. In this respect, the fall of comets could have played a crucial role by inject-

P. Ehrenfreund et al. (eds.), Astrobiology: Future Perspectives, 179–203.

ing organic — and possibly prebiotic — molecules (Oró 1961; Thomas et al. 1997). This possibility, which has been the object of passionate debates for a long time, relies on the answers to two questions:

- Do indeed comets contain molecules of a prebiotic interest?
- Could such molecules survive cometary impacts on Earth?

The object of the present Chapter is to try to answer the first question. Our knowledge of the chemical nature of cometary volatiles has considerably evolved during the last two decades with the advent of space exploration and of modern spectroscopic techniques. About two dozen species are now known, including the main constituents (H_2O, CO, CO_2, CH_4, CH_3OH, H_2CO, NH_3, H_2S)[1], and a number of trace species. We note that several species among identified molecules are "building blocks" for prebiotic molecules, i.e. they may react spontaneously in liquid water to form molecules like amino acids or nucleic acids. These are H_2CO, HCN, HC_3N, CH_3CN... Fortunately, comets have not evolved, so that the study of comets nowadays informs us on the nature of comets which could have seeded the Earth in the early times.

To the second question, Nature has offered opposite answers. On the one hand, the collision of comet Shoemaker-Levy 9 on Jupiter in July 1994 was followed by a surge of new molecular species in the atmosphere of the giant planet. These species were not the original molecules contained in the comet, rather the result of intense chemical processing of cometary material with the planetary atmosphere (Crovisier 1996; Lellouch 1996). The memory of the chemical content of the comet was lost. On the other hand, the analysis of meteorites — especially of carbonaceous chondrites, which are presumed to be akin to cometary nuclei — does show the presence of many organic molecules which survived the impact (Berzelius 1834; Botta & Bada 2002).

Next section will explain the various techniques (remote sensing or in situ analysis) which can be used to probe the chemical composition of comets. Section 3 will assess our present knowledge of the molecular content of comets, with an accent on molecular complexity and on the comet-to-comet variations. Section 4 will deal with future prospects: Which molecules could we expect to detect with new techniques?

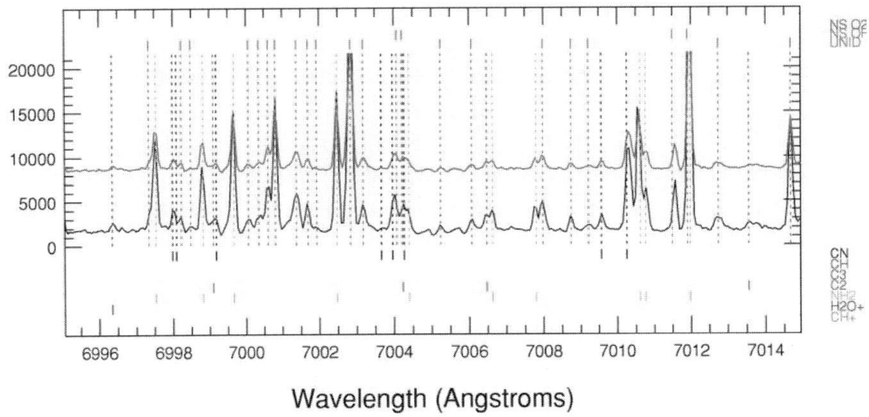

Figure 8.1. Excerpt of the high-resolution spectrum of comet 122P/de Vico. One can see lines of CN, C_2, NH_2, and many unidentified features. (Adapted from Cochran & Cochran 2002.)

2. The chemical composition of comets: methods of investigation

Awaiting for the results of ongoing or future space missions with in situ analyses and the return of samples from cometary nuclei, information of the chemical composition of cometary ices can only be obtained from remote sensing of the coma formed by matter released during cometary activity. Volatile molecules coming from the sublimation of cometary ices — the so-called *parent molecules* — are readily photo-processed to form secondary products, radicals, ions, atoms — the so-called *daughter molecules*.

2.1 Visible spectroscopy

Visible spectroscopy has been at work in astronomy for nearly one century and a half. Its application to comets (Fig. 8.1) is intimately linked to the beginnings of astrophysics and to laboratory spectroscopy. It took time to identify the features... Only dissociation or ionization products can be observed in this spectral domain, so that guesses had to

[1]Listed here are the species with abundances $\geq 1\%$ relative to water.

be made to infer what are their parent molecules. In the 1950's, from the observation of daughter molecules (CH, CO^+, NH, NH_2, OH, C_2, CN), only a handful of parents (H_2O, CO_2, CH_4, NH_3) could be proposed (e.g., Whipple, 1950). These guesses were basically right! However, it was not yet possible to have access to the actual complexity of the composition of cometary ices. One had to wait for the development of spectroscopy in other domains to directly identify parent molecules.

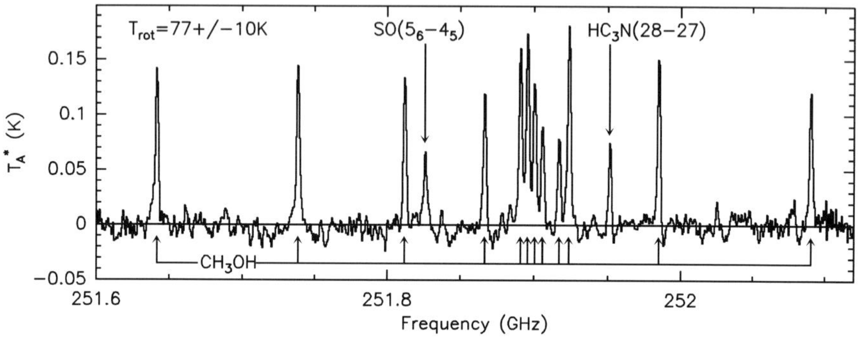

Figure 8.2. A millimetric spectrum of comet C/1995 O1 (Hale-Bopp) observed at the *Caltech Submillimeter Observatory* on 21 March 1997. This spectrum is chosen for its complexity. It covers a total frequency range of about 1.2 GHz (in this receiver setup, the spectrum is a mixture of the lower side-band — whose frequency scale is indicated in the figure — and of the upper side-band which is shifted by 5 GHz). The spectrum shows a series of lines of methanol for which the corresponding rotational temperature is indicated, and the first observations of SO and HC_3N (at 244.700 GHz in the upper side-band) in a comet. (From Lis et al. 1999.)

2.2 Radio and infrared spectroscopy

In the last two decades, radio and infrared spectroscopy (and to a lesser extent ultraviolet spectroscopy) led to the identification of about two dozen cometary volatiles (Figs 8.2 & 8.3). Radio spectroscopy (rotational lines) is more sensitive than IR spectroscopy (vibrational bands) for species with a significant dipolar moment[2]. Non-polar species (such as symmetric hydrocarbons) can only be investigated by IR spectroscopy.

[2]Species with abundances as small as 2×10^{-4} relative to water were observed in the radio spectra of comet Hale-Bopp. Species observed in the infrared had abundances of several 10^{-3}.

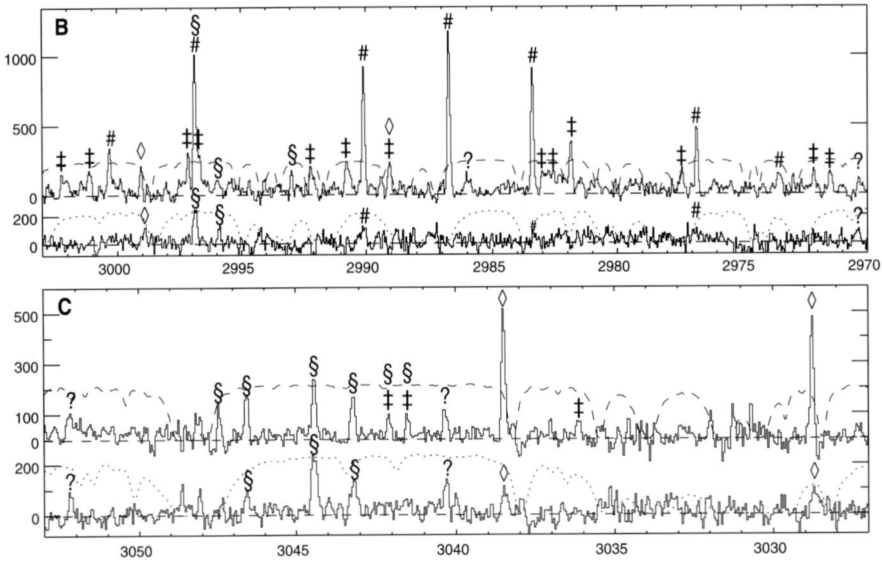

Figure 8.3. Excerpts of the infrared spectra of comets C/1999 H1 (Lee) (top of each frame) and C/1999 S4 (LINEAR) (bottom), observed with the Keck telescope. One can see lines of CH_3OH (‡), C_2H_6 (♯), CH_4 (□) and OH (§). Note the presence of several unidentified lines (?). The dashed and dotted lines show telluric transmission. (Adapted from Mumma et al. 2001.)

2.3 Surface reflectance spectroscopy

Cometary nuclei are very dark (albedo ≈ 0.03) and red (but not as red as some Centaurs and Kuiper belt objects). Reflectance spectroscopy of cometary nuclei could inform us on the composition of their surface. Such an observation at a large distance is, however, a formidable challenge because of the weakness of the signal, which could be dominated by the much more intense emission by the dust coma when the comet is active. It was successfully applied to the investigation of large Kuiper belt objects and Centaurs (de Bergh 2004). It could also be performed on the nucleus of comet 19P/Borrelly during its fly-by by *Deep Space 1*. Surprisingly, no water ice signature was revealed. The near-IR spectrum was featureless, except for an unidentified feature at 2.39 μm (Soderblom et al. 2004; Fig. 8.4).

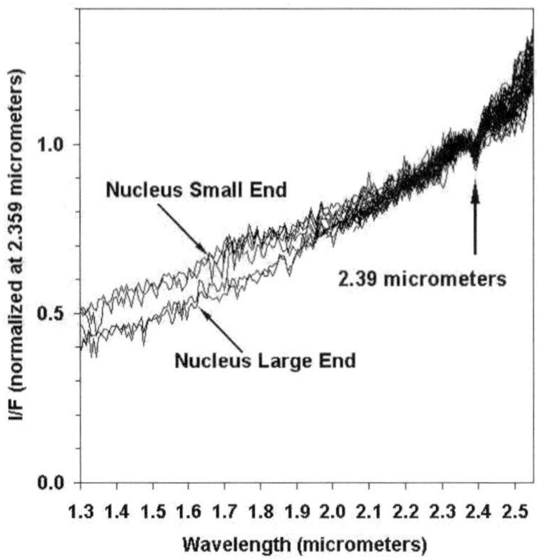

Figure 8.4. Near-infrared spectra of the nucleus of comet 19P/Borrelly observed by *Deep Space 1*. Note the absence of spectral features, except for an unidentified absorption at 2.39 μm. (From Soderblom et al. 2004.)

2.4 Mass spectroscopy

In situ mass spectroscopy is a very powerful mean of investigation. However, up to now it was only exercised in comet Halley (Altwegg et al. 1999; Eberhardt et al. 1999; Fig. 8.5). It was then seriously hampered by mass ambiguity due to its limited spectral resolution. The *ROSINA* instrument designed for *Rosetta* (Balsiger et al. 2003) has a much better resolution than the mass spectrometers of *Giotto*.

Another problem is that this technique is only sensitive to ions — either ions already present in the coma, or coming from the ionization or protonization of cometary neutrals by the apparatus. Thus chemical modelling is necessary to retrieve information on the parent molecules from the ionized species effectively observed by the mass spectrometer. This problem may be as difficult as the guess of parents from the observed daughters mentioned above about visible spectroscopy. It may be pointed out that in the interpretations of the mass spectroscopy data from comet Halley, most of the *correct* identifications and abundance determinations were made a posteriori, after the way had been cleared by spectroscopic remote sensing.

The time-of-flight mass spectrometer *PUMA* on board *VEGA 1* revealed several high molecular mass compounds in comet Halley's dust grains (Kissel & Krueger 1987; also Kissel et al. 1997), that the authors

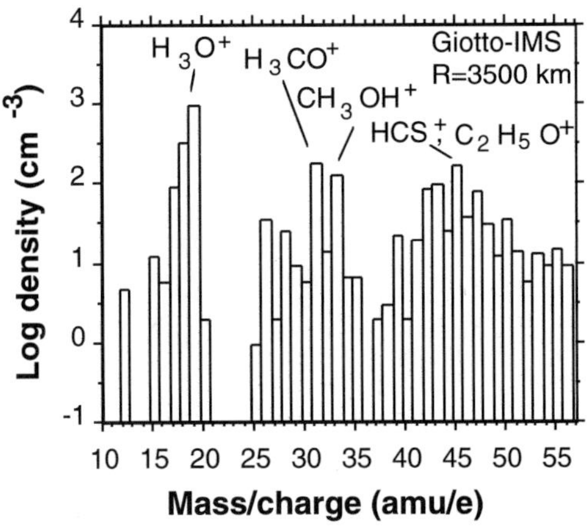

Figure 8.5. The mass spectrum of gas observed in 1P/Halley by the *IMS* instrument. (From Altwegg et al. 1999.)

attributed to specific organic species such as adenine, pyrimidine, and their compounds. These highly speculative hypotheses will have to be confirmed in the future using more selective identification techniques.

2.5 Chromatography

Chromatography is another powerful in situ mean of investigation, especially when associated with mass spectroscopy. Space gas chromatography was successfully used on Mars (*Viking* probes) and Venus (*Pioneer-Venus*, *Venera* and *Vega* probes). It will investigate the atmosphere of Titan (*Cassini-Huygens*). It will be used on the *Rosetta* lander to probe the nucleus of comet 67P/Churyumov-Gerasimenko (the *COSAC* experiment: Rosenbauer et al. 1999; Thiemann et al. 2001; Szopa et al. 2003; Fig. 8.6). Szopa et al. estimate that species with abundances 10^{-5}–10^{-6} relative to water could be detected with *COSAC* (for those species that can be sampled with the dedicated chromatographic columns). Semi-refractory species (e.g. glycine) will not be released; they could be studied by pyrolysis and derivatisation. Chirality could be investigated with specific chromatographic columns.

2.6 Fetching cometary material

Ultimately, the return of a sample of cometary nucleus would allow us to perform its study at leisure in Earth laboratories. Several scenarios of *Comet Nucleus Sample Return* missions have been studied and

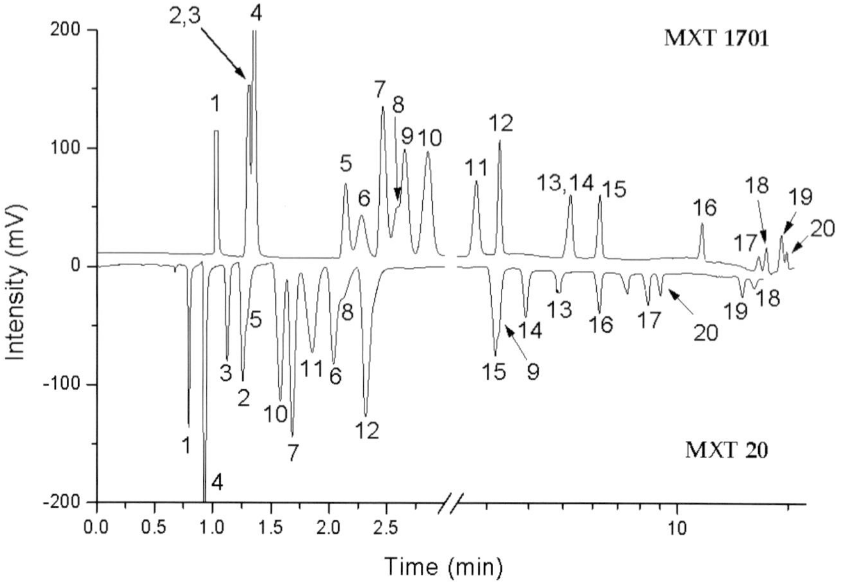

Figure 8.6. Example of calibration chromatograms obtained with two different columns in preparation of the *COSAC* experiment on the lander of *Rosetta*. The peaks correspond to 1 = nitrogen, 2 = water, 3 = methanol, 4 = benzene, 5 = formamide, 6 = pyridine, 7 = toluene, 8 = pentanol, 9 = octane, 10 = ethyl benzene, 11 = m-xylene, 12 = p-xylene, 13 = cyclohexanone,14 = o-xylene, 15 = nonane, 16 = propyl benzene, 17 = trimethyl benzene, 18 = decane, 19 = p-cymene, 20 = indene. (From Szopa et al. 2003.)

proposed[3]. There is presently no firmly scheduled mission to return a genuine sample of cometary nucleus, due to complexity and cost. The *Stardust* mission (Brownlee et al. 1997) towards comet 81P/Wild 2 (fly-by in January 2004) is expected to return samples of dust grains to Earth in January 2006. These grains will not have retained any icy component. But perhaps they will still contain organic refractories — they could be similar to some interplanetary dust particles collected in the Earth stratosphere which are believed to be of cometary origin.

2.7 Home delivery of cometary material

Why bother going at a comet nucleus and back to fetch samples, or studying it painstakingly at distance? Large amounts of cometary

[3] Among them, earlier versions of *Rosetta* and *Champollion* (an aborted NASA mission similar to *Rosetta*).

Table 8.1. Molecules, radicals, ions and atoms observed in cometary comae, with their means of detection: radio (rad.), infrared (IR), visible (vis.), ultraviolet (UV) spectroscopy, or mass spectrometry[a] (MS).

	rad.	IR	vis.	UV	MS		rad.	IR	vis.	UV	MS
H_2O	x	x			x	N				x	
H_2O^+			x			N_2^+			x		
H_3O^+	x					NH_3	x	x			x
OH	x	x		x		NH			x		
H			x	x		NH_2		x	x		
H_2				x		HCN	x	x			x
O			x	x		HNC	x				
O^+				x		CN	x	x	x		
CO	x	x		x	x	CH_3CN	x				x
CO_2		x			x	HC_3N	x				
CO^+	x		x	x		HNCO	x				
CO_2^+			x			NH_2CHO	x				
C				x		H_2S	x				x
C_2		x	x			CS	x			x	
C_3			x			SO	x				
CH_4		x				SO_2	x				
C_2H_2		x			x	OCS	x	x			
C_2H_4					x	H_2CS	x				
C_2H_6		x			x	NS	x				
C_4H_2		x				S_2				x	
CH		x	x			S				x	
CH^+			x			Na			x		
CH_2					x	K			x		
H_2CO	x	x				Ar				?	
HCO^+	x					metals[b]			x		
CH_3OH	x	x			x						
HCOOH	x										
CH_3CHO	x										
$HCOOCH_3$	x										

[a] The species detected by mass spectroscopy of comet 1P/Halley are taken from the compilation of Altwegg et al. 1999 (Table II); also listed as "probably detected" were CS, OCS, CH_3CHO, C_3H_2 and C_2H_5CN (ibid., Table III).
[b] Various metal atoms were only observed in the sungrazing comet C/1965 S1 (Ikeya-Seki).

material are continuously delivered to Earth. This is the very mechanism which is the motivation of the present Chapter, albeit a delivery rate much lower nowadays than it was in the early times!

Meteor showers are attributed to the shedding of dust by cometary nuclei. The spectroscopy of meteoroids, although a challenging obser-

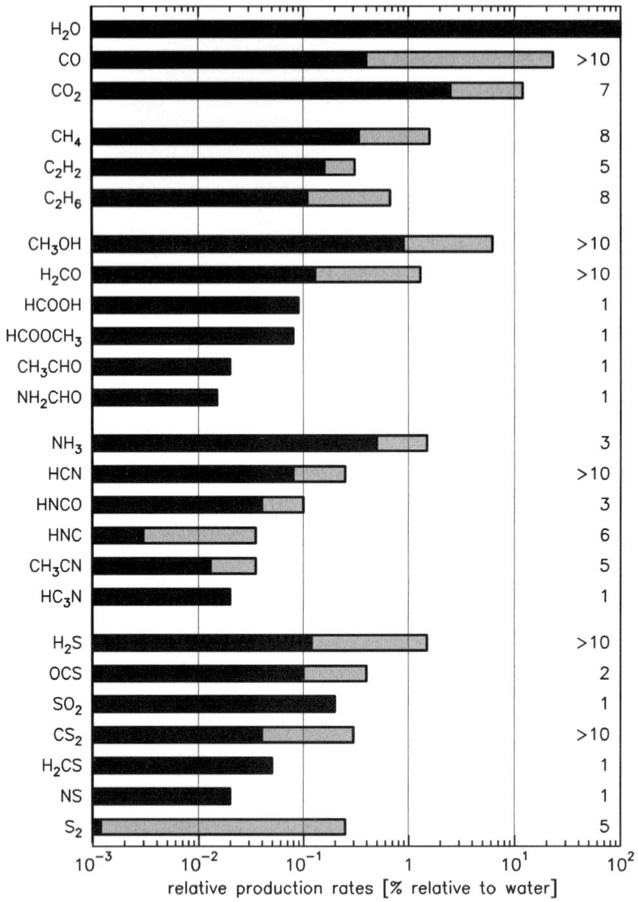

Figure 8.7. Relative production rates of cometary volatiles and their comet-to-comet variations[a]. These rates are believed to trace the relative abundances in cometary ices. The grey part of each bar indicates the range of variation from comet to comet. On the right, the number of comets in which the species was detected is indicated. (From Bockelée-Morvan et al. 2005.)

[a] Some species reported in this figure appear to come (in part) from extended sources, not directly from nucleus ices (See Section 3.3). CS_2 is not observed directly: its production rate is estimated from its daughter CS. NS is a radical for which no plausible parent has yet been proposed. The origin of S_2 is not well understood.

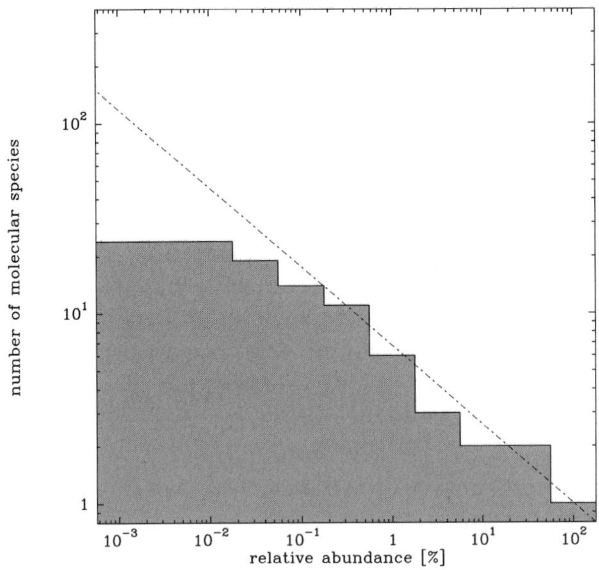

Figure 8.8. The cumulative histogram of molecular parent species detected in comet Hale-Bopp with production rates (in percent relative to water) larger than a given value. The observed distribution departs from the true distribution for the less abundant species because of sensitivity issues. The dashed line is a guess of the true distribution. (From Bockelée-Morvan & Crovisier 2002.)

vation, would inform us on their chemical composition (e.g., Abe et al. 2002). IDP's collected in the upper atmosphere as well as micrometeorites are supposed to contain a significant fraction of cometary dust particles (Molster 2004).

In situ analysis and the study of returned samples of cometary dust will provide ground truth for evaluating the relation between chondritic porous IDP's and cometary material. It has been argued for a long time whether some meteorites (among CI and CM carbonaceous chondrites) could actually be parts of cometary nuclei (Campins & Swindle 1998; Lodders & Osborne 1999), but direct clues are still lacking.

3. The chemical composition of comets: present status

Although this is still small number statistics (for both the number of comets investigated and the number of molecules detected), our knowl-

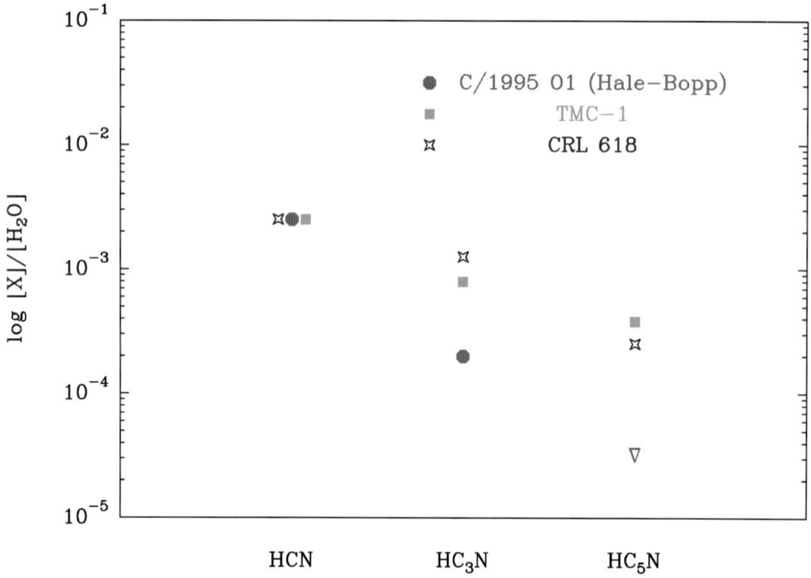

Figure 8.9. The relative abundances of linear cyanopolyynes in comet Hale-Bopp (Bockelée-Morvan et al. 2000; Crovisier et al. 2004), in the dark cloud TMC–1 (Irvine et al. 1987) and in the proto-planetary nebula CRL 618 (Cernicharo et al. 2001). For the comet, the abundances are plotted relative to water and the triangle indicates an upper limit. For the interstellar sources, the abundances are normalized to that of HCN.

edge of the composition of cometary material is rapidly expanding. It is the topic of several recent reviews (Crovisier 1998; Crovisier & Bockelée-Morvan 1999; Bockelée-Morvan et al. 2000; Irvine et al. 2000; Despois et al. 2002; Bockelée-Morvan et al. 2005). It is summarized in Fig. 8.7. Detailed information is only available for a very small number of comets (1P/Halley, C/1996 B2 (Hyakutake), C/1995 O1 (Hale-Bopp), C/1999 H1 (Lee), C/1999 S4 (LINEAR), 153P/Ikeya-Zhang). About 7 species were observed in comet Hale-Bopp only.

3.1 Molecular complexity

Altogether, 44 molecular species, radicals or molecular ions are identified in comets (Table 8.1). This is to be compared with about 130 species (excluding isotopologues) detected in the interstellar medium. However, we must have in mind that these *interstellar molecules* are not all detected in a single source[4].Among them are 25 stable volatile molecules, likely to have sublimated from nucleus ices (Fig. 8.7), but upper limits are set on many other species[5].

Fig. 8.8 shows the cumulative distribution of the number N of identified cometary parent molecules as a function of their relative production rates X observed in comet Hale-Bopp[6]. We probably now know the most abundant species down to a level of about 1%. The distribution is obviously stalling for abundances $< 0.1\%$, which reveals that less abundant (trace) species are still sparsely known at this level. The dashed line (which corresponds to a power law $N \propto X^{-0.4}$) is a rash extrapolation of the distribution to small abundances. If realistic, it would imply we should expect ≈ 100 species with abundances $> 0.001\%$ relative to water (isotopologues are excluded from this study).

Cometary volatiles have very different equilibrium sublimation temperatures[7]. Thus, fractionation at sublimation may occur, so that the observed relative production rates of cometary volatiles do not directly correspond to their relative abundances in cometary ices (Huebner & Benkhoff 1999; Biver et al. 2002a). For instance, the $[CO]/[H_2O]$ ratio is observed to increase with heliocentric distance, due to the higher volatility of CO; no water is produced at $r_h > 4$ AU whereas CO was still observed up to 15 AU in comet Hale-Bopp (and could explain alone cometary activity at these distances). But in a high sublimation regime at $T > 180$ K dominated by water, such as that which occurs at $r_h < 1$ AU, all volatiles sublime alike and little fractionation is expected. Another object of concern is the possibility of chemical synthesis: could some of the observed trace species be the result of chemical processes rather than coming from nucleus ices? According to the calculations of Rodgers & Charnley (2001), this is not the case, even in a very productive comet such as Hale-Bopp. This is due to the cold temperature of the coma, its low ionisation state, and the short time spent by molecules in the collision zone.

---

[4]For instance, the longest carbon-chain molecules are only observed in dark clouds (TMC–1); complex CNO molecules are only observed in hot cores (Galactic centre, Orion IRc2); some hydrocarbons were only detected in circumstellar envelopes (IRC+10216 or CRL 618). (See e.g., Millar 2004.)

[5]For instance, ketene (CH_2CO) $<$ 0.03%, ethanol (C_2H_5OH) $<$ 0.1%, glycine (NH_2CH_2COOH) $<$ 0.15%, dimethyl ether (CH_3OCH_3) $<$ 0.5%, cyanodiacetylene (HC_5N) $<$ 0.003%, methanimine (CH_2NH) $<$ 0.03% relative to water (Crovisier et al. 2004).

[6]It would be interesting to plot similar histograms for the relative molecular abundances in typical interstellar medium objects, or for biological bodies, or for those resulting from chemical models, to see whether a power law $N \approx X^{-\alpha}$ is emerging with a similar or different exponent.

[7]The sublimation temperatures range from 25–30 K for CO and CH_4, to 70–100 K for CO_2, NH_3, HCN, CH_3OH, up to 150–180 K for H_2O.

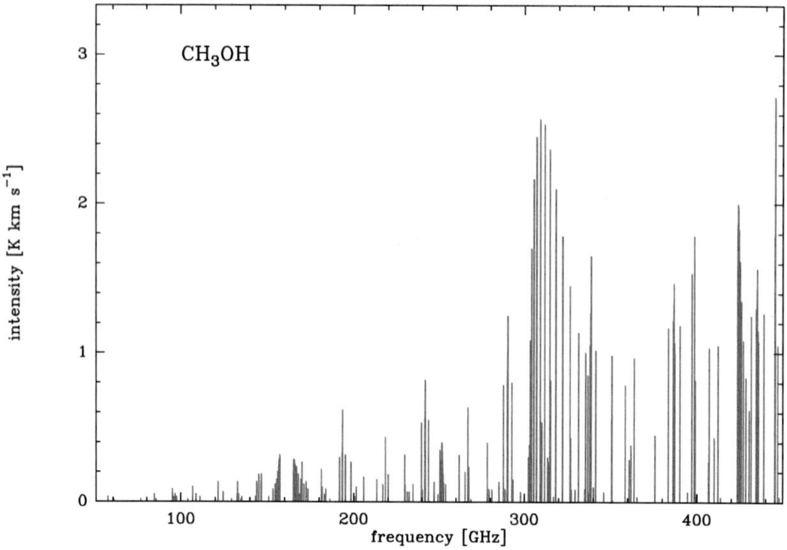

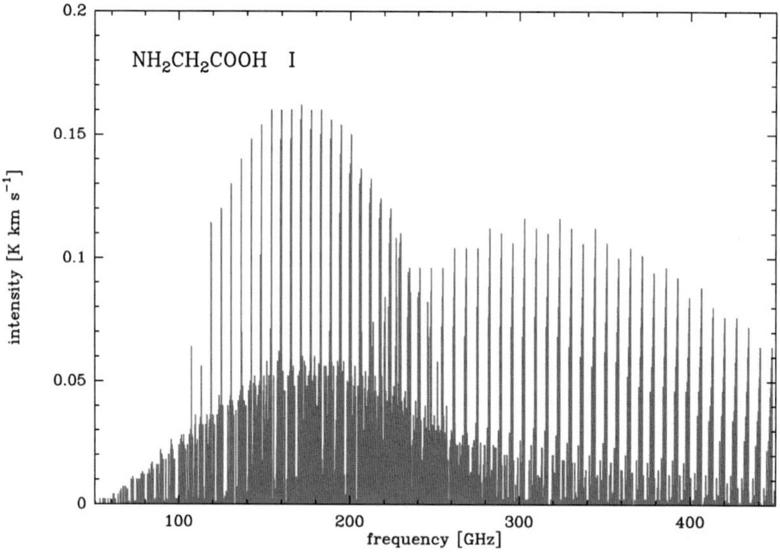

Figure 8.10. The microwave spectra expected for cometary methanol (top) and glycine (conformer I, bottom). They were computed for the following conditions: comet at 1 AU from Sun and Earth, observed with a 30-m telescope, coma at a temperature of 70 K and 0.8 km/s expansion velocity, molecular production rates of 10^{28} s^{-1} for each species. Note the much more intense lines of methanol. Many lines of CH_3OH were observed in several comets. Glycine is still undetected. Molecular data are taken from the JPL molecular data base.

Obviously, abundances decrease with increasing molecular complexity. Fig. 8.9 shows that the relative abundance of cometary cyanopolyynes ($HC_{2n+1}N$) decrease more rapidly than in interstellar clouds[8]. From the upper limits determined in comet Hale-Bopp (Crovisier et al. 2004), the ratio [ethanol]/[methanol] is also rather low ($< 1/25$). This may not be the same for all cometary molecules in *homologous series*. For instance, [ethane]/[methane], now observed in several comets (Gibb et al. 2003), is close to one[9].

Cyanopolyynes are simple, linear molecules with a strong dipolar moment. Therefore, their radio spectra consist in a simple series of strong lines. They can be observed with very low abundances. In contrast, the upper limits on complex molecular species are not good. For instance the upper limit is only 0.15% in number relative to water for glycine[10](Crovisier et al. 2004). This is due to the high value of the partition function of such molecules which have many rotational levels populated even at low temperatures, which have additional degrees of freedom due to internal rotation, and which may have several low-energy conformers. Their spectra are complex, with a forest of lines (Fig. 8.10). Such species should be searched for by in situ analysis rather than by remote sensing.

3.2 Chemical diversity

According to the current cosmogonic theories, all comets formed at the very beginning of the history of the Solar System, from accretion of planetesimals in the proto-planetary disc beyond the *snow line*[11]. But they followed different fates, depending on where they were formed. Beyond Neptune, comets constituted part of the Kuiper belt. In the inner Solar System, newly formed comets in the Jupiter–Neptune region were readily ejected by gravitational interaction with the giant planets: they either left the Solar System, or formed a spherical, distant cloud, the Oort cloud.

There are thus two reservoirs of comets. Further perturbations provoke changes of orbits that may bring back comets from both reservoirs

[8]Cronin (1998) has noted that for carboxylic acids analysed in meteorites, the abundance decreases by $\approx 60\%$ for each additional C-chain atom.

[9]In this specific case, this could be due to a difference of volatility (sublimation equilibrium temperatures of 31 K for CH_4, 44 K for C_2H_6) rather than to a chemical effect.

[10]This corresponds to about 0.3% in mass in nucleus ices. This is to be compared with the abundance of $\approx 10^{-5}$ in mass measured for the whole amino acids in carbonaceous chondrites (Botta and Bada 2002).

[11]The snow line is the heliocentric distance beyond which water could condense. This is ≈ 5 AU (about the Jupiter's heliocentric distance) in the present Solar System.

to the inner Solar System. Comets from the Kuiper belt give Jupiter-family comets, with short periods and low inclinations on the ecliptic plane. Comets from the Oort cloud may have all kinds of periods and random inclinations. One would expect different compositions for these two main dynamical classes of comets, depending on where they were initially formed. For instance, comets formed at large heliocentric distances in the Kuiper belt could have retained species more volatile than the Oort cloud comets.

A first attempt of a taxonomy of comets based upon their chemical composition was made by A'Hearn et al. (1995) from the relative production rates of daughter specie (OH, CN, C_2, C_3, NH). Two classes of comets were identified: *typical comets*, and *carbon-depleted comets*, on the basis of the $[C_2]/[CN]$ ratio. Half of the Jupiter-family comets appear to be carbon-depleted, whereas most of the Oort cloud comets are typical. How does this variety in relative abundances of daughters translate into variety in the abundances of parents is unknown, because of, e.g., the large number of precursors for C_2.

There is now a significant sample of comets (about twenty) for which the main volatile components (parent molecules) have been investigated. No "typical comet" could be proposed with respect to chemical composition. A large chemical diversity is revealed (radio observations: Biver et al. 2002b; infrared observations of hydrocarbons: Mumma et al. 2002; Fig. 8.7). Although the $[HCN]/[H_2O]$ ratio is fairly constant, CH_3OH, H_2CO and H_2S show large variations. Carbon monoxide is still more variable (by about a factor fo 50).

The relation of chemical composition with dynamical classes of comets and their site of formation — *Oort cloud* (long-period) comets versus *Kuiper belt* (Jupiter-family) comets — is unclear. This relation may be blurred by the fractionation in Jupiter-family comets expected from repeated passages close to the Sun. More observations are needed to increase the number of comets (especially the weaker Jupiter-family comets) and of molecules.

Of which dynamical class were the comets which seeded the Earth in the early times? Discussing the possible external delivery of water to Earth by small bodies impacts, Morbidelli et al. (2000) estimate that the Earth benefited from cometary impacts all along its formation. First, from comets from the Jupiter–Saturn region (i.e., the same comets as those which were ejected to the Oort cloud), when the Earth was still less than half its present mass. Last, from comets originating from the Uranus–Neptune region and the Kuiper belt[12].

According to Binzel et al. (2002), up to 30% of the near-Earth objects — which are objects with orbits crossing the Earth orbit and therefore likely to impact the Earth — are extinct or dormant comets[13].

3.3 The mystery of the "extended sources"

Extended sources of molecules are needed to explain the production, distribution, and evolution with heliocentric distance of cometary CO, H_2CO and HNC/HCN.

The evidences for extended sources (see review in Bockelée-Morvan & Crovisier 2002) first came from in situ exploration of 1P/Halley, then from long-slit spectroscopy in the IR and mapping in the radio (both by single dishes and interferometers). Extended sources are suggested for CO, H_2CO, HNC, OCS. The most clear-cut case is that of H_2CO, which might have no significant nuclear source at all. The most important extended source is that observed in comet Hale-Bopp, for which about half of the 20% (relative to water) production of CO was coming from the extended source (DiSanti et al. 2001). This extended source appears to have turned on at $r_h < 1.5$ AU. However, this analysis is complicated by the presence of rotating jets of CO (Henry et al. 2002) in that comet. The extended source of CO was observed to be relatively much weaker in comet C/1996 B2 (Hyakutake) (DiSanti et al. 2003) and no extended source could be found in comet 153P/Ikeya-Zhang (DiSanti et al. 2002).

Admittedly, extended sources are expected from icy grains and from the photolysis of parent molecules (CO_2, H_2CO and CH_3OH produce CO). However, the scale length observed for the extended source of CO ($\approx 10^4$ km) does not fit: it is too long for sublimating icy grains, too short for photolysed parents. Furthermore, the production of extended CO is much larger in comet Hale-Bopp than that observed for plausible CO-parents.

Such sources could be CHON refractories (or semi-refractories) and/or polymers on cometary grains. From the elemental composition[14] of the dust grains of 1P/Halley, Fomenkova (1999) estimated that $\approx$ 50% in

[12]But they estimate that the bulk of the water presently on Earth was coming from a few planetary embryos originally formed in the outer asteroid belt.

[13]*Dormant comets* are cometary nuclei which have built up a protective crust and are therefore inactive even at 1 AU. Their inner nucleus has the same composition as other comets. *Extinct comets* have exhausted their volatiles and are no longer active. They could however still contain semi-refractory CHON material.

[14]Measured by the dust-impact time-of-flight mass spectrometers *PUMA* on *VEGA* and *PIA* on *Giotto*.

mass of the grains were organic refractories. Of course, the exact composition of these organics cannot be unambiguously assessed. HCN polymers and polyoxymethylene (H_2CO polymer) have been specifically invoked. The release mechanisms of light molecules from refractories is still ill-understood. Laboratory studies are necessary, such as those made on the photo- and thermo-degradation of polyoxymethylene by Cottin et al. (2001). The steep variation with heliocentric distance of the production of molecules from extended sources rather suggests that a temperature-dependent mechanism, such as thermo-degradation, is at work.

The nature of cometary refractory organics will be investigated in situ by *Stardust* and *Rosetta*. But it is dubious that the true composition of such complex species could be revealed. After all, even terrestrial material such as tholins, kerogen or caramel is not so well chemically characterized from laboratory analyses.

Being less sensitive to destruction during impacts, such organics could have been a significant source of prebiotic material to the early Earth!

4. Future prospects: sensitivity and confusion issues

A gain in sensitivity is clearly needed to make significant progresses, but the ultimate limit will be provided by the confusion limit in the spectra.

4.1 Confusion in cometary spectra

4.1.1 Confusion in radio spectra. Avery (1987) analysed the radio spectra of interstellar medium objects He found that the number of lines with antenna temperature between T_A and $T_A + \Delta T_A$ follows the law[15] $T_A^{-2} \Delta T_A$. A similar law might apply to cometary spectra.

A first advantage of comets over interstellar sources is the small width of their molecular lines: 2 km s^{-1}, which is to be compared to interstellar medium objects: 50 km s^{-1} for the Galactic centre, 5 km s^{-1} for the Orion IRc2 hot core, 30 km s^{-1} for the circumstellar envelope IRC+10216, and 10 km s^{-1} for the bipolar flow L 1157). Cold dark clouds (e.g., TMC–1) have narrow lines comparable to those of comets (or even narrower), but their gas-phase content is not so rich because most molecules are frozen into ices. Another advantage of comets is

[15]Note that this law does not converge when integrated to low T_A's: it leads to an infinite continuum (Olbers' paradox). If all species had similar spectroscopic properties, this would convert into a similar law for the column densities of species; the integration would similarly lead to an infinite total column density.

the relatively colder temperature of their molecules (≈ 100 K for comet Hale-Bopp at $r_h \approx 1$ AU — less for less productive comets or comets at larger heliocentric distances), which is to be compared with $T_{rot} \approx 250$ K in Orion IRc2. More than a factor of ten more lines (and thus ten times more species) could ultimately be detected in comets before being limited by confusion.

4.1.2 Confusion in infrared spectra. The *VIRTIS* instrument (Coradini et al. 1998) designed for *Rosetta* has a limited resolution, similar to that of the *ISO* short-wavelength spectrometer ($R \approx 2000$). This barely allows us to resolve the rotational structure of the vibrational bands. The number of detectable species will be severely limited given this resolution. Could deuterated species be detectable (HDO, CH_3D... Gibb et al. 2002; Kawakita et al. 2003) with such instrumentation? Examples of what could be confused infrared spectra is given by the Earth atmosphere spectra obtained by limb sounding with the *ATMOS* experiment ($R \approx 500\,000$; Farmer & Norton 1989). The cometary spectra recently measured from the ground by *IRTF/CSHELL* and *Keck/NIRSPEC* have a much higher resolution ($R \approx 20\,000$), but telluric lines may be a problem (Mumma et al. 2003).

4.1.3 Confusion in chromatograms and mass spectra. For an experiment such as *COSAC*, confusion caused by coelution also limits the number of species that could be detected with a single chromatographic column to about 20–30 (Fig. 8.6). At the detection limit of *COSAC* (10^{-5}–10^{-6} relative to water), one would expect more than 100 cometary species (Fig. 8.8). Confusion rather than sensitivity is thus a problem. The use of several chromatographic columns, complementary data from mass spectroscopy together with sophisticated retrieval methods (Pietrogrande et al. 2003) are needed to get the best from the data.

4.2 Unidentified lines

The numerous lines in cometary spectra which are not yet identified strongly suggest that many more species are potentially present.

The visible spectrum of comets is still far from being fully understood. For instance, in their high-resolution spectral atlas of comet 122P/de Vico, Cochran & Cochran (2002) identified 12 219 emission lines and reported no less than 4 219 unidentified lines[16] (Fig.8.1). In the far UV, the *FUSE* satellite observatory has observed several unidentified lines, presumably due to atoms or ions (Feldman et al. 2002). In the infrared, high-resolution spectra (Fig. 8.3; Mumma et al. 2003) many unidentified lines have been noted in the 3–5 μm region. Methanol is known to be the main contributor of the emission observed in the 3.2–3.5 μm region. But a definite evaluation of this contribution is still awaiting a detailed model of the methanol infrared spectrum. PAH's are probably contributing to the 3.28 μm emission feature observed in some comets. Other contributors could be CHO species, hydrocarbons, radicals or even molecular ions. Even in the radio domain, unidentified lines have been noted (Crovisier et al. 2004).

Solving these problems require further spectroscopic laboratory work and the building of comprehensive molecular data bases. This is especially the case for NH_2 in the visible and CH_3OH in the infrared: these species have so complex spectra that cometary spectra must be cleaned from their contribution to assess the presence of genuinely new species.

4.3 Future prospects

What are the future molecules to be detected? All likely simple molecules have now been detected (H_2O, CO, CO_2, NH_3, CH_4), except N_2 and O_2[17]. A similarity between the composition of cometary ices and that observed in the interstellar hot cores has been noted (Bockelée-Morvan et al. 2000). This does not necessarily mean that comets have directly incorporated unprocessed interstellar material, but that processes leading to the formation of molecules were similar in the interstellar medium and in the protosolar nebula. Going on with this analogy would suggest that further *interstellar molecules* (e.g. dimethyl ether CH_3OCH_3, acetic acid CH_3COOH, glycolaldehyde $HOCH_2CHO$, ethylene glycol $HOCH_2CH_2OH$) are to be identified in comets[18].

[16] A significant part of them could be attributed to NH_2 (M. Vervloet, *personal communication*).

[17] N_2 and O_2 do not have easily observable spectral signatures, so that no very stringent upper limit could be put on their abundances (but these species, which are very volatile, may have not been retained in cometary ices). The presence of N_2 has been suggested from the identification of a N_2^+ band in the near UV. However, this identification has been questioned by Cochran et al. (2002) who failed to detect this band in their high-resolution spectra of recent comets. H_2 has been detected in the UV with *FUSE* (Feldman et al. 2002), but it is presumably a daughter species coming from the dissociation of water and hydrocarbons.

[18] Ethylene glycol has recently been identified in archive radio spectra of comet Hale-Bopp (Crovisier et al. 2004, Astron. Astrophys. 418:L35–L38). This is the most complex organic molecule firmly identified in a comet to date. [Note added in proof]

Concerning remote sensing investigations, the breakthrough performed in comets Hyakutake and Hale-Bopp has been, at best, repeated for some molecules in other bright comets which appeared since that time. For new progresses, we need sensitive new instruments, but also new bright comets. The real advantages of the new generation ground-based high-resolution infrared spectrometers (*IRTF/CSHELL*, *Keck/NIRSPEC* and *VLT/CRIRES* which will be soon available) have not yet been fully exploited due to line identification issues. The *ALMA*[19] radio telescope, to be completed in 2010, will achieve a gain in sensitivity of more than an order of magnitude over present instruments, when operating in the autocorrelation mode. The *Herschel Space Observatory* (to be launched in 2007) will give us access to a still shorter wavelength range, which cannot be explored from the ground. It will be adequate for probing light hydrides, but not for heavy molecules which have strong rotational lines at longer wavelengths.

Concerning space missions, the next step after *Stardust* and *Rosetta* should be the return of a cometary ice sample. An interesting technical challenge for the next generations.

5. Conclusion

There is now little doubt that *prebiotic* molecules are present in cometary ices. Evidences for more complex organic molecules are also present and indeed, we still have no idea of the limit in complexity of the cometary material.

There are also converging clues to the existence of a (semi-)refractory organic material composing a large fraction of the cometary grains. This material, of ill-defined composition, is likely to deliver lighter organic molecules through thermo-degradation.

Such organic matter probably does not survive giant cometary impacts. However, experimental simulations have shown that in specific conditions, amino acids could be delivered (Blank et al. 2001). Some of the simplest molecules could also be recomposed in the shock chemistry following impact.

On the other hand, cometary grains and small nucleus fragments could survive their delivery to Earth. They constitute part of the IDP's collected in the stratosphere, of the micrometeorites collected on the ground or in Antarctic ices. Indeed, they do contain complex heavy organic molecules. They have lost most of their volatiles, however.

---

[19] ALMA is a radio interferometer array of 64 × 12-m antennas to operate in the millimetre-submillimetre range.

Remote sensing observations will continue to assess the gross composition of cometary ices and its diversity from comet to comet. It will be limited by sensitivity and confusion, however, so that really complex trace species will escape to this technique.

In situ analyses could be more sensitive, but will be limited to a small number of objects. The analysis of the refractory organic cometary macro-molecules could be difficult because such material is expected to have an ill-characterized composition.

An expected return of cometary space exploration is to provide ground truth for discriminating cometary material in the extraterrestrial material available on Earth.

References

A'Hearn, M.F., Millis, R.L., Schleicher, D.G., Osip, D.J. and Birch, P.V. (1995). The ensemble properties of comets: results from narrowband photometry of 85 comets, 1976–1992. *Icarus*, 118:223–270.

Abe, S., Yano, H., Ebizuka, N., et al. (2002). First results of OH emission from meteor and after glow: search for organics in cometary meteoroids. In *Asteroids, Comets, Meteors, ACM 2002*, ESA SP-500, 213–216.

Altwegg, K., Balsiger, H., and Geiss, J. (1999). Composition of the volatile material in Halley's coma from in situ measurements. In *ISSI workshop* Composition and Origin of Cometary Material. *Space Scie. Rev.*, 90:3–18.

Avery, L.W. (1987). Radio and millimetre observations of larger molecules. In *Astrochemistry*, IAU Symp. No 120, M.S. Vardya & S.P. Tarafdar edts, Reidel, Dordrecht, 187–197.

Balsiger, H., Altwegg, K., Gliem, F., Fieth, et al. (2003). Rosetta orbiter spectrometer for ion and neutral analysis ROSINA. *ESA SP-1165*, in press.

Berzelius, J.J. (1834). Über Meteorstein, 4. Meteorstein von Alais. *Ann. Phys. Chem.*, 33:113–123.

Binzel, R.P., Stuart, J.S., Rivkin, A.S., et al. (2002). Exploring the comet component within the Near-Earth Object population. *Bull. Amer. Astron. Soc.*, 34:840.

Biver, N., Bockelée-Morvan, D., Winnberg, A., et al. (2002a). The 1995-2002 long-term monitoring of comet C/1995 O1 (Hale-Bopp) at radio wavelengths. *Earth, Moon, Planets*, 90:5–14.

Biver, N., Bockelée-Morvan, D., Crovisier, J., et al. (2002b). Chemical composition diversity among 24 comets observed at radio wavelengths. *Earth, Moon, Planets*, 90:323–333.

Blank, J.G., Miller, G.H., Ahrens, M.J., and Winans, R.E. (2001). Experimental shock chemistry of aqueous amino acid solutions and the cometary delivery of prebiotic compounds. *Origins Life Evol. Biosphere*, 31:15–51.

Bockelée-Morvan, D., and Crovisier, J. (2002). Lessons of comet Hale-Bopp for coma chemistry: observations and theory. *Earth Moon Planets*, 89:53–71.

Bockelée-Morvan, D., Crovisier, J., Mumma, M.J., and Weaver, H.A. (2005). The composition of cometary volatiles. In *Comets II*, M. Festou, H.U. Keller & H.A. Weaver edts, Univ. Arizona Press, Tucson, USA, in press.

Bockelée-Morvan, D., Lis, D.C., Wink, et al. (2000). New molecules found in comet C/1995 O1 (Hale-Bopp). Investigating the link between cometary and interstellar material. *Astron. Astrophys.*, 353:1101–1114.

Botta, O., and Bada, J.L. (2002). Extraterrestrial organic compounds in meteorites. *Surv. Geophys.*, 23:411–467.

Brownlee, D.E., Tsou, P., Burnett, D., et al. (1997). The STARDUST mission: returning comet samples to Earth. *Meteor. Planet. Scie.*, 32:A22.

Campins, H., and Swindle, T.D. (1998). Expected characterisitics of cometary meteorites. *Meteoritics Planet. Scie.*, 33:1201–1211.

Cernicharo, J., Heras, A.M., Tielens, A.G.G.M., et al. (2001). Infrared Space Observatory's discovery of C_4H_2, C_6H_2, and benzene in CRL 618. *Astrophys. J.*, 546:L123–126.

Cochran, A.L., Cochran, W.D. (2002). A high spectral resolution atlas of comet 122P/de Vico. *Icarus*, 157:297–308.

Cochran, A.L., Cochran, W.D., and Barker, E.S. (2000). N_2^+ and CO^+ in comets 122P/1995 S1 (de Vico) and C/1995 O1 (Hale-Bopp). *Icarus*, 146:583–593.

Coradini, A., Capaccioni, F., Drossart, P., et al. (1997). VIRTIS: An imaging spectrometer for the ROSETTA mission. *Planet. Space Scie.*, 46:1291–1304.

Cottin, H., Gazeau, M.C., Benilan, Y., and Raulin, F. (2001a). Polyoxymethylene as parent molecule for the formaldehyde extended source in comet Halley. *Astrophys. J.*, 556:417–420.

Cronin, J.R. (1998). Clues from the origin of the Solar System: meteorites. In *The Molecular Origins of Life*. A. Brack edt. Cambridge University Press, Cambridge, UK, 119–146.

Crovisier, J. (1996). Observational constraints on the composition and nature of comet D/Shoemaker-Levy 9. In *The Collision of Comet Shoemaker-Levy 9 and Jupiter*, K.S. Noll, H.A. Weaver and P.D. Feldman edts, Cambridge University Press, 31–54.

Crovisier, J. (1998). Physics and chemistry of comets: recent results from comets Hyakutake and Hale-Bopp. Answers to old questions and new enigmas. *Faraday Discuss.*, 109:437–452.

Crovisier, J., and Bockelée-Morvan, D. (1999). Remote observations of the composition of cometary volatiles. In *ISSI workshop* Composition and Origin of Cometary Material. *Space Scie. Rev.*, 90:19–32.

Crovisier, J., Bockelée-Morvan, D., Colom, P., et al. (2004). The composition of ices in comet C/1995 O1 (Hale-Bopp) from radio spectroscopy. Further results and upper limits on undetected species. *Astron. Astrophys.*, 418:1141–1157.

de Bergh, C. (2004). Kuiper Belt: water and organics. *This volume.*

Despois, D., Crovisier, J., Bockelée-Morvan, D., and Biver, N. (2002). Comets and prebiotic chemistry: the volatile component. In *Proceedings of the Second European Workshop on Exo/Astrobiology*, ESA SP-518, 123–127.

DiSanti, M.A., Dello Russo, N., Magee-Sauer, K. et al. (2002). CO, H_2CO, and CH_3OH in comet 202 C1 Ikeya-Zhang. In *Asteroids, Comets, Meteors, ACM 2002*, ESA SP-500, 571–574.

DiSanti, M.A., Mumma, M.J., Dello Russo, N., and Magee-Sauer, K. (2001). Carbon monoxide production and excitation in comet C/1995 O1 (Hale-Bopp): isolation of native and distributed CO sources. *Icarus*, 153:361–390.

DiSanti, M.A., Mumma, M.J., Dello Russo, N., Magee-Sauer, K., and Griep, D.M. (2003). Evidence for a dominant native source of carbon monoxide in Comet C/1996 B2 (Hyakutake). *J. Geophys. Res.* 108(E6):5061.

Eberhardt, P. (1999). Comet Halley's gas composition and extended sources: results from the neutral mass spectrometer on Giotto. In *ISSI workshop* Composition and Origin of Cometary Material. *Space Scie. Rev.*, 90:45–52.

Farmer, C.B., and Norton, R.H. (1989). A high-resolution atlas of the infrared spectrum of the Sun and the Earth atmosphere from space. *NASA RP-1224*.

Feldman, P.D., Weaver, H.A., and Burgh, E.B. (2002). Far Ultraviolet Spectroscopic Explorer observations of CO and H_2 emissions in comet C/2001 A2 (LINEAR). *Astrophys. J.*, 576:L91–L94.

Fomenkova, M.N. (1999). On the organic refractory component of cometary dust. *Space Scie. Rev.* 90:109–114.

Gibb, E.L., Mumma, M.J., Dello Russo, N., DiSanti, M.A., and Magge-Sauer, K. (2003). Methane in Oort Cloud comets. *Icarus*, 165:391–406.

Gibb, E.L., Mumma, M.J., DiSanti, M.A., Dello Russo, N., and Magee-Sauer, K. (2002). An infrared search for HDO in comets. In *Asteroids, Comets, Meteors 2002*, ESA SP-500, 705–708.

Henry, F., Bockele-Morvan, D., Crovisier, J., and Wink, J. (2002). Observations of rotating jets of carbon monoxide in comet Hale-Bopp with the Plateau de Bure Interferometer. *Earth Moon Planets*, 90:57–60.

Huebner, W.F., and Benkhoff, J. (1999). From coma abundances to nucleus composition. In *ISSI workshop* Composition and Origin of Cometary Material. *Space Scie. Rev.*, 90:117–130.

Irvine, W.M., Goldsmith, P.F., and Hjalmarson, Å. (1987). Chemical abundances in molecular clouds. In *Interstellar Processes*, D.J. Hollenbach & H.A. Thronson, Jr edts, Reidel Publishing Company, 561–609.

Irvine, W.M., Schloerb, F.P., Crovisier, J., Fegley, B., and Mumma, M.J. (2000). Comets: a link between interstellar and nebular chemistry. In *Protostars and Planets IV*, V. Mannings, A. Boss and S. Russell edts, Univ. Arizona Press, Tucson, Arizona, 1159–1200.

Kawakita, H., and Watanabe, J. (2003). Fluorescence efficiencies of monodeuteriomethane in comets: toward the determination of the deuterium/hydrogen ratio in methane. *Astrophys. J.*, 583:534–539.

Kissel, J., and Krueger, F.R. (1987). The organic component in dust from comet Halley as measured by the PUMA mass-spectrometer on board VEGA 1. *Nature* 326:755–760.

Kissel, J., Krueger, F.R., and Roessler, K. (1997). Organic chemistry in comets from remote and in situ observations. In *Comets and the Origin and Evolution of Life*, P.J. Thomas, C.F. Chyba and C.P. McKay edts, Springer-Verlag, New York, 69–109.

Lellouch, E. (1996). Chemistry induced by the impacts: Observations. In *The Collision of Comet Shoemaker-Levy 9 and Jupiter*, K.S. Noll, H.A. Weaver and P.D. Feldman edts, Cambridge University Press, 213–242.

Lis, D.C., Mehringher, D., Benford, D., et al. (1999). New molecular species in comet C/1995 O1 (Hale-Bopp) observed with the Caltech Submillimeter Observatory. *Earth Moon Planets*, 78:13–20.

Lodders, K., and Osborne, R. (1999). Perspectives on the comet-asteroid-meteorite link. *Space Scie. Rev.*, 90:289–297.

Millar, T.J. (2004). Large organics in the interstellar gas and their evolution. *This volume.*

Molster, F. (2004). IDPs and implications for astrobiology. *This volume.*

Morbidelli, A., Chambers, J., Lunine, J.I., et al. (2000). Source regions and timescales for the delivery of water to Earth. *Meteoritics Planet. Scie.*, 35:1309–1320.

Mumma, M.J., Dello Russo, N., DiSanti, M.A., et al. (2001). Organic composition of C/1999 S4 (LINEAR): a comet formed near Jupiter? *Science*, 292:1334–1339.

Mumma M. J., DiSanti M. A., Dello Russo N., et al. (2003). Remote infrared observations of parent volatiles in comets; a window on the early Solar System. *Adv. Space Res.*, 31:2563–2575.

Oró, J. (1961). Comets and the formation of biochemical compounds on the primitive earth. *Nature*, 190:389–390.

Pietrogrande, M.C.,Tellini, I., Szopa, C., et al. (2003). Interpretation of chromatographic data recovered from space missions: decoding of complex chromatograms by Fourier analysis. *Planet. Space Scie.*, 51:581–590.

Rodgers, S.D. and Charnley, S.B. (2001a). Organic synthesis in the coma of comet Hale-Bopp? *Month. Not. Roy. Astron. Soc.*, 320:L61–L64.

Rosenbauer, H., Fuselier, S.A., Ghielmetti, A., et al. (1999). The COSAC experiment on the lander of the Rosetta mission. *Adv. Space Res.*, 23(2):333–340.

Soderblom, L.A., Britt, D.T., Brown, R.H., et al. (2004). Short-wavelength infrared (1.3–2.6 μm) observations of the nucleus of comet 19P/Borrelly. *Icarus*, 167:100–112.

Szopa, C., Sternberg, R., Raulin, F., and Rosenbauer, H. (2003). What can we expect from the in situ chemical investigation of a cometary nucleus by gas chromatography: First results from laboratory studies. *Planet. Space Scie.*, 51:863–877.

Thiemann, W.H.-P., Rosenbauer, H., and Meierhenrich, U.J. (2001). Conception of the "chirality-experiment" on ESA's mission Rosetta to comet 46P/Wirtanen. *Adv. Space Res.*, 27(2):323–328.

Thomas, P.J., Chyba, C.F., and McKay, C.P. (Edts) (1997). *Comets and the Origin and Evolution of Life*. Springer, New York.

Whipple, F.L. (1950). A comet model. I. The acceleration of comet Encke. *Astrophys. J.*, 111:375–394.

Chapter 9

KUIPER BELT: WATER AND ORGANICS

Catherine de Bergh
LESIA Observatoire de Paris, Meudon, France
catherine.debergh@obspm.fr

Abstract

The recent discovery of a large number of Solar System bodies that orbit the Sun beyond Neptune has opened new possibilities to study primordial matter and processes in the early solar nebula. Indeed, Kuiper Belt objects are the most primitive solid bodies in the solar system. Due to their intrinsic faintness and remoteness, these objects are very difficult to study, and progress is slow. I will review some of the current knowledge about these bodies, with emphasis on their water and organics contents, and the various processes that could have altered their surfaces. Comparisons with other small Solar System bodies will be made. Some of the brigthest objects in the Belt, which are the only ones for which detailed physical studies can be made, have been discovered only recently, and it is postulated that many more await discovery. Furthermore, major on-going or planned ground-based surveys as well as planned space missions should bring important new information on Kuiper Belt Objects. Prospects are therefore very good, and major progress is expected in the near future.

Keywords: Kuiper Belt Objects, Centaurs, Surface Composition, Ices, Spectroscopy

1. INTRODUCTION

The Kuiper Belt is the region of the solar system where small bodies called Kuiper Belt Objects (also called Edgeworth-Kuiper Belt Objects or Transneptunian Objects) are now discovered. These bodies orbit the Sun beyond Neptune. The existence of a belt of objects beyond Neptune had been suspected for a long time. There were two main reasons for that: first, it was hard to imagine that the solar nebula would end at Neptune, secondly a reservoir for Jupiter-family comets different from the Oort cloud was needed. Jupiter-family comets have

P. Ehrenfreund et al. (eds.), Astrobiology: Future Perspectives, 205–243.

low orbital inclinations that could not be explained if the comets had originated in the Oort cloud, the postulated reservoir for long-period comets, which must have a spherical shape.

The first Kuiper Belt object was discovered in 1992 (Jewitt and Luu 1993). More than 850 objects have now (Jan. 2004) been identified. The diameters of these objects range from about 50 km to about 1,200 km. From its orbital properties, it is now clear that Pluto, with its very eccentric and strongly inclined orbit relative to the ecliptic plane actually belongs to this recently identified population of Solar System objects.

As these objects are located in very cold regions of the Solar System, they have been much less modified since their origin than objects closer to the Sun. Furthermore, as they are small bodies, they are much less prone to internal evolution. It is thus hoped that their study can provide important information about the original composition and conditions in the solar nebula at large distances from the Sun, and about the formation and evolution of the solar system (accretion, collisions, chemical processes,...), and, in particular, the formation of the giant planets. Furthermore they may have contributed to bringing water and other biogenic material to the Earth.

A decade of photometric and spectroscopic studies of these objects has revealed that Kuiper Belt Objects (KBOs) have very diverse surfaces, something which was completely unexpected. These studies have also shown the presence at their surfaces of two very important ingredients for life on Earth, namely water ice and organics. Water ice is not always detected, however. Concerning organics it is still difficult to know which ones are present. It is important to find out which material is pristine and which material is processed. And where should we look to find the most pristine material? State of the art and comparison with other small bodies of the solar system are presented. In particular, I include the Centaurs which are most probably escaped bodies from the Kuiper Belt and that are now on unstable orbits between Jupiter and Neptune. I will then consider what progress can be expected in the next decade.

2. STRUCTURE AND EXTENT OF THE BELT

Although Kuiper Belt Objects were searched for since 1988 (as soon as large format and very sensitive CCDs became available), the first one was discovered only in 1992 (Jewitt and Luu 1993). These objects are difficult to find because they are small and located far away from the Sun. Furthermore, they appear to have dark surfaces. Their magnitudes are in general higher than 19. It is not sufficient to observe the objects once to be able to find them again. Indeed their motion is so slow that a series of observations spread over several months and repeated for at least two oppositions (an object is at opposition when, as seen from the Earth, it is opposite to the Sun) are required to secure their orbits. Objects for which the follow-up is not done properly are lost. More than 850 have now been identified (diurnally updated lists of the objects can be found at the web addresses http:cfa-www.harvard.edu/cfa/ps/lists/TNOs.html and http:cfa-www.harvard.edu/cfa/ps/lists/Centaurs.html).

Given the sky coverage achieved so far at the required magnitudes to detect such objects, it has been estimated that there could exist 100,000 objects with diameter larger than 100 km and maybe 800 million objects with diameter larger than 5 km (Jewitt 1999). The population of Kuiper Belt Objects must therefore exceed that of asteroids in the main-belt. Nine binary objects (besides Pluto-Charon) are now known.

Orbits of some of the objects identified so far are shown in Figure 9.1.

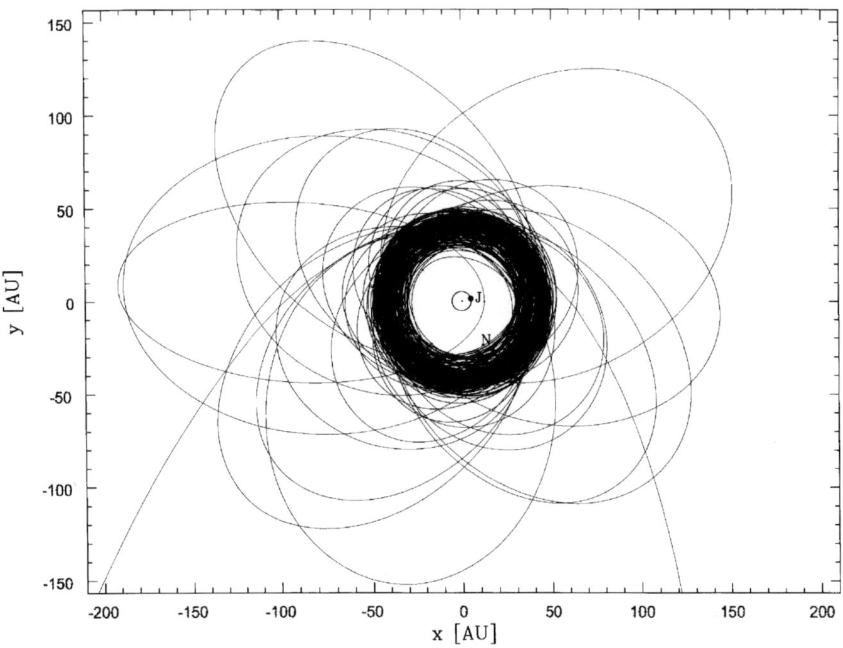

Figure 9.1. Orbits of some KBOs projected on the ecliptic plane. The objects with the most eccentric orbits are Scattered Disk Objects. The orbits of Jupiter (J) and Neptune (N) are also indicated (from D. Jewitt's web page: `http://www.ifa.hawaii.edu/faculty/jewitt/`).

There are at least three main dynamical classes of objects. The so-called **Resonant Objects** are located in some major mean motion resonance with Neptune (Figure 9.2). They are mostly present in the 3:2 resonance zone (semi-major axis: 39.4 AU), as is Pluto. These objects cover two orbits around the Sun while Neptune (which has an orbital semi-major axis of 30 AU) covers three orbits. Such objects are called Plutinos. Other resonances are also populated: the 4:3, 5:3 and 2:1, in particular, (but also the 5:2 and 7:4 resonances). Most resonant objects have very stable orbits. Their eccentricities can be quite large.

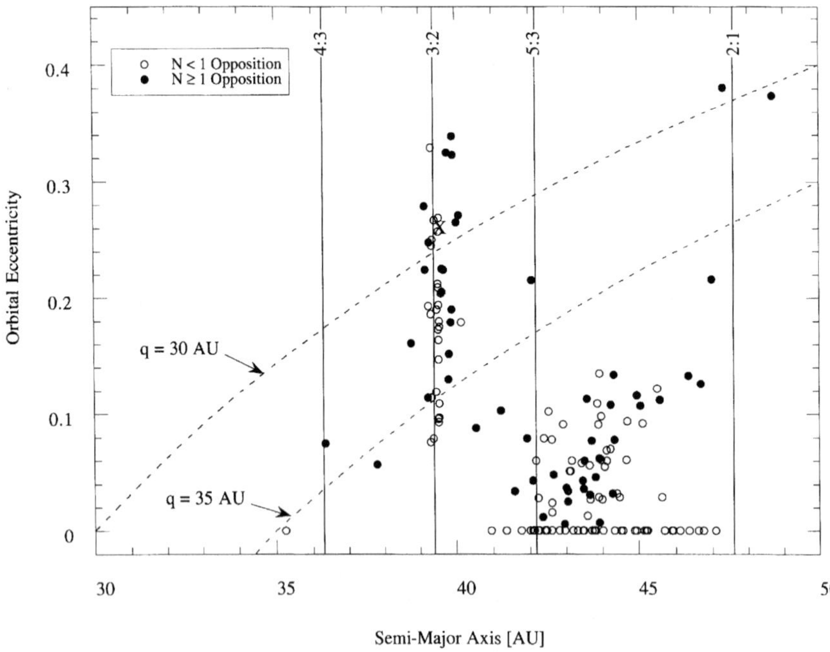

Figure 9.2. Orbital semi-major axis versus eccentricity for Classical and Resonant Objects (Jewitt and Luu 2000). Copyright 2000 The Arizona Board of Regents. Reprinted by permission of the University of Arizona Press.

The **Classical Objects** are objects with nearly circular or moderately eccentric orbits that are not in resonance with Neptune. Their orbital semi-major axes range between 30 and 50 AU (Figure 9.2). As they have relatively small eccentricities, they always stay far from Neptune, even at perihelion. They are a priori the objects that best represent remnants of the solar nebula accretion disk. However, some have quite

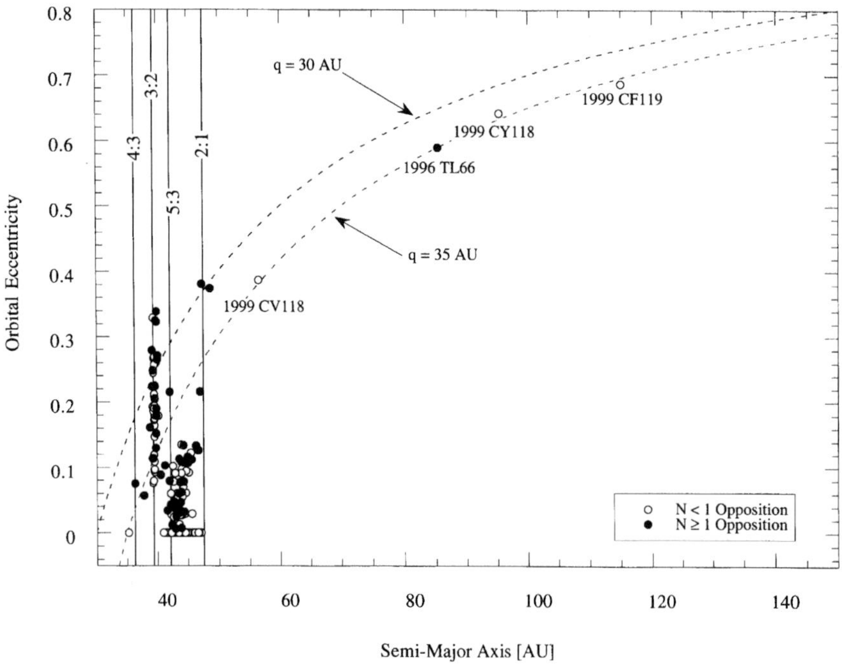

Figure 9.3. Orbital semi-major axis versus eccentricity for Scattered Disk Objects (Jewitt and Luu 2000). Copyright 2000 The Arizona Board of Regents. Reprinted by permission of the University of Arizona Press.

high orbital inclinations, which is somewhat contradictory. The so-called **Scattered Disk Objects** have very eccentric (Figure 9.3) and inclined orbits. They have perihelia close to that of Neptune (at 30 AU). The objects that constitute the Scattered disk (objects that are still weakly interacting with Neptune) most probably originated in the Classical Belt. Their orbits must have been nearly circular at their formation, and then evolved dynamically to their current configuration. Aphelia of the objects discovered so far can be as high as several hundreds of AU (415 AU for 2001 FP_{185}, which has a perihelion at 32 AU). The first one was discovered only in 1996.

Other objects are found now that do not belong to any of these three categories. This is the case, for instance, of a few objects (currently five) with semi-major axes higher than 50 AU and highly eccentric orbits that are not considered as part of the scattered disk because their orbital parameters indicate that they never had a close encounter with Neptune (they have perihelia outside the domain controlled by Neptune). 2000

CR 105 (semi-major axis: 225 AU, perihelion: 45 AU) is one of them. These objects are called "extended scattered disk objects" (Gladman et al. 2002). According to Morbidelli et al. (2004), they may have formed much closer to the Sun. A newly discovered object, 2003 VB_{12} (provisionally called Sedna), which has a very large perihelion distance (about 76 AU) and an aphelion around 990 UA, may also belong to this category, although this object could also be considered as a member of the inner Oort cloud (Brown et al. 2004). 2003 VB_{12} (currently at 90 AU from the Sun) is most probably the largest known object in the outer Solar System after Pluto.

No Classical objects have been found with semi-major axes in excess of 50 AU, although their detection should be feasible. It is therefore believed that there is an edge around 50 AU. But it may be just a hole. The Belt could extend much further outward. Furthermore, the Classical Belt is not at all flat. Indeed, as already mentionned, many of the objects have quite large orbital inclinations.

To explain the formation of bodies as large as Pluto and the other large KBOs at such large distances from the Sun, it is generally assumed that the Kuiper Belt was much more massive in the past (see, e.g., Stern and Colwell 1997). Another argument in favor of this hypothesis is that the extrapolation of the surface density of solids incorporated in the giant planets to the Kuiper Belt leads to a much higher mass than currently estimated in the Belt (about 0.1 the mass of the Earth, Luu and Jewitt 2002). Collisions would therefore have been much more frequent at that time. The current populations in the Classical Kuiper Belt and Scattered Disk may correspond only to about 1 per cent of the original populations. Loss of mass could have occurred from the loss of large objects due to dynamical instabilities or through collisions after the mass had been converted to small dust grains and then dragged toward the Sun by Poynting-Robertson effect or by the solar wind, or chased outwards by radiation pressure.

The dynamical structure of the Kuiper Belt presents characteristics that are much more complex than expected. The gravitational influence of Neptune and, to a lesser degree, the other giant planets, is not sufficient to account for the observed dynamical structure, and, in particular, the high orbital eccentricities and inclinations of many objects, a possible edge around 50 AU, and the existence of "extended scattered disk objects". Other factors must have played a role, such as: the migration of the giant planets, and particularly Neptune, during their formation, the invasion by large objects that were remnants of giant planet formation, or close stellar encounters if the Sun was formed in a cluster or association of stars (which is very likely). Saturn, Uranus and Neptune would have migrated outwards, Jupiter inwards. Neptune's migration could have been as large as 8 AU. Collisions must have also participated in the sculpting of the Belt (see Morbidelli et al. 2004 for a review). A major test for understanding the primordial evolution of the outer Solar System will be to explain how objects with large perihelia such as 2000 CR_{105} and 2003 VB_{12} have acquired their strange orbits (Morbidelli and Levison 2004).

3. RELATED POPULATIONS OF OBJECTS: CENTAURS AND JUPITER-FAMILY COMETS

The study of Kuiper Belt Objects is very difficult due to their faintness. This is why it is important also to consider populations that most probably originated in the Kuiper Belt and are more accessible to observations. This is the case, in particular, of the Centaurs. The Centaurs are objects that orbit among the giant planets. Dynamical studies indicate that most of them probably come from the Kuiper Belt, while the others may come from the Oort cloud. It is worth mentionning here that Oort cloud comets were not formed in situ, but probably originated in the region located between Uranus and Neptune.

The best-studied Centaur object is 2060 Chiron. Chiron was discovered in 1977. Its orbit crosses Saturn's orbit (Figure 9.4). Chiron's perihelion is at 8.45 AU, while its aphelion is at 19 AU (just inside Uranus orbit). Orbital and physical characteristics of some Centaurs are listed in Table 9.1. Chiron was first classified as an asteroid, then as a comet when a coma was detected around the object. Emissions due to CN and maybe CO have been seen (Bus et al. 1991, Womack and Stern 1999). As Centaurs are not trapped in resonance with a planet, their orbits are unstable. Lifetimes of these objects probably do not exceed a few million years. About 50 are currently known (following the Minor Planet Centaur classification for Centaurs). There exists three more objects (29P/SW1, 39P/Oterma and C/NEAT(2001 T4)) that are also classified as Centaurs (as well as comets) but are not part of this list. Statistics based on existing sky surveys indicate that there could be about 100 Centaurs with diameters larger than 100 km and 10 million with diameters larger than 2 km (Sheppard et al. 2000).

Other dynamical studies have shown that the orbital characteristics and the number of Jupiter-family comets (which are comets with orbital periods less than 20 years gravitationally controlled by Jupiter) that we detect now could be explained by an origin in the Kuiper Belt. Centaurs may in fact constitute an intermediate family between Kuiper Belt

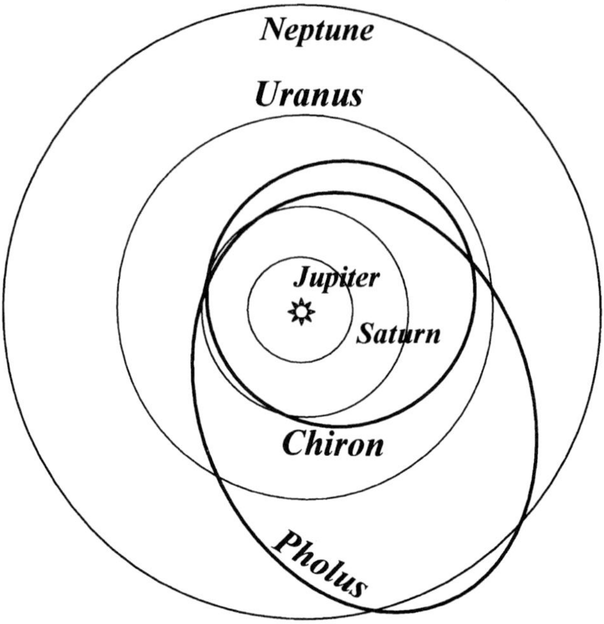

Figure 9.4. Orbits of the Centaurs Pholus and Chiron.

Table 9.1. Orbital and physical characteristics of some Centaurs (adapted from Barucci et al. 2003)

Name	q(AU)	Q(AU)	e	i(deg)	Diameter (km)	Albedo
2060 Chiron	8.44	18.85	0.38	6.9	180±10[1]	0.15±0.05[1]
					180±10[2]	0.15±0.05[2]
5145 Pholus	8.70	32.12	0.57	24.7	190±22[3]	0.04±0.03[3]
7066 Nessus	11.77	37.21	0.52	15.6	~75	–
8405 Asbolus	6.84	29.00	0.62	17.6	66±4[2]	0.12±0.03[2]
10199 Chariklo	13.07	18.64	0.17	23.4	302±30[1]	0.045±0.010[1]
					273±19[4]	0.055±0.008[4]
10370 Hylonome	18.86	31.38	0.25	4.1	~150	–
52872 Okyrhoe	5.81	11.00	0.31	15.6	~35	–
54598 2000 QC_{243}	13.14	19.80	0.20	20.8	~190	–

[1]Jewitt and Kalas (1998)
[2]Fernandez et al. (2002)
[3]Davies et al. (1993)
[4]Altenhoff et al. (2001)

Objects and Jupiter-family comets, although the Centaurs that have been discovered so far appear in general larger than nuclei of Jupiter-family comets. Jupiter-family comets (dynamical lifetime: about 10^5 years) could be small primitive KBOs or collisional remnants of larger KBOs, and would be detected only when they come closer to us. Times to change from a KBO orbit to a Jupiter-family orbit are of the order of 1 to 10 million years (the average Centaur lifetime).

Being closer to the Sun than Kuiper Belt Objects, these related populations are in principle easier to study. However, comets are much more evolved bodies, due to their closer proximity to the Sun, and their nuclei (which are what we would like to probe) are not so easily accessible to Earth-based investigations as they are very small and relatively far away when not surrounded by a coma. Furthermore, the surfaces of cometary nuclei would have been affected by past activity, and may not be as representative of KBOs surfaces. However, observations of their gaseous and dusty emission can provide very useful information on interiors of KBOs. Dedicated space missions to Jupiter-family comets have just been achieved, are in progress or are planned, so our knowledge on the composition of these comets should considerably improve in the near future (see Crovisier 2004).

4. ALBEDOS, SIZES, ROTATIONAL PERIODS, AND SHAPES OF KUIPER BELT OBJECTS

The knowledge of the surface albedo (reflecting power of the surface) is very important not only to get information about the nature of the surface material but also to estimate the sizes of the objects. Direct size measurements have been obtained for only one object, Quaoar, using the Hubble Space Telescope (Brown and Trujillo 2004a). Its diameter (1260 $\pm$190 km), is about that of Charon (Charon's diameter is of the order of 1200 km).

The albedo measurements require combined measurements of the light reflected by the object and of its thermal radiation. High albedo objects are optically bright, but do not emit much thermal radiation. Low albedo objects are optically dark but emit more thermal radiation. Some model is used to convert thermal radiation to effective surface temperature (Spencer et al. 1989). Given the expected surface temperature of these objects (30-50 K) the peak of the thermal blackbody radiation is around 60-80 μm, a wavelength inaccessible to Earth-based telescopes.

The direct size measurement of Quaoar seems to be corroborated by a measurement of the thermal radiation of the object using the IRAM radiotelescope (Bertoldi, private communication). There have been only very few albedo measurements of KBOs, as thermal radiation measurements for such objects are difficult. Jewitt et al. (2001) measured an albedo (in the R-band) of $0.07^{+0.03}_{-0.017}$ for Varuna (corresponding to a diameter of 900 $^{+129}_{-145}$ km), while Lellouch et al. (2002) got slightly lower albedo values (R albedo of $0.049^{+0.029}_{-0.013}$, and V albedo of $0.038^{+0.022}_{-0.01}$) for the same object, corresponding to a diameter of 1060 $^{+180}_{-220}$ km. For 2002 AW_{197}, Margot et al. (2002) measured an albedo around 10% (with a corresponding diameter of 886^{+115}_{-131} km). These diameters have been subsequently slightly revised by Altenhoff et al. (2004). Altenhoff et al. also obtained an albedo of about 5% for 1999 TC_{36}, leading to a diameter of 609^{+93}_{-47} km for this object if it was single. However, as this object is a binary with a difference in brightness of about 2 magnitudes between the two components, the components' diameters are estimated to be about 566 km and 225 km, respectively (Altenhoff et al. 2004).

When albedo measurements are not available, an albedo of 0.04 typical of Centaurs presumably inactive (Table 9.1) and of cometary nuclei is used to estimate the sizes of the objects. Albedos of cometary nuclei are of the order of 0.02-0.06 in the V band (Lamy et al. 2004). Pluto has an albedo of 0.6, Charon an albedo of 0.38. But this does not seem to be the case for other KBOs which appear to be darker. The recent measurements have shown however that some KBOs can be brighter than cometary nuclei (see above).

Rotation periods are obtained by studying variations with time of the luminosity by photometry, very often R photometry since this is the most favorable wavelength (less sky contamination, and, in general, more flux than in V because of the redness of the objects). Photometric lightcurves have been obtained for a number of the brightest objects. Some of the large objects lightcurves have very high amplitude variations, much larger than for asteroids. All lightcurves seem to be due to irregular shapes rather than to albedo variations. According to Sheppard and Jewitt (2002), only 1999 KR_{16} could have a lightcurve affected by surface albedo variations.

The large amplitude variations indicate that objects in the Kuiper Belt can have quite elongated shapes. Assuming Jacobi ellipsoids, objects like Varuna, 2000 GN_{171} and 1998 BU_{48} have a:b:c of the order of 1.5-1.8:1:0.7 (where a, b and c are the three axes of the ellipsoid, with $a>b>c$) (Sheppard and Jewitt, 2002).

Measured rotation periods range from about 6 to about 15 hours. In many cases, it has not been possible to measure them because the amplitude in magnitude is very small, which must indicate a regular shape (spherical, or, possibly, ellipsoidal if viewed pole-on). Measured rotational periods of KBOs are in the same range as those of Centaurs (see Barucci et al. 2004). The large KBOs for which rotational periods could be measured have quite rapid rotation periods.

5. MASS, DENSITIES, AND INTERIOR MODELS OF KBOs

What indication do we have on their bulk composition and interior structure? Size combined with mass provide density. However, most of these parameters are still unknown for KBOs.

Nine binaries have now been detected, in addition to Pluto-Charon. The pairs have characteristics quite different from pairs observed in the asteroid belt, with individual bodies of more comparable sizes and more spatially separated. Mass estimates would be possible from the study of the binaries. However, the first binary has been discovered only in 2001, and good trajectories are needed to get mass estimates, which requires many observations.

What kind of density could we have for these objects? McKinnon (2002) has made the following predictions for KBOs, based on what we know about Pluto and Charon and solar nebula models. Pluto-Charon and Triton densities are of the order of 2000 kg/m^3. This corresponds to a rock/ice ratio of 70/30. This is in agreement with models of the outer solar nebula, assuming that all C is in CO. But not all nebular C is in CO. In particular, Halley's comet observations have shown that about 80% of the carbon is in refractory solids (CHON) and 20% is in the gas (dominated by CO and CO_2). Therefore, the rock/ice ratio in the solid matter in the Kuiper Belt must have been less than 70/30, and, according to McKinnon and others, Pluto-Charon and Triton would have lost ices. KBOs should be rock + organics rich. McKinnon estimates that they could have a rock/ice ratio of 70/30, and, with 20% total organic fraction, the rock/ice+organics would be 58/42. This leads to a solid or grain density of 1640 kg/m^3.

What can we infer from our current knowledge about these objects? Lower limits on the densities of a few objects and an estimate of the density of Varuna have been obtained from their shapes and rotation periods (Sheppard and Jewitt 2002; Jewitt and Sheppard 2002). According to Jewitt and Sheppard, the most plausible explanation for the large amplitudes of the lightcurves (0.4 mag. for Varuna) and their rapid rotations (period of about 6 hours for Varuna) is that the objects are elongated by their rapid rotation. Assuming that the objects are in equilibrium (Jacobi or McLaurin ellipsoids) the density can be derived. The estimated density for Varuna is 1000 kg/m^3. Since it is very unlilely that

Varuna is made of pure ice, but must rather be made of a mixture of ice and rocks, Varuna must be porous (volume-average porosity of 20-25%, for a cosmochemically plausible rock-ice mixture).

According to McKinnon (2002), Merk and Prialnik (2004), Choi et al. (2002), De Sanctis et al. (2001), and De Sanctis et al. (2002), the largest KBOs could have gone through some thermal evolution. Their thermal evolution would be dominated by radiogenic heating (either the short-lived radioactive isotope ^{26}Al and/or the long-lived ones, mostly ^{238}U, ^{235}U, ^{232}Th and ^{40}K). As an example, the models developed by McKinnon, who favors evolution through heating by long-lived radionuclides, indicate that the large KBOs could have gone through massive conversion of interior amorphous ice to crystalline ice, a transition which is strongly exothermic. This would have led to central porosity reduction and maybe some early geologic activity. The largest bodies could have remained geologically active for quite a long time, maybe hundreds of millions of years. And these bodies could have been subjected to partial differentiation, outgassing of very volatile material, and, maybe even cryogenic volcanism. On the contrary, bodies with diameters less than about 340 km (but this critical size strongly depends on the thermal conductivity, which is a poorly constrained parameter, and on the amount of radiogenic elements initially present) would have kept their cosmochemically primordial surfaces. According to McKinnon, large KBOs would be "cousins to midsized icy satellites, not just big comets". Indeed, many of the icy satellites of Saturn and Uranus have densities near that of water ice and may be porous. The smallest bodies, including those that probably form the Jupiter-family comets (and are not collisional fragments of larger KBOs), would be instead primordial.

Thermal models developed by Choi et al. (2002), taking into account the combined effects of solar insolation and radiogenic heating (by short-lived and long-lived isotopes), indicate that long-term evolution of bodies with diameters ranging between 20 and 1000 km would lead to a stratified internal composition (as also found by De Sanctis et al. 2001 and De Sanctis et al. 2002), with the outer layers being less affected than the inner part by evolution. In these models, the initial composition is assumed to be a porous mixture of ices (H_2O, CO and CO_2) and dust. Choi et al. conclude that the large KBOs have probably lost all volatiles that sublimate below 40-50 K, and may have partially lost less volatile ices. Their models indicate also a porosity that varies with depth, but, maybe, not monotonically, with some "weak" regions (regions of sharp density variations) at depths ranging from a few hundred meters to 1 km in size. They suggest that this could explain the small size of Jupiter-family comets, which are possible fragments of large KBOs.

6. THE COLORS OF CENTAURS AND KUIPER BELT OBJECTS

Colors are obtained from broad-band photometric measurements made with different filters (B, V, R, I, J, H, K). The colors (B–V, V–R, R–I...) are then compared to solar colors, as the radiation received from these objects in the visible and near-infrared ranges is solar radiation reflected by the objects' surfaces. For example, a "red" object is an object which has a color redder than that of the Sun.

A much larger sample of objects have been studied by photometry than by spectroscopy. For the large majority of these objects, only visible photometry exists, because the objects are too faint for good infrared magnitude measurements. These studies indicate a large diversity of colors (see, e.g., Luu and Jewitt, 1996, Tegler and Romanishin 2000, Doressoundiram et al 2002., McBride et al. 2003). This has come as a big surprise as it was expected that all objects, being formed in more or less the same region of the solar system, and having spent all their existence in the same space environment, would have similar colors.

The reddest objects in the solar system are found among Centaurs and KBOs. One does not find such red colors on cometary nuclei. And there is a comparable range of colors for KBOs and Centaurs, which, according to some, supports a Kuiper Belt origin for Centaurs. However, the fact that one does not find very red objects for cometary nuclei (but only few of them have been observed) poses a problem concerning the origin of short period comets. But as the comets have come much closer to the Sun, they are more evolved objects. Furthermore, we are sampling small objects with comets. No such small Centaurs have been detected, although they certainly exist. The Centaurs and KBOs can be much redder than D and P-type asteroids, and they present a much larger diversity in color. Peculiar objects in the Kuiper Belt are therefore the very red objects. Objects with colors similar to those of more neutral Centaurs and KBOs can be found in other populations. But we do not know if they have the same albedo. So, until we have a more complete data base on albedo measurements, we will not be able to say how similar they are.

Tegler and Romanishin claimed in 1998 and then again in 2000, and later in 2003 (Tegler and Romanishin, 1998, 2000, 2003) that there exist two classes of objects, with one class of neutral-moderately red objects and another class of more red objects. But this does not seem to be confirmed by more recent work (Figure 9.5). It may be because Tegler and Romanishin included both Centaurs and KBOs in their sample. However, such a dichotomy may be true for Centaurs (Peixinho et al. 2003).

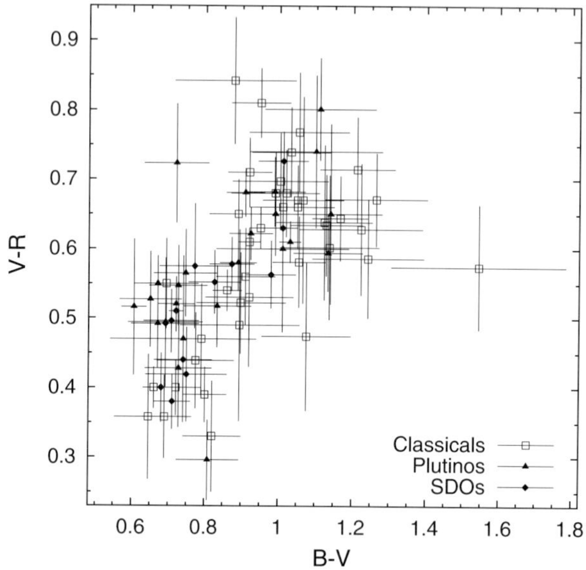

Figure 9.5. Color indices of KBOs from observations at CFHT (Doressoundiram et al. 2002) and ESO–VLT (Peixinho et al. 2003).

Some of this color diversity could be biased because of surface heterogeneities. We know that Pluto has a very heterogeneous surface. In addition, photometric variations have been found for some objects for which repeated observations were made. As an example, 1995 SM_{55} shows V–R variability (Doressoundiram et al. 2002). Furthermore, in a very limited number of cases, spectra of a given object taken at different times look different, as we will see further (we see differences between spectra recorded by different teams at different times as well as differences between spectra recorded by the same team at different times), and this is most probably due to surface heterogeneities. In other cases, photometric changes have been seen that are certainly due to sporadic comet-like activity (case of Chiron) or may be due to sporadic activity (case of 1996 TO_{66}; Hainaut et al. 2000). However, for most objects, repeated observations provided the same photometric results (see Doressoundiram et al. 2002). So the important diversity of colors of the objects is real, it is not due to important surface heterogeneities.

Do these objects have different original composition because they come from different regions of the solar nebula? For this to occur in these low temperature regions of the solar system, widely different formation regions would be required to have a significant effect on the composition. And this is not very likely. However, it may be true for some objects, and efforts to retrace the past dynamical history of the different classes of objects are essential. Or could different objects have suffered sufficiently different surface alterations that the resulting colors could be so dissimilar? We will get back to this issue further.

7. SPECTROSCOPY AND SURFACE MODELLING

Only a limited number of objects can potentially be explored by spectroscopy with the instrumentation at our disposal. Indeed, only a few Centaurs and KBOs are brighter than 21st magnitude (in V). Large telescopes are needed, as well as very sensitive spectrometers operating at low or medium resolution. This is why the list of telescopes and instruments used so far for this type of study is also limited, as well as the list of teams involved. A group of Europeans have joined to propose a Large Programme at the ESO Very Large Telescope (four 8-m telescopes) in Chile for both photometry and spectroscopy in the visible (FORS instrument) and near infrared (ISAAC instrument) ranges, and most of the spectroscopic work done so far in the visible range comes from this team. The NIRC instrument at the Keck 10-m telescope (Hawaii) has provided a large portion of the infrared spectra (but not all of them have been published). Other instruments that have been used for infrared spectroscopy are CISCO at the 8-m Subaru telescope (Hawaii) and NICS at the 4-m TNG telescope (Canary Islands). The infrared spectra are currently limited to wavelengths shorter than 2.5 μm. Many of them are low S/N spectra. In only very few cases (5145 Pholus, 8405 Asbolus, 2060 Chiron, 2000 EB_{173}, ...), the same object has been observed by different teams.

When spectroscopic studies of KBOs started, people expected to see many signatures of ices, as had been discovered in spectra of Pluto, and Triton, a satellite of Neptune (which may be a captured KBO). The Pluto spectrum is very rich (Figure 9.6) and dominated by absorptions of CH_4 ice. Ices of N_2, CO, and probably also H_2O, have also been detected (Owen et al. 1993, Cruikshank et al. 1997, Grundy and Buie 2002; see also Roush and Cruikshank 2004).

In addition, recent studies extending the spectrum of Pluto beyond 2.5 μm have revealed still unexplained absorptions in the 2.8-3.1 μm

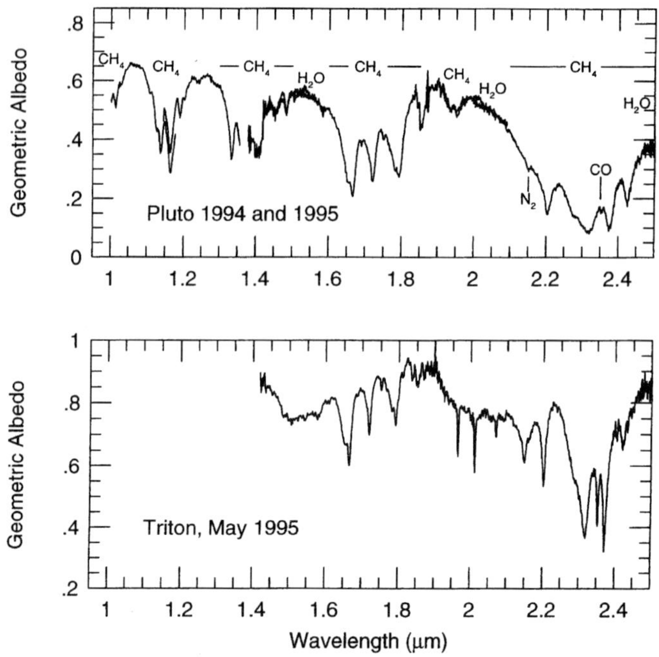

Figure 9.6. Near-infrared spectrum of Pluto (top) and Triton (bottom). The molecular identifications shown for Pluto are the same as for Triton, except that Triton shows three sharp bands of CO_2 in the region 1.96 – 2.07 μm (Cruikshank et al. 1998b). Reproduced with kind permission of Kluwer Academic Publishers.

range and the 3.6-4.1 μm range. The former could be due to some water ice + tholin mixture, while the later could be due to CO_2 and/or SO_2 ices (Grundy et al. 2002). But there are other possibilities.

There are a lot of similarities between Pluto and Triton. The same ices, N_2 ice, CH_4 ice, CO ice and H_2O ice, have been detected at the surface of both objects. In addition, CO_2 ice has been seen on Triton only (Figure 9.6; Cruikshank et al. 1993). There are also unidentified broad absorption features further in the infrared (Grundy et al. 2002).

Spectra of Charon are not as rich in signatures. It is essentially only H_2O ice which has been detected. From the prominence of a feature at 1.65 μm, it is inferred that the water ice at the surface of Charon is mostly in the crystalline state (Figure 9.7; Brown and Calvin 2000). In addition, a small feature has been seen at 2.2 μm (only on part of the satellite) that could be due to a combination of ammonia and ammonia hydrate ice (Brown and Calvin 2000; Brown 2002; see also the discussion in Roush and Cruikshank 2004).

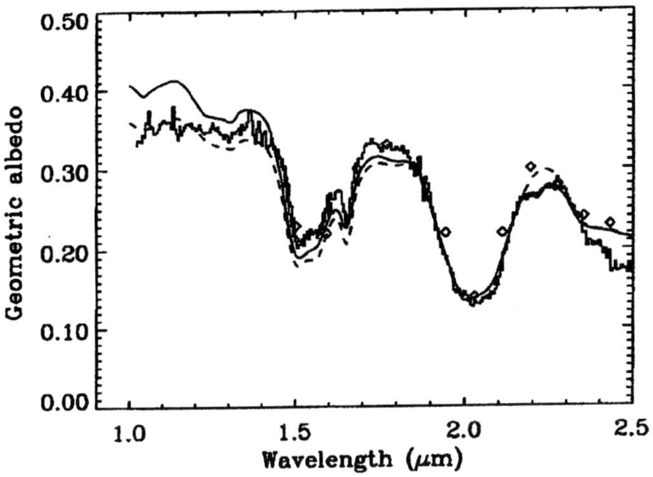

Figure 9.7. Near-infrared spectrum of Charon (Brown and Calvin 2000). Reprinted, with permission, from Science, 287, 2000. Copyright 2000 AAAS.

For other KBOs or Centaurs, until very recently, only very few spectral features had been detected. Detected features were essentially water ice, maybe methanol (or a similar molecule) ice for the Centaur Pholus (Cruikshank et al. 1998a), possibly olivine (also for the Centaur Pholus), and, maybe, hydrated silicates at the surface of two Plutinos (Resonant Kuiper Belt Objects) and one Scattered Disk Object, from signatures in the visible and in the near-infrared (see further). A set of unpublished spectra presented at a meeting (M.E. Brown) present in addition a continuum absorption beyond 2.2 μm, as is seen for other objects like comet Hale-Bopp when it was still at 7 AU from the Sun (Davies et al. 1997).

A very recent work on the brightest known KBO, 50000 Quaoar, reveals a quite rich near infrared spectrum with weak absorptions that may be due to methane ice, in addition to the more prominent absorptions of water ice (Brown and Trujillo 2004b). Quite remarkably (see below), the water ice seems to be, as for Charon, mostly in the crystalline state. In addition, there are a few weak unidentifed features in this spectrum that would require higher signal-to-noise observations for confirmation (and, eventually, identification).

Many of the spectra of Centaurs and KBOs are featureless, but they have variable spectral shapes. Many of the near-infrared spectra have quite low signal-to-noise ratios, however. This is why spectrophotometric measurements, which are available for a much larger sample of objects, have also been modelled to get information on surface properties (see Cruikshank and Dalle Ore 2004).

Why are the spectra so different from those of Pluto (and Triton)? Probably because Pluto and Triton are much larger bodies and are therefore more able to retain a subliming atmosphere. The surfaces of Pluto and Triton are continuously recovered by volatile compounds. Volatile transport is enhanced due to important seasonal effects (Yelle et al. 1995, Spencer et al. 1997).

The general shapes of the spectra provide additional constraints. Different types of radiative transfer models have been used to model the spectra. To get the best constraints from spectral shapes, it is very important to cover the widest possible spectral range. And this has been possible so far for only a few objects (see Figure 9.8).

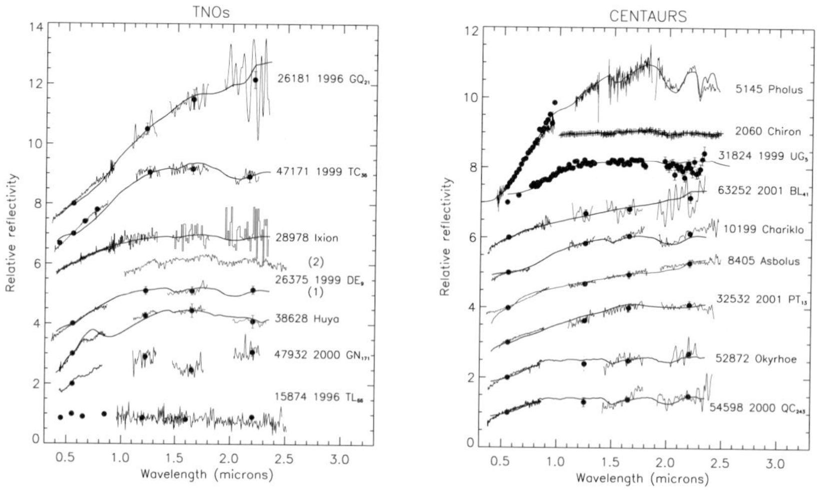

Figure 9.8. Combined visible and near-infrared spectra of TNOs (or KBOs) and Centaurs for which models have been made (from Dotto et al. 2003a).

Different radiative scattering models have been used: essentially Hapke's model, a slightly modified Hapke's model (Douté and Schmitt 1998), and Shkuratov's model (see Chapter by Roush and Cruikshank for more details). Very often, one needs to include in the models grains of very small sizes to improve the fits. And this violates Hapke's model domain of validity. In these cases, Shkuratov model is preferred, although it has other limitations. And different types of mixtures are also used in the models (areal, intimate, ...). This is also discussed in Roush and Cruikshank (2004). Details on the components included in the models presented in Figure 9.8, and references, can be found in Dotto et al. 2003a.

We have evidence for surface heterogeneities revealed by spectroscopic studies. The most obvious cases are: the Centaurs 2001 PT_{13} (Barucci et al. 2002) and 1999 UG_5 (Bauer et al. 2002) for which very different spectra were obtained when different parts of the objects were observed. The Centaur 10199 Chariklo also seems to have an heterogeneous surface (Dotto et al. 2003b), as is the case also for the KBO 1996 TO_{66} (Brown et al. 1999).

8. ORGANICS

As we have seen, only very few distinct features that could be due to carbon compounds are detected. Methane ice was detected on Pluto and Charon, and may be also present on Quaoar (Brown and Trujillo, 2004b). In addition, there is an absorption at 2.27 μm in the spectrum of Pholus (Figure 9.8) that is attributed to frozen methanol and/or a photolytic product of methanol with small molecular weight.

A large part of our current ideas on the surface composition of these objects, and particularly on their organic content, therefore relies on spectral slopes, which does not provide very strong constraints. In the visible range, colors vary from neutral to very red. In the infrared, the slope may continue to rise or may flatten when going to longer wavelengths.

People have looked for mixtures of plausible constituents (ices, organics, silicates) that could account for the observed slopes. The spectral slopes of red objects have been reproduced by including organics such as tholins or kerogens in mixtures. Amorphous carbon, which is spectrally neutral (Figure 9.9), has also been used to lower the albedo.

Some tholins have been included in models of Pluto to account for a slight red slope in the visible, and have been considered to explain the dark regions on Pluto (Grundy and Fink 1996). In addition, tholins (Titan, Triton or ice tholins) have been used to model the general shapes of the spectra of the Centaurs 5145 Pholus, 1999 UG_5, 8405 Asbolus, 2001 PT_{13}, and 10199 Chariklo, as well as spectra of the TNOs 1999 TC_{36}, 28978 Ixion, and 38628 Huya, while kerogens have been used to model the general shapes of the spectra of the Centaurs 52872 Okyrhoe and 2000 QC_{243} (Figure 9.8).

Why these choices? By analogy with dark asteroids (such as D-type asteroids), with meteorites or with planetary satellites. The tholins that have been used in the models are either Titan tholins, Triton tholins, or ice tholins, which are the only ones for which optical constants are

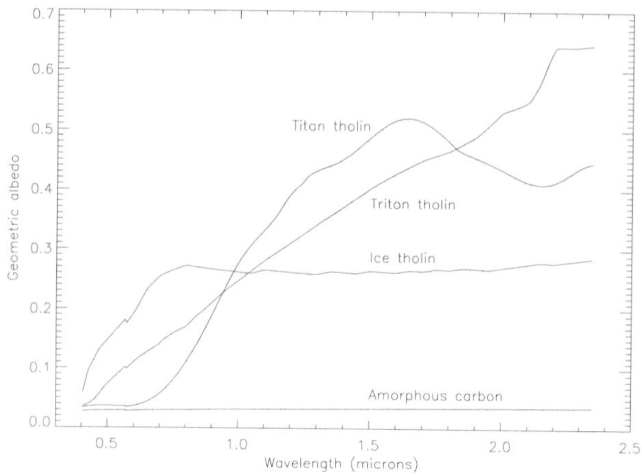

Figure 9.9. Spectra of tholins and amorphous carbon (adapted from Romon 2002).

available (Khare et al. 1984, McDonald et al. 1994, Khare et al. 1993, McDonald et al. 1996). Titan tholins are obtained by plasma discharge of a gas mixture of 90% N_2 and 10% CH_4. Triton tholins are formed in the laboratory by plasma discharge of a gaseous mixture more rich in N_2 (99.9% of N_2, and 0.1% of CH_4). Ice tholins I are formed by irradiating an ice mixture made of 86% H_2O and 14% C_2H_6. Spectra of Titan tholins, Triton tholins and Ice tholin I are shown in Figure 9.9.

Terrestrial kerogens are complex insoluble carbonaceous materials resulting from the decomposition of organisms at the bottom of oceans or lakes. Kerogen-like (or macro-molecular carbon) materials are found in carbonaceous chondrites (see Roush and Cruikshank 2004). They contain, in addition to carbon, elements such as H, N, O, and S. Spectra of D-type asteroids and of the dark side of Iapetus have been successfully modelled with kerogens (Gradie and Veverka 1980). It does not mean that such compounds have been identified, but just that the compounds present have spectral similarities with kerogen-type material. But few materials have been studied in the laboratory. Especially few plausible organic materials have been irradiated (see Roush and Cruikshank 2004).

We have difficulties in reproducing the slopes of red objects in the visible (see Figure 9.8). In particular, most of the visible slopes for red objects have convex or straight shapes, which cannot be explained by the presence of red organics which tend to have concave shapes, as explained by Moroz et al. (2004a,b). Moroz et al. demonstrate that it is better to mix organics with some silicates (and ices) to improve fits, even if no obvious signatures are present. According to Moroz et al. (2004a,b), KBOs would be mixtures of ices, silicates and organics, and their spectra would depend on which of the constituents optically dominates the surfaces. For a surface optically dominated by ices or silicates, space weathering (see further) would cause the reddening and darkening of the surface. A surface dominated by complex organics may have red colors which become gray (and darkens) due to aging. Collisional break-up of such gray bodies would expose fresh red fragments. This could explain the diversity of colors, but in one case the red surfaces would be the oldest ones, while in the other case they would be the youngest ones. However, much more work remains to be done in the laboratory to go further. There is work in progress in the laboratory by L. Moroz and colleagues from Catania.

In the same "vein", Cruikshank et al. (2001) have demonstrated that to explain moderately red surfaces like that of the Jupiter Trojan asteroid Hektor (but many KBOs have surfaces of equivalent redness), you do not need to include organics in the models. Some unirradiated silicates could produce the moderately red slopes. This is the case, in particular, of Mg-rich pyroxene.

9. WATER ICE

Water ice has been detected at the surface of only few objects. This is surprising as, given the temperatures in these cold regions of the solar system, water ice must be an abundant component of Centaurs and KBOs. Water ice has been detected with certainty on the Centaurs 2060 Chiron, 5145 Pholus (Figure 9.8), 10199 Chariklo (and maybe on 1999 UG_5 and 2001 PT_{13}). It has also been detected with certainty on the KBOs 1996 TO_{66}, 1999 DE_9, 1999 TC_{36}, 50000 Quaoar (and maybe on 20000 Varuna; Licandro et al. 2001).

In the case of the Centaur 8405 Asbolus, an absorption was seen around 1.6 μm in a spectrum recorded with the HST (Kern et al. 2000), but it has not been confirmed by other spectroscopic studies, and, in particular, a study that surveyed the whole surface of this Centaur (Romon-Martin et al. 2002).

In the case of 2060 Chiron, water ice was seen only when the object had no detectable activity. Water ice was not seen in 1993, when it was active. Then it was seen in 1996 and 1999 (Figure 9.10), which corresponded to inactive periods (Luu et al. 2000). In 2001, Chiron was observed following its rotation, and no water ice absorption was detected (Romon-Martin et al. 2003). It was soon realized that, in fact,

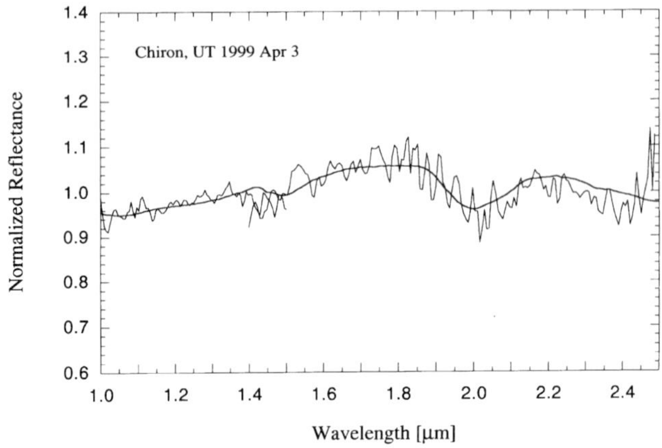

Figure 9.10. Spectrum of the Centaur Chiron obtained during a period of low activity with a model (full line) that incorporates water ice (from Luu et al. 2000).

the object was very active at that time (Chiron activity is sporadic, and hard to predict). Therefore, there seems to be a correlation between the non-detection of water ice and the object cometary-type activity. This could be due to obscuration by a gravitationally bound atmosphere.

Water ice can exist in different states. On average, one gets amorphous ice for low temperatures (T less than about 90 K), and crystalline ice for higher temperatures. But it is more complicated than that. Indeed, the lattice order of the ice depends not only on condensation temperature and rate, but also on temperature history and radiation environment (Hansen and McCord 2004). Radiation causes disorder in crystalline ice.

What do we expect for KBOs? A priori it should be amorphous ice, given the low temperatures of the objects. The hypothesis that the ice in KBOs must have been initially amorphous is supported by the fact that short-period comets emit very volatile species like CO (Choi et al. 2002). Indeed, amorphous ice is very efficient at trapping volatiles. The surface water ice seems to be amorphous on 10199 Chariklo (formerly known as 1997 CU_{26}) and 1996 TO_{66} (which has very deep ice bands) (Brown et al. 1998; Brown et al. 1999), but higher quality data are needed to be completely sure. However, it is predominantly crystalline on Charon, and also on Quaoar. Why? The answer is not obvious. Is it in relation with the size of the objects, with a possible massive conversion from amorphous ice to crystalline ice only inside the larger KBOs?

Surface heating and irradiation by high energy particles could also have played a role. According to Mastrapa and Brown (2001) a small impulse of heat due to a collision could convert amorphous ice to crystalline ice at the surface of a KBO. Furthermore, you could have different states of water ice at the different depths probed by different bands. This is the case for instance of Europa, for which it has been suggested (Hansen and McCord 2004) that ice at the surface (probed by the 3.1 μm band) was amorphous, while at a depth of about 1 mm (which is supposed to be the level probed by the weaker 1.65 μm band) the ice is crystalline, although you would have expected predominantly crystalline ice on Europa's surface (temperatures of 120-130 K). On the contrary, the ice on Callisto is uniformly crystalline at the surface and mostly crystalline at depth (Hansen and McCord 2004). But radiation is strong in Jupiter's environment which may lead to some peculiar effects that would not apply to KBOs.

For other Centaurs and KBOs, we do not know what is the state of water ice at the surface because the data are not of high enough quality that we can discriminate. The spectra are not of high enough quality either that we can measure the surface temperature from the depth and shape of the absorptions (particularly the 1.65 μm absorption in crystalline ice which is very sensitive to temperature), as has been done for giant planets icy satellites (Grundy et al. 1999). It would be very important to get better spectra in the H and K bands, where water ice bands are present, in order to retrieve some information on the lattice order of the water ice at the surface of these objects. Progress in laboratory work and in thermal modelling is also needed.

How much surface water ice is there? This very much depends on the model used. The same spectrum can give rise to widely different interpretations. As an example, models of Pholus spectrum with Hapke's model can be interpreted with 6% water ice in intimate mixture with other compounds, while models with Shkuratov's model require 42% water ice (Poulet et al. 2002). Why are the band depths different for the different objects? Is there a correlation with the albedo? Why don't we see water ice on all objects? Brown (2000) has found no correlation with the albedo when comparing spectra of Centaurs and irregular planetary satellites. If water ice is, at the surface of these bodies, mixed with a dark component, even in very small amounts, then one cannot detect it. The detection or non-detection of water ice depends on how it is mixed with the other components, and on the size of the grains. Furthermore, it depends on how much irradiation it has received and for how long.

The surface ice is not as "fresh" as on the Jupiter Galilean satellites. However, Phoebe, an irregular satellite of Saturn, which is very dark (albedo of 0.08-0.1) has ice absorptions somewhat similar to those observed on Chiron. It has been suggested that Phoebe is a captured object. It could have been formed at much larger distances from the Sun. On Uranian satellites, the water ice bands are quite subdued, but they are however deeper than generally seen on KBOs (1996 TO_{66} appears to be an exception). The "dirty" ice at the surface of these satellites could be ice contaminated by energetic particles from Uranus magnetosphere. Nereid, an irregular satellite of Neptune, has an albedo and a near-infrared spectrum quite similar to that of an Uranian satellite, Oberon. However, one should be careful when doing this type of comparison, as the depths of water ice absorptions depend on many parameters.

10. IS AQUEOUS ALTERATION POSSIBLE FOR THESE OBJECTS?

Although we do not have yet clear evidence for it, the presence of aqueous altered minerals on KBOs has been suggested by several authors. As it is something particularly interesting, we will briefly discuss it here.

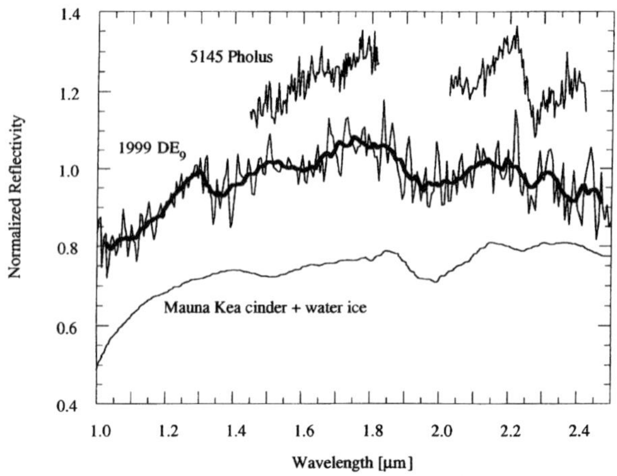

Figure 9.11. Spectrum of the KBO 1999 DE_9 with a model. Absorptions due to water ice (2μm), silicates (1μm) and the OH stretch (1.4 and 2.25 μm) are detected (Luu and Jewitt 2002). Reprinted, with permission, from the Ann. Rev. of Astron. Astrophys., vol. 40. Copyright 2002 by Annual Reviews.

Lederer, Vilas and colleagues have been looking specifically for absorptions due to such compounds in the visible range (Lederer et al. 2001, 2002). In 2000, Jewitt and Luu (2001) recorded at the Subaru telescope a spectrum of 1999 DE_9, and tentatively attributed two weak absorptions at 1.4 and 2.25 μm to the presence of Al or Mg in minerals incorporating OH within their structure (Figure 9.11).

More recently, weak bands were detected in visible spectra (ESO-Very Large Telescope) of two KBOs 2000 GN171 and 38628 Huya (formerly known as 2000 EB_{173}) around 0.6 – 0.7 μm (Lazzarin et al. 2003). However, de Bergh et al. (2004) acquired new spectra of these objects one year later to confirm them, and did not detect anything. As explained in de Bergh et al. (2004), this may be due to a heterogeneous composition of the surface, as has been found for other objects. If the features are real (which still needs to be confirmed by new observations), the only possible explanation found by de Bergh et al. is that they would be due to the presence of hydrated minerals over their surface, as exists for asteroids. The asteroids that present weak absorption bands in their spectra are dark asteroids, mostly C-type asteroids (see Roush and Cruikshank 2004). Another possible indication for the presence of hydrated minerals in the Kuiper Belt is the 2.2 μm feature detected in a spectrum of Charon that could perhaps be due to hydrated NH_3 ice (see above).

Hydrated silicates are present in carbonaceous meteorites, in Interplanetary Dust Particles (IDP), in micrometeorites, and, maybe, in comets. It is generally assumed that the aqueous alteration observed in meteorites occurred inside their parent bodies, the asteroids. Aqueous alteration in comets could have been facilitated by the hydrocryogenic mechanism known to exist in terrestrial permafrost and which can occur at temperatures below 273 K (Rietmeijer and Mackinnon 1987).

How could aqueous alteration of minerals be possible in the very cold environment of the Kuiper Belt? Two processes could be invoked to heat KBOs to temperatures sufficient for low temperature alteration by liquid water or hydrocryogenic alteration: collisions and internal heating. Internal heating would be possible only for the large TNOs. de Bergh et al. also considered as a possibility that the hydrated minerals were formed in the solar nebula, as has been suggested before to explain the presence of hydrous minerals in IDPs, meteorites, and, maybe, comets. In this hypothesis, the aqueous alteration would have occurred on grains when the nebula started to cool down.

These surfaces would probably have to be quite young for the hydrated minerals at the surface not to have been destroyed by irradiation.

11. HOW CAN WE EXPLAIN THE LARGE DIVERSITY IN SURFACE CHARACTERISTICS?

We have seen that colors are very diverse and that spectra, although available only for a limited number of objects, are also very diverse. Why? Could such a diversity be explained by differences in the formation region? Could some objects have been formed closer to the Sun, for instance, and brought later to the Kuiper Belt? Or could some have come from further away?

Are there any trends? Any relation with size, heliocentric distance or orbital parameters? Different teams have been looking for trends in their homogeneous data sets. Others have combined all existing data sets. A complete review on the subject can be found in Doressoundiram (2004). In short, the current conclusion is that there are no correlations between colors and size or heliocentric distance for any of the KBOs populations or for Centaurs. No clear correlation with other orbital parameters has been found either for Plutinos, Scattered objects or Centaurs. Some trends have been found for Classical objects, however. Classical objects with perihelion distances at and beyond 40 AU are mostly red (Tegler and Romanishin 2000, Doressoundiram et al. 2002). In addition, Classical objects with semi-major axis between the 2:3 and 1:2 resonance zones which have high eccentricities and inclinations tend to be neutral or slightly red. A red color-low inclination cluster of KBOs has been found in the Classical Belt (Trujillo and Brown 2002, Doressoundiram et al. 2002). All large classical objects have high orbital eccentricities. They are what is called “dynamically hot”. But these trends need to be verified by additional observations.

Dynamical models are in progress, with the hope that they will explain some of the color features. New work by Gomes (2003) shows that, if the initial mass of the Kuiper Belt was large enough, you could now have two groups of objects in the Classical Belt that would have been formed at different locations in the solar system, which could perhaps lead to differences in their surface compositions, and therefore their colors. One group, the dynamically cold objects would have been formed in situ or would have been moderately transported to the Kuiper Belt region by resonance sweeping. The other group, the dynamically hot ones, would have been formed in the 15-28 AU region and then expelled outwards, and some would have become permanently trapped in the Kuiper Belt. However, in Gomes’ model, one would also have two groups of objects among Plutinos, a group from the inner disk (scattered) and a group from the outer disk (trapped).

It is very important to pursue such models. At the same time, such scenarios, if confirmed, would not explain everything, and it is clear that a lot of the differences we see now must be due to different surface evolutions. What are the processes that could have affected the surfaces? There are four main mechanisms: irradiation, collisions, cometary-like activity and internal evolution. Some would uniformly affect the surfaces, others could lead to surface heterogeneities.

An airless body in space is subjected to what is called "space weathering". This term includes the combined effects of: solar UV, solar wind plasma, heliospheric cosmic rays, galactic cosmic rays, interplanetary dust, and interstellar neutral gas. In the case of KBOs, the main source of irradiation is galactic cosmic rays. Contrary to general belief, the irradiation, which occurs over very long timescales, could have been different depending on where the objects were formed (Cooper et al. 2004). It may therefore have played a role in the color diversity. Indeed, an Anomalous Cosmic Ray (ACR) component discovered recently, and which, at 40 AU, has an order of magnitude more H^+ flux than the normal cosmic ray H^+ component, could produce an increase in intensity when going from 40 AU to the very local interstellar medium. As discussed by Cooper et al. (2004), Scattered Disk Objects with aphelia near the heliosheath (which is located around 85 AU), as well as Oort cloud comets, would be strongly irradiated by this anomalous cosmic rays component, while Classical Objects near 40 AU would be much less irradiated. In addition, the slight gradients in intensity of ACR ions within the Classical Belt itself could have induced some color variations for objects belonging to this population, but this effect is probably minor.

By irradiating mixtures of ices, you may form new ices. Irradiated mixed ices (H_2O, CO_2, CO, N_2) tend to redden at moderate dosages. And laboratory experiments have shown that there is a flattening of the spectrum towards the near-infrared (Thompson et al. 1987). If surfaces were made initially of ices containing carbon, red surfaces would be due to the formation of more complex hydrocarbons. When you irradiate even more, the surfaces may become neutral again (but darker), as already mentioned. If this is true, neutral surfaces initially covered by such ices could be young or old. If you have a mixture dominated by silicates, irradiation would provide a reddening and a darkening of the surface. It is important to note that the effect will vary as a function of depth, as different types of irradiation penetrate to different depths (see, e.g., Strazulla 1998). Spectroscopy in the visible and near infrared could provide information on different depths, depending on the strength of the absorptions.

We have seen that occasional collisions between objects could have been very important in the past when the Belt was much more populated. Heterogeneous surfaces are expected, at least for large KBOs. Model results of Durda and Stern (2000) show that small KBOs would have completely reworked surfaces over 3.5 Gyr. The surface would be a regolith of impact products instead of a solid ice.

Irradiation and collisions are two competing mechanisms. Collisions would bring some "fresher", unirradiated material to the surface by excavating craters or producing impacts debris covering nearby regions. Cooper et al. (2004) estimate that time scales for creating an appreciable irradiation mantle on the surface of KBOs are comparable or even shorter than disruptive processes like collisions. One of the first proposed explanation for the color diversity was the collisional-resurfacing mechanism (Luu and Jewitt 1996). This model would naturally lead to a continuous range of colors. Initial work by Tegler and Romanishin, which tended to show that there are two groups of colors, was in contradiction with this mechanism. More recent works (see, e.g., Doressoudiram et al. 2002, Peixhino et al. 2004) are in favor of a continuous range of colors for KBOs. However, the collisional-resurfacing model initially suggested by Luu and Jewitt has been later questionned by that same team (Jewitt and Luu 2000). Indeed, they have argued that observed color differences are too extreme to be explained by this model. Furthermore, they claim that, if this mechanism was responsible for the color diversity, heterogeneous surfaces should be more common.

Concerning Centaurs, when plotting existing colors of Centaurs, one finds however that there are two main groups of objects: one group of red objects and another group of more neutral objects (Peixhino et al. 2003). It has been suggested that the red objects, like Pholus, would have an old surface because they would have been expelled relatively recently from the Kuiper Belt. The more neutral objects would have spent more time closer to the Sun and their irradiation mantle formed in the Kuiper Belt would have been progressively destructed or buried (Cruikshank et al. 1998a). But the current sample of objects is not very large (about 20 objects). We need to observe more objects to improve statistics.

Is cometary activity common for KBOs and Centaurs? Pluto is active. This is why we see essentially the most volatile ices at the surface. The Centaur Chiron is also active. It may also be the case for 1996 TO66 for which an important change in its light-curve has been observed (Hainaut et al., 2000). Cometary activity has been observed in distant comets, up to 20 AU (Meech and Hainaut 2001). At such large distances from the Sun, the activity is not due to H_2O but rather to CO or CO_2, as these ices can sublimate at much colder temperatures. Is activity common? We do not know. Observations of comae around very faint objects are extremely difficult to achieve, and activity is hard to predict for such objects. The Centaurs Chiron and Asbolus have about the same orbital properties, but only Chiron appears to be active. Why? Is it because of very different past histories? Centaurs are certainly more susceptible than KBOs to cometary-type activity, but the possibility of cometary activity for the largest KBOs must be considered when trying to explain the color diversity.

A modelling attempt to include all three processes (irradiation, collisions, cometary activity) to explain the color distributions of KBOs and Centaurs is in progress (Delsanti et al. 2004).

As we have seen before, internal heating could have occurred for the largest bodies. And this would have had consequences on the surfaces early in their history. It is possible that small bodies would have kept their primordial surfaces (with further evolution due to space weathering), while the largest bodies would have gone through some chemical and structural evolution, which could have affected their surfaces early in their evolution. According to McKinnon (2002), part of the color diversity could be due to differences in internal evolution which are very strongly size dependent. So this possibility must be kept in mind when trying to explain the color diversity.

Given all possible sources of surface alteration, the problem is quite complex and far from being solved. In addition, one should be very careful in not going too far in the interpretation. Indeed, for two given objects you could have the same color but very different albedos, which would mean very different surface compositions. Albedo measurements are essential to discriminate between models, and to eventually be able to determine which surfaces are young and which ones are old. Furthermore, it is very important, when trying to classify colors, to have the most possible complete range of colors, and going to the infrared, as we know that some objects have similar colors in the visible but widely different colors in the infrared.

12. DISSEMINATION OF KUIPER BELT MATERIAL ELSEWHERE IN THE SOLAR SYSTEM

The current mass in the Belt in the 40-50 AU region is of the order of 0.1 terrestrial mass. We have seen that there are several good arguments for a much more massive early Kuiper Belt, and, consequently, an intense collisional activity at that time. One should mention that this has been recently questioned by two different groups of dynamicists, however. One group argues that KBOs binaries (which seem to be quite common) could not have survived in such an intense collisional environment. The other group is currently working on a model that does not require a massive Belt at the start to explain the complex current dynamical structure. However, it remains to be seen how the other lines of evidence could be explained.

If the Belt was indeed 100 times more massive, where has the large part of the mass gone? Part of the mass initially present in the Classical Belt was probably transfered to the Scattered Disk. And it is not easy to estimate how much is still there since Scattered Disk objects, with their very long orbital periods, are most of the time inaccessible to observations. But current estimates indicate that the mass of objects that could have gone to the Scattered Disk would not be sufficient. We have seen that some bodies currently located outside of the Kuiper Belt, Centaurs and short-period comets, could be escaped KBOs or collisional remnants of KBOs. As discussed above, dynamical studies have shown that Centaurs could constitute an intermediate population between KBOs and Jupiter-family comets. Some of the Jupiter-family comets escape Jupiter's attraction and may constitute an important source of volatiles (including water and organics) to the Earth and the other terrestrial planets. Centaurs that are not scattered close to the Sun could have impacted a planet (or the Sun) or could have been expelled out of the solar system (Jewitt and Luu 2000).

Other objects could have originated in the Kuiper Belt. One Neptune Trojan (2001 QR322) has been discovered recently, and it has been suggested that Neptune Trojans are captured objects from the Kuiper Belt, and that they may be more numerous than Jupiter Trojans (Chiang et al. 2004). Triton, with its retrograde orbit and similarity to Pluto, could also be a captured Kuiper Belt Object. All irregular satellites of the giant planets, which are satellites located at large distances from the central planet, and that usually have highly inclined and eccentric (and often retrograde) orbits, could have originated elsewhere and then be captured by the central planet. They have diverse surface properties. One irregular satellite of Jupiter is very red, much redder than the other Jupiter satellites and the Trojans (Grav et al. 2003). Phoebe, an irregular satellite of Saturn, which has a very dark surface and is covered with water ice, presents analogies with KBOs. These bodies could therefore have been formed in the Kuiper Belt.

However, a lot of dust could have been produced during the intense collisional episode, as well as because of impacts of interstellar dust on KBOs (Stern 1996, Kenyon and Luu 1999). The dust would not stay there forever, and a very important mass loss could be explained by the loss of small dust particles in very large amounts. A large part of the grains would be ejected from the Solar System. but some would be carried towards the interior of the Solar System by the Poynting-Robertson effect and plasma drag forces, as discussed before. How much of this Kuiper Belt dust could have reached the Earth? This is a very controversial topic. But we need to discuss it for astrobiology.

Pioneer 10 detected dust impacts out to about 18 AU, and Pioneer 11 detected dust out to about 13 AU. But Voyager 1 and 2 instruments have detected dust beyond Neptune's orbit. However, the dust coming from the Kuiper Belt can be collisionally destroyed en-route to the Earth (see Liou et al. 1996). The Liou et al. dynamical study indicates that, in spite of that, Kuiper Belt dust grains of diameter between about 1 and 9 μm could still constitute a large portion of the small IDPs collected in the Earth stratosphere. These IDPs contain distinctive types of primitive materials (Bradley et al. 1988, Clemett et al. 1993). They have therefore been considered as a source of the first organic materials to the Earth and could have contributed to the origin of life.

On the contrary, a more recent dynamical study (Moro-Martin and Malhotra, 2003) indicates that Kuiper Belt dust would permeate only essentially the outer solar system. According to these authors, Kuiper Belt dust would be the source of micrometeoroid impacts on outer solar system bodies, and the source of large particles in the local ISM, but, because of important scattering by Jupiter and Saturn, almost none of the dust would reach the Earth. Further studies on the ultimate fate of Kuiper Belt dust that spirals towards the Sun are needed before one can assess its role for bringing primitive materials to the Earth.

Another interesting approach to this problem deals with analyses of micrometeorites collected in Antartica. It has been suggested that some of the early micrometeorites (which have dimensions larger than 50 μm) could have played an important role in the origin of the Earth's hydrosphere (see, e.g., Maurette et al. 2000). Maurette et al. further suggest that, based on time-scale arguments, the early micrometeorites that fell in abundance on Earth during the period of heavy bombardment (around 3.8-3.9 Gyr ago) must have come for a large part from Kuiper Belt comets rather than from asteroids.

13. CONCLUSIONS AND PROSPECTS

Contrarily to all expectations, objects in the Kuiper Belt appear to have surfaces of widely different colors. Very few components have been identified at the surface of these objects, but useful information can be obtained from the shape of their spectra. A large number of Kuiper Belt Objects are covered by very red material that seems to be peculiar to this class of objects. Although KBOs are the most primitive bodies in the solar system, some of them appear to have been subjected to important evolution. Quite surprisingly, water ice, which must have been present in abundance in the solar nebula when the objects were formed, is detected at the surface of only a few of the studied objects.

To better constrain surface models of KBOs, one would need spectra with higher signal-to-noise ratio in the near-infrared. Spectra at longer wavelengths would also be important. If such spectra cannot be obtained because of the faintness of the objects, one should have at least spectrophotometric data over the widest possible spectral range. Simultaneous measurements in the visible and near-infrared are essential as some of the objects may have heterogeneous surfaces, unless rotationally-resolved data are obtained in both spectral ranges. Besides, it is very important to continue looking for surface heterogeneites as this would provide some clues on which processes are dominant in providing the color diversity.

It is also important to look for more objects and to obtain accurate orbits. This will help in retracing the dynamical evolution of the different categories of objects and in looking for color trends. A significant number of large ("bright") objects for which physical studies can be carried out must be present in the Kuiper Belt that await discovery. Indeed, according to Trujillo and Brown (2004), 1 to 2 objects the size of Pluto, 10 objects with diameter larger than 1000 km, and about 125 brighter than R = 21 may be present in the Belt. Many more binaries must also exist. Their studies are important for mass determination.

Only a handful of KBOs and Centaurs have had their albedos measured. Albedo measurements are essential.

What are the prospects? Various sky surveys are currently carried out. In particular a survey devoted to the brightest objects (limiting R magnitude of 20.7) is operated at Palomar using a 1.22-m diameter Schmidt telescope (Trujillo and Brown 2004). It is estimated that, by January 2005, about 100 bright KBOs (sky coverage: north of about - 30° latitude) will have been found by this survey. The European astrometric satellite GAIA, which will be able to observe objects down to a magnitude of about 20 everywhere in the sky, should also play an important role. In addition, there are plans for a dedicated ground-based survey facility called Pan Starrs that will have much higher sensitivity than the 3.6-m CFHT and 8-m Subaru surveys (which are currently the most sensitive ones) and should revolutionize completely the field (Jewitt 2004).

According to Jewitt (2004), in 10 years from now, thanks to the ground-based facilities in development, one should have discovered about 20,000 KBOs with multi-opposition orbits (compared to only 200 now). And about 1000 binaries will have been detected. Spectacular color families will have been defined, and one will have been able to observe objects in the Classical Belt beyond the "edge" (around 50 AU).

The infrared Spitzer Space Telescope (formerly SIRTF, the Space Infrared Telescope Facility), which was launched in August 2003, should provide albedo measurements for about 100 KBOs. Other instruments in preparation that should also provide albedo data on the brightest KBOs are the international Atacama Large Millimeter Array (ALMA) of 64 antennas of 12 meters diameter in Chile and the 3.5-m Herschel sub-millimeter space telescope of ESA.

Last, but not least, the planned NASA New Horizons space mission to Pluto-Charon and Kuiper Belt Objects that will include instruments operating in the ultraviolet, visible and radio ranges (Stern and Spencer 2004) should revolutionize our understanding of these bodies, and, in particular, of their surface composition and of possible internal evolution and surface alteration processes.

14. REFERENCES

Altenhoff, W.J., Bertoldi, F., and Menten, K.M. 2004, Size estimates of some optically bright KBOs, Astron. Astrophys., 415, 771.

Altenhoff, W.J., Menten, K.M., and Bertoldi, F. 2001, Size determination of the Centaur Chariklo from millimeter-wavelength bolometer observations, Astron. Astrophys., 366, L9.

Barucci, M.A., Boehnhardt, H., Dotto, E., Doressoundiram, A., Romon, J., Lazzarin, M., Fornasier, S. et al. 2002, Visible and near-infrared spectroscopy of the Centaur 32532 (2001 PT_{13}). ESO Large Programme on TNOs and Centaurs: First spectroscopy results, Astron. Astrophys. 392, 335.

Barucci, M.A., Cruikshank, D.P., Mottola, S., and Lazzarin, M. 2003, Physical properties of Trojan and Centaurs asteroids, in "Asteroids III", W.F. Bottke, A. Cellino, P. Paolicchi, and R.P. Binzel eds, Univ. Arizona Press, 273.

Barucci, M.A., Doressoundiram, A., and Cruikshank, D.P. 2004, Physical Characteristics of TNOs and Centaurs, in Comets II, Univ. Arizona Press, M. Festou et al. eds, in press.

Bauer, J.M., Meech, K.J., Fernandez, Y.R., Farnham, T.L., and Roush, T.L. 2002, Observations of the Centaur 1999 UG_5: evidence of a unique outer solar system surface, PASP, 114, 1309.

Bradley, J.P., Sandford, S.A., and Walker, R.M. 1988, Interplanetary dust particles, in "Meteorites and the Early Solar System", J.F. Kerridge and M.S. Matthews eds, Univ. Arizona Press, 861.

Brown, M.E. 2000, Near-infrared spectroscopy of Centaurs and irregular satellites, Astron. J., 119, 977.

Brown, M.E. 2002, Pluto and Charon: Formation, Seasons, Composition, Ann. Rev. Earth Planet. Sci., 30, 307.

Brown, R.H., Cruikshank, D.P., Pendleton, Y., and Veeder, G.J. 1998, Identification of water ice on the Centaur 1997 CU_{26}, Science, 280, 1430.

Brown, R.H., Cruikshank, D.P., and Pendleton, Y.J. 1999, Water ice on Kuiper Belt object 1996 TO_{66}, Astrophys. J. Lett. 519, L101.

Brown, M.E., and Calvin, W.M. 2000, Evidence for crystalline water and ammonia ices on Pluto's satellite Charon, Science, 287, 107.

Brown, M.E., and Trujillo, C.A. 2004a, Direct measurement of the size of the large Kuiper Belt Object (50000) Quaoar, Astron.J., in press.

Brown, M.E., and Trujillo, C.A. 2004b, Ices on the surface of (50000) Quaoar, Astrophys.J. Letters, in press.

Brown, M.E., Trujillo, C.A., and Rabinowitz, D. 2004, Discovery of a candidate inner Oort cloud planetoid, submitted to Ap.J. Letters.

Bus, S.J., A'Hearn, M.F., Schleicher, D.G.,Bowell, E. 1991, Detection of CN emission from (2060) Chiron, Science, 251, 774.

Chiang, E.I., Millis, R.L., Buie, M.W., Wasserman, L.H., and Meech, K.J. 2004, Resonant and Secular Families of the Kuiper Belt, Earth, Moon and Planets, in press.

Choi, Y.-J., Cohen M., Merk R., and Prialnik, D. 2002, Long-term evolution of objects in the Kuiper Belt Zone - Effects of insolation and radiogenic heating, Icarus, 160, 300.

Clemett, S.J., Maechling, C.R., Zare, R.N., Swan, P.D., and Walker, R.M. 1993, Identification of complex aromatic molecules in individual interplanetary dust particles, Science, 262, 721.

Cooper, J.F., Christian, E.R., Richardson, J.D. and Wang, C. 2004, Proton irradiation of Centaurs, Kuiper Belt, and Oort cloud objects at plasma to cosmic ray energy, Earth, Moon and Planets, in press.

Crovisier, J. 2004, The molecular complexity of comets, this volume.

Cruikshank, D.P., Roush, T.L., Owen, T.C., Geballe, T.R., de Bergh, C., Schmitt, B., Brown, R.H., and Bartholomew, M.J. 1993, Ices on the surface of Triton, Science, 261, 742.

Cruikshank, D.P., Roush, T.L., Moore, M., Sykes, M., Owen, T.C., Bartholomew, M.J., Brown, R.H., and Tryka, K.A. 1997, The surfaces of Pluto and Charon, in "Pluto and Charon", S.A. Stern and D.J. Tholen eds, Univ. Arizona Press, 221.

Cruikshank, D.P., Roush, T.L., Bartholomew, M.J., Moroz, L.V., Geballe, T.R., Pendelton, Y.J., White, S.M., Bell, J.F., Davies, J.K., Owen, T.C., de Bergh, C.,Tholen, D.J., Bernstein, M.P., Brown, R.H., Tryka, K.A., and Dalle Ore, C.M. 1998a, The composition of Centaur 5145 Pholus, Icarus, 135, 389.

Cruikshank, D.P, Roush, T.L., Owen, T.C., Quirico, E., and de Bergh, C., 1998b, The surface compositions of Triton, Pluto, and Charon, in "Solar System Ices", B. Schmitt, C. de Bergh, and M. Festou eds, Kluwer Academic Publishers, 655.

Cruikshank, D.P., Dalle Ore, C.M., Roush, T.L., Geballe, T.R., Owen, T.C., de Bergh, C., Cash, M.D., and Hartmann, W.K. 2001, Constraints on the composition of Trojan asteroid 624 Hektor, Icarus, 153, 348.

Cruikshank, D.P., and Dalle Ore, C.M. 2004, Spectral models of Kuiper Belt Objects and Centaurs, Earth, Moon, and Planets, in press.

Davies, J., Spencer, J., Sykes, M., Tholen, D., and Green, S., 1993, 5145 Pholus, IAU Circular, 5698.

Davies, J.K., Roush, T.L., Cruikshank, D.P., Bartholomew, M.J., Geballe, T.R., Owen, T.C., and de Bergh, C. 1997, The detection of water ice in comet Hale-Bopp, Icarus 127, 238.

de Bergh, C., Boehnhardt, H., Barucci, M.A., Lazzarin, M., Fornasier, S., Romon-Martin, J., Tozzi, G.-P., Doressoundiram, A., and Dotto, E. 2004, Aqueous altered silicates at the surface of two Plutinos?, Astron. Astrophys., 416, 791.

Delsanti, A., Hainaut, O., Jourdeuil, E., Meech, K.J., Boehnhardt, H., and Barrera, L. 2004, Simultaneous visible-near IR photometric study of Kuiper Belt Object surfaces with the ESO/Very Large Telescopes, Astron. Astrophys., 417, 1145.

De Sanctis, M.C., Capria, M.T., and Coradini, A. 2001, Thermal evolution and differentiation of Edgeworth-Kuiper Belt Objects, Astron. J., 121, 2792.

De Sanctis, M.C., Capria, M.T., and Coradini, A. 2002, Thermal evolution models of objects at large heliocentric distances, Proceedings of Asteroids, Comets, Meteors, (ACM 2002), ESA-SP-500, 39.

Doressoundiram, A. 2004, Colour properties and trends in trans-Neptunian objects, Earth, Moon, and Planets, in press.

Doressoundiram, A., Peixinho, N., de Bergh, C., Fornasier, S., Thebault, Ph., Barucci, M.A., and Veillet, C. 2002 The color distribution of the Kuiper Belt, Astron. J., 124, 2279.

Dotto, E., Barucci, M.A., and de Bergh, C. 2003a, Surface composition of TNOs and Centaurs: visible and near-infrared spectroscopy, Comptes Rendus Physique, Acad. Sciences - Paris, 7, 775.

Dotto, E., Barucci, M.A., Leyrat, C., Romon, J., de Bergh, C., and Licandro, J. 2003b, Unveiling the nature of 10199 Chariklo: near-infrared observations and modeling, Icarus, 164, 122.

Douté, S., and Schmitt, B. 1998, A mulilayer bidirectional reflectance model for the analysis of planetary surface hyperspectral images at visible and near-infrared wavelengths, J. Geophys. Res., E103, 31367.

Durda, D.D., and Stern S.A. 2000, Collision rates in the present-day Kuiper Belt and Centaur regions: applications to surface activation and modification on comets, Kuiper Belt Objects, Centaurs, and Pluto-Charon, Icarus, 145, 220.

Fernandez, Y.R., Jewitt, D.C., and Sheppard, S.S., 2002, Thermal properties of Centaurs Asbolus and Chiron, Astron. J., 123, 1050.

Gladman, B., Holman, M., Grav, T., Kaavelars, J.J., Nicholson, P., Aksnes, K. and Petit, J.M. 2002, Evidence for an extended scattered disk, Icarus 157, 269.

Gomes, R.S. 2003, The origin of the Kuiper Belt high-inclination population, Icarus, 161, 404.

Gradie, J., and Veverka, J. 1980, The composition of the Trojan asteroids, Nature, 283, 840.

Grav, T., Holman, M.J., Gladman, B.J. and Aksnes, K. 2003, Photometric Survey of the Irregular Satellites, Icarus, 166, 33.

Grundy, W.M., and Buie, M.W. 2002, Spatial and compositional constraints on non-ice components and H_2O on Pluto's surface, Icarus, 157, 128.

Grundy, W.M., and Fink, U. 1996, Synoptic CCD Spectrophotometry of Pluto over the Past 15 Years, Icarus, 124, 329.

Grundy, W.M., Buie, M.W., and Spencer, J.R. 2002 Spectroscopy of Pluto and Triton at 3-4 microns: possible evidence for wide distribution of nonvolatile solids, Astron. J,. 124, 2273.

Grundy, W.M., Buie, M.W., Stansberry, J.A., and Spencer, J.R. 1999 Near-infrared spectra of outer solar system surfaces: remote determination of H_2O ice temperatures, Icarus, 142, 536.

Hainaut, O.R., Delahodde, C.E., Boehnhardt, H., Dotto, E., Barucci, M.A., Meech, K.J., Bauer, J.M., West, R.M., and Doressoundiram, A. 2000 Physical properties of TNO 1996 TO_{66}; Lightcurve and possible cometary activity, Astron. Astrophys., 356, 1076.

Hansen, G.B. and McCord, T.B. 2004, Amorphous and crystalline ice on the Galilean satellites: a balance between thermal and radiolytic processes, JGR-Planets, in press.

Jewitt, D.C. 1999, Kuiper Belt Objects, Ann. Rev. Earth Planet. Sci., 27, 287.

Jewitt, D.C. 2004, Project Pan-STARRS and the outer Solar System, Earth, Moon, and Planets, in press.

Jewitt, D., and Kalas, P. 1998, Thermal observations of Centaur 1997 CU_{26}, Astroph. J., 499, L103.

Jewitt, D.C. and Luu, J. 1993, Discovery of the candidate Kuiper Belt Object 1992 QB1, Nature 362, 730.

Jewitt, D.C., and Luu, J. 2000, Physical nature of the Kuiper Belt, in "Protostars and Planets IV", V. Mannings, A. Boss, and S. Russell eds, Univ. Arizona Press, Tucson, 1201.

Jewitt, D.C., and Luu, J. 2001, Colors and spectra of Kuiper Belt Objects, Astron. J., 122, 2099.

Jewitt, D.C., Aussel, H., and Evans, A. 2001, The size and albedo of the Kuiper-belt object (20000) Varuna, Nature, 411, 446.

Jewitt, D.C., and Sheppard, S.S. 2002, Physical properties of Trans-Neptunian object (20000) Varuna, Astron. J., 123, 2110.

Kenyon, S.J., Luu, J.X. 1999, Accretion in the Early Kuiper Belt. II. Fragmentation. Astron. J. 118, 1101.

Kern, S.D., McCarthy, D.W., Buie, M.W., Brown, R.H., Campins, H., and Rieke, M. 2000, Compositional variation on the surface of Centaur 8405 Asbolus, Astrophys. J., 542, L155.

Khare, B.N., Sagan, C., Arakawa, E.T., Suits, R., Callcott, T.A., and Williams, M.W. 1984, Optical constants of organic tholins produced in a simulated Titanian atmospehre: from soft x-ray to microwave frequencies, Icarus 60, 127.

Khare, B.N., Thompson, W.R., Cheng, L., Chyba, C., Arakawa, E.T., Meisse, C., and Tuminello, P. 1993, Production and optical constants of ice tholin from charged particle irradiation of (1:6) C_2H_6/H_2O at 77 K, Icarus, 103, 290.

Lamy, P.L., Toth, I., Fernandez, Y.R., and Weaver, H.A. 2004, The sizes, shapes, albedos, and colors of cometary nuclei, in Comets II, Univ. Arizona Press, Festou et al. eds, in press.

Lazzarin, M., Barucci, M.A., Boehnhardt, H., Tozzi, G.P., de Bergh, C., and Dotto, E. 2003, ESO Large Programme on Physical Studies of Trans-Neptunian Objects and Centaurs: Visible Spectroscopy, Astron. J., 125, 1554.

Lederer, S.M., Vilas, F., Jarvis, K.S., and French, L. 2001, Colors and compositional characteristics of Kuiper Belt Objects and Centaurs, BAAS, 33, 1046.

Lederer, S.M., and Vilas, F. 2002, Spectrophotometry of Kuiper Belt Objects and Centaurs, BAAS, 34, 846.

Lellouch, E., Moreno, R., Ortiz, J.L., et al., 2002, Coordinated thermal and optical observations of trans-neptunian object (20000) Varuna from Sierra Nevada, Astron. Astrophys., 391, 1133.

Licandro, J., Oliva, E., Di Martino, M. 2001, NICS-TNG infrared spectroscopy of thransneptunian objects 2000 EB_{173} and 2000 WR_{106}, Astron. Astrophys. 373, L29.

Liou, J.C., Zook, H.A., and Dermott, S.F. 1996, Kuiper Belt dust grains as a source of interplanetary dust particles, Icarus, 124, 429.

Luu, J.X., and Jewitt, D. 1996, Color diversity among the Centaurs and Kuiper Belt Objects, Astron. J. 112, 2310.

Luu, J.X., and Jewitt, D. 2002, Kuiper Belt Objects: relics from the accretion disk of the Sun, Ann. Rev. Astron. Astrophys. 40, 63.

Luu, J., Jewitt, D., and Trujillo, C. 2000, Water ice in 2060 Chiron and its implications for Centaurs and Kuiper Belt Objects, Ap.J., 531, L151.

Margot, J.L., Trujillo, C., Brown, M.E., and Bertoldi, F., 2002, The size and albedo of KBO 2002 AW_{197}, AAS–DPS meeting, Bul. Amer. Astron. Soc., 34, 871.

Mastrapa, R.M.E., and Brown, R.H. 2001, Modelling the effects of ion radiation on crystalline water ice, Lunar and Planetary Science XXXII, 1381.

Maurette, M., Gounelle, M., Duprat, J., Engrand, C., and Matrajt, G. 2000, The early micrometeorites accretion scenario and the origin of the Earth's hydrosphere, in "A new Era in Bioastronomy", ASP Conference Series (G. Lemarchand, K. Meech, eds), 213, 263.

McBride, N., Green, S.F., Davies, J.K., Tholen, D.J., Sheppard, S.S., Whiteley, R.J., and Hillier, J.K. 2003 Visible and infrared photometry of Kuiper Belt Objects: searching for evidence of trends, Icarus, 161, 501.

McDonald, G.D., Thompson, W.R., Heinrich, M., Khare, B.N., and Sagan, C. 1994, Chemical investigation of Titan and Triton tholins, Icarus, 108, 137.

McDonald, G.D., Whited, L.J., DeRuiter, C., Khare, B.N., Patniak, A., and Sagan, C. 1996, Production and chemical analysis of cometary ice tholins, Icarus, 122, 107.

McKinnon, W.B. 2002, On the initial thermal evolution of Kuiper Belt Objects, in Proceedings of Asteroids, Comets, Meteors (ACM 2002), ESA-SP-500, 29.

Meech, K.J., and Hainaut, O.R. 2001, in "Collisional processes in the solar system", M. Marov and H. Rickman eds, Kluwer Academic Publishers, 163.

Merk, R., and Prialnik, D. 2004, Early thermal and structural evolution of small bodies in the trans-Neptunian zone, Earth, Moon, and Planets, in press.

Morbidelli, A., Brown, M.E., and Levison, H.F. 2004, The Kuiper Belt and its primordial sculpting, Earth, Moon and Planets, in press.

Morbidelli, A., and Levison, H.F. 2004, Scenarios for the origin of the orbits of the Trans-Neptunian Objects 2000 CR_{105} and 2003 VB_{12}, submitted to Astrophys. J.

Moro-Martin, A., and Malhotra, R. 2003, Dynamical models of Kuiper Belt dust in the inner and outer Solar System, Astron. J., 125, 2255.

Moroz, L.V., Baratta, G., Distefano, E., Strazzulla, G., E. Dotto, and M.A. Barucci 2004a, Ion irradiation of Asphaltite: optical effects and implications for Trans-Neptunian Objects and Centaurs, Earth, Moon, and Planets, in press.

Moroz, L., Starukhina, L., Strazzulla, G., Baratta, G., Dotto, E., Barucci, A., Arnold, G. 2004b, Optical alteration of complex organics induced by ion irradiation: II. Implications for primitive small bodies of the Solar System, submitted to Icarus.

Owen , T.C., Roush, T.L., Cruikshank, D.P., Elliot, J.L., Young, L.A., de Bergh, C., Schmitt, B., Geballe, T.R., Brown, R.H., and Bartholomew, M.J. 1993, Surface ices and atmospheric composition of Pluto, Science, 261, 745.

Peixinho, N., Doressoundiram, A., Delsanti, A., Boehnhardt, H., Barucci, M.A., and Belskaya, I. 2003, Reopening the TNOs color controversy: Centaurs bimodality and TNOs unimodality, Astron. Astrophys., 410, L29.

Peixinho, N., Boehnhardt, H., Belskaya, I., Doressoundiram, A., Barucci, M.A., and Delsanti, A. 2004, ESO Large Program on Centaurs and TNOs: Visible Colors - Final Results, Icarus, in press.

Poulet, F., Cuzzi, J.N., Cruikshank, D.P., Roush, T., and Dalle Ore, C.M. 2002, Comparison between Shkuratov and Hapke scattering theories for planetary surfaces: application to the surface composition of two Centaurs, Icarus 160, 313.

Rietmeijer, F.J.M., and Mackinnon, I.D.R. 1987, in Symposium on the diversity and similarity of comets, ESA SP-278, 363.

Romon, J. 2002, Observations et modélisation des états de surface d'objets de Kuiper et de noyaux cométaires, PhD thesis, Univ. Paris XI, France.

Romon-Martin, J., Barucci, M.A., de Bergh, C., Doressoundiram, A., Peixinho, N., and Poulet, F. 2002, Observations of Centaur Asbolus: searching for water ice, Icarus, 160, 59.

Romon-Martin, J., Delahodde, C., Barucci, M.A., de Bergh, C., and Peixinho, N. Photometric and spectroscopic observations of 2060 Chiron, 2003, Astron. Astrophys., 400, 369.

Roush, T.L., and Cruikshank, D.P. 2004, Observations and laboratory data of planetary organics, this volume.

Sheppard, S.S., Jewitt, D.C., Trujillo, C.A., Brown, M.J., and Ashley, M.C.B. 2000, Astron. J. 120, 2687.

Sheppard, S.S. and Jewitt, D.C. 2002, Time-resolved photometry of Kuiper Belt Objects: rotations, shapes and phase functions, Astron. J., 124, 1757.

Spencer, J.R., Lebofsky, L.A., and Sykes, M.V. 1989, Systematic biases in radiometric diameter determinations, Icarus 78, 337.

Spencer, J.R., Stansberry, J.A., Trafton, L.M., Young, E.F., Binzel, R.P., and Croft, S.K. 1997, Volatile transport, seasonal cycles, and atmospheric dynamics on Pluto, in "Pluto and Charon", S.A. Stern, D.J. Tholen eds, Univ. Arizona Press, 435.

Stern, S.A. 1996, On the Collisional Environment of the Massive, Primordial Kuiper Belt, Astron. J. 112, 1203.

Stern, S.A. and Colwell, J.E. 1997, Accretion in the Edgeworth-Kuiper Belt: forming 100-1000 km radius bodies at 30 AU and beyond, Astron. J., 114, 841.

Stern, S.A. and Spencer, J.R. 2004, New Horizons: the First Reconnaissance Mission to Bodies in the Kuiper Belt, Earth, Moon, and Planets, in press.

Strazzulla, G. 1998, Chemistry of Ice induced by Bombardment with Energetic Charged Particles, in "Solar System Ices", B. Schmitt, C. de Bergh, and M. Festou eds, Kluwer Academic Publisher, 281.

Tegler, S.C., and Romanishin, W. 1998, Two distinct populations of Kuiper Belt Objects, Nature, 392, 49.

Tegler, S.C., and Romanishin, W. 2000, Extremely red Kuiper Belt Objects in near-circular orbits beyond 40 AU, Nature, 407, 979.

Tegler, S.C., and Romanishin, W. 2003, Resolution of the Kuiper Belt object color controversy: two distinct color populations, Icarus, 161, 181.

Thompson, W.R., Murray, B.G., Khare, B.N., and Sagan, C. 1987, Coloration and darkening of methane clathrate and other ices by charged particle irradiation - applications to the outer solar system, JGR, 92, 14933.

Trujillo, C.A., and Brown, M.E. 2002, A Correlation between Inclination and Color in the Classical Kuiper Belt, Astrophys. J. 566, 125.

Trujillo, C.A., and Brown, M.E. 2004, The Caltech Wide Area Sky Survey, Earth, Moon, and Planets, in press.

Womack, M., and Stern, S.A. 1999, The Detection of Carbon Monoxide Gas Emission in (2060) Chiron, Astronom. Vestnik, 33, 187.

Yelle, R.V., Lunine, J.I., Pollack, J.B., and Brown, R.H. 1995, Lower atmospheric structure and surface-atmosphere interactions on Triton, in Neptune and Triton, ed. D.P. Cruikshank, Univ. Arizona Press, 1031.

Chapter 10

INTERPLANETARY DUST PARTICLES AND ASTROBIOLOGY

F.J. Molster
ESTEC/ESA
Keplerlaan 1
2201 AZ Netherlands
frank.molster@esa.int

Abstract Interplanetary Dust Particles are amongst the most pristine materials of the Solar System that can be studied here on Earth. The study of these primitive particles gives a lot of information about the evolution of our Solar system and about the delivery of (pre-)biotic material on Earth. Although the sample size of IDPs is small, typically 10^{-9} gram, this does not prevent the study of them and several techniques are available. At the moment the possibilities for detailed astrobiology research are limited. But with the present day evolution of the different instruments, the time for detailed astrobiology research of interplanetary dust particles is near.

Keywords: Astrobiology, Interplanetary Dust Particles

1. Introduction

Stratospheric dust particles are collected by NASA U2 planes (see Fig 10.1). Among the particles collected are particles with an extraterrestrial origin. These particles, better known as interplanetary dust particles (IDPs), are small ($< 50\mu$m) dust particles (Fig 10.2) originating from the tails of active Oort-cloud and Kuiper-belt comets and the collisions of asteroids with each other or with (small) meteorites. The IDPs from comets are expected to be among the most pristine materials available for scientific research. They contain (sometimes relatively large amounts of) material that has not experienced significant thermal or aqueous processing.

P. Ehrenfreund et al. (eds.), Astrobiology: Future Perspectives, 245–265.

Some of these IDPs are amongst the most chemically and isotopically primitive meteoritic materials that can be studied in laboratories here on Earth. Several IDPs contain bulk ratios of D/H which are far beyond the average solar system ratios. High spatial resolution investigation show that these anomalous D/H ratios are often concentrated in hot-spots located in organic matter. The large D/H ratios (δD of 50,000 per mil, i.e. 50 times solar) found by Messenger et al. 2000 and Keller et al. 2000 in an IDP are only observed in the cold molecular clouds from which new stars form. This is clear evidence that some IDPs contain pre-solar material (that has not been exposed to high temperatures). Looking at the D/H ratios, Aléon et al. (2001) found three distinct classes of organic matter within an IDP with different D/H and C/H ratios. These three different phases do not only give important information about the D/H ratio evolution, but at the same time also about the carbonaceous matter evolution in the proto-solar nebula. There is more evidence that IDPs contain materials from different phases in the proto-solar nebula evolution. Besides the evidence for material that only experienced low temperature, IDPs often contain also clear evidence for materials that have undergone high temperature processing, like crystalline silicates. All these different components in a single IDP provide valuable insight in the grain dynamics and chemistry in the (early) solar system and in the pre-solar interstellar and circumstellar environment.

There are several ways to classify IDPs. An often-used classification is based on the difference between anhydrous and hydrous IDPs, the last ones show (some) evidence for aqueous alteration, in the form of layer silicates, like smectite and saponite, while the former ones are often dominated by olivine and/or pyroxene. A second distinction often made is between the cluster and non-cluster particles. Cluster particles are so fragile that they break into multiple fragments upon impact with the collection surface, while non-cluster particles seem to survive the capture as one piece. Isotopic anomalies are common in cluster IDPs.

There are some statistics about the morphology of an IDP and its origin, but in general it is quite difficult to indicate for an individual IDP where it came from based on its appearance. The chondritic smooth (CS) IDPs often contain hydrated (layer-lattice) silicates (smectite, saponite or serpentine) and show compositional and spectroscopic similarities with the main-belt C-type asteroids. The members of the chondritic porous (CP) group contain anhydrous silicates (pyroxene and olivine), although it should be noted that layer silicates have been found in chondritic porous IDPs (Rietmeijer and Mackinnon, 1985). The origin of the CP IDPs is sought in the primitive P and D type asteroids and comets, where postaccretional (aqueous) alteration likely did not

Figure 10.1. The U2 plane used for the collection of IDPs high in the stratosphere. The dust collectors are located at the tip of the wing. Photo credit: NASA.

take place. For a further discussion of the mineralogical classification of IDPs see: Brownlee et al., 1982; Sandford & Walker, 1985; Schramm et al., 1989.

Another distinction between IDPs from comets and those from asteroids can be made based on their entry velocity. During the entry in the Earth atmosphere IDPs get heated. IDPs coming from asteroids will have in general a lower entry velocity than those having a cometary origin. This is due to the fact that, particles originating from the asteroid belt are more or less in the same plane as the Earth and slowly spiral towards the Sun due to the Poynting-Robertson drag, and the velocity difference with the Earth is relatively low (about the escape velocity of the Earth). Cometary IDPs on the other hand often cross the Earth orbit with a high inclination, and the relative velocities are therefore much higher. In general the IDPs coming from comets experience a peak heating which is roughly 300K higher than those of similar size and mass coming from the asteroid belt (Love and Brownlee 1991; 1994). Solar wind implanted gas release experiments can tell us something about the entry velocity and therefore about their origin. By measuring the temperature at which gasses are released from the IDP in the lab, one can make an estimate of the entry velocity.

The preservation of solar flare tracks in the silicate crystals of many IDPs shows already that the maximum temperature experienced by many of the particles was less than 900 K, which is the estimated pyrolysis temperature of the organic matter. Calculations of the maximum temperature arrived during entry in the Earth atmosphere, also confirm that many particles with the size and structure of a typical IDP will not heat up above 900 K (Flynn et al. 2000). So, pre-biotic molecules have a good chance to survive the delivery to Earth.

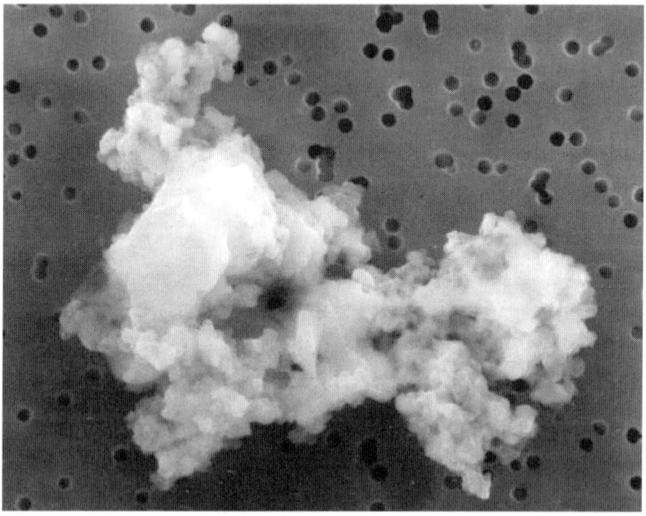

Figure 10.2. IDP U2012C11, is a typical cluster IDP. The size of the IDP is roughly 10μm. Photo credit: NASA.

Most IDPs are aggregates of diverse materials. As noted before, it is even possible that you will find high-temperature processed materials close to materials that still have evidence of low temperature processing. The main components of IDPs are I) silicates: amorphous silicates, crystalline anhydrous and hydrous silicates, II) carbonaceous material: amorphous carbon, nano-diamonds and organic material (macromolecular material with aromatic and aliphatic nature), graphite (although this is in general an artefact of atmospheric entry heating), III) other materials: oxides, metal-sulphides. The abundance (or even absence) of the different materials give indications about the dust processing in the (early) Solar system. For example, the apparent absence of nano-diamonds in IDPs associated with comets and their omni presence in meteorites, which come from the asteroid belt, suggests that (most)

Table 10.1. Limited overview of the analytical techniques which can be performed on IDPs. BMCA = Bulk and Mineral Compositional Analyses, MAS = Mineralogy and Atomic Structure. Partially implies that due to the sample preparation the particle is partially destroyed, but that the sample is still useful for other investigations.

Technique	Used for	Sample mass	Destructiveness
Light-Optical Techniques	Imaging	ng	non-destructive
Micro Raman Spectroscopy	Organics	ng-mg	non-destructive
Scanning Electron Microscopy/ Energy Dispersive Spectrometry	Imaging	ng	non-destructive
(Synchrotron) X-ray Diffraction/ Fluorescence tomography	MAS & BMCA	ng	non-destructive
Transmission and Reflectance Visual to FIR spectroscopy	Organics & MAS	ng-mg	non-destructive to partially
(Atomic/Magnetic) Force Microscopy	Imaging & MAS	ng	partially
Transmission Electron Microscopy/ Analytical Electron Microscopy	Imaging & MAS	ng	partially
Electron Energy Loss Spectroscopy	MAS	ng	partially
Scanning Transmission X-ray Microscopy	Organics & MAS	ng	partially
Ion Mass Spectrometry	Imaging & MAS & BMCA	ng	destructive/ partially
Laser desorption/laser ionisation mass spectroscopy	Organics & MAS	ng	destructive/ partially

nano-diamonds do not have a pre-solar origin, but are formed inside our Solar system (Dai et al. 2002). Mineralogical information can be used to trace back the origin of an IDP, and if related with pre-biotic material also the origin of this pre-biotic material can be traced back (and maybe the formation history and further evolution).

Roughly 30,000 tons of IDPs reach the Earth each year (Love & Brownlee, 1993), and this number was likely higher in the early days of the Earth. So, if the delivery of pre-biotic molecules from space was important for the origin of life, IDPs might have had an important rôle in this process. Since the material that IDPs deliver here on Earth is more or less similar (although the ratios of pristine and altered matter might differ) to what has been delivered 3.5-4.5 billion years ago, it is very worthwhile to investigate them in the field of astrobiology.

This paper is as organised as follow. First the different techniques to study the IDPs will be discussed. Secondly we will discuss the link with astrobiology, and we will end with some conclusions.

2. How to study IDPs

As noted above, IDPs contain very interesting information about the history of the solar system, and might even provide clues about the origin of life on Earth. However, there is one drawback studying them. IDPs are very small and have therefore only a very small sample mass (typically 10^{-9}g). Furthermore, IDPs are very fine grained, with typical grain sizes of (much) less than 1 μm. This is very different from the most primitive anhydrous meteorites, which have grain sizes at least an order of magnitude larger. Fortunately, even with the small overall mass of an IDP and the even tinier individual components in an IDP a lot of useful research can be done and has been done.

In Table 10.1 an (incomplete) overview is given of the different types of investigations that are possible with such small amounts of material and for which purpose it is used. Also indicated is whether the technique is destructive or not (either the technique itself, or due to the necessary sample preparation). In the following the above-mentioned techniques will be discussed. This overview is not meant to be complete, but only to explain a few of the most used and useful methods available to study the IDPs.

Light Optical Techniques

IDPs are large enough that they can be visualized in a good microscope, which very much aids the handling of these particles. It also gives a first impression of the particle and helps to focus other instruments. However due to the small size of the individual components of an IDP, its usefulness for other investigations is quite limited.

Micro Raman Spectroscopy

Raman spectroscopy provides information about the molecular bonds of the constituent components. Raman spectroscopy can be used to determine on a micrometer scale the different C bonds in a material and to distinguish between carbon-bonds in diamond, graphite, carbonates, benzene-rings or kerogen like material. This makes it in principle a very powerful method to look for organic matter. Wopenka (1988) was one of the first who performed micro-Raman spectroscopy on IDPs. She found that deuterium depletions were connected with the most ordered carbonaceous material, while deuterium enhancements were found in carbonaceous material with a degree of order, which was significantly lower. She also noted that no traces were found of silicate Raman bands, and attributes this to the fact that these silicates might be coated with carbonaceous material, which also explains why the particles look black in the visual.

Scanning Electron Microscope and Energy Dispersive Spectrometry

From every particle that has been catalogued a Scanning Electron Microscope (SEM) image has been taken. Also a preliminary investigation of its atomic composition is done with the aid of Energy Dispersive Spectrometry (EDS). This non-destructive technique not only allows the identification of the extraterrestrial material it also serves as a guideline for a first classification of the IDP. This information is stored in a catalogue and used for the distribution of the samples. Figure 10.2 is a typical example of such an image.

Bulk chemical composition EDS measurements by Schramm et al. (1989) of 200 IDPs show that chondritic porous IDPs form a close match to the bulk CI chondrites (which are supposed to be the best match to the solar atomic ratios). On the other hand the chondritic smooth IDPs are systematic depleted in Ca and Mg and contain stoichiometric excess oxygen. This is consistent with the presence of hydrous phases, and correlates well with a CM or CI asteroidal origin of the CS IDPs.

Thomas et al. (1995) found after an EDS study of more than 100 IDPs that chondritic IDPs (hydrated and anhydrous) in general have C contents which are on average 2-3 times CI. Some IDPs have the highest bulk C abundances (sometimes the weight in C is more than in any other element) of any extraterrestrial sample.

Synchrotron X-ray Diffraction and X-ray fluorescence Tomography

Due to the small size of IDPs X-ray diffraction and X-ray fluorescence tomography studies should be done with an intense synchrotron light source.

X-ray diffraction studies are a non-destructive technique, by which it is possible to identify the different minerals in an IDP based on the diffraction pattern. This method can clearly distinguish between the hydrous and anhydrous IDPs (see for example Nakamura et al. 1999t) and is therefore a very useful tool to learn more about the origin of an IDP.

X-ray fluorescence (micro)tomography is another non destructive method that has been applied to IDPs (see for example Sutton et al. 2000). With this method it is possible to form a 3D image of the heavy elements (above S) through the whole IDP. Elements between Si and S might still be "imaged" in the smaller IDPs, but below that is not possible due to the limited escape depth for the K-line fluorescence X-rays. Indirectly one can still get some information about the lighter elements by looking at the heavier trace elements that follow the lighter ones (for example Sr is likely to follow Ca).

It can be concluded that these synchrotron X-ray analysing methods are very useful tools to study the minerals and can give important information about the mineralogical evolution of the IDP. However in the light of astrobiology, it might be of less direct relevance.

Transmission and Reflectance spectroscopy from UV to FIR

Transmission and reflection spectroscopy is in principle a very powerful technique to identify not only minerals but also molecules. However problems still exist and the sample preparation technique can be destructive.

Ultravioltet-Visible (UV-VIS). UV absorption measurements have been attempted, but were not really successful up to now (see for example Gezo et al 2000). The small sample size and the special requirements for the optics make it difficult to succeed, but it appears to be possible. A disadvantage of transmission measurements in the UV is that the IDPs have to be cut in thin slices ($< 1\mu$m), on the other hand the measurement itself is non-destructive.

Reflection studies have been performed from the UV to the near-IR. Bradley et al. (1996) found that this non-destructive technique in principle works, but that not many details could be investigated. Still, different classes of IDPs can be distinguished based on spectral trends. Around 1μm also information about the silicate minerals can be derived. Strongly heated IDPs (during atmospheric entry), which developed a magnetite rim, are detectable. Unfortunately in the UV the small components in an IDP cause spectral artefacts. Bradley et al. also found that the chondritic IDPs are dark objects in the visual. This is very likely caused by the carbonaceous matter that encompasses everything, although the presence of nanophase Fe-metal grains in GEMS can also not be excluded as source for this darkening.

IR radiation. Another powerful and non-destructive tool to investigate the global material properties of the IDP is infrared (IR) transmission spectroscopy. Many molecular and solid state transitions can be found in this wavelength range. In principle this can give you the mineralogical and the molecular structure of your IDP. Infrared transmission spectra of IDPs have already been taken for almost 20 years (Sandford & Walker, 1985), and do require neither the use of a synchrotron light source nor destruction of the IDP. However if one wants to go to wavelengths beyond 20μm the intense beam of a synchrotron light source becomes inevitable. Currently, there exist synchrotron light sources that allow absorption spectra of IDPs to be measured far beyond 30μm (see Fig. 10.3).

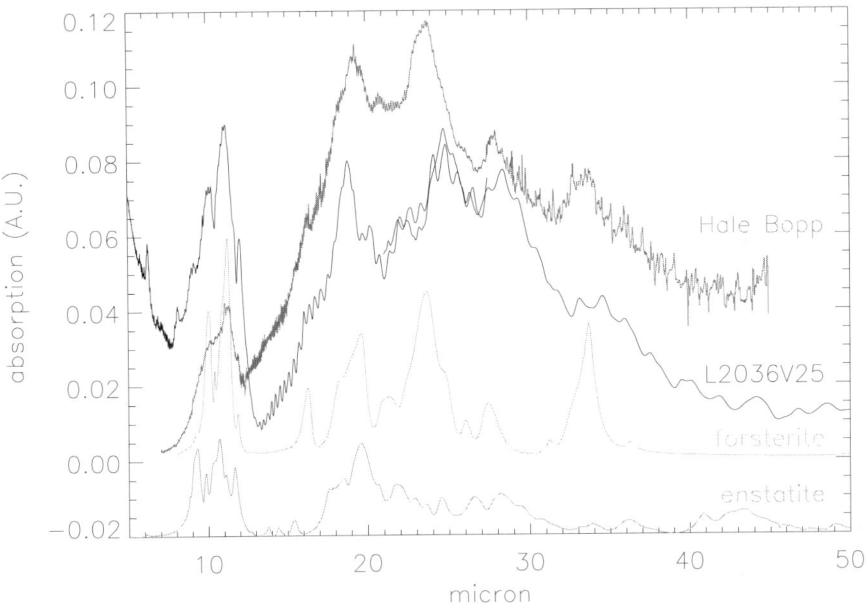

Figure 10.3. The first far-IR transmission spectra of an IDP (IDP L2036V25), compared with the laboratory spectra of enstatite and forsterite, and the ISO (emission) spectrum of Hale Bopp. Note the shift to longer wavelengths in the most prominent peak positions of the IDP. This is an indication for a slightly different composition of the crystalline silicates. Figure taken from Molster et al. 2003.

Because of the high beam intensity, these synchrotron facilities also make it possible to measure the absorption spectra of ultra-microtomed slices (thickness less than 100nm) of an IDP. These slices are thin enough that they can also be analysed in a transmission electron microscope (TEM). So, in principle characteristic absorption features can now be traced back to individual components. In Figure 10.4 one can see the spectra of several slices of one IDP (IDP U220A19). The absence or presence of features in a single slice can be combined with the compositional information determined by TEM analysis. In principle, this is also a great way to identify the infrared spectral characteristics of components for which no laboratory spectra exist yet.

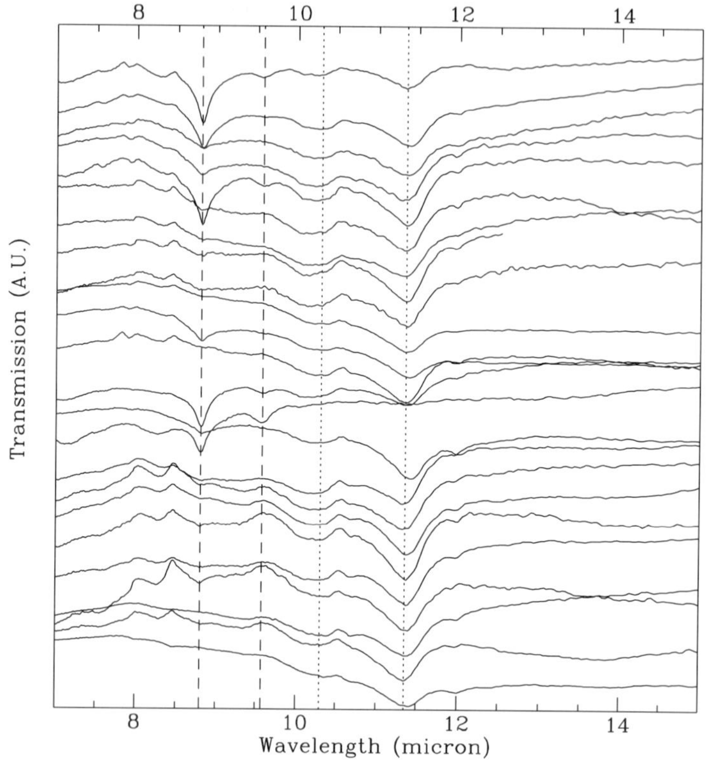

Figure 10.4. IR transmission spectra of different ultra-microtomed slices of IDP U220A19 (the same IDP as in Fig 10.6). Note the differences in the individual spectra (which are shifted with respect to each other for clarity). With the aid of TEM measurements of the individual slices they can be traced back to individual components. For example, the absorption features at 10.3 and 11.3μm (see the dotted lines) are due to olivines (the broad feature around 10.3 micron has also a contribution from pyroxenes) and the features at 8.8 and 9.6 μm (see the dashed lines) are due to residual silicon oil.

Infrared transmission spectroscopy is a great tool for the investigation of minerals. But it is not limited to minerals, it can also be used to study the organic compounds in an IDP as can be seen in Fig 10.5. However, many organic molecules show rather similar features in the infrared region, because they are all based on the same kind of vibrations (i.e. C-H, C-C, C-O and C-N). Differences between the different molecules are still beyond our reach to be detected by IR-transmission spectroscopy. Future developments are likely to improve this situation. In any case IR spectroscopy is very powerful to investigate the mineralogical contents (and history) of the individual IDPs.

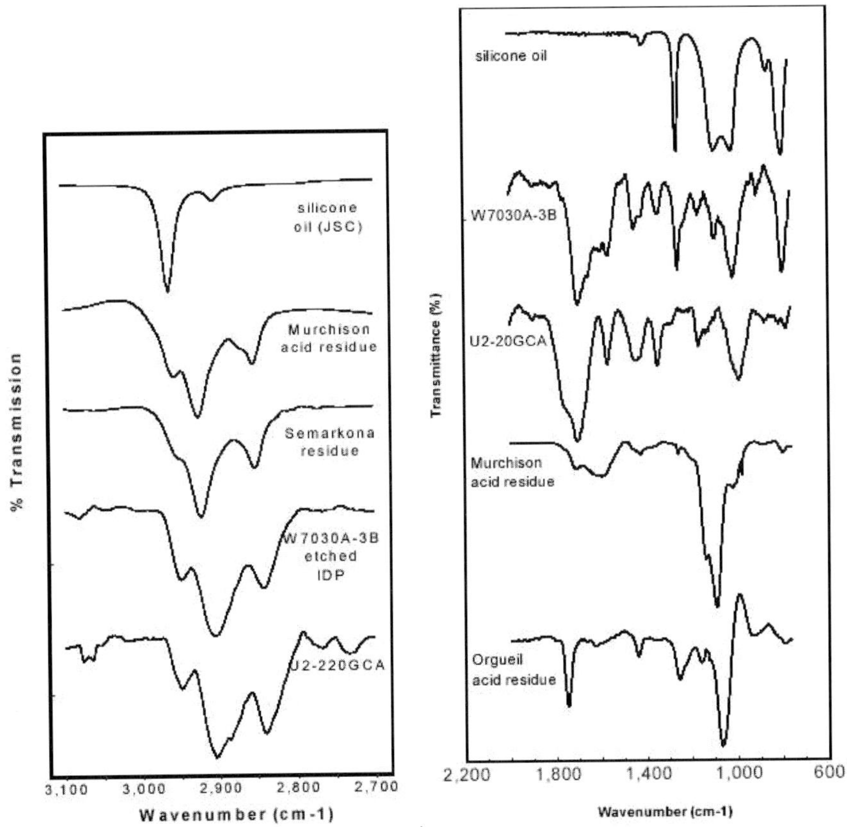

Figure 10.5. IR transmission spectra of different samples, including meteorites and IDPs showing the organic bonds. Pictures taken from Flynn et al. (2002)

Atomic/Magnetic Force Microscopy

Atomic force microscopy has been performed on IDPs (Nakamura et al., 1999k) mostly as a test for MIDAS onboard ESA's cometary mission: Rosetta (Schwehm, 1995). This technique is able to achieve high-resolution (up to nm scale) 3D images of the surface structure. It is therefore able to detect small surface interactions due to space weathering or entry in the Earth atmosphere. In tapping mode it is also possible to see differences between different materials. It is not possible yet to trace these differences back to material identifications, but indications for carbonaceous coatings might be found in this way.

Magnetic force microscopy is able to detect small magnetic domains inside particles. This can be very important for the dust coagulation in the early solar system (see for example Nübold et al. 2003) and thus for the dust evolution (and shielding of carbonaceous materials). In principle both techniques do not have to be destructive.

(Scanning) Transmission Electron Microscope

Sub-micron details about the mineralogical composition and atomic abundances of the IDP are best investigated with the aid of a (Scanning) Transmission Electron Microscope ((S)TEM), equipped with an EDS and/or EELS. This also requires the cutting of the IDP in very thin slices, and only the 2-D structure remains, however the investigation of several slices next to each other, can still give some 3D information. An example of a TEM image can be seen in Figure 10.6. From the same IDP, even from the same section as from which this image is part of, also the IR transmission characteristics have been analysed (see Figure 10.4). For small-scale investigations of the compositions and atomic bonds the TEM remains the most used instrument. It is also very useful to see solid-state evolutionary marks in the different crystals and the interactions between materials and different kinds of irradiation.

Electron Energy Loss Spectroscopy

Electron energy loss spectroscopy (EELS), in general attached to a TEM, is used to determine the composition and atomic binding state of the material. For example, differences between diamond and graphitic carbon are easily seen with this method. With EELS it is also possible to determine the oxidation state of Fe. Furthermore, EELS is also used to study the implanted hydrogen in GEMS, by studying the OH bond. Advances in the technology make it now possible to map also the light elements (C, O, N etc.). Because of the use of electrons, the investigated area can be very small.

This technique requires the cutting of the IDP in very thin slices and is therefore partially destructive, although it can of course still be used for imaging with the TEM, or absorption spectroscopy.

Scanning Transmission X-ray Microscopy

A Scanning Transmission X-ray Microscope (STXM) can be used to detect K & L-edge absorption of the different elements up to a spa-

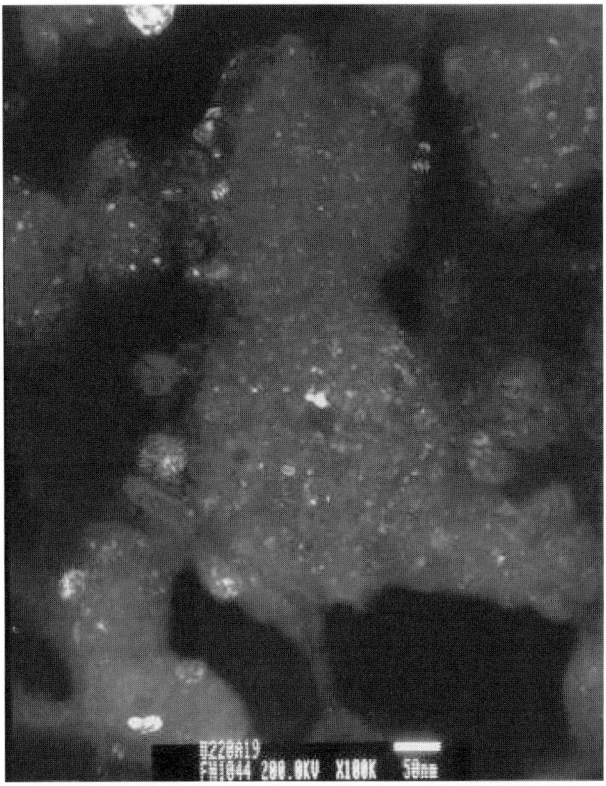

Figure 10.6. A TEM backscattering image of a part of the IDP U220A19. The part shown here contains predominantly GEMS (Glass with embedded metals and sulphides) The bright areas are metal(-sulphides), while the grey area is a glassy silicate. In Figure 10.4 one can see the IR transmission spectroscopy data from this IDP.

tial resolution of 50nm. This can be used to estimate the elemental abundance and its bonding state. If X-ray absorption near edge structure (XANES) spectroscopy is used, one can detect absorptions that correspond to photon induced transitions from core level electrons into various bound and virtually excited states. Variations in the electron density surrounding these photo-excited states leads to absorption band profiles, corresponding to different functional groups of the investigated element (see for example Figure 10.7 for carbon).

This technique has now been used for several years for IDPs (Chapman et al. 1995) starting with C-XANES. Nowadays, the energies of the XANES are diagnostic of the type of bonding for C, N and O. It therefore

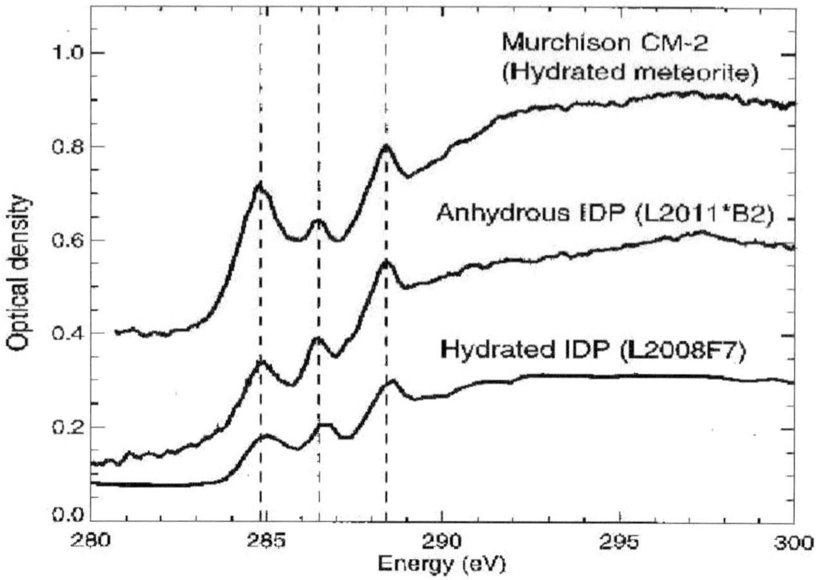

Figure 10.7. C-Xanes spectra of the carbon rich regions of two IDPs, one hydrated and the other anhydrous compared to the Murchison carbonaceous meteorite. The peak at 285 eV is characteristic for the C-ring bonding, while the peak at 288.5 eV is due to the carbonyl bond. Picture taken from Flynn et al. (2003).

allows the identification of the functional groups in organic matter on sub-micron scale. So, although this does not give information about the exact molecules present, it does give an indication of the building blocks of these molecules. For example, the C-XANES structure at $\approx$285 eV is a characteristic of the C-ring while at $\approx$288.5 eV there is a feature which is characteristic for the C=O bond (carbonyl). The C-ring structure is found both in elemental (graphitic or amorphous) carbon and in organic carbon (for example PAHs). The presence of the carbonyl absorption in IDPs (Flynn et al. 2001) indicates that IDPs contain some organic carbon. The height of the C-edge, i.e. the increase in absorption from 282 to 300 eV is a direct measure of the total mass of carbon, while the area of the C=O absorption is proportional to the abundance of this functional group. Based on these results it has been estimated that at least 1-2 wt-% of the total particle mass is bound in the C=O functional group. These results are confirmed by O-XANES measurements where also the C=O absorption has been found (Flynn et al. 2001; 2003).

This technique also requires ultramicrotomed thin sections, and is therefore destructive. Fortunately, as stated above these thin sections can still be used for other observations.

Ion Mass Spectroscopy

With an ion microprobe a beam of focussed ions is used to sputter the first atomic layers of the sample. These sputtered atoms are then analysed. Often they are ionised, filtered according to their energy, selected in mass and finally counted by electron multipliers. This process is called secondary ion mass spectroscopy (SIMS). With this technique it is even possible to detect different isotopes of the same element. Since you sputter the top layers away, in principle, this a destructive process, however since it is only a few atomic layers, the samples can often still be used for other experiments.

With this method it is in principle possible to investigate the whole range of atomic and isotopic compositions, however in reality it is difficult to measure very different atoms at the same time, due to the set-up of the instruments. The focus of IDP studies has been primarily on the light elements H, C, N, O and sometimes Si, S and Mg. Due to the sputtering the bonding structure is lost in this process.

Nano-SIMS. The Nano-SIMS is a recently developed secondary ion mass spectrometer to work below the 100nm scale. It is very suitable to measure isotopic ratios of the different elements on very small scales. Recently Messenger et al. (2003a) discovered the presence of small silicate particles which have clear non-solar oxygen isotopic ratios. From three of the discovered grains the mineralogical structure could be identified with TEM analysis. Two grains were identified as GEMS (Glass with Embedded Metals and Sulphides; Bradley, 1994), of which one is displayed in Figure 10.8, and the third one was a forsterite crystal. GEMS are often found in IDPs and in many aspects resemble the interstellar medium (ISM) grains (Bradley et al., 1999). This lead to the speculation that GEMS have a pre-solar origin. The findings of Messenger et al. give this claim even more ground. This kind of research proves the usefulness of multiple ways to investigate the same sample.

Gas release experiments. Several experiments have been performed to measure the noble gas release from IDPs at different temperatures. In general an oven is used, to step-heat the sample. After each step in temperature the release of the noble gasses is measured by an ion mass spectrometer (see for example, Nier & Schlutter, 1992; Brownlee et al. 1993; Pepin et al., 2000). These measurements give an estimate for

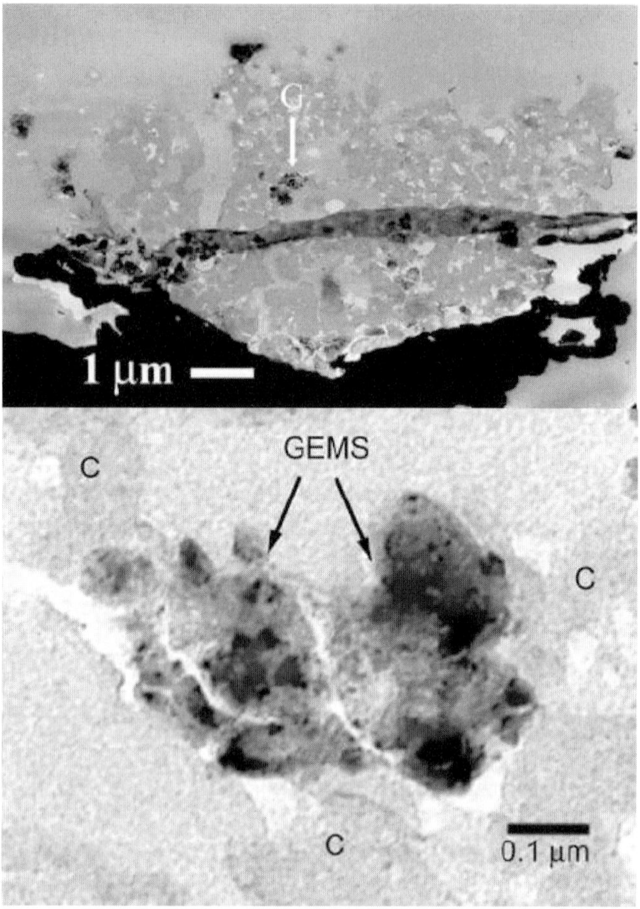

Figure 10.8. TEM images of IDP L2036 C14 showing the location of a presolar GEMS grain (G). The lacy grey material surrounding this grain is abundant indigenous carbonaceous matrix. The black material at the bottom is the gold from the sample mount. A fold in the slice is also apparent. The Figure is taken from Messenger et al. (2003b).

the maximum temperature experienced by the IDP during entry in the Earth atmosphere (i.e. the point where the release of noble gases suddenly increases), and are therefore an indication of the origin of the IDP. As long as the sample is not heated to temperatures where significant alteration is expected, it is a non-destructive method and the samples can be used for other experiments, like morphological and mineralogical studies.

Laser desorption/laser ionisation mass spectroscopy

This technique is more or less similar to ion microprobe technique. However, heating via a laser beam now does the ablation of the top layers. In principle similar results can be achieved. The spatial resolution is comparable to a normal SIMS (not a nano-SIMS). With this technique Clemett et al. (1993) detected PAHs with zeptomole (10^{-21} mole) sensitivity. The bulk of the PAHs found have an atomic mass between 200 and 450 amu, and there is also some contribution around 60 amu. The virtual absence of the low mass PAHs, such as alkyl-benzenes, alkyl-naphthalenes, and alkyl-phenanthrenes on these IDPs is consistent with evaporation during prolonged exposure in space, as well as the loss by heating during atmospheric entry.

Kehm et al. (2002) used laser heating to release the noble gases. They found a rough correlation between the Zn/Fe ratios and the He content. Zn-poor particles generally have lower He contents than other IDPs. This suggests that both elements were lost by heating during atmospheric entry. It confirms the view that Zn can serve as an entry-heating indicator in IDPs (Kehm et al. (2002). The determination of the Zn/Fe ratio by EDS thus provides a non-destructive method to get an estimate of the entry velocity and therefore the origin.

The laser desorption spectroscopy is naturally destructive, although it can be limited to the top layers. The use of a laser in the gas release experiment does not have to be destructive but alteration due to the high temperatures is expected, especially, since there is no good control of the temperature reached in the IDP.

3. IDPs and astrobiology

It is still not well known how life emerged here on Earth. However there is quite some evidence that if pre-biotic matter formed in space it can have efficiently been transported to Earth via IDPs (Anders 1989).

IDPs can contain a very significant fraction of carbon. The carbonaceous material can hold up to 90 volume% of an IDP. In Fig. 10.9 one can see an IDP before and after HF etching. During this etching procedure most components are etched away (especially the abundant silicates). The only left over material is the carbonaceous material and metal sulphides as can be seen in the bottom part in the graph of Fig. 10.9. It is clear from these kind of experiments that the carbonaceous matter is the glue that holds the IDP together (see also Fig. 10.10). This carbonaceous material is not only badly ordered graphite, but can also contains quite a significant fraction of organic matter. Due to the relative small thermal heating during atmospheric entry, this material has

Figure 10.9. The result of HF etching of an IDP. The top image shows the original IDP. The middle image shows the IDP after HF etching. The bottom graph shows the atomic composition before and after HF etching. Pictures courtesy of Don Brownlee

a good change to survive the way to the surface of the Earth. So, if life emerged from pre-biotic molecules brought from space, the IDPs might have been (and still are) a very important carrier.

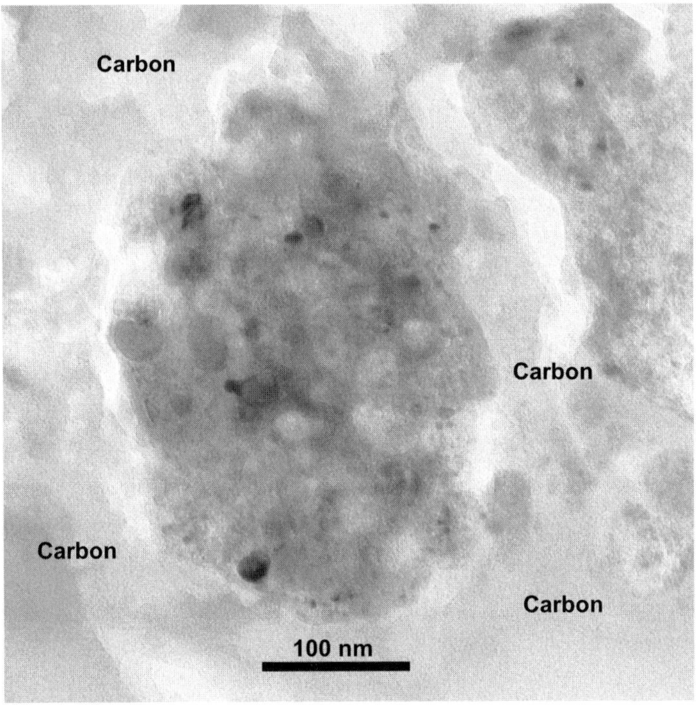

Figure 10.10. TEM image of a GEMS (Glass with embedded metals and sulphides) in an IDP, note the way carbon is acting as the glue of this rubble pile

It can be concluded that, IDPs form unique and interesting samples for the study of extraterrestrial pre-biotic matter and are therefore of great significance for astrobiology. Unfortunately, although a lot of tools exist to study these particles, there is at the moment not a tool available to extract and to uniquely identify the very small amounts of organic molecules. A lot of research can still be done and information can be retrieved about the possible formation conditions and environments of these pre-biotic molecules. And the instrumental possibilities are still growing and will soon reach the level that individual molecules can be uniquely defined at those levels. Finally, it should not be forgotten: IDPs have the whole history of the Solar system written in their mineralogical structure and therefore also indirectly the history of the pre-biotic matter.

4. Conclusions

It can be concluded that IDPs are very relevant to the study of the materials which has been delivered to the Earth in its early days. They contain materials dating back from the beginning or even before the formation of the solar system, and at the same time also processed materials. So, in principle the study of the IDPs would enable us to get important information about all the different kinds of materials that has been delivered to Earth during its lifetime. However there is one problem: The sample size. IDPs are very small ($< 50\mu m$) and not so abundant and therefore only a limited amount of material (typically in the order of a few nanogram) is available for scientific research. Furthermore, IDPs are aggregates of diverse materials, which makes the analysis even more difficult. Fortunately, it is not impossible to investigate these particles. Several techniques described in this paper have been employed for the research of these particles. At the moment a lot of work is put into new instrumentation. So, although it seems not possible yet to identify individual organic molecules, this likely will become possible in the near future. Important in this aspect is also the Stardust mission. This mission will bring cometary material back to Earth in 2006 and has triggered instrumentation development in several fields. So, maybe limited at the moment to indirect investigations, the direct investigation possibilities for astrobiology are expected to grow in the near future, and a close eye should therefore be kept on the developments in the field of IDP research.

Acknowledgments

The author wish to thank Dr. J.P. Bradley, Dr. D. Brownlee and Dr. G.J. Flynn for the use of their figures and an anonymous referee for the useful comments.

References

Aléon, J., Engrand, C., Robert, F. & Chaussidon, M., 2001, *Geochim. Cosmochim. Acta* **65**, 4399
Anders, E., 1989, *Nature* **342**, 255
Bradley, J.P., 1994, *Science* **265**, 925
Bradley, J.P., Keller, L.P., Brownlee, D.E. & Thomas, K.L., 1996, *Meteoritics & Planetary Science* **31**, 394
Bradley, J.P., Keller, L.P., Snow., T.P., et al., 1999, *Science* **285**, 1716
Brownlee, D.E., Olszewski, E. & Wheelock, M.M., 1982, *LPSC* **13**, 71
Brownlee, D.E., Joswiak, D.J., Love S.G., et al., 1993 *Meteoritics* **28**, 332
Chapman, H.N., Bajt, S. Flynn, G.J. & Keller, L.P., 1995 *Meteoritics* **30**, 496
Clemett, S.J., Maechling, C.R., Zare, R.N., 1993, *LPSC* **24**, 309

Dai, Z.R., Bradley, J.P., Joswiak, D.J., et al. 2002, *Nature* **418**, 157
Flynn, G.J., Keller, L.P., Jacobsen, C. & Wirick, S., 1998, *Meteoritics & Planetary Science* **33**, A50
Flynn, G.J., Feser, M., Keller, L.P., et al., 2001, *LPSC* **31**, abstract nr: 1603
Flynn, G.J., Keller, L.P., Joswiak, D. & Brownlee, D.E., 2002, *LPSC* **33**, abstract nr: 1320
Flynn, G.J., Keller, L.P., Wirick, S. & Jacobsen, C., 2003, *LPSC* **34**, abstract nr: 1652
Gezo, J.C., Bradley, J.P., Brownlee, D.E., et al., 2000, *LPSC* **31**, abstract nr: 1816
Kehm, K., Flynn, G.J., Sutton, S.R. & Hohenberg, C.M., 2002, *Meteoritics and Planetary Science* **37**, 1323
Keller, L.P., Messenger, S., Flynn, G.J. et al., 2000, *Meteoritics & Planetary Science* **35**, A86
Love, S.G. & Brownlee, D.E., 1991, *Icarus* **89**, 26
Love, S.G. & Brownlee, D.E., 1993, *Science* **262**, 550
Love, S.G. & Brownlee, D.E., 1994, *Meteoritics* **29**, 69
Messenger, S., Keller, L.P., Stadermann, F.J., et al. 2003a, *Science* **300**, 105
Messenger, S., Keller, L.P. & Walker, R.M., 2003b, *LPSC* **34**, abstract nr: 1630
Messenger, S., 2000, *Nature* **404**, 968
Molster, F.J., Demyk, K., d'Hendecourt, L., et al., 2003, *LPSC* **34**, abstract nr: 1148
Nakamura, K., Klöck, W., Romstedt, J., et al., 1999k, *Meteoritics and Planetary Science* **34**, A85
Nakamura, T., Nakamuta. Y. & Tanaka, M., 1999t, *Meteoritics and Planetary Science* **34**, A86
Nier, A.O. & Schlutter, D.J., 1992, *Meteoritics* **27**, 166
Nübold, H., Poppe, T., Rost, M., et al., 2003, *Icarus* **165**, 195
Pepin, R.O., Palma, R.L., Schlutter, D.J., 2000, *Meteoritics and Planetary Science* **35**, 495
Rietmeijer, F.J.M. & Mackinnon, I.D.R., 1985, *LPSC* **16**, 149
Sandford, S.A. & Walker, R.M., 1985, *Astrophys. J.* **291**, 838
Schramm, L.S., Brownlee, D.E. & Wheelock, M.M., 1989, *Meteoritics* **24**, 99
Schwehm, G., 1995, *ESA-SP* **1179**, 28
Sutton, S.R., Flynn, G.J., Rivers, M., et al., 2000, *LPSC* **31**, abstract nr: 1857
Thomas, K.L., Keller, L.P. & McKay, D.S., 1995, *Meteoritics* **30**, 587
Wopenka, B., 1988, *Earth & Planetary Science Letters* **88**, 221

Chapter 11

THE PREBIOTIC ATMOSPHERE OF THE EARTH

Franck Selsis
Centro de Astrobiología (CSIC-INTA)
Ctra de Ajalvir km 4 - 28850 Torrejón de Ardoz - Madrid - Spain
selsis@obs.u-bordeaux1.fr

Abstract Because of the paucity of geologic samples dating back to the early Earth, trying to unveil the nature of the prebiotic terrestrial environment is as frustrating as it is fascinating. An understanding of the characteristics of this period in our planet's history is, however, crucial to studies of the origin of Life. Recent progress in astrophysics, geochemistry and simulation of planetary accretion provide some new and precious constraints but also question some of the previously admitted "facts". The aim of this chapter is to highlight some of the important open-ended questions about the environment of the prebiotic Earth (for a more comprehensive review, see Kasting and Catling, 2003). The implications of recent theories of terrestrial planets formation on the origin and early evolution of the atmosphere are discussed, in particular, along with the various influences of our faint but active young Sun during this period.

Keywords: early Earth, primitive atmosphere, young Sun, origins of life

1. Introduction: The formation of the Earth

Within the last decade, a coherent scenario has arisen for the formation of terrestrial planets from an initial dusty proto-planetary disk. Much of our insight is the result of numerous N-body simulations of solar system formation (Chambers and Wetherill, 1998; Levison et al., 1998; Chambers, 2001). The evolution from dust to planet can be divided into three successive phases. The first is the coagulation of micrometer-sized grains into kilometer-sized *planetesimals*, a process which may occur in 10^4-10^5 yrs (Weidenschilling and Cuzzi, 1993). In this first step, grains

P. Ehrenfreund et al. (eds.), Astrobiology: Future Perspectives, 267–286.

are assumed to stick together if an impact occurs at a critical, threshold velocity; this value is composition dependent (Kouchi et al., 2002). As this occurs, planetesimals decouple from the gaseous disk and start to interact gravitationally with each other. This catalyses the second phase: a *runaway growth* leading to the formation of planetary *embryos* with masses 1-10% of of that of Earth (Makino et al., 1998). These embryos accrete material locally and form a dense population distributed throughout the solar system. Modeling has these embryos separated typically by $1/100^{\text{th}}$ of an AU. While only 10^4 yrs more are needed to form the embryos, the final mass of the Earth is reached within 50-150 Myrs according to numerical simulations ((Chambers, 2001) or less than 30 Myrs based on constraints of core segregation determined using Hf-W dating (Kleine et al., 2002). This discrepancy can be removed by considering realistic merging of two differentiated bodies during a collision (Halliday, 2004). Indeed, simulations of the Moon-forming impact show a quasi-instantaneous merging of the two cores (Canup and Asphaug, 2001). Thus, the 30 Myrs obtained from Hf-W dating can be interpreted as a mean differentiation time for the embryos that made the Earth. These embryos can incorporate into a planet several AU from their formation region. This last step of the planetary accretion makes the novelty of the scenario: the Earth is no longer formed continuously by accreting planetesimals formed around 1 AU. The final mean composition of the planet is not a simple function of its orbital distance: It results from the mixture of a small number of embryos coming from different heliocentric regions. In this model, the formation of a water-rich planet at 1 AU is a question of luck.

The origin of water

Scenarios assuming a continuous and local accretion of the Earth did not provide a consistent theory for the origin of water. The correlation between the water content and the orbital distance of asteroids (assuming that their present orbital distance does not differ too much from where they formed) suggests that material formed at 1 AU was dry (Morbidelli et al., 2000). This implies a late veneer of chondritic or cometary material to feed the water and volatile reservoir of the Earth. Several authors proposed the comets as a late source of water (Chyba, 1987; Delsemme, 2000) but both deuterium constraints and dynamical models limit the contribution of comets to less than 15% of the terrestrial water content (Morbidelli et al., 2000). Abundances of noble gases and metals may provide an even more drastic limit on the cometary con-

tribution (Dauphas and Marty, 2002), though the abundance of these elements in comets is still derived from laboratory studies and models. A chondritic origin of the water fits the D/H ratio of Earth's oceans but would require the delivery of a "super veneer" of about 1% of the terrestrial mass, which contradicts recent re-estimate of the bombardment (see section 11.2.0).
The existence of very old zircons also requires enough liquid water on Earth to form continental crust as early as -4.4 Gyr (Wilde et al., 2001). Finally, one would have to explain why the late delivery was inefficient on Mars. Indeed the upper limit on the initial relative water and volatile contents are much lower than on Earth and are difficult to explain only by scaling the impact rate or by escape processes (Lammer et al., 2004).
In the "embryos model", where the Earth is the result of the fusion of 10 to 100 large embryos, the final water content depends on the nature and thus the region of formation of these embryos. For instance, a unique Moon-sized embryo formed at 3 AU and made of 10% of structural water would give the Earth 5 times its present water content. Wiechert et al. (2001) showed that the identical isotope fractionation of oxygen on the Earth and the Moon implies a similar composition of the Moon-forming impactor "Theia" and the proto-Earth. Assuming that the oxygen fractionation is a signature of the heliocentric distance of formation, most of the embryos that formed the Earth and Theia should thus have originated from a narrow circumsolar region. A large contribution of "wet" embryos coming from more distant regions would have modified this signature. In addition to numerical simulations (Chambers, 2001; Raymond et al., 2004), this is another argument suggesting that the delivery of water to the telluric planets is a stochastic process relying on a small number of collisions with wet embryos.
The relatively small initial Martian water reservoir can be understood in this context: At the orbital distance of Mars, planetary formation is less efficient because of the influence of Jupiter and Mars is probably a remaining dry embryo (or the result of a very small number of dry embryos) formed locally and on which water was only brought by the bombardment (Lunine et al., 2003). The accretion of the Earth being fed by distant formation regions, a late delivery of water is no longer necessary. Such delivery certainly contributed somewhat, and can even be required to explain some anomalies of siderophiles or noble gas content of the Earth, but the bulk of the terrestrial water can be present from the "beginning", that is before -4.45 Gyr.

Giant impacts and atmospheric erosion

One of the reasons often given for the need of a late veneer of volatile-rich material is the atmospheric loss due to the giant impacts. Chen and Ahrens (1997) estimated that such impacts produce ground velocities above the escape velocity resulting in the escape of almost all the atmosphere. However, the question was recently revisited by Genda and Abe (2003): they found that, in a collision similar to the Moon-forming impact (i.e. a collision betweeen a proto-Earth and a Mars-sized planet), less than 30% of the atmosphere of both bodies is lost to space. Therefore, giant impacts can result in a net delivery of gases to the growing proto-Earth (in the case of volatile-rich impactors) and are unlikely to reset to zero the atmospheric content, because:
- 30% is the extreme upper limit for the escaping fraction estimated by these authors,
- impact with Mars-sized bodies are among the biggest ones experienced by proto-planets, according to accretion models (Chambers, 2001),
- a significant fraction of the volatiles can be kept in the mantle (especially CO_2 and H_2O, but also N_2 depending on the oxidation state reached by the mantle at that stage, see section 11.3.0). This is demonstrated by the survival of near solar isotopic ratios of Ne in the deep mantle.

2. Dating the prebiotic Earth

We focus here on the period just before the emergence of Life on Earth. The prebiotic Earth is the place where the last steps of chemical evolution led to the first biological processes. This concept becomes of course very fragile if one considers the possibility of panspermia (see for instance Napier, 2004). It is usually assumed that the emergence of the terrestrial Life occurred on Earth, panspermia being considered as a more exotic hypothesis. The assumption of a terrestrial origin, however, is not clearly supported by scientific arguments. In order not to eliminate any possibility, let us consider that the prebiotic Earth was the place where the conditions allowed life to evolve, whether its origin is endogenous or extraterrestrial.

The last sterilizing impact

The impact history of the Earth between the end of the "embryo accretion" (before 4.45 Gyr ago) and the for mation of the Lunar basins (4.0-3.8 Gyr ago) is a controversial topic. Previous models describing the evolution of the impacts (in rate and impactor sizes) used to extrapolate the gradual decrease observed between -3.8 and -3.5 Gyr back to the Moon-forming impact that occcured at -4.5 Gyr. Scaled to the Earth, this extrapolation yields huge impactor fluxes among which a small fraction are bigger than 500 km. These large impactors are able to vaporize the entire terrestrial ocean and melt the remaining rocky surface into a magma surrounded by a H_2O-CO_2 atmosphere of more than 300 bars (Zahnle and Sleep, 1997). It is reasonable to consider that such an impact not only sterilizes the planet but destroys its organic content (although some authors estimate that ejecta containing bacteria can orbit the "boiling" Earth and return later, reseeding the home planet after its recovery - Wells et al., 2003). Therefore, our most ancient terrestrial ancestor must appear (or reseed the Earth) after the last sterilizing event. Several authors used this extrapolation of the Lunar impacts to estimate the rate of such catastrophes and the horizon for ancient life on Earth (Maher and Stevenson, 1988; Oberbeck and Fogelman, 1989; Sleep et al., 2001). However, Ryder (2003, but also other authors) re-interpreted the Lunar crater record. He concluded that, first, the bombardment in the period 4.4 - 3.9 Gyr ago has been exaggerated in most studies, and, second, a flux within an order of magnitude of the present one, with an increase at 3.9 Gyr ago (the "Lunar cataclysm"), is presently the most consistent with the impact history recorded on the Moon.

Lunar Basins were typically formed by 20 km sized impactors with relative velocities of about 30 km/s (Morbidelli et al., 2001) with kinetic energies of 5×10^8 Megatons. Scaled to the Earth, about 20 times more impacts should have occurred, allowing statistically bigger bodies to collide the planet. However this scaling does not imply the existence of 500 km bodies (3×10^{12} Megatons) required to evaporate the entire oceans nor the 200 km bodies (10^{11} Megatons) able warm it up to 100°C. A low impact rate between the end of the embryo accretion and the peak at -3.9 Gyr can explain the moderate temperatures, the presence of liquid water and continental crust attested by detrital zircons (Valley et al., 2002). Note that, in this "quiet and cool" scenario, the effective delivery of extraterrestrial organics is not necessarily lower. Indeed, the flux can be integrated over a much longer period in the absence of late vaporizing events resetting the chemical content of the superficial Earth. The controversy is still very active and the point here is clearly not to abandon the concept of an heavy bombardment frustrating the origins of life but to realize that there is a possibility that this phase ended with the main accretion of the Earth about 4.45 Gyr ago.

Interestingly, while the horizon for early life is moving backward in time, another debate questions the oldest isotopic and morphological evidences

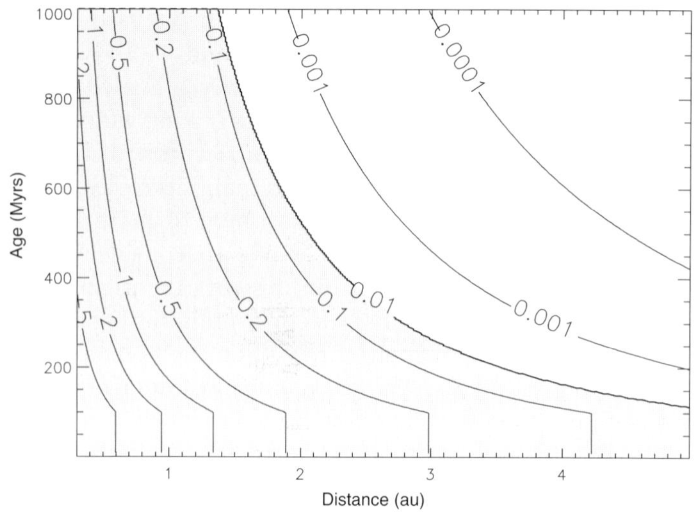

Figure 11.1. Time integrated X-EUV flux received by the Earth. The dashed curve shows the evolution of the integrated energy received in the Ly-α line and XUV range [1-1000Å], from 4.54 Gyr ago (dissipation of the protoplanetary disk) to t. These wavelengths drive the escape of H (XUV and Ly-α) and the production of H atoms through photolysis (Ly-α). Thus, they power the oxidation of the Earth (see section 11.3.0). Due to the steep decrease of this energy flux at early ages, the efficiency of the oxidation and the chemical evolution of the Earth is highly dependent on the time of incorporation of the bulk of water (t_{H_2O}). The solid lines give the fluxes integrated between (t_{H_2O}) and t. To underline the importance of t_{H_2O}, it is shown (dotted lines and arrows) that a same amount of energy usable for oxidation is received in the 340-Myrs interval from 4.54 to 4.2 Gyr ago and in the 700-Myrs interval from 4.4 to 3.7 Gyr ago.

for ancient life found in the interval 3.85-3.45 Gyr ago (Brasier et al., 2004).

3. Insights into the composition of the prebiotic atmosphere

The bulk of the terrestrial water was likely brought to the Earth by one or several wet embryos, formed beyond 2-3 AU. The formation of these embryos from km-sized objects takes typically 10,000 yrs. To radiate their internal energy ($> 3GM^2/5R$) during this period of time implies a mean cooling rate exceeding by orders of magnitude the 300 Wm^{-2} runaway greenhouse threshold. This runaway produces a "magma ocean" phase, during which a dense steam atmosphere equilibrates with a melt rocky surface (Sleep and Zahnle, 2001). This phase lasts 0.5 to 4 Myrs for embryos with masses between 0.01 and 0.1 M_{Earth} which is shorter or comparable to the typical lifetime for protoplanetary nebulae (Lyo et al., 2003; Armitage et al., 2003). This has interesting consequences: First, embryos may capture directly a solar-composition atmosphere from the nebula while the opacity of the disk provide a protection against the solar radiation able to induce gravitational escape of the atmosphere[1]. Second, during the magma-ocean phase, a nebular component incorporates into the embryo through the melted surface, which is consistent with the isotopic and elemental abundances of noble gases in the terrestrial mantle and crust (Dauphas, 2003).

A faint but active Early Sun

While the total luminosity of the young Sun was lower than today, its high-energy emission was higher. No consistent model is available to describe the evolution of the solar spectrum in the X and extreme UV range but observations of Sun-like stars at different ages provide an indirect estimate. In their *Sun in Time* program, Guinan and Ribas (2003) used a sample of solar proxies representing most of the Sun's main-sequence lifetime from ~ 130 to 8 Gyr to derive a power-law relationship (11.1) for the time dependence of the emission in the range [1-1000 Å]. Lammer et al. (2003) used HST observations of some of the stars from this sample to derive a similar reationship (11.2) for the emission in the Lyman-α line (1215 Å).

$$\begin{cases} F_{\mathrm{XUV}}(t) = 6.13\,t^{-1.19} f_{\mathrm{XUV}} & \text{if } t > 0.1 \text{ Gyr,} \\ F_{\mathrm{XUV}}(t) = F_{\mathrm{XUV}}(0.1) & \text{if } t < 0.1 \text{ Gyr} \end{cases} \tag{11.1}$$

$$\begin{cases} F_{\alpha}(t) = 3.17\,t^{-0.75} f_{\alpha} & \text{if } t > 0.1 \text{ Gyr,} \\ F_{\alpha}(t) = F_{\alpha}(0.1) & \text{if } t < 0.1 \text{ Gyr} \end{cases} \tag{11.2}$$

where t is the age in Gyr, $f_{\mathrm{XUV}} = 3.4 \times 10^{-3}$ W.m^{-2} and $f_{\alpha} = 5.7 \times 10^{-3}$ W.m^{-2} are respectively the Sun XUV and Lyman-α fluxes at 1AU. Note that for an age t=4.5 Gyr, Equ. (11.1) and (11.2) recover the solar values. The high fluxes emitted at early ages (associated with the strong solar wind mentioned above), played a major role in the atmospheric loss processes from the primitive atmosphere and the oxidation of the Earth. The estimates presented here (Fig. 1 and 2) are only preliminary, and a more detailed study, including also photochemical effects, is in progress.

Survival of reduced atmospheres and early oxidation of the Earth

During the accretion of embryos by the growing Earth (before -4.4 Gyr), such magma-ocean phases must have occurred (at least once the water abundance reached a certain value). But in this case, the disk no longer exists and the atmosphere is submitted to the strong X and Extreme UV (XUV) radiation from the young Sun. If the core segregation is still not complete (until at least 30 Myr from Hf-W dating) water reacts with iron in the mantle, releasing large amounts of molecular hydrogen in addition of the captured nebular atmosphere (Zahnle, 1998). As shown on Fig. 11.3.0, the lifetime of such an atmosphere is extremely short due to the hydrodynamic escape driven by the XUV heating: a mass of hydrogen equivalent to the mass of the present Earth's atmosphere is lost in less than 1 Myr during the first 150 Myrs. Therefore, the existence of a primitive H_2-dominated atmosphere depends on the production rate of H_2 by the oxidation of iron. In an atmosphere where H_2 is not the main compound, the escape rate of hydrogen can still be of the same order (Hunten et al., 1989). Indeed, during the first 100 Myrs, the XUV solar flux would produce exospheric temperatures above the critical temperature at which H atoms have a mean kinetic energy above their gravitational energy ($\sim$ 7500 K for an Earth mass). Gravitational escape from an initial nebular atmosphere is consistent with the observed fractionation of residual atmospheric xenon and neon.
At these epochs, the Lyman-α irradiation was also a major source of H through the photodissociation of H_2O or CH_4, able to photolyse the equivalent of several meters of water every Myr. The result of this production of rapidly escaping hydrogen is a fast oxidation of the atmosphere and the upper mantle. This oxidation started as early as the incorporation of water into the Earth (before-4.4 Gyr from zircons and numerical accretion models).
This fast oxidation is consistent with the early degassing of N_2 (before -4.3 Gyr, Tolstikhin and Marty, 1998) inferred from its isotopic distribution. Because such degassing is incompatible with reducing conditions (Libourel et al., 2003), the upper mantle should have reached its present oxidation state as early as 4.3 Gyr ago. It is likely that the fast and early oxidation of the upper mantle was made possible by the chemical action of water combined with the activity of the young Sun. If the bulk of water would have been delivered later, when the Sun was less active and the loss of atmospheric hydrogen less efficient, the fate of the Earth might have been different (Fig. 11.2.0). The fact that atmospheric neon and xenon are highly fractionated while nitrogen seems not sug-

gests that these noble gases were submitted even earlier to an intense fractionation through atmospheric hydrodynamic escape This also requires a very reducing mantle able to keep nitrogen in the silicates and thus an incomplete core segregation when the bulk of noble gases was lost. The relatively short period of time during which nitrogen was kept in the silicates may then be of high importance by protecting nitrogen from gravitational escape. A secondary nitrogen component may have been brought later by comets (Owen and Bar-Nun, 2001) although this contribution is constrained to low values by D/H constraints. A limited chondritic delivery is also possible.

The partial presure of N_2 in the prebiotic atmosphere was probably similar to its present one. Indeed, the upper mantle contains a reservoir of N_2 smaller than 15% of the atmospheric N_2 (Zhang and Zindler, 1993). Therefore, the partial pressure could have been slightly higher but probably not lower because the recycling of atmospheric N_2 into the mantle is certainly more efficient nowadays because of the biological fixation of nitrogen (Marty and Dauphas, 2003a, 2003b).

The partial pressure of CO_2 is a subject of debate and is discussed in section 11.4.0. After giant impacts producing a runaway greenhouse, the partial pressure of CO_2 (and H_2O) was only limited by its reservoir and its solubility in magma. Out of these extreme conditions, it seems reasonable to say that it was in the range [10-200 mbars], despite a huge reservoir in the crust and upper mantle respectively equivalent to 60 and 150 bars (Zhang and Zindler, 1993).

The composition in other gases is much more speculative: CO is often pointed as a major prebiotic gas because of its formation during impact and its abundance in the protosolar nebula but no quantitative study about its sources and sinks is available. Moderate levels of CH_4 are plausible (see section 11.4.0). N- bearing prebiotic compounds like HCN and NO can be produced by lightning and impacts at a rate depending critically on the level of CO_2 and CH_4 (Navarro-Gonzalez et al., 2001; Commeyras et al., 2004a). O_2 was kept to very low levels ($< 10^{-6}$) by the release of reduced volcanic gases despite its production by CO_2 photolysis and H_2O photolysis associated with H escape (Selsis et al., 2002).

4. The faint young Sun and climate

Standard models for the evolution of the Sun like Baraffe et al. (1998) give a solar luminosity equal to 73% the present one 4 Gyr ago. Assuming an atmosphere with fixed radiative properties (an obviously weak assumption) results in a frozen Earth during more than the first half of

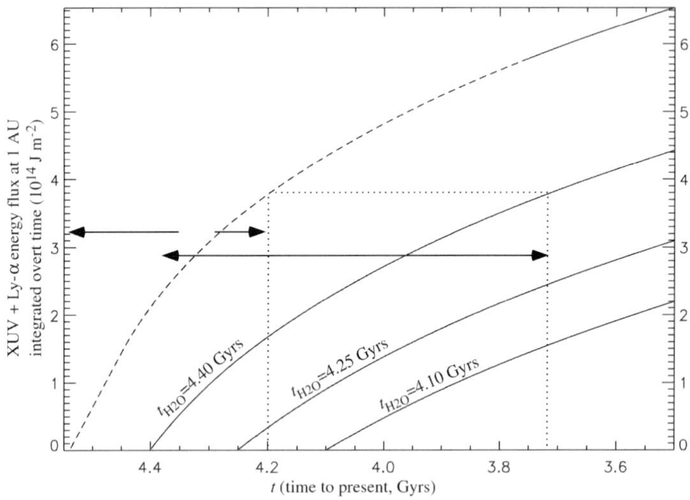

Figure 11.2. Mass loss (in M_{atm}/Myr) from H_2-rich atmospheres due to the solar X-ray and extreme UV heating. M_{atm} is the mass of the present terrestrial atmosphere (5×10^{14}kg). H_2-dominated atmospheres generated by capture or degassing are highly unstable once the protecting protosolar nebula has vanished (which occurred during the first 10 Myrs). This graph gives the escape rate due to the XUV heating (additional solar-wind induced mass loss would enhance these rates). Solar XUV and Lyman-α are from Equ. (11.1) and (11.2) The grey area is where energy-limited hydrodynamic escape occurs, while the white area is the Jeans escape domain.

Earth's history, until about 2 Gyr ago. Discussing this so-called *faint young Sun problem* (Sagan and Mullen, 1972; Kasting and Grinspoon, 1991) provides another interesting approach to investigate the prebiotic conditions.

The temperature of the early Earth

As described in section 11.1, the formation of the Earth implies huge impacts by planetary embryos with masses between 1% and 10% of Earth's mass. Post impact conditions induced a magma ocean surrounded by an atmosphere of several hundreds bar of H_2O and CO_2. During the cooling and condensation of this atmosphere, a hot ocean (230°C) can be found in equilibrium with a 300 bars atmosphere. The thermal evolution between this extreme phase and the present conditions is still very obscure. Knauth and Lowe (1978) used the isotopes of oxygen ($\delta^{18}O$) in cherts to infer the temperature of the early ocean. However, the high temperatures they found (80-100°C, 3.5 Gyr ago) are now interpreted as the temperature in hydrothermal systems (Pinti et al., 2001), which is not related to the mean ocean temperature.
The coldest times in Earths history are the *Snowball Earth* events characterized by an ice cover down to the equator. At the beginning of these events, the runaway ice-albedo feedback makes the global mean temperature drop to -50°C for a few tens of thousands of years (Schrag et al., 2002). This temperature drop is followed by a period of a few million years during which the mean temperature is around -10°C. The most ancient known snowball events occurred near the beginning of the Proterozoic (2.45 - 2.22 Gyr ago), and another one may have occurred 2.7-2.8 Gyr ago (Crowley, 1983). These ancient low-latitude glaciations may be a consequence of the first release of biological oxygen (Selsis, 2002). The absence of other major glacial event suggests a mean surface temperature well above the freezing point of water and thus an efficient greenhouse warming. However, after the emergence of life, the climate can be strongly influenced by biological activity. In particular, methanogenesis is known as a very primitive metabolism, able to sustain a warm climate even under low solar irradiation (Pavlov et al., 2000). Therefore, high temperatures recorded in the Proterozoic or even the Archean should not be extrapolated to prebiotic periods, for which our knowledge of the thermal conditions still rely on climate models.

Climate regulation by the carbonate-silicate cycle

Walker et al. (1981) proposed the attractive hypothesis that the level of atmospheric CO_2 is regulated in such way that the temperature is

always above and close to the freezing point of water, except for geologically brief periods. If, for any reason, the greenhouse warming becomes unable to maintain such temperature, surface water freezes making ineffective the removal of atmospheric CO_2 into carbonates. In the absence of a CO_2 sink, the release of volcanic gases restores the level of CO_2. Thanks to this feedback mechanism, a global glaciation is not an irreversible trap. Assuming such regulation throughout Earth's history, and a constant albedo, the minimum level of CO_2 can be expressed as a function of time from the evolution of the luminosity (Kasting and Grinspoon, 1991). 3 Gyr old paleosols seem to indicate that the level of CO_2 was below the freezing limit (Rye et al., 1995), implying another greenhouse gas. The nature of these geological samples and how they should be interpreted in terms of CO_2 level is very debated but the problem can be solved anyway by the production of methane by the primitive biosphere (Pavlov et al., 2000), without questioning the role of the carbonate-silicate cycle in abiotic conditions.
Recently, however, Sleep and Zahnle (2001) developed a detailed model for CO_2 cycling on the early Earth. They find CO_2 levels lower than the warming limit, except for brief post-impact periods. For these authors, other efficient sinks for atmospheric CO_2 (mainly the basalt carbonatization) make a dense CO_2 atmosphere unstable. They conclude that life certainly emerged on a frozen planet. The model is however sensitive to several parameters affected by large uncertainties, like the impact rate and the quantitative coupling of the different carbon reservoirs.

Abiotic methane

The existence of a globally reduced prebiotic atmosphere has been generally abandoned during the last decades, but there is still room for methane as a minor (but maybe precious) atmospheric gas. Indeed, a small production of methane (small compared to the biogenic one) exists in present hydrothermal systems. This production is due to the reaction of hot water with iron associated to a production of H_2 and, then, CH_4 in the presence of CO_2. To contribute to the surface warming the abiotic production of CH_4 had to be orders of magnitude higher than today but this is supported by the geological context on the prebiotic Earth (Shock et al., 2000). Furthermore, the Ly-α emission, which drives the photo-oxidation of CH_4 into CO_2 and H_2O was an order of magnitude higher than its present one after the formation of the Earth, requiring enhanced sources.
The production of methane after large iron-rich asteroid impacts has

recently been studied (Sekine et al., 2003), mostly for its implications on HCN production and prebiotic chemistry. The effect on the climate is negligible considering the other consequences of the large impacts considered. Impacts by comets, which contain a significant fraction of methane (see the Chapter by Crovisier in the present volume), can also sporadically deliver CH_4, but only a continuous production of this gas can sustain a sufficient level in the atmosphere because of the short photochemical lifetime of this molecule.

A fat young Sun?

Whitmire et al. (1995) proposed the idea that the Sun lost a fraction of its mass through solar wind and was thus initially more massive. An increased mass (by only a few percent) results in an initial insolation of the planets comparable to the present one. This is due not only to the increase of the intrinsic luminosity but also to the slightly smaller planetary orbital distances. Heliosismic constraints authorize a massive enough young Sun (up to 1.07 times its present mass) provided that the decrease of the mass loss to its present value was slow (Sackmann and Boothroyd, 2003). On the other hand, indirect determination of the stellar wind of solar-type stars suggest on the contrary a fast decrease of the mass loss following the power law $\dot{M} \propto t^{(-2.0\pm0.5)}$ (Wood et al., 2002). The highest initial mass that can be obtained from this relation is 1.04 M_{sol} (considering no evolution during the first 100 Myrs: the saturation phase). This is enough to provide an initial insolation comparable to the present one, but contradicts the slow decrease required to fit heliosismic constraints. It also implies that 90% of the mass excess is lost within the very early history of the solar system (the first 200-300 Myr) with no long-lasting effect on the planetary climates.

Early Cloudiness

The effective temperature of a planet considered as a black body is given by

$$T_{\text{eff}} = \left(\frac{S(1-A)}{4\sigma}\right)^{\frac{1}{4}}$$

where S is the solar flux (in W.m^{-2}) received at the orbital distance, A is the albedo (the fraction of energy reflected to space), σ the Stefan-Boltzmann constant. The factor 4 is due to the distribution of the solar energy over the whole planetary surface. With the present albedo of

the Earth (close to 0.3), the effective temperature is 255 K, while the observed mean surface temperature of the Earth is 288 K. The 33 K difference is provided by the IR opacity of the atmosphere. An Earth without clouds would have an effective temperature close to 273 K because of the low albedo of oceans and lands. An Early Earth without clouds (A=0.1) would have a similar effective temperature as the Earth today (A=0.3).
Shaviv (2003) used this argument and the controversial relationship between cloudiness and the cosmic ray flux (CRF) to solve the faint young Sun problem. A growing literature exists on the effect of the CRF on climate (see for instance Carslaw et al., 2002). This debated effect can be summarized as follows: by inducing condensation along their path, ionizing particles increase the cloudiness and thus the albedo. On the other hand, the solar wind lowers the CRF and consequently warms the surface. While the Sun evolves, its bolometric luminosity increases and its magnetic activity and mass loss decrease: according to Shaviv, taking into account these two trends lead to a more or less constant effective temperature of the Earth (or at least less variations). However, even assuming that the present CRF effect on the cloudiness is real, its impact must depend on the nature of the atmosphere. In particular, the contribution of the CRF to the cloudiness in an anoxic early atmosphere could be much lower. Indeed, the abundance of OH (the main cleaning agent in present day atmosphere - Brasseur and Salomon, 1984) was considerably lower while the sources of aerosols were higher: volcanic ashes and dusts, interplanetary dust particles (see previous chapter) and even, to some extent, organic hazes produced by CH_4 photolysis. As a consequence, the nucleation sites were likely to be more abondant, providing favorable conditions for condensation and cloud formation. In this context, the additional role of the CRF seems less significant. Yet, the relation between CRF and climate (if confirmed) could have played a role in the second part of Earth's history.

The origin of Life on a frozen Earth?

New models addressing the faint young Sun paradox all aim to provide a scenario that seems to be motivated by an undisputed need for a mean surface temperature above 273 K for the origin of life. However, a frozen surface also offers attractive properties. First it still allows the existence of liquid water: close to volcanic hot spots; after impacts (Bada et al., 1994); or as a thin upper layer around the substellar point (at noon and low latitude). The latter case provides a unique site to concentrate the

compounds deposited from the atmosphere (organics, dust) and a daily dry-humid alternation required by some prebiotic scenarios (Commeyras et al., 2004b). Moreover, the fresh water produced by melted ice is more suitable to polymerization processes (Monnard et al., 2002).
In addition, the existence of hydrothermal systems and black smokers is not affected by a cold climate and is a possible site for prebiotic organic synthesis (Baross and Hoggman, 1985).

5. Conclusions and future perspectives

Until a few years ago, the interval of time within which life arose was estimated to be very short by geological standards, bounded by the last "sterilizing impact event" (4.0-3.9 Ga, Sleep et al., 2001) and the oldest indirect evidence for life (3.85 Ga, Mojzsis et al., 1996). Now, both of these boundaries are contended strongly and, as a consequence, the window for the evolution of life has expanded. There is now plenty of time between the end of terrestrial accretion ($\sim$ 4.45 Gyr ago) and the oldest recognized microfossils (cf., Westall, this volume). Unfortunately, the face of the Earth during this interval of time may remain indeterminate because tectonics and biological activity have obscured the ancient records. And, while the exploration of our neighboring planets, Venus and Mars, may provide new insights into the origins of atmosphere and biosphere on terrestrial bodies, the surface of Venus has experienced recent resurfacing (Turcotte et al., 1999) and Mars could well have a very different history than the Earth (Lunine et al., 2003). A fascinating perspective is offered by the search of ejecta from the early Earth on the Moon or even on Mars (Armstrong et al., 2002) but it is more probable that more pieces of the puzzle will come from the study of extrasolar planetary systems.
The detection and spectroscopic analysis of terrestrial planets with a space observatories such as the Darwin/TPF programs of ESA and NASA (Volonte et al., 2000; Beichman et al., 1999) will certainly provide a better understanding of how atmospheres form and evolve. Observing solar-type stars at different ages has allowed us, to some degree, to recreate the history of our Sun from its protostellar state to his final fate (subgiant, giant, white dwarf...). Thanks to these observations, our knowledge of the evolution of solar activity, luminosity and mass loss is growing rapidly. The capability to characterize planetary atmospheres of terrestrial planets in nearby systems at different stages of evolution should yield similar improvements.

Acknowledgments

The author wish to thank Jennifer Blank, Didier Despois and Tobias Owen for their helpful comments. This chapter also benefitted from precious discussions with A. Morbidelli, B. Marty, H. Lammer, I. Ribas.

Notes

1. Here it is assumed that the embryos orbiting within the first few AU are embedded in the proto-planetary nebula though this cannot be confirmed with the present spatial resolution of astronomical observations.

References

Armitage, P. J., Clarke, C. J., and Palla, F. (2003). Dispersion in the lifetime and accretion rate of T Tauri discs. *MNRAS*, 342:1139-1146.

Armstrong, J. C., Wells, L. E., and Gonzalez, G. (2002). Rummaging through Earth's Attic for Remains of Ancient Life. *Icarus*, 160:183-196.

Bada, J. L., Bigham, C., and Miller, S. L. (1994). Impact melting of frozen oceans on the early Earth: Implications for the origin of life. *Proc. Natl. Acad. Sci.*, 91:1248-1250.

Baraffe, I., Chabrier, G., Allard, F., and Hauschildt, P. H. (1998). Evolutionary models for solar metallicity low-mass stars: mass-magnitude relationships and color-magnitude diagrams. *A&A*, 337:403-412.

Baross, J. A. and Hoggman, S. E. (1985). Submarine Hydrothermal Vents and Associated Gradient Environments as Sites for the Origin and Evolution of Life. *Origins of Life*, 15:327.

Beichman, C. A., Woolf, N. J., and Lindensmith, C. A., editors (1999). *The Terrestrial Planet Finder (TPF): a NASA Origins program to search for habitable planets*. JPL Publications.

Brasier, M., Green, O., Lindsay, J., and Steele, A. (2004). Earth's Oldest (3.5 Ga) Fossils and the 'Early Eden Hypothesis': Questioning the Evidence. *Origins of Life and Evolution of the Biosphere*, 34:257-269.

Brasseur, G. and Solomon, S. (1984). *Aeronomy of the middle atmosphere: Chemistry and physics of the stratosphere and mesosphere*. Dordrecht, D. Reidel Publishing Co., 1984, 457 p.

Canup, R. M. and Asphaug, E. (2001). Origin of the Moon in a giant impact near the end of the Earth's formation. *Nat*, 412:708-712.

Carslaw, K. S., Harrison, R. G., and Kirkby, J. (2002). Cosmic Rays, Clouds, and Climate. *Science*, 298:1732-1737.

Chambers, J. E. (2001). Making More Terrestrial Planets. *Icarus*, 152:205-224.

Chambers, J. E. and Wetherill, G. W. (1998). Making the Terrestrial Planets: N-Body Integrations of Planetary Embryos in Three Dimensions. *Icarus*, 136:304-327.

Chen, G. Q. and Ahrens, T. J. (1997). Erosion of terrestrial planet atmosphere by surface motion after a large impact. *Physics of the Earth and Planetary Interiors*, 100:21-26.

Chyba, C. F. (1987). The cometary contribution to the oceans of primitive Earth. *Nat*, 330:632-635.

Commeyras, A., Boiteau, L., Vandenabeele-Trambouze, O., and Selsis, F. (2004a). Peptide Emergence, Evolution and Selection on the Primitive Earth. II. The Primary Pump Scenario. In *Adv. Astrobiology and Biogeophysics Series, in press, Springer.*

Commeyras, A., Taillades, J., Collet, H., Boiteau, L., Vandenabeele-Trambouze, O., Pascal, R., Rousset, A., Garrel, L., Rossi, J., Biron, J., Lagrille, O., Plasson, R., Souaid, E., Danger, G., Selsis, F., Dobrijvic, M., and Martin, H. (2004b). Dynamic Co-evolution of Peptides and Chemical Energetics, a Gateway to the Emergence of Homochirality and the Catalytic Activity of Peptides. *Origins of Life and Evolution of the Biosphere*, 34:35-55.

Crowley, T. J. (1983). The geologic record of climatic change. *Rev. Geophys. Space Phys.*, 21:828-877.

Dauphas, N. (2003). The dual origin of the terrestrial atmosphere. *Icarus*, 165:326-339.

Dauphas, N. and Marty, B. (2002). Inference on the nature and the mass of Earth's late veneer from noble metals and gases. *J. Geophys. Res.*, 107:12-1.

Delsemme, A. H. (2000). 1999 Kuiper Prize Lecture: Cometary Origin of the Biosphere. *Icarus*, 146:313-325.

Genda, H. and Abe, Y. (2003). Survival of a proto-atmosphere through the stage of giant impacts: the mechanical aspects. *Icarus*, 164:149-162.

Guinan, E. F. and Ribas, I. (2002). Our Changing Sun: The Role of Solar Nuclear Evolution and Magnetic Activity on Earth's Atmosphere and Climate. In *ASP Conf. Ser. 269: The Evolving Sun and its Influence on Planetary Environments*, page 85.

Halliday, A. N. (2004). Mixing, volatile loss and compositional change during impact-driven accretion of the Earth. *Nat*, 427:505-509.

Hunten, D. M., Donahue, T. M., Walker, J. C. G., and Kasting, J. F. (1989). Escape of atmospheres and loss of water. In *Origin and Evolution of Planetary and Satellite Atmospheres*, pages 386-422.

Kasting, J. F. and Catling, D. (2003). Evolution of a Habitable Planet. *ARA&A*, 41:429-463.

Kasting, J. F. and Grinspoon, D. H. (1991). The faint young Sun problem. In *The Sun in Time*, pages 447-462. Univ. Arizona Press.

Kleine, T., Mnker, C., Mezger, K., and Palme, H. (2002). Rapid accretion and early core formation on asteroids and the terrestrial planets from Hf-W chronometry. *Nat*, 418:952-955.

Knauth, L. P. and Lowe, D. R. (1978). Oxygen isotope geochemistry of cherts from the Onverwatch Group (3.4 billion years), Traansval Group, South Africa, with implications for secular variations in the isotopic composition of cherts. *J. Geol.*, 41:209-222.

Kouchi, A., Kudo, T., Nakano, H., Arakawa, M., Watanabe, N., Sirono, S., Higa, M., and Maeno, N. (2002). Rapid Growth of Asteroids Owing to Very Sticky Interstellar Organic Grains. *ApJ Lett.*, 566:L121-L124.

Lammer, H., Selsis, F., Penz, T., Amerstorfer, U. V., Lichtenegger, H. I. M., Kolb, C., and Ribas, I. (2004). Atmospheric evolution and the history of water on Mars. In *Water on Mars*, pages 25-44. Springer.

Lammer, H., Selsis, F., Ribas, I., Guinan, E. F., Bauer, S. J., and Weiss, W. W. (2003). Atmospheric Loss of Exoplanets Resulting from Stellar X-Ray and Extreme-Ultraviolet Heating. *ApJ Lett.*, 598:L121-L124.

Levison, H. F., Lissauer, J. J., and Duncan, M. J. (1998). Modeling the Diversity of Outer Planetary Systems. *AJ*, 116:1998-2014.

Libourel, G., Marty, B., and Humbert, F. (2003). Nitrogen solubility in basaltic melt. Part I. Effect of oxygen fugacity. *Geochim. Cosmochim. Acta*, 67:4123-4135.

Lunine, J. I., Chambers, J., Morbidelli, A., and Leshin, L. A. (2003). The origin of water on Mars. *Icarus*, 165:1-8.

Lyo, A.-R., Lawson, W. A., Mamajek, E. E., Feigelson, E. D., Sung, E., and Crause, L. A. (2003). Infrared study of the ? Chamaeleontis cluster and the longevity of circumstellar discs. *MNRAS*, 338:616-622.

Maher, K. A. and Stevenson, D. J. (1988). Impact frustration of the origin of life. *Nat*, 331:612-614.

Makino, J., Fukushige, T., Funato, Y., and Kokubo, E. (1998). On the mass distribution of planetesimals in the early runaway stage. *New Astronomy*, 3:411-416.

Marty, B. and Dauphas, N. (2003a). "Nitrogen isotopic compositions of the present mantle and the Archean biosphere": Reply to comment by P. Cartigny and M. Ader. *Earth Planet. Sci. Lett.*, 216:433-439.

Marty, B. and Dauphas, N. (2003b). The nitrogen record of crust-mantle interaction and mantle convection from Archean to Present. *Earth Planet. Sci. Lett.*, 206:397-410.

Mojzsis, S. J., Arrhenius, G., McKeegan, K. D., Harrison, T. M., Nutman, A. P., and Friend, C. R. L. (1996). Evidence for life on Earth before 3,800 million years ago. *Nat*, 384:55-59.

Monnard, P., Apel, C. L., Kanavarioti, A., and Deamer, D. W. (2002). Influence of Ionic Inorganic Solutes on Self-Assembly and Polymerization Processes Related to Early Forms of Life: Implications for a Prebiotic Aqueous Medium. *Astrobiology*, 2:139.

Morbidelli, A., Chambers, J., Lunine, J. I., Petit, J. M., Robert, F., Valsecchi, G. B., and Cyr, K. E. (2000). Source regions and time scales for the delivery of water to Earth. *Meteoritics and Planetary Science*, 35:1309-1320.

Morbidelli, A., Petit, J.-M., Gladman, B., and Chambers, J. (2001). A plausible cause of the late heavy bombardment. *Meteoritics and Planetary Science*, 36:371-380.

Napier, W. M. (2004). A mechanism for interstellar panspermia. *MNRAS*, 348:46-51.

Navarro-Gonzalez, R., McKay, C. P., and Nna Mvondo, D. (2001). A possible nitrogen crisis for Archaean life due to reduced nitrogen fixation by lightning. *Nat*, 412:61-64.

Oberbeck, V. and Fogelman, G. (1989). Impacts and the origin of life. *Nat*, 339:434.

Owen, T. C. and Bar-Nun, A. (2001). Contributions of Icy Planetesimals to the Earth's Early Atmosphere. *Origins of Life and Evolution of the Biosphere*, 31:435-458.

Pavlov, A. A., Kasting, J. F., Brown, L. L., Rages, K. A., and Freedman, R. (2000). Greenhouse warming by CH_4 in the atmosphere of early Earth. *J. Geophys. Res.*, 105:11981-11990.

Pinti, D. L., Hashizume, K., and Matsuda, J.-i. (2001). Nitrogen and argon signatures in 3.8 to 2.8 Ga metasediments: clues on the chemical state of the archean ocean and the deep biosphere. *Geochim. Cosmochim. Acta*, 65:2301-2315.

Raymond, S. N., Quinn, T., and Lunine, J. I. (2004). Making other Earths: Dynamical simulations of terrestrial planets formation and water delivery. *Icarus*, 168:1-17.

Ryder, G. (2003). Bombardment of the Hadean Earth: Wholesome or Deleterious? *Astrobiology*, 3:3.

Rye, R., Kuo, P. H., and Holland, H. D. (1995). Atmospheric carbon dioxide concentrations before 2.2 billions years ago. *Nat*, 378:603-605.

Sackmann, I.-J. and Boothroyd, A. I. (2003). Our Sun. V. A Bright Young Sun Consistent with Helioseismology and Warm Temperatures on Ancient Earth and Mars. *ApJ*, 583:1024-1039.

Sagan, C. and Mullen, G. (1972). Earth and Mars: Evolution of atmospheres and surface temperature. *Science*, 177:52-56.

Schrag, D. P., Berner, R. A., Hoffman, P. F., and Halverson, G. P. (2002). On the initiation of a snowball Earth. *Geochemistry, Geophysics, Geosystems*, page 1.

Sekine, Y., Sugita, S., Kadono, T., and Matsui, T. (2003). Methane production by large iron meteorite impacts on early Earth. *J. Geophys. Res.*, page 6.

Selsis, F. (2002). Occurrence and detectability of O_2-rich atmosphere in circumstellar "habitable zones". In *ASP Conf. Ser. 269: The Evolving Sun and its Influence on Planetary Environments.*

Selsis, F., Despois, D., and Parisot, J.-P. (2002). Signature of life on exoplanets: Can Darwin produce false positive detections? *A&A*, 388:985-1003.

Shaviv, N. J. (2003). Toward a solution to the early faint Sun paradox: A lower cosmic ray flux from a stronger solar wind. *J. Geophys. Res.*, page 1437.

Shock, E. L., Amend, J. P., and Zolotov, M. Y. (2000). The Early Earth vs. the Origin of Life. In *Origin of the Earth and Moon*, pages 527-543. University of Arizona Press.

Sleep, N. H. and Zahnle, K. (2001). Carbon dioxide cycling and implications for climate on ancient Earth. *J. Geophys. Res.*, 106:1373-1400.

Sleep, N. H., Zahnle, K., and Neuhoff, P. S. (2001). Initiation of clement conditions on the earliest Earth. *PNAS*, 98:3666-3672.

Turcotte, D. L., Morein, G., Roberts, D., and Malamud, B. D. (1999). Catastrophic Resurfacing and Episodic Subduction on Venus. *Icarus*, 139:49-54.

Valley, J. W., Peck, W. H., King, E. M., and Wilde, S. A. (2002). A cool Early Earth. *Geology*, 30:351-354.

Volonte, S., Laurance, R., Whitcomb, G., Karlsson, A., Fridlund, M., Ollivier, M., Gondoin, P., Guideroni, B., Granato, G. L., Amils, R., and Smith, M. (2000). Darwin: the infrared space interferometer. Technical report, ESA.

Walker, J. C. G., Hays, P. B., and Kasting, J. F. (1981). A negative feedback mechanism for the long-term stabilization of Earth's surface temperature. *J. Geophys. Res.*, 86:9776-9782.

Weidenschilling, S. J. and Cuzzi, J. N. (1993). Formation of planetesimals in the solar nebula. In *Protostars and Planets III*, pages 1031-1060.

Wells, L. E., Armstrong, J. C., and Gonzalez, G. (2003). Reseeding of early earth by impacts of returning ejecta during the late heavy bombardment. *Icarus*, 162:38-46.

Whitmire, D. P., Doyle, L. R., Reynolds, R. T., and Matese, J. J. (1995). A slightly more massive young Sun as an explanation for warm temperatures on early Mars. *J. Geophys. Res.*, 100:5457-5464.

Wiechert, U., Halliday, A. N., Lee, D.-C., Snyder, G. A., Taylor, L. A., and Rumble, D. (2001). Oxygen Isotopes and the Moon-Forming Giant Impact. *Science*, 294:345-348.

Wilde, S. A., Valley, J. W., Peck, W. H., and Graham, C. M. (2001). Evidence from detrital zircons for the existence of continental crust and oceans on the Earth 4.4 Gyr ago. *Nat*, 409:175-178.

Wood, B. E., Mller, H., Zank, G. P., and Linsky, J. L. (2002). Measured Mass-Loss Rates of Solar-like Stars as a Function of Age and Activity. *ApJ*, 574:412-425.

Zahnle, K. (1998). Origins of Atmospheres. In *ASP Conf. Ser. 148: Origins*, pages 364-391.

Zahnle, K. and Sleep, N. H. (1997). Impacts and the early evolution of life. In *Comets and the origin and evolution of life*, pages 175-208. Springer: New York.

Zhang, Y. and Zindler, A. (1993). Distribution and evolution of carbon and nitrogen in Earth. *Earth and Planetary Science Letters*, 117:331-345.

Chapter 12

EARLY LIFE ON EARTH: THE ANCIENT FOSSIL RECORD

Frances Westall

Centre de Biophysique Moleculaire, CNRS, Rue Charles Sardon, 45071 Orléans Cedex 02, France

Abstract: The evidence for early life and its initial evolution on Earth is linked intimately with the geological evolution of the early Earth. The environment of the early Earth would be considered extreme by modern standards: hot (50-80°C), volcanically and hydrothermally active, anoxic, high UV flux, and a high flux of extraterrestrial impacts. Habitats for life were more limited until continent-building processes resulted in the formation of stable cratons with wide, shallow, continental platforms in the Mid-Late Archaean. Unfortunately there are no records of the first appearance of life and the earliest isotopic indications of the existence of organisms fractionating carbon in ~3.8 Ga rocks from the Isua greenstone belt in Greenland are tenuous. Well-preserved microfossils and microbial mats (in the form of tabular and domical stromatolites) occur in 3.5-3.3 Ga, Early Archaean, sedimentary formations from the Barberton (South Africa) and Pilbara (Australia) greenstone belts. They document life forms that show a relatively advanced level of evolution. Microfossil morphology includes filamentous, coccoid, rod and vibroid shapes. Colonial microorganisms formed biofilms and microbial mats at the surfaces of volcaniclastic and chemical sediments, some of which created (small) macroscopic microbialites such as stromatolites. Anoxygenic photosynthesis may already have developed. Carbon, nitrogen and sulphur isotopes ratios are in the range of those for organisms with anaerobic metabolisms, such as methanogenesis, sulphate reduction and photosynthesis. Life was apparently distributed widely in shallow-water to littoral environments, including exposed, evaporitic basins and regions of hydrothermal activity. Biomass in the early Archaean was restricted owing to the limited amount of energy that could be produced by anaerobic metabolisms. Microfossils resembling oxygenic photosynthesisers, such as cyanobacteria, probably first occurred in the later part of the Mid Archaean (~2.9 Ga), concurrent with the tectonic development of suitable shallow shelf environments. The development of an oxygenic metabolism allowed a

P. Ehrenfreund et al. (eds.), Astrobiology: Future Perspectives, 287–316.

considerable increase in biomass and increased interaction with the geological environment.

Key words: early Earth, extreme environment, early habitat, early life, prokaryotes.

1. INTRODUCTION

This review concerns principally the earliest evidence for life and its early evolution over the time period of the early Achaean (4-3.2 Ga). It briefly addresses the further evolution of life to the very important stage of oxygenic photosynthesis, which probably took place in the late Mid Archaean (~3.0 - 2.8 Ga).

The Earth upon which life first appeared was very different to the Earth of today. It was the epitomy of an extreme habitat. Therefore, in order to place early life on Earth in its environmental context, a short summary of the conditions existing on Earth during the first one and a half billion years of its history is relevant. The first fossil evidence for life dates back to 3.5 billion years ago (Dunlop et al., 1978; Walsh, 1992; Hofmann et al., 1999; Westall et al., 2001), *i.e.* one billion years after the formation of the Earth. However, the characteristics of the microfossils indicate that they represent life forms that were already relatively evolved, as will be described below. Life obviously appeared much earlier and had already undergone a number of important evolutionary phases, from the formation of the first cellular membranes containing (a) primitive reproductive and energy-obtaining mechanism(s), to the organisms characterised by the type of genetic reproduction in force today, as well as many (but not all) of the metabolic pathways found in modern organisms. Nevertheless, signatures of life have been described from older rocks, ranging in age between 3.65 to more than 3.8 Ga, from the highly metamorphosed terrains of the Isua and Akilia greenstone belts in Greenland (Pflug, 1979, 2001; Pflug and Jaeschke-Boyer, 1979; Schidlowski et al., 1983; Robbins, 1987; Schidlowski, 1988, 2001; Robbins and Ibberall, 1991; Mojzsis et al., 1996; Rosing, 1999). Recent studies, however, have invalidated all but one of these signatures (a - 18‰ $\delta^{13}C$ from a 3.8 Ga metasedimentary unit, Rosing, 1999; van Zuilen et al., 2002).

When could life potentially have arisen on Earth? We assume that life requires liquid water, the earliest evidence for which is 4.3-4.4 Ga. This evidence is based on the occurrence of many zircon crystals of this age trapped in younger rocks (Wilde et al., 2001; Mojzsis et al., 1999). The formation of large amounts of these crystals requires the fractionation of hydrated basaltic crust. Moreover, the $\delta^{18}O_2$ isotope values obtained from the zircons is consistent with a mantle source altered by low-temperature hydrothermal fluids (Wilde et al., 2001). Water-rich fluids could have supported the origin of life at this early time in the Hadean Eon (4.56-4.0 Ga). [In the following, the period of Earth's history between 4.56 and 4.0 Ga is termed the Hadean, that between 4.0 and 3.2 Ga the Early Archaean, the 3.2 to 2.8 Ga period is the Mid Archaean, and the 2.8-2.5 Ga period is the Late Archaean]. On the other hand, analysis of the lunar impact record suggests that the inner planets were subjected to an intense impact flux early in their histories (Melosh et al., 1993), and there maybe have been a peak in bombardment rate between 4-3.85 Ga ago (Ryder, 2002). The most pessimistic impact scenarios envision the annihilation of any life existing at the surface of the planet (Maher and Stevenson, 1988; Sleep et al., 1989), whereas the most optimistic hypotheses envision that only the uppermost few hundred meters of ocean water boiled off, thus avoiding global sterilisation (Ryder, 2002, 2003). To date, we have no means of finding out when life arose because we have no direct record since most rocks older than about 3.5 Ga, have either been destroyed by plate tectonic processes or severely altered by metamorphism.

2. EARLY ENVIRONMENTAL HABITATS

The environment of the early Earth differs from the present day Earth in a number of respects. Early continental masses were represented by volcanic rocks in the form of volcanoes or oceanic ridges (such as Iceland today), or even thickened portions of oceanic crust (such as the Ontong Java Plateau in the Pacific Ocean) that presented a variety of potential habitats for early life, ranging from deep-water to subaerial environments. The deep ocean floor, with its hydrothermal vents, was also a potential habitat, as was the oceanic water column. The lack of broad, shallow continental shelves, typical of modern continents, during this early period in Earth's history had important consequences on the development of life, as we will see below; the biomass of early life was limited until the evolution of oxygenic photosynthesising that could develop and expand in the suitable, sunbathed environments provided by the shelves.

Apart from a small amount of partially fractionated rocks (*i.e.* those richer in silica), most of the rocks from the Early Archaean are basic to ultrabasic in composition. The composition of the rocks on early Earth is important since it is assumed that life is surface-based and the first organisms needed a subsurface to support their growth. Furthermore, the earliest forms of life, according to the tree of life, probably used a chemoautolithic metabolism (the most primitive form of metabolism), whereby they obtained both their energy and their carbon from non-organic sources. The surfaces of the volcanic rocks are highly reactive chemically and, thus, can provide an ideal substrate for chemoautotrophic microorganisms. Temperatures on the early Earth, both in the mantle and at the surface, were probably higher than they are today (Arndt, 1994; Knauth and Lowe, 2003). Higher mantle temperatures would have resulted in faster tectonic recycling and greater volcanic and hydrothermal activity. Knauth and Lowe (2003) calculated an average ocean water temperature of more than 50°C, and possibly as high as 70-80°C. Any early life form at this temperature, by definition, would have been thermophilic, although not necessarily hyperthermophilic as has been suggested (see Wiegel and Adams, 1998, and articles therein; for a discussion on a thermophilic versus a hydrothermophilic origin of life see Forterre et al., 1995). The more active volcanism on the early Earth entrained with it greater hydrothermal activity, although the nature of the hydrothermal activity is poorly constrained (see Paris et al., (1985) and Knauth and Lowe (2003) for divergent views).

The composition of the ocean water is unknown, as are its pH and salinity. One theoretical model hypothesises an early "soda" ocean based on comparisons with modern, alkaline, volcanic lakes (Kempe and Degens, 1985). Others suggest that the pH may have been slightly acidic, about 5, owing to the high partial pressure of CO_2 in the early atmosphere (Grotzinger and Kasting 1993). Holland (1984), on the other hand, proposed a neutral pH, similar to that of the modern ocean. Salinity measurements on fluid inclusions from early Archaean rocks also produce mixed results: Touret (2003) noted normally saline ocean waters whereas de Ronde and Ebbessen (1996) calculated that Early Archaean salinities were slightly higher than those of the present day.

The composition of the early atmosphere was probably dominated by CO_2, although other gases would have been present, such as H_2O vapour, N_2, and CH_4 (Kasting, 1993, Pavlov et al, 2001). Radiation from the young Sun was probably up to 30% less intense than it is today (Sagan and Mullen, 1976; Sagan and Chyba, 1997). Therefore, in order to keep the planet warm and the oceans liquid, either the atmospheric partial pressures of CO_2 would have had to have been very high (above 10 bars) or another greenhouse gas, such as CH_4, was present (Kasting, 1993, Pavlov et al, 2001).

There is much debate concerning the amount of oxygen in the early atmosphere and the rise in the Earth's history of free oxygen (O_2 gas). Ohmoto (1997, 1999), for example, maintains that oxygen levels were already high in the Early Archaean atmosphere, whereas Holland (1984) contests that they were very low. Mineralogical studies of ancient sediments (Rye et al., 1995; Rasmussen et al, 1999), as well as sulphur isotope data (Farquhar et al., 2000), support the theory that there was little oxygen in the atmosphere. However, the presence of jasperites and banded iron formations (BIFs) in the early Archaean requires a certain amount of O_2. The oxygen necessary for the production of these chemical precipitates could have been generated in minor amounts by photolysis of H_2O vapour in the atmosphere and in the upper layers of the oceans. However, it probably represented concentrations below 1% of present atmospheric levels (Kasting, 1993). Low quantities of O_2 would have important implications for early life. First it would suggest that early life on Earth was anaerobic and second, a protective ozone layer may not have existed, in which case deleterious UV radiation would have reached habitats in exposed environments such as in the uppermost layers of water, the littoral zone, and land surfaces (Cockell, 2001). On the other hand, the early atmosphere was probably very thick with dust and aerosols resulting from the abundant volcanic activity as well as the higher occurrence of impactors hitting the planet. The identification of many layers of spherulites with an extraterrestrial Cr isotope signal in 3.2-3.5 Ga-old rocks is consistent with the theory that the flux of extraterrestrial impacts continued to be higher than today (Lowe and Byerly, 1986; Lowe et al., 2003; Kyte et al., 2003).

Other environmental factors that could have affected early life are the shorter day (the planets were spinning faster) and higher tides since the Moon was closer to the Earth than it is today (Sonett et al., 1996).

In summary, the Earth upon which life appeared and first flourished was hot, volcanically and hydrothermally active, subjected to impacts from extraterrestrial bodies, had an anoxic atmosphere and suffered from high UV radiation. Moreover, the ocean water may not have had the same pH and salinity as today. The geological adage that "the present is the key to the past" cannot be used with respect to the early Earth.

Despite the extreme environment, the early Earth offered a wide range of potential habitats available for colonization, albeit somewhat more restricted and less diverse than today because the geological evolution of the planet had not yet reached its present tectonic development, including the formation of broad continental shelves that is characteristic of the Earth today. Moreover, the higher ambient temperatures (Knauth and Lowe, 2003) would probably have precluded the existence of temperate or cold conditions, thus further limiting the range of habitats.

3. EARLY LIFE

If one accepts the assumption that life originated at the surface of the Earth, then few ancient terrains have rocks relevant to the study of ancient life. The relevant rocks include extruded volcanic material, sedimentary rocks derived from eroded volcanic material, precipitated minerals, and minerals of evaporative origin. Of these, the oldest are found in the >3.7 Ga Isukasia/Akilia greenstone belts in southwest Greenland, followed by the 3.2- >3.5 Ga Pilbara and Barberton greenstone belts in northwest Australia and eastern South Africa, respectively (Figure 1).

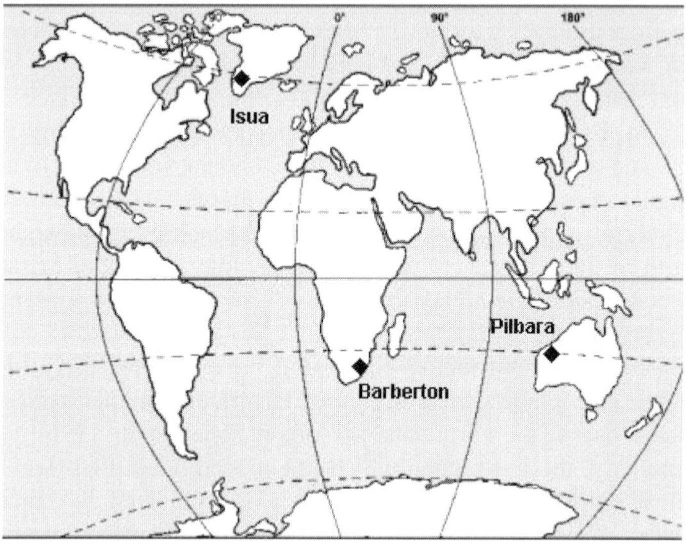

Figure 12.1. Location of the early Archean terrains if Isua, Barberton and the Pilbara.

3.1 The Isua/Akilia Greenstone Belts

Greenland hosts a number of early Archaean terrains, two of which have been the subject of early life investigations: (i) the Isukasia area to the north east of Nuuk, the capital, and (ii) the Akilia association, including Akilia Island to the south west of Nuuk. These terrains consist of highly metamorphosed supracrustal rocks (volcanics and volcaniclastic and chemical sediments) and are older than 3.75 Ga (Nutman, 1986; Appel et al., 1998; Appel and Moorbath, 1999). Owing to the degree of alteration, there is some debate as to whether these terrains actually include rocks that were originally sediments (Fedo and Whitehouse, 2002; Myers, 2003).

These rocks have been analyzed using a number of different methods in an effort to identify evidence of life. Thin section study and scanning and transmission electron microscopy (SEM and TEM) observations of carbonaceous, acid residue materia l revealed bacteriomorph structures that

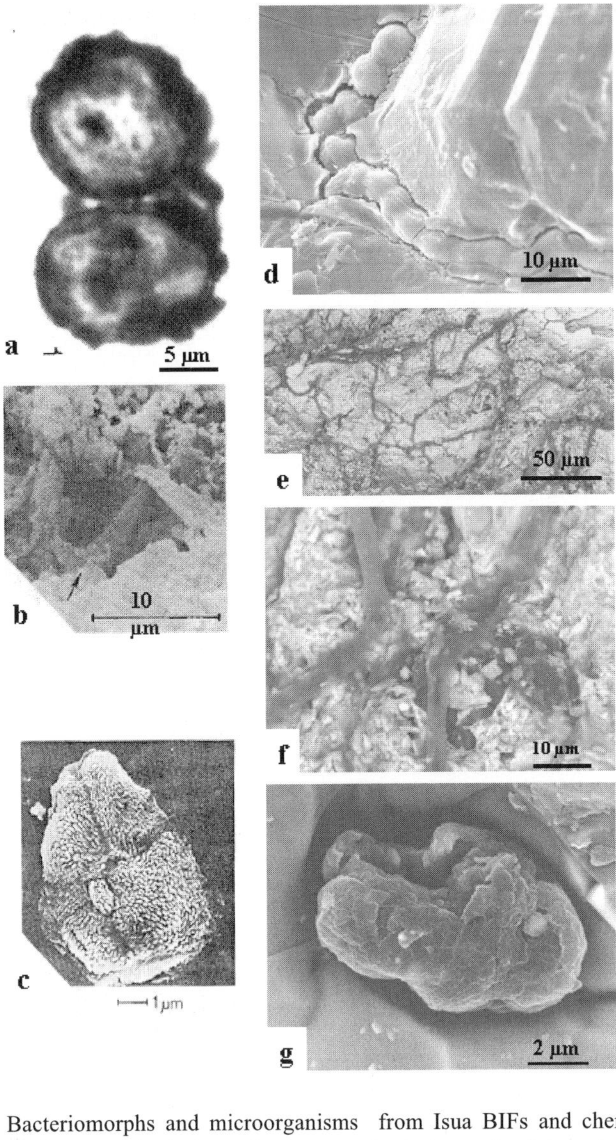

Figure 12.2. Bacteriomorphs and microorganisms from Isua BIFs and cherts. (a) TEM micrograph of large, spherical, carbonaceous bacteriomorph interpreted as fossil cyanobacteria, liberated by acid maceration (Pflug, 2001). (b) SEM micrograph of filaments liberated by acid maceration (Robbins and Iberall, 1991). (c) Apelella ferrifera, liberated by acid maceration (Robbins and Iberall, 1991). (d) Recent, endolithic cyanobacteria in a BIF; compare with (a) (Westall and Folk, 2003). (e,f) Recent endolithic fungal hyphae on a fracture surface in a BIF; compare with (b) (Westall and Folk, 2003). (g) Clump of generic kerogen of possible exogenous origin from a BIF; compare with (c) (Westall, unpub. data).

include yeast-like cells (*Isuasphera isua* Pflug) (Pflug and Jaeschke-Boyer, 1979), structures similar to cyanobacteria (Pflug, 1979, 2001) (Figure 2a), as well as limonitic and hematitic bacteriomorphs, such as *Apellela ferrifera* (Robbins, 1987)(Figure 12.2c), all of which were interpreted as representing microbial colonies, and iron coated filaments (Robbins et al., 1987; Robbins, 1987; Robbins and Iberall, 1991)(Figure 12.2b).These structures were later re-interpreted as possible weathering products, fluid inclusions, or the result of modern contamination (Roedder, 1981; Bridgewater et al., 1981; Appel et al., 2003; Westall and Folk, 2003). Westall and Folk (2003), for instance, showed that recent endolithic microorganisms, such as cyanobacteria and fungi, inhabit cracks and fissures in these rocks along with carbonaceous remains washed into cracks by rainwater (Figures 12.2d-g).

Carbon isotope investigations of BIFs and cherts from this area also revealed the presence of ^{12}C-depleted material, which suggests that organisms using a metabolism that fractionated the lighter ^{12}C against the heavier ^{13}C isotope existed at the time that the protoliths of the Isukasia and Akilia rocks were formed (Schidlowski et al., 1983; Schidlowski, 1988; 2001). On the basis of bulk sample analyses, these authors measured a -22 to -28 ‰ [PDB] value that was attributed to biological fractionation processes, and, in particular, to oxygenic photosynthesis. Later, Mojzsis et al. (1996) made *in situ* investigations of graphite inclusions in apatite crystals. They found a range of isotope values from - 21 to - 49 ‰, consistent with the photosynthetic activity today found among methanogens. An additional study by Rosing (1999) on metasediments (turbidites) in the same area produced a result of -11.4 to -20.2‰, which he interpreted as also reflecting fractionation due to life processes.

Alternative explanations for the isotope record have been put forward. Fedo and Whitehouse (2002) hypothesise an abiogenic origin via Fischer Tropsch type reactions for hydrocarbons exhibiting a $\delta^{13}C$ value less than - 20 ‰ formed via Fischer-Tropsch type reactions. van Zuilen et al. (2002) suggest that these reactions could have taken place in a CO_2 rich environment caused by metamorphic decarbonation of magnetite that formed by the serpentinisation of ultramafic rocks. Furthermore, they demonstrated that the graphite in the rocks could also have been produced abiogenically by thermal decomposition of siderite. Using stepped combustion techniques, they documented the presence of a contaminating carbon isotope signal (*e.g.* the results of Lepland et al., 2002, and the contaminating microorganisms and microfossils found by Westall and Folk, 2003). However, van Zuilen et al. (2002) also found an apparently original stable carbon isotope signature in one sample. The sample in question comes from what were originally described to be turbiditic metasediments (Rosing, 1999; van Zuilen et al., 2002. This result is intriguing since Myers (2003) has interpreted the rock as a mylonite and not a metasediment.

The metamorphic rocks from Isua were also the subjects of a stable nitrogen isotope study (Pinti et al., 2001). Measuring biotites and iron oxides from what they believed to have been metasediments, these authors obtained a - 1.7 ‰ $\delta^{15}N$ value that they interpreted as possibly being of biogenic origin, although they recognised that this hypothesis needed to be confirmed by further investigation.

3.2 The Barberton and Pilbara Greenstone Belts

As we have seen above, the degree of alteration of the > 3.7 Ga rocks from Isua and Akilia is so great that it is often difficult to clearly interpret their origin. There are, however, two other ancient terrains that have conserved well-preserved sedimentary rocks. These are the 3.2-3.5 Ga greenstone belts of Barberton, South Africa, and the Pilbara Australia (Figure 1). These two terrains consist of intrusive and extrusive volcanic rocks, volcaniclastic and chemical sediments, and granitic intrusions. Considering the general lack of preservation of Early Archaean supracrustal sequences, it is somewhat surprising that these two cratons survived. Their conservation appears to be due to several factors related to the gradual tectonic evolution of the planet. Continued fractionation in the lower protocontinental crust resulted in silica-rich granitoid intrusions and a more rigid crust upon which continental sedimentary basins could develop. Continental growth occurred by lateral accretion (cratonisation) while underplating of the protocontinental masses contributed to their stabilisation.

The Pilbara block has undergone only very mild alteration in its ~3.5 Ga history (prehnite-pumpellyite facies), whereas the Barberton block was subjected to a Late Archaean metamorphic event that resulted in a slightly higher-grade metamorphic facies (greenschist). Nevertheless, the volcanic and sedimentary rocks in these terrains are remarkably well preserved. The sedimentary beds, consisting of volcaniclastic and chemical sediments, tend to be thin intercalations between thick piles of volcanic flows. The sediments represent deposition during a lull in volcanic activity over relatively long periods of time, whereas the volcanics appeared to have been extruded very rapidly (Lowe and Byerly 1999). These protocontinental masses provided a wide variety of potential habitats for early life, ranging from deeper water (at least below storm base), through shallow waters to the littoral and even subaerial environment.

There have been numerous investigations for evidence of life in these sediments. Schopf and Walter (1983) and Schopf (1993) concluded that the vast majority of the microfossils previously described (see Schopf and Walter, 1983, for the abundant list of references) were actually artefacts, dubiofossils, or contaminants. Recently, a number of filamentous structures superficially resembling fossil cyanobacteria that were originally described by Schopf and Packer (1987) and Schopf (1993), have been shown to be, themselves, carbonaceous precipitations in a hydrothermal chert vein (Brasier et al., 2002). On the other hand, numerous carbon isotope investigations of carbon-rich cherts from the Pilbara and Barberton document $\delta^{13}C$ ratios ranging from -14‰ to 42‰, consistent with biogenic fractionation (Hayes et al., 1983; Walsh, 1992; Walsh and Lowe, 1999; Westall et al., 2001; Ueno et al., 2001a-c; Brasier et al., 2002). Brasier et al (*op.cit.*), however, have raised doubts as to the value of such data in cases where abiogenic production of carbon with a fractionated signal could be a possible alternative explanation. Moreover, Walsh and Westall (unpub. observations) have noted that the cherts, especially the carbon-rich samples (the carbon-poor cherts pose less of a problem), are susceptible to infiltration by younger microorganisms in cracks, thus potentially biasing any measurements of an isotopic signal. This could also be the case for other nitrogen and possibly sulphur isotope studies (see below).

Obviously, great care needs to be taken in the search for evidence of life in these most ancient sediments. It is not the scope of this paper to make a detailed critique of the criteria that are important for identifying signs of life, but this has been done by previous authors (Schopf and Walter, 1983; Buick, 1990; Schopf, 1993; Westall, 1999; Westall, 2003a). An important point to bear in mind is the fact that any investigation of past life needs to take into account a variety of information ranging from macroscopic to microscopic observations, as well as a variety of observational and biogeochemical methods. Careful study of the geological context is of utmost importance since the presence of potential evidence for life must be coherent with the environment of deposition (cf. Brasier et al., 2002). Other criteria include (i) morphological characteristics (size, shape, cell wall texture, evidence for cell division, evidence for flexibility in the case of filaments), (ii) colonial characteristics (colony formation, presence of consortia, production of polymer - extracellular polysaccharide substances or EPS), and their propensity to form biofilms at sediment and mineral surfaces, (iii) carbonaceous composition (*N.B.* modern oxidising environments, however, do not preserve organic carbon), (iv) isotope signals compatible with life processes, and (v) organic biomarkers. New studies also show that the mapping of elements related to microfossils (Philippot et al., 2003), as well as REE trace elements (van Kranendonk et al., 2003) provides information that is useful in identifying signs of life.

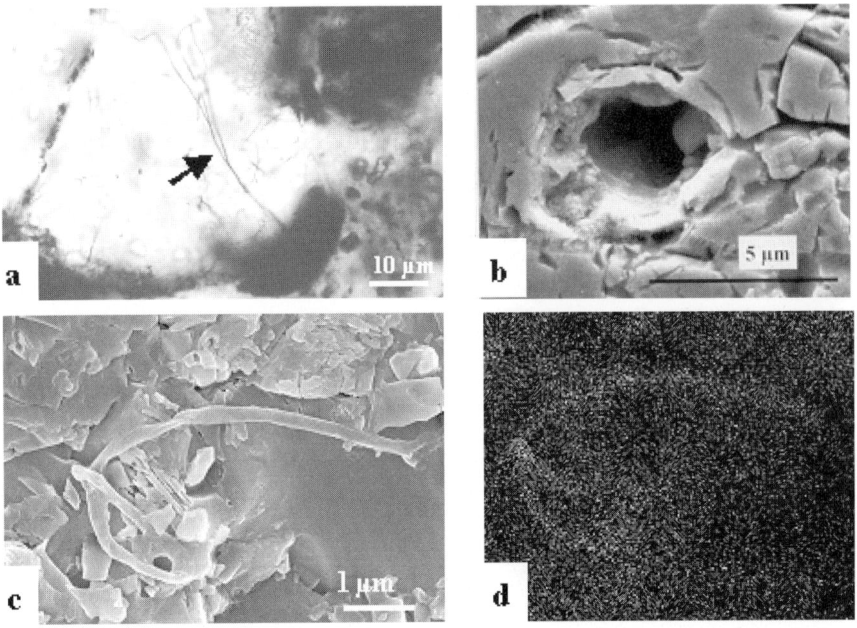

Figure 12.3. Filaments from Early Archaean formations in the Pilbara and Barberton.(a) Large, carbonaceous filaments (arrow) protruding perpendicularly (or at an angle) from a thick, carbonaceous, biofilm substrate. The filaments are embedded in chert. Kromberg Formation, Barberton (Westall and Walsh, 2000). (b) Cross section of a large filament exhibiting a silica-impregnated, finely granular outer envelope, embedded in chert. Acid-etched thin section. Kromberg Formation, Barberton (Westall and Walsh, 2000). (c) Smaller, silica-impregnated filament embedded in chert. Acid-etched cut surface. Coppin Gap Formation, Pilbara (Westall et al., 2002). (d) EDX mapping illustrating the carbon enrichment in the microfossil (Westall, unpub. data).

The following review of the evidence for life in the 3.2-3.5 Ga sediments from the Pilbara and Barberton is based on field observations, thin section study, scanning and transmission electron microscope (SEM, TEM) investigations, and biogeochemical analyses (EDX/microprobe evaluation of carbonaceous composition, and carbon, nitrogen and sulphur isotope studies). No organic biomarker studies have yet been undertaken, indeed, the organic carbon is relatively highly mature (Rouzaud, pers.comm. 2002). This review is based on the microfossil and microbial mat (stromatolite) investigations of Dunlop et al. (1978), Byerly et al., (1986), Robert (1989), Strauss and Morre (1992), Walsh (1992), Walsh and Lowe (1999), Rasmussen (1999), Hofmann et al. (1999), Ueno et al., (2001a-c), Westall and Walsh (2000, 2003), Westall et al. (2001, 2002), Westall (2003a,b),

Walsh and Westall (2003), and the isotope investigations of Schidlowski et al. (1983), Walsh (1992), Walsh and Lowe (1999), Beaumont and Robert (1999), Farquhar et al., (2000), Pinti and Hashizumi (2001), Pinti et al., 2001; Westall et al., (2001), Shen et al. (2001), Ueno et al. (2001a-c), Kakegawa (2001).

The evidence for Early Archaean life is hotly contentious and the following discussion is based on the author's own investigations (also contested). Obviously future studies will provide more data that will either confirmed or disprove present understanding regarding the nature of Early Archaean life.

4. MICROFOSSILS AND MICROBIAL MATS

The authors listed above identified a variety of microfossils, including:

- large, sinuous, carbonaceous filaments, up to 5 μm in width and sometimes over 100 μm in length, that form and are anchored to, wavy, carbonaceous laminations resembling microbial mats (Walsh, 1992; Rasmussen, 1999; Westall and Walsh, 2000; Ueno, 2001c; Walsh and Westall, 2003) (Figure 12.3a,b);
- smaller, sinuous carbonaceous filaments, between 0.25 – 0.5 μm in width and up to many tens of micrometers in length that form thick, robust microbial mats (Figure 5b), or looser, more delicate mats together with coccoidal microfossils (Westall at al., 2002; Westall, 2003a,b) (Figures 12.3c,d);
- carbonaceous, coccoidal microfossils that occur in carpet-like, aggregated colonies of hundreds of individuals (Figures 12.4a,b) in chain-like associations (Figure 12.4c), or in looser associations (Figure 4d). Average sizes vary from colony to colony but are generally between 0.4- 0.5 μm or between 0.8-1 μm (Westall et al., 2001; Westall 2003b);
- rod-shaped bacteriomorphs, 1 μm in length and 0.5 μm wide (Figure 4e; Westall unpub. data), or longer rods of about 10 μm in length and 1-2 μm diameter (Walsh, 1992).
- - vibroid-shaped microfossils that form small colonies and are about 1 μm wide with lengths ranging from 2-3.8 μm (Westall et al., 2001) (Figure 12.4b).

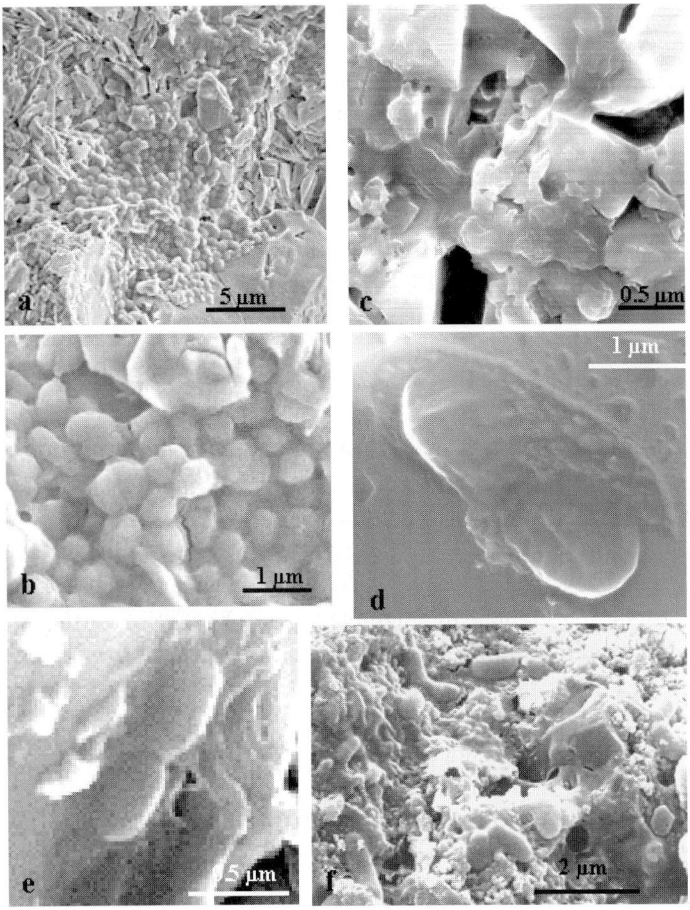

Figure 12.4. Coccoidal and vibroid-shaped microfossils from Early Archaean formations in the Pilbara and Barberton. (a) Carpet-like colony (arrow) of small, carbonaceous coccoidal cells embedded in chert. The detail (b) shows two size classes of coccoids (large and small arrows), many of which are undergoing cell division. Acid-etched cut surface. Coppin Gap Formation, Pilbara (Westall, unpub. data). (c) Carbonaceous coccoids in chain formation, embedded in chert. Acid-etched cut surface. Coppin Gap Formation, Pilbara (Westall et al., 2002). (d) Dividing coccoids still partially embedded in the matrix quartz. The surfaces of the microfossils are rugged and they exhibit a deflated aspect, indicating that the micoorganisms were already dead and lysed before fossilisation. Acid-etched cut surface, Barberton, Kromberg Formation (Westall, unpub.data). (e) Short, rod-shaped microfossils, Barberton, Kromberg Formation (Westall and Gerneke, 1998). (f) Small colony of vibroid-shaped microfossils surrounded by a polymer film, embedded in chert. Acid-etched fracture surface, Kromberg Formation, Barberton (Westall et al., 2001).

Microfossils usually occur in colonial associations consisting of more than one morphological types, for instances coccoids and filaments, or two different species of coccoids (different size classes). The numbers of individuals in the colonies range from just a few to many hundreds. Colonies may be loose or relatively compact associations. They may form delicate biofilms or thick, robust mats at the surfaces of volcaniclastic and chemical sediments (Figure 12.5b).These mats are sometimes macroscopically visible as laterally extensive films of finely sinuous layers, domical structures (Figures 12.5a,c,d), or laterally extensive stratiform layers.The mats investigated formed at the surfaces of shallow water and littoral sediments. In the latter case, the association with evaporite facies crystals, together with desiccation cracks (Figure 12.5b) testifies to their formation in a subaerial environment. To date, not enough work has been done to establish whether mats also formed at the surfaces of deeper water sediments (although certainly detrital remnants of mats occur in deeper water facies).

There appears to be an association between many of the environments where the mats formed and the presence of hydrothermal activity. However, the nature of the association(s) need(s) to be better constrained: (i) were the mats formed on the vicinity of an active vent or hot spring, or (ii) were they simply impregnated with hot, silica-rich fluids emanating from hydrothermal vents that may have been at some distance, but whose effects would have still been strongly felt in a marine setting where the average seawater temperatures and mineral concentrations were, anyway, high (cf. Knauth and Lowe, 2003)?

The Early Archaean microfossils, biofilms, and mats investigated appear to have been preserved preferentially through silification (cf., Westall et al., 1999, 2000, 2001; Westall, 1999). In this process, silica-rich seawater (cf., Knauth and Lowe, 2003) or hydrothermal fluids chelated to the surfaces of the organic matter and there polymerised on the organic template, thus acting as a chemical fixative. The organisms were mineralised (or "fixed") in the state in which they occurred, *i.e.*, if an organism were in a living, healthy state, it would therefore be preserved as a turgid structure. On the other hand, if the organism were dead and degraded, it would consequently have a lysed, deflated and often very "ragged" appearance. Upon continued degradation, it would simply become part of the generic polymer of the biofilms, thus becoming unrecognisable. These early Archaean biofilms therefore provide an important record of a whole spectrum of living to dead states, such as occurs in modern microbial communities. Other mineral ions occurred in solution in the early Archaean seawater, such as Fe, Ca and Ba (of hydrothermal origin), with which microorganisms could also have been permeated but, to

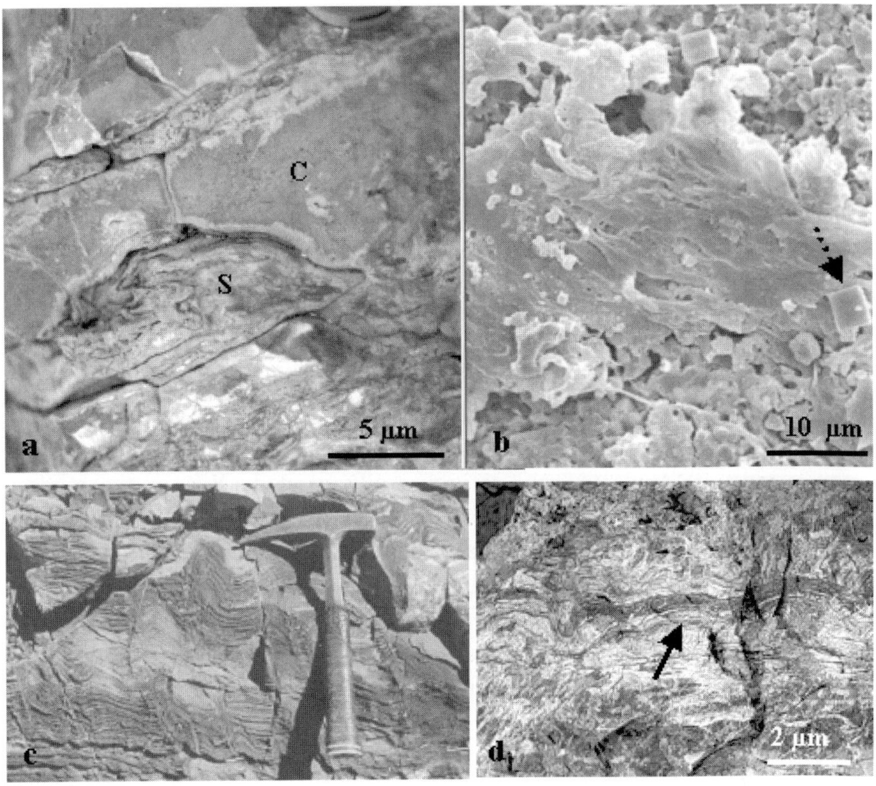

Figure 12.5. Tabular and domal stromatolites from the Pilbara and Barberton. (a) Small domal stromatolite (S) with deformed internal layering, lying on top of a layer of hydrothermal chert and coated ("squashed") by another layer of chert. This example occurs near a hydrothermal vent exit. Field photograph, North Pole Dome, Pilbara (Westall et al., 2001). (b) Thick, robust, desiccated microbial mat formed of thin, flow-laminated filaments (individual filament arrowed). Evaporite minerals (dotted arrow) are embedded within it and on it. Kromberg Formation, Barberton (Westall et al., 2001, 2002). (c) Domical stromatolites. Strelley Pool Formation, Pilbara (Westall, 2003a). (d) Low amplitude domical stromatolites (arrowed). Fig tree Formation, Barberton (Westall 2003a).

date, reports of such mineralisations are rare. For example, Rasmussen (2000) described pyritised, probable microbial filaments from a hydrothermal setting in 3.235 Ga volcanogenic massive sulphide deposits from the eastern Pilbara craton. One of the microbial mats described by Westall et al. (2001, 2002) appears to have been both silicified, as well as calcified (Figure 12.5b).

There are a number of reports of stromatolites from Early Archaean formations in Barberton and the Pilbara. Domical stromatolites were described from the North Pole region/ Strelley Pool Chert area in the Pilbara by Dunlop et al. (1978), Schopf and Walter (1983), Lowe (1983), Hofmann et al., (1999) and Westall et al. (2002) (see also Figure 5a), and from the southern part of the Barberton greenstone belt by Byerly et al. (1986) (observation confirmed by Westall, unpub. data). However, Lowe (1994) noted that some of the domical stromatolitic structures could be abiogenic in origin, reflecting hydrothermal deposits rather than microbial mats. Indeed, in some cases, there appears to be some relationship between the occurrence of the "stromatolitic" structures and hydrothermal activity (*e.g.*, Westall et al., 2002). On the other hand, this does not necessarily imply that either a biogenic or an abiogenic situation. Microbial mats can form on chemically precipitated sediments and, conversely, can act as a nucleation point for mineral precipitation. Interactive microbial growth of this type was described by Westall et al. (1999). However, more work needs to be done to establish (i) the biogenicity of the domical stromatolites, (ii) whether such an association with hydrothermal activity really existed (*N.B.* in the modern environment, stromatolites are common around hot springs, cf. Cady and Farmer (1996), (iii) whether the association was simply coincidental, and (iv) if the association really existed, how common was it?

5. BIOGEOCHEMICAL DATA

The Early Archaean chert sediments from the Pilbara and the Barberton greenstone belts contain varying amounts of organic carbon (kerogen). Measurements of bulk samples range from < 0.01 to >1.4% (Hayes et al., 1983; Walsh, 1992; Walsh and Lowe, 1999; Westall et al., 2001). More detailed spot analyses and mapping of specific microfossils using electron diffraction and microprobe analysis (SEM and TEM) has demonstrated the carbon composition of the fossiliferous structures (Figure 12.3 and Westall, in prep.). Carbon isotope analyses were made on bulk chert samples containing from 0.01 to 1% organic carbon. As noted above, those samples richer in carbon are particularly susceptible to modern contamination in the form of infiltration in cracks. However, no contamination has been observed in samples with lower carbon contents. The ratios measured range from -14‰ to 42‰ (Hayes et al., 1983; Walsh, 1992; Strauss and Moore, 1992; Walsh and Lowe, 1999; Westall et al., 2001; Ueno et al., 2001a-c; Brasier et al., 2002). These are consistent with biological fractionation, although it was also noted above that abiogenic Fischer Topsch synthesis can apparently produce a similar phenomenon. Moreover, Hayes (1983) and Robert (1989) showed that post depositional processes could alter the original isotopic record. Many of the samples analysed in the studies noted above were obtained from bedded, sedimentary horizons and are thus presumably not hydrothermal in origin, but some came from hydrothermal veins (*e.g.* Ueno, et al., 2001c; Brasier et al., 2002). The isotope ratios measured cover a variety of microbial metabolisms ranging from methanogenesis to anoxygenic photosynthesis. It also covers the oxygenic photosynthetic range but, as noted below, there is no morphological or mineralogical evidence for oxidising conditions associated with the shallow water to littoral environment mats.

Brasier et al.'s (2002) investigation of the samples from the Chinaman Creek, Marble Bar site of Schopf and Packer (1987) and Schopf (1993) demonstrated their hydrothermal vein provenance (inconsistent with oxygenic photosynthesising cyanobacteria). They also made measurements of the carbon isotopes, obtaining values of -16 to –26‰. Although these values could reflect biogenic fractionation, Brasier et al. suggest that they were more likely to have been produced by Fischer Tropsch synthesis of the organic matter. There is, however, an alternative explanation: Microbial biofilms appear to be common on the surfaces of shallow water sediments in the Early Archaean formations of the Pilbara. Some of the horizons traversed by the hydrothermal vein (from which Schopf's (1993) sample was obtained) contain bedded volcaniclastic sediments. Would it have been possible for the hydrothermal fluids to have leached biogenic carbon from the country rock that had been trapped in fossil microbial mats, depositing it higher up in the vein? Ueno et al. (2001a-c) provide possible support for this hypothesis. They document $\delta^{13}C$ values of -40 to –31‰ (from their own data and also that of Hayes et al., 1983) for kerogen from bedded cherts from a 3.49 Ga chert/barite unit in the North Pole Dome area of the Pilbara. The kerogen is described as forming very large filamentous bacteriomorphs that the authors compared with the fossil "cyanobacteria" of Schopf (1993). They also investigated the organic carbon in crosscutting chert veins. The vein carbon is widely distributed and also formed bacteriomorph filaments similar to those previously described by Schopf (1993). Their $\delta^{13}C$ ratios ranged from -42 to -32 ‰. The authors interpreted these data as resulting from an active autotrophic microbial population living in a hydrothermal habitat. This interpretation needs to be re-evaluated on the basis of the Brasier et al. (2002) study. However, the overlap between the isotope values for the vein chert and those of the bedded chert/barite units, through which the veins traversed, could also suggest some causal relationship, as suggested above (*i.e.*, that the kerogen was leached out of the bedded sediments by the hydrothermal fluids, and subsequently redeposited higher up in the veins).

There is a further conundrum: considering the fact that there was much hydrothermal activity in the early Archaean (cf. Paris, 1985; de Wit and Hart, 1993; Westall et al., 2002), would it be possible to distinguish between organic carbon formed by abiotic Fisher Tropsch reactions and deposited in bedded chemical sediments from biogenic organic carbon (microbial mats) formed at the surface of the chemical sediments? However, the latter would probably contain morphologically recognisable remains unless the mats were in such a highly degraded state that individual microfossils are no longer distinguishable - in which case identification would, indeed, be problematic.

$\delta^{15}N$ values from organic matter from the Archaean carbonaceous cherts differ from those of modern carbonaceous sediments owing to changes in atmospheric chemistry over geological time (Pinti and Hashizume, 2001). Metabolic processes involving reduced N probably dominated the N cycle in the Archaean because of the lack of oxygen in the atmosphere. Thus, the isotopic ranges for biogenically formed carbon would have been between -9 and - 4‰ (Pinti and Hashizume, 2001), whereas the isotopic shifts in $\delta^{15}N$ for organic carbon in the modern ocean are positive. The negative nitrogen isotope ratios obtained from carbonaceous cherts from the Pilbara and Barberton are therefore interpreted to indicate the presence of chemosynthetic bacteria using inorganic NH_4^+ from hydrothermal fluids (Beaumont and Robert, 1998; Pinti and Hashizumi, 2001; Pinti et al., 2001; Ueno et al., 2001c).

The interpretation of the sulphur isotopes ratios from Early Archaean sediments is much debated. Shen et al. (2001) interpreted the results of their analyses of microscopic sulphides in Early Archaean barite deposits from the North Pole Dome in the Pilbara (21.3 to 7.4 ‰) as indicating the presence of sulphate reducing bacteria at the time of deposition of the sediments. A similar conclusion was reached regarding the isotope results from variety of sediments from early Mid Archaean formation in the Barberton greenstone belt by Kakegawa (2001). However, the results of Shen et al. (2001) were based on the hypothesis that the barite sediments replaced, original, primary, bedded gypsum. Other studies conclude that the barites are primary hydrothermal exhalations (Nijman et al., 1999; Farquhar et al., 2000; Runnegar, 2002) and that the isotope results largely reflect the influence of UV photochemistry on sulphur species in the atmosphere.

6. CONCLUSIONS REGARDING EARLY ARCHEAN LIFE

The morphological, colonial and mat-building characteristics of the microfossils from the early Archaean formations are remarkably similar to those of modern prokaryotes. These characteristics, taken altogether with their carbonaceous composition and the various C, N, and S isotope ratios, suggest that the organisms that formed these microfossils and fossil biofilm/mats were single celled prokaryotes. Considering the fact that these organisms represent the earliest evidence of life on Earth, they exhibit characteristics that are remarkably evolved compared to what we imagine the first cells to have been like, *i.e.* simple, more or less spherical cells, possibly restricted to a single habitat, such as around deep sea vents. These Early Archaean organisms already display a variety of morphologies, ranging from spherical cells, vibroids, filaments, and possibly rods (Westall, 2003b). Missing are the spiral morphologies characteristic of some modern prokaryotes, as well as some of rectangular, or multi-angular to almost amorphous morphologies of some of the organisms found in hot vent environments today. Moreover, evidence for cell differentiation or trichome formation, morphological peculiarities typical of microorganisms such as some cyanobacteria, has not been observed. Lack of evidence in the geological record does not necessarily imply that such organisms did not exist. However, biofilms containing well-preserved microfossils that do not exhibit cyanobacterial characteristics are abundant in the kind of environments in which such (oxygenic) photosynthesising organisms would be expected. It could be argued that the non-cyanobacteria microorganisms in such biofilms were preferentially preserved, whereas the cyanobacteria, if they were originally present, were destroyed. This argument, however, is inconsistent with the fact that cyanobacteria, with their polysaccharide-rich sheaths and glycocalyx (also a polysaccharide)- coated colonies, are far more susceptible to fossilisation than other types of microorganisms. I therefore conclude that cyanobacteria probably did not exist during the Early Archaean.

Certain environmental characteristics provided by the general geological data, as well as more detailed sedimentological investigations, can furnish additional information regarding the mode of life of the microorganisms. In the first place, the generally elevated temperatures (cf. Knauth and Lowe, 2003) indicate that the organisms must have been thermophilic. Secondly, all the documented mats formed in very shallow water environments and even in the littoral environment. This implies that they must have had some resistance to UV radiation (for example, effective gene repair mechanisms as well as physical protection, such as is found in thick mats formed in modern exposed environments that have an outer layer of dead cells protecting the underlying layers from desiccation as well as UV radiation). The occurrence of mats in evaporitic environments suggests that the organisms building those mats were probably halotolerant, if not halophilic. Finally, given the seemingly widespread occurrence of mats in shallow-water environments, is it possible that the organisms had developed some kind of phytosynthesis? Could the occurrence of stromatolites in the Early Archaean formations be taken as conclusive proof of photosynthetic activity? As noted above, there is still debate as to the biogenicity of the domal structures, although it is difficult to explain the plastically deformed layers in the domed structures in Figure 12.5a as having an abiogenic origin. Moreover, the fine filaments making up the mats formed in exposed littoral environments (Figure 12.5b) (Westall et al., 2001; Westall et al., 2002; Westall et al., 2003a,b) are similar in appearance to modern anoxygenic photosynthetic bacteria, such as *Chloroflexus*. If photosynthesis had already developed by the Early Archaean, it was probably anoxygenic photosynthesis since there is no mineralogical evidence for oxidised minerals associated with those mats that may have been produced by photosynthesis, on the contrary, reduced minerals, such as pyrite, can occur in these mats. Moreover, as noted above, none of the microfossils found in these shallow water to subaerial mats resembles organisms that use oxygenic photosynthesis, such as cyanobacteria. There is a tendency in palaeontology to automatically associate stromatolites with cyanobacteria, because this is the case today and was common in the Late Archaean and Proterozoic Eons. However, stromatolites can also be formed by other non-oxygenic photosynthetic organisms (Krumbein, 1983; Gerdes and Krumbein, 1987).

Given the time-consuming nature of these kinds of studies and the limited number of investigations undertaken to date, it is not yet possible to obtain a full overview of the distribution of life in the Early Archaean, and its importance in sedimentological processes. However, the preliminary data, if confirmed, suggest that life was widespread in shallow water habitats, as well as in the littoral environment, and formed macroscopic constructions in the form of stratiform and domical mats. The biofilms and microbial mats produced by these early organisms would have had the same stabilising effect on sediment surfaces as they do today, thus influencing sediment transport dynamics. Thick piles of microbialite sediments, especially carbonates, are a common product of bioprecipitation and biomediated precipitation from the Late Archaean onwards. However, such features were largely the result of organisms using oxygenic photosynthesis. Early Archaean microbialites are more restricted because there was less biomass owing to the fact that energy produced by anaerobic metabolisms, including anaerobic photosynthesis, is less efficient than that produced by oxygenic photosynthesis (Des Marais, 2000). In fact, microbialites in the Early Archaean are only represented by thin layers of tabular mats (stratiform stromatolites) and by relatively small domical stromatolites, if the latter are indeed biogenic. In terms of mineral precipitation, the microorganisms, biofilms and mats offered suitable nucleation sites for minerals, such as silica, carbonates and sulphides, which subsequently polymerised in and on the organic matrix, thus preserving the organisms and their biofilms. Work remains to be done to determine the potential role of biogenic activity in the formation of other chemical deposits, such as BIFs and barite, from this time.

7. MID TO LATE ARCHEAN LIFE

The Mid to Late Archaean marks a critical period of continued geological evolution that was characterised by cratonisation, consolidation and stabilisation of continental masses, and the widespread development of broad, shallow, continental platforms surrounding the newly formed continents (Veizer, 1994). True arenaceous, continental sediments appeared for the first time. The platforms provided huge areas of sunlight-bathed, carbonate-rich habitats that did not exist previously (although shallow water environments were common in restricted basins in the Early Archaean) (cf. Grotzinger, 1994). It is in this period of the Earth's history that we have the first record of cyanobacterial microfossils implying, as a consequence, the development of oxygenic photosynthesis. However, direct evidence for life in the Mid Archaean is sparse. Noffke et al. (2003) describe the oldest known occurrence of cyanobacteria-like filaments, extensive films formed by bacteria on arenaceous sediment surfaces from the 2.9 Ga Mozaan Group of the Pongola Supergroup in South Africa (Figure12.6).The sediments were deposited on a shallow, storm influenced, shelf.

Chemical biomarker investigations of 2.5 to >2.7 Ga formations in the Pilbara of NW Australia provide confirmation of the evolution of cyanobacteria (Summons et al., 1999; Brocks et al., 1999; 2003). The shales investigated contained macromolecules, such as 2-methylhopanoids, that are the degradation products of hopane molecules, specific to lipids in the cell walls of cyanobacteria. Brocks et al. (1999) also found another type of macromolecule, a sterane (derivative of sterol) that was interpreted as being derived from the cell walls of eukaryote organisms. This would have put the evolution of eukaryotes far earlier than previously thought. It has since been recognised, however, that these particular macromolecule derivatives can be

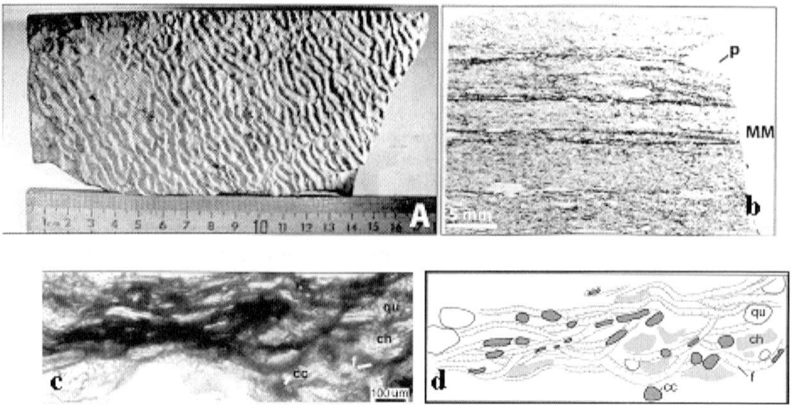

Figure 12.6. Mid Archaean microbial mat formed of layers of entwined filaments that resemble cyanobacteria from the Pongola Supergroup, South Africa (Noffke et al., 2003).

found in the walls of certain prokaryote organisms, hence, they are not specific to eukaryotes. Indeed, arguments for the oldest evidence for eukaryotes are based on the observation of some carbonaceous compressions in the 1.9 Ga Negaunee Formation in Michigan (Han and Runnegar, 1992).

The development of organisms utilising oxygenic photosynthesis, which is far more efficient than even anaerobic photosynthesis, opened up the possibility for a huge increase in the amount of biomass on Earth, and therefore a greater interaction with geological processes. It is in the Mid to Late Archaean that we start seeing the widespread development of stromatolites and the bioprecipitation of vast amounts of carbonate sedimentation that characterises much of the latter half of the Precambrian (cf. Walter, 1976, 1983; Awramik and Sprinkle, 1999). Furthermore, the oxygen produced by the cyanobacteria, a poison for most of the other microorganisms co-existing at the time, was essential to the further evolution of life and the appearance of eukaryotes. Additional indications of an increase on oxygen in the environment come from the massive oxidation of ferrous iron, which precipitated as thick BIFs, as well as deposits of reduced sulphide and sulphate on the continental shelves.

As a postscript to this résumé of early life, it is interesting to note that the rise of O_2 in the atmosphere, due both to microbial output and to the burial of carbon in the mantle by plate tectonic processes (Lindsay and Brasier, 2002), may have triggered the first major glaciation on Earth in the Palaeoproterozoic (Krischvink et al., 2000).

8. SUMMARY

The Earth upon which life appeared was a very different planet to that of today. It was, in fact, the extreme environment *per se*. The planet was characterised by strong volcanic and hydrothermal activity; tectonic turnover was rapid but large continents with platforms did not exist, subaerial land masses being represented by exposed ridges, volcanoes, and thickened portions of oceanic crust; the atmosphere was anoxic, consisting mostly of CO_2; it was probably also very "dirty" owing to the amount of gaseous and dusty aerosols from volcanic activity and impact debris; there was no ozone protection from UV radiation (although the dirty atmosphere would have offered some protection); the oceans would not have had the same chemistry as today and may have been more saline, although hypotheses regarding the pH vary from slightly acidic, neutral to alkaline; extraterrestrial bolide impacts were more frequent and larger; last but not least, days were shorter and tides were greater because the Moon was closer to the Earth.

It was on this Earth that life first appeared. There is no record of its first steps. The evidence that life existed by 3.75-3.8 Ga-ago is tenuous. However, if the conjectures of Westall (2003) regarding its seemingly widespread distribution and relatively evolved level of development by 3.4-3.5 Ga are correct, life must have already been in existence by 3.75-3.8 Ga. The oldest direct evidence for life shows that it consisted of colony- and biofilm-forming, prokaryote-like organisms that already displayed a variety of morphologies (coccoid, rod, vibroid, filament) and created stratiform, domical and coniform stromatolites. The biofilms formed on the surfaces of shallow-water sediments, as well as in evaporitic environments. These organisms must have been thermophilic. They also included halo-tolerant if not halophilic forms. There is some evidence that anaerobic photosynthesis may already have developed but not oxygenic photosynthesis. Although life was apparently widespread, microbial biomass was limited because of the limitations of the anaerobic metabolisms. The effects of the biomass on sedimentary processes were therefore of limited extent, resulting in sediment stabilisation and the formation of thin layers of biomediated, precipitated sediments.

Further evolution in the form of oxygenic photosynthesis and the invasion of a newly evolved habitat, *i.e.* broad continental shelves, appeared in the Mid- to Late-Archaean when organisms similar to modern cyanobacteria evolved. They contributed to geological processes by producing vast amounts of carbonate sediments and stromatolitic structures, trapping detritus with the large amounts of sticky saccharide polymers produced and forming oxygen as a by-product of their metabolism. The latter acted as a poison on co-existing anaerobic organisms but, eventually, produced the environmental conditions necessary for the appearance of eukaryotic life.

ACKNOWLEDGEMENTS

M. de Wit, W. Nijman, D. Marchesini and S. de Vries are thanked for their help in the field and I am indebted to S. Awramik for his insightful criticism of this manuscript.

REFERENCES

Appel, P.W.U., Moorbath, S. (1999). Exploring Earth's oldest geological record in Greenland. *EOS*, **80**, 257-264.

Appel, P.W.U., Fedo, C.M., Moorbath, S., Myer, J.S. (1998). Recognizable primary volcanic and sedimentary features in a low-strain domain of the highly deformed, oldest known (~3.7-3.8 Gyr) greenstone belt, Isua, West Greenland. *Terra Nova*, **10**, 57-62.

Appel, P.W.U., Moorbath, S., Myers, J.S. (2003). Isuasphaera isua (Pflug) revisited. *Precambrian Res*. **126**, 309-312.

Arndt, N.T. (1994). Archean komatiites, in *Archean crustal evolution*, ed. K.C. Condie, p. 11-44, Elsevier, Amsterdam.

Awramik, S.M., Sprinkle, J. (1999). Proterozoic stromatolites: the first marine evolutionary biota. *Historical Biology*, **13,** 241-253

Beaumont, V., Robert, F. (1999). Nitrogen isotope ratios of kerogens in Precambrian cherts: a record of the evolution of atmospheric chemistry? Precambrian Res., 96, 63-82.

Brasier, M.D., Green, O.R., Jephcoat, A.P., Kleppe, A.K., van Kranendonk, M., Lindsay, J.F., Steele, A., Grassineau, N. (2002). Questioning the evidence for Earth's oldest fossils. *Nature*, **416**, 76-81.

Bridgewater, D., Allaart, J.H., Schopf, J.W., Kelin, C., Walter, E.S., Strother, A.H., Gorman, B.E. (1981). Microfossil-like objects from the Archaean of Greenland: a cautionary note. *Nature*, **289**, 51-53.

Brocks J.J., Logan G.A., Buick R., Summons R.E. (1999). Archean molecular fossils and the early rise of eukaryotes. *Science*, **285,** 1033-1036.

Brocks, J.J. Love, G.D., Snape, C.E., Logan, G.A., Summons, R.E., Buick, R. (2003). Release of bound aromatic hydrocarbons from late Archean and Mesoarchean kerogens via hydropyrolysis. *Geochim. Cosmochim. Acta.*, **67**, 1521-1530.

Buick, R., (1990). Microfossil recognition in Archean rocks: an appraisal of spheroids and filaments from a 3500 m.y. old chert-barite unit at North Pole, Western Australia. *Palaios*, **5**, 441-459.

Byerly G. R., Walsh, M.M., Lowe, D.L. (1986). Stromatolites from the 3300-3500 Myr Swaziland Supergroup, Barberton Mountain Land, South Africa. *Nature*, **319**, 489-491.

Cady, S.L., Farmer, J.D. (1996). Fossilization processes in siliceous thermal springs: trends in preservation along the thermal gradient. In: *Evolution of hydrothermal ecosystems on Earth (and Mars)*, ed. G.R. Bock and J.A. Goodie, Ciba Symposium 202, John Wiley, Chichester, pp 150-173.

Cockell, C.S. (2001). A photobiological hoistory of Earth. In: *Ecosystems, Evolution and Ultraviolet Radiation*, eds. C.S. Cockell and A.R. Blaustein, Springer, Berlin, pp 1-35.

de Ronde, C.E.J., Ebbesen, T.W. (1996). 3.2 billion years of organic compound formation near sea-floor hot springs. *Geology*, **24**, 791-794.

de Wit, M.J., Hart, R.A. (1993). Earth's earliest continental lithosphere, hydrothermal flux and crustal recycling. *Lithos*, **30**, 309- 336.

Des Marais, D.J. (2000). When did photosynthesis emerge on Earth? Science, **289**, 1703-1705.

Dunlop, J.S.R., Muir, M.D., Milne, V.A., Groves, D.I. (1978). A new microfossil assemblage from the archeaen of Western Australia. *Nature*, **274**, 676-678.

Farquahr, J., Bao, H., Thiessen, M. (2000). Atmospheric influence of Earth's earliest sufur cycle. *Science*, **289**, 756-758.

Fedo, C.M. (2000). Setting and origin for problematic rocks from the > 3.7 Ga Isua Greenstone Belt, southern west Greenland: Earth's oldest coarse sediments. *Precambrian Res.*, **101**, 69-78.

Fedo, C.M., Whitehouse, M.J. (2002). Metasomatic origin of quartz-pyroxene rock, Akilia, Greenland, and implications for Earth's earliest life. *Science*, **296**, 1448-1452.

Forterre, P., Confalonieri, F., Charbonnier, F, Duguet, M. (1995). Speculations on the origins of life and thermophily: review of available information on reverse gyrase suggests that hyperthermophilic procaryotes are not so primitive. *Origins Life Evol. Biosphere*, **25**, 235-249.

Gerdes, G., Krumbein, W.E. (1987). *Biolaminated deposits*. In, Lecture Notes in Earth Sciences (S. Bhattacharji et al.). Springer, Berlin.

Grotzinger, J.P. (1994). Trends in Precambrian carbonate sediments and their implication for understanding evolution. In *Early life on Earth*, ed. S. Bengtson, Nobel Symp. 84, Columbia Univ. Press, N.Y., pp.245-258.

Grotzinger, J.P., Kasting, J.F. (1993). New constraints on Precambrian ocean composition. *J. Geol.*, **101**, 235-243.

Han, T.-M., Runnegar, B. (1992). Megascopic eukaryotic algae from the 2.1-billion-year-old Negaunee Iron-Formation, Michigan. *Science*, **257**, 232-235.

Hayes, J.M., Kaplan I.R., Wedeking K.W. (1983). Precambrian organic chemistry, preservation of the record. In: *Earth's earliest biosphere*, ed. J.W. Schopf, p. 93-134, Princeton Univ. Press.

Hofmann, H. J., M. Masson, (1994). Archean stromatolites from Abitibi greenstone belt, Quebec, Canada. *Geol. Soc. Am. Bull.*, **106**, p. 424-429.

Hofmann, H.J., Grey, K., , A.H., Thorpe, R.I. (1999). Origin of 3.45 Ga coniform stromatolites in Warrawoona Group, Western Australia. *Geol. Soc. Am. Bull.*, **111**, 1256-1262.

Holland, H. D. (1984). The chemical evolution of the atmosphere and oceans. Princeton Univ. Press, Princeton.

Kakegawa, T. (2001). Isotopic signatures of early life in the Archean oceans: influence from submarine hydrothermal activities. In: *Geochemistry and the Origin of Life*, Eds. S. Nakashsima, S. Maruyama, A. Brack, and B.F. Windley, pp. 237-249., Universal Acad. Press, Tokyo

Kasting, J.F. (1993). Earth's early atmosphere. *Science*, **259**, 920-926

Kempe, S. Degens, E.T. (1985). An early soda ocean? *Chem. Geol.*, **5**, 95-108.

Kirschvink, J.L. Gaidos, E.J., Bertani,L.E., Beukes, N.J., Gutzmer,J., Maepa, l.N., Steinberge, R.E. (2000). Paleoproterozoic snowball Earth: Extreme climatic and geochemical global change and its biological consequences. *Proc. Natl. Acad. Sci. USA*, **97**, 1400-1495.

Knauth, L.P., Lowe, D.R. (2003). High Archean climatic temperature inferred from oxygen isotope geochemistry of cherts in the 3.5 Ga Swaziland Supergroup, South Africa. *Geol. Soc. Am. Bull.*, **115**, 566-580.

Krumbein, W.E. (1983) Stromatolites - the challenge of a term in space and time. *Precambrian Res.*, **20**, 493-531.

Kyte, F.T., Shukolyukov, A., Lugmaor, G.W., Lowe, D.R., Byerly, G.R. (2003). Early Archean spherule beds: Chromium isotopes confirm origin through multiple impacts of projectiles of carbonaceous chondrite type. *Geology*, **31**, 283-286.

Lepland, A., Arrhenius, G., Cornell, D. (2002) Apatite in early Archean Isua supracrustal rocks, southern West Greenland: its origin, association with graphite and potential as a biomarker. *Precambrian Res.*, **118**, 221-241.

Lindsay, J.F., Brasier, M.D. (2002). Did global tectonics drive early biosphere evolution? Carbon isotope record from 2.6 to 1.9 Ga carbonates of Western Australian basins. *Precambrian Res.*, **114**, 1-34.

Lowe, D.R. (1983). Restricted shallow water sedimentation of zearly Archean stromatolitic and evaporitic strata of the Strelley Pool Chert, Pilbara Block, Western Australia. *Precambrian Res.*, **19**, 239-283.

Lowe, D. R., 1994. Abiological origin of described stromatolites older than 3.2 Ga. *Geology*, **22**, 287-390.

Lowe, D.R., Byerly, G.R. (1986). Early Archean silicate spherules of probable impact origin, South Africa and Western Australia, *Geology*, **14**, 83-86.

Lowe, D.R., Byerly, G.R. (1999). Geologic evolution of the Barberton greenstone belt, South Africa. *Geol. Soc. Am. Spec. Paper* **329**.

Lowe, D.R., Byerly, G.R., Kyte, F.T., Shukulyukov, A., Asaro, F., Krull, A. (2003). Spherule beds 3.47-3.24 billion years old in the Barberton Greenstone Belt, South Africa: a record of large meteorite impacts and their influence on early crustal and biological evolution. *Astrobiology*, **3**, 7-48.

Maher, K.A., Stevenson, D.J. (1988). Impact frustration of the origin of life. *Nature*, **331**, 612-614.

Melosh, H.J., Vickery, A.M., Tonks, W.B. (1993) Impacts and the early environment and evolution of the terrestrial planets. In *Protostars and planets III*, Ed. E.H. Levt and J.L. Lunine. Univ. Arizona Press, Tucson, pp. 1339-1370.

Mojzsis, S.J., Arrhenius, G., McKeegan, K.D., Harrison, T.M., Nutman, A.P., Friend, C.R.L. (1996). Evidence for life on Earth before 3800 millio-n years ago. *Nature*, **384**, 55-59.

Mojzsis, S.J., Harrison, T.M., Pidgeon, R.T. (2001). Oxygen-isotope evidence from ancient zircons for liquid water at the Earth's surface 4,300 Myr ago. *Nature*, **409**, 178-181

Myers, J.S.,(2003). Isua enigmas: illusive tectonic, sedimentary, volcanic, and organic features of the > 3.8 - >3.7 Ga Isua greenstone belt, Southwest Greenland. *Geophys. Res. Abstracts*, **5**, 13823.

Nijman, W., de Bruijne, K.H. Valkering, M. (1999). Growth fault control of Early Archaean cherts, barite mounds and chert-barite veins, North Pole Dome, Eastern Pilbara, Western Australia. *Precambrian Res.*, **95**, 247-274.

Noffke, N., Hazen, R., Nhieko, N. (2003). Earth's earliest microbial mats in a siliciclastic marine environment (2.9 Ga Mozaan Group, South Africa). *Geology*, **31**, 673-676.

Nutman, A.P. (1986). The early Archaean to Proterozoic history of the Isukasia area, southern West Greenland. *Greenland Geol. Surv. Bull.* **154**, Copenhagen.

Ohmoto, H. (1997). Evidence in pre-2.2 Ga paleosols for the early evolution of atmospheric oxygen and terrestrial biota. *Geology*, **24,** 1135-1138.

Ohmoto, H. (1999). Redox state of the Archean atmosphere: Evidence from detrital heavy minerals in ca. 3250-2750 Ma sandstones from the Pilbara craton, Australia: Comment. *Geology*, **27,** 1151-1152.

Paris, I., Stanistreet, I.G., Hughes, M.J. (1985). Cherts of the Barberton greenstone belt interpreted as products of submarine exhalative activity. *J. Geol.*, **93**, 111-129.

Pavlov, A.A., Kasting, J.F., Brown, L.L., Rages, K.A., Freedman, R. (2001). Greenhouse warming by CH4 in the atmosphere of early earth. *J. Geophys. Res.*, **105**, 11981-11990.

Pflug, H.D. (1979). Archean fossil finds resembling yeasts. *Geol. Palaeontol.*, **13**, 1-8.

Pflug, H.D., (2001). Earliest organic evolution, Essay to the memory of Bartholomew Nagy. *Precamb. Res.*, **106**,79-92.

Pflug, H.D., Jaeschke-Boyer, H. (1979). Combined structural and chemical analysis of 3,800-Myr-old microfossils. *Nature*, **280**, 483-486.

Philippot, P., Foriel, J., Cauzid, J., Susini, J., Dumas, P., Somogyi, A., Khodia, H., Ménez, B., Fouquet, Y., Moreira, D., Garcia-Lopez, P. (2003). High resolution synchrotron-based imaging of sulphur oxidation states in individual microfossils and contemporary microbial filaments. *Goldschmidt Conf. Abst.* A379.

Pinti, D.L., Hashizume, K. (2001). 15N-depleted nitrogen in Early Archean kerogens: clues on ancient marine chemosynthetic-based ecosystems? *Precambrian Res.*,**105**, 85-88.

Pinti, D.L., Hashizume, K., Matsuda, J.-I. (2001). Nitrogen and argon signatures in 3.8 to 2.8 Ga metasediments: Clues on the chemical state of the Archean ocean and the deep biosphere. *Geochim. Cosmochim. Acta*, **65**, 2301-2315.

Rasmussen, B. (2000). Filamentous microfossils in a 3,235-million-year-old volcanogenic massive sulphide deposit. *Nature*, **405**, 676-679.

Rasmussen, B., Buick, R., Holland, H.D. (1999). Redox state of the Archean atmosphere: evidence from detrital heavy minerals in ca. 3250-2750 Ma sandstones from the Pilbara Craton, Australia: reply. *Geology*, **27**, 1152.

Robert, F. (1988). Carbon and oxygen isotope variations in Precambrian cherts. *Geochim. Cosmochim. Acta*, **52**, 1473-1478.

Robbins, E.I. (1987). *Appelella ferrifera*, a possible new iron-coated microfossil in the Isua Iron-Formation, Southwestern Greenland. In: *Precambrian Iron Formations* eds., Appel, P.W.U., LaBerge, G.L., Theophrastes, Athens, pp. 141-154.

Robbins, E.I., Iberall, A.S. (1991). Mineral remains of early life on Earth? On Mars? *Geomicrobiol. J.*, **9**, 51-66.

Roedder, E. (1981). Are the 3,800-Myr-old Isua objects microfossils, limonite-stained fluid inclusions, or neither? *Nature*, **293**, 159-162.

Rosing M.T. (1999). 13C depleted carbon microparticles in >3700 Ma seafloor sedimentary rocks from West Greenland. *Science*, **283**, 674-676.

Runnegar, B. (2002). Archean sulphates from Western Australia: implications for Early Archean atmosphere and chemistry. *Goldschmidt Conf.,* Abstr. 3859.

Ryder, G. (2002). Mass influx in the ancient Earth-Moon system and benign implications for the origin of life on Earth. *J. Geophys. Res.*, **107**, 10.1029/2001JE001583.

Ryder, G. (2003). Bombardment of the Hadean Earth: wholesome or deleterious. *Astrobiology*, **3**, 3-6.

Rye, R., Kuo, P.H., Holland, H.D. (1995). Atmospheric carbon dioxide concentrations before 2.2 billion years ago. *Nature*, **378**, 603-605.

Sagan, C., Mullen, G. (1972). Earth and Mars: Evolution of atmospheres and surface temperatures. *Science*, **177**, 52-56.

Sagan, C., Chyba, C. (1997). The early sun paradox: organic shielding of ultraviolet-labile greenhouse gases. *Science*, **276**, 1217-1221.

Schidlowski, M. (1988). A 3800 million-year isotopic record of life from carbon in sedimentary rocks. *Nature*, **333**, 313-318.

Schidlowski, M., (2001). Carbon isotopes as biogeochemical recorders of life over 3.8 Ga of Earth history: evolution of a concept. *Precambrian Res.*, **106**, 117-134.

Schidlowski, M., Hayes, J.M., Kaplan, I.R. (1983). Isotopic inferences of ancient biochemistries: carbon, sulfur, hydrogen and nitrogen. In: *Earth's earliest biosphere: its origin and evolution*, ed. Schopf, J.W., Princeton Univ. Press, Princeton, pp. 149-186.

Schopf, J.W. (1993). Microfossils of the Early Archean Apex Chert: new evidence of the antiquity of life. *Science*, **260**, 640-646.

Schopf, J.W., Walter, M.R. (1983). Archean microfossils: new evidence of ancient microbes, in *Earth's earliest biosphere*, ed. J.W. Schopf, p.214-239, Princeton Univ. Press, Princeton.

Schopf, J.W., Packer, B.M. (1987). Early Archean (3.3 billionto 3.5 billion-year-old) microfossils from Warawoona Group, Australia. *Science*, **237**, 70-73.

Shen, Y., Buick, R., Canfield, D.E. (2001). Isotopic evidence for microbial sulphate reduction in the early Archaean era. *Nature*, **410**, 77-81.

Sleep, N.H., Zahnle, K.J., Kasting, J.F., Morowitz, H.J. (1989). Annihilation of ecosystems by large asteroid impacts on the early Earth. *Nature*, **342**, 139-142.

Sonett, C.P., Kvale, E.P., Zarkharian, A., Chan, M.A., Demko T.M. (1996). Late Proterozoic and Paleozoic tides, retreat of the Moon, and rotation of the Earth *Science* **273**, pp. 100-104.

Strauss, H. , Moore, T.B. (1992). Abundances and isotopic compositions of carbon and sulfur species in whole rock and kerogen samples. In: *The Proterozoic Biosphere: A multidisciplinary study,* Eds. J.W. Schopf, J.W., C. Klein, p.709-798, Camdridge Univ. Press, Cambridge.

Summons R.E., Jahnke L.L., Hope J.M., Logan J.H. (1999). 2-Methylhopanoids as biomarkers for cyanobacterial oxygenic photosynthesis. *Nature*, **400**, 554-7.

Touret, J. (2003). Remnants of early Archaean hydrothermal methane and brines in pillow-breccia from the Isua Greenstone Belt, West Greenland. *Precambrian Res*., **126**, 219-233.

Ueno, Y., Maruyama, S., Isozaki, Y., Yurimoto, H., (2001a). Early Archean (ca. 3.5 Ga) microfossils and 13C-depleted carbonaceous matter in the North Pole area, Western Australia: Field occurrence and geochemistry. In: *Geochemistry and the Origin of Life*, Eds. S. Nakashsima, S. Maruyama, A. Brack, and B.F. Windley, p. 203-236, Universal Acad. Press, Tokyo

Ueno, Y., Isozaki, Y., Yurimoto, H., Maruyama, S. (2001b). Carbon isotope signatures of individual Archean microfossils (?) from Western Australia. *Intl. Geology Review*, **43**, 196-212.

Ueno, Y., Yoshioka, H., Maruyama, S., Isozaki, Y. (2001c). Carbon and nitrogen isotope geochemistry of kerogen-rich ssilica dikes in the ca. 3.5 Ga North Pole area, Western Australia: sub-seafloor biosphere in the Archean. *International Archean Symposium extended abstracts*, K.F. Cassidy, J.N. Dunphy, and M.J. van Kranendonk (Eds.), AGSO-Geoscience Australia **2001/37**, 99-101.

Ueno, Y., Yurimoto, H., Yoshioka, H., Komiya, T., Matuyama, S. (2002). Ion microprobe analysis of graphite from ca. 3.8 Ga metasediments, Isua suprcrustal belt, West Greenland: relationship between metamorphism and carbon isotope composition. *Geochim. Cosmochim. Acta*, **66**, 1257-1268.

Van Kranendonk, Webb, G.E. Kamber, B.S. (2003) Geological and trace element evidence for a marine sedimentary environment of deposition and bogenicity of 3.45 Ga carbonates in the Pilbara, and support for a reducing Archean ocean. *Geobiology*, **1**, 91-108.

Van Zuilen, M., Lepland, Arrhenius, G. (2002). Reassessing the evidence for the earliest traces of life. *Nature*, **418**, 627-630.

Veizer, J. (1994). The Archean-Proterozoic transition and its environmental implications. In: *Early life on Earth*, ed. S. Bengtson, Nobel Symp. 84, Columbia Univ. Press, N.Y., pp. 208-219.

Walsh, M.M. (1992). Microfossils and possible microfossils from the Early Archean Onverwacht Group, Barberton Mountain Land, South Africa. *Precambrian Res.*, **54**, 271-293.

Walsh, M.M., Lowe, D.R. (1999). Modes of accumulation of carbonaceous matter in the early Archaean: A petrographic and geochemical study of carbonaceous cherts from the Swaziland Supergroup. In: *Geologic evolution of the Barberton greenstone belt, South Africa*, eds. DR. Lowe, G.R. Byerly, Geol. Soc. Am Spec. Paper, 329, 115-132,.

Walsh, M.M., Westall, F. (2003). Archean biofilms preserved in the 3.2-3.6 Ga Swaziland Supergroup, South Africa. In *Fossil and Recent Biofilms* (ed. W.E. Krumbein, T. Dornieden, and M. Volkmann), Kluwer, Amsterdam, in press.

Walter, M.R. (1976). Stromatolites, Springer verlag, Berlin.

Walter, M. R. (1983). Archean stromatolites: evidence of the Earth's earliest benthos. In, *Earth's Earliest Biosphere* (J. W. Schopf, eds.). Princeton University Press, Princeton. p. 187-213.

Westall, F. (1999). The nature of fossil bacteria. *J. Geophys. Res.*, **104**, 16,437-16,451.

Westall, F. (2003a). The geological context for the origin of life and the mineral signatures of fossil life. In *The Early Earth and the origin of Life*, H. Martin, M. Gardaud, G. Reisse, B. Barbier (Eds). Springer, Berlin, in press.

Westall, F. (2003b). Stephen Jay Gould, les procaryotes et leur évolution dans le contexte géologique. *Palevol*, **2** , 485-501.

Westall, F., Gerneke, D. (1998). Electron microscope methods in the search for the earliest life forms on Earth (in 3.5-3.3 Ga cherts from the Barberton greenstone belt, South Africa): applications for extraterrestrial studies. SPIE: Instruments, Methods and Missions for Astrobiology, **3441**, 158-169.

Westall, F., Walsh, M.M. (2000). The diversity of fossil microorganisms in Archaean-age rocks. In, *Journey to Diverse Microbial Worlds*, Ed. J. Seckbach, Kluwer, Amsterdam, 15-27.

Westall, F., Walsh, M.M. (2003). Fossil biofilms and the search for life on Mars. In *Fossil and Recent Biofilms,* eds. W.E. Krumbein, D. Patterson, G. Zavarzin, Kluwer, Amsterdam, in press.

Westall, F., Folk, R.L. (2003). Exogenous carbonaceous microstructures in Early Archaean cherts and BIFs from the Isua greenstone belt: Implications for the search for life in ancient rocks. *Precambrian Res.*, **126**, 313-330.

Westall, F., De Wit, M.J., Walsh, M.M., Folk, R.L., Chafetz, H., Gibson, E.K. (1999). An Early Archean, organic carbon-rich microbialite (3.3-3.4 Ga) from the Barberton greenstone belt. *South Africa. Intl. Soc. Study of the Origin of Life (ISSOL*), Abstr.

Westall, F., Steele, A., Toporski, J. Walsh, M., Allen, C., Guidry, S., Gibson, E., Mckay, D., Chafetz, H., (2000). Polymeric substances and biofilms as biomarkers in terrestrial materials: Implications for extraterrestrial samples. *J. Geophys. Res. Planets*. **105**, 24,511-24,527.

Westall, F., De Wit, M.J., Dann, J., Van Der Gaast., S., De Ronde., C., Gerneke., D., (2001). Early Archaean fossil bacteria and biofilms in hydrothermally-influenced, shallow water sediments, Barberton greenstone belt, South Africa. *Precambrian Res.*, **106**, 93-116.

Westall, F., Brack, A., Barbier, B., Bertrand, M., Chabin, A. (2002). Early Earth and early life: an extreme environment and extremophiles - application to the search for life on Mars. *Proceedings of the Second European Workshop on Exo/Astrobiology Graz, Austria, 16-19 September 2002*, **ESA SP-518**, pp. 131-136.

Westall, F., Hofmann, B, Brack, B. (2004). Searching for fossil microbial biofilms on Mars: a case study using a 3.46 billion-year old example from the Pilbara in Australia.

Proceedings of the Third European Workshop on Exo/Astrobiology Madrid, Spain 18-20 2003 **ESA Spec. Pub. 545**, in press.

Wiegel, J., Adams, M.W. (1998). *Thermophiles: The Keys to Molecular Evolution and the Origin of Life?* Taylor & Francis, N.Y.

Wilde, S.A., Valley, J.W., Peck, W.H., Graham, C.M. (2001). Evidence from detrital zircons for the existence of continental crust and oceans on the Earth 4.4 Gyr ago. *Nature*, **409**, 175-178.

Chapter 13

HIGHLY ALTERED ORGANIC MATTER ON EARTH: BIOSIGNATURE RELEVANCE

Beda A. Hofmann
Natural History Museum Bern

Abstract: On earth, organic matter of biological origin is subjected to various alteration processes, dominated by oxidation and thermal metamorphism due to deep burial, typically a result of plate-tectonic processes. Other alteration processes unrelated to the oxidizing atmosphere and plate tectonics include impact metamorphism, irradiation and thermal degradation of dissolved organic species. The implications of these transformation processes as observed on earth for the use of carbon forms as biosignatures on other planetary bodies are discussed. Carbon minerals, such as graphite and diamond, are potential biosignatures, their major advantage being a high degree of stability.

Key words: organic matter, graphite, diamond, carbonado, oxalate, whewellite

1. INTRODUCTION

Terrestrial organic matter in the broadest sense (compounds containing reduced carbon) is of dominantly biogenic origin, but a minor fraction is of abiotic origin. The terrestrial geological record contains various forms of more or less altered types of organic matter much of which is likely of biological origin, based mainly on the geological settings of occurrences and carbon isotope signatures. The most common type of organic matter on earth is kerogen, the insoluble breakdown product of microbial and higher living tissue. The typical final product of kerogen maturation on earth is graphite, with possible intermediate release of carbon dioxide and mobile hydrocarbons. Graphite has lost all structural information related to

P. Ehrenfreund et al. (eds.), Astrobiology: Future Perspectives, 317–331.

precursor materials, but it's carbon isotope signature may be used as an indicator of a biogenic origin. The oldest report of a possible biosignature on earth is of this type (Mojzsis, Arrhenius et al. 1996). The difficulty with the interpretation of such data is demonstrated by the discussions related to a possible nonbiogenic origin of this graphite following the original publication (Fedo and Whitehouse 2002; Mojzsis, Harrison et al. 2002; van Zuilen, Lepland et al. 2002). Due to the stability even under surface conditions on earth, graphite is largely stable during weathering and flakes are found in all types of sediments, including oxidizing red-beds. During metamorphism, the characteristic carbon isotope signature of organogenic reduced carbon may be lost, due to equilibration with carbonate minerals.

The dominant formation process of graphite on Earth is thermal metamorphism resulting from deep burial of sediments rich in organic matter in the course of plate-tectonic processes. The relevance of graphite as a potential biomarker on other planetary bodies lacking plate tectonics may thus be questioned. An investigation of processes having the potential to transform carbon into very stable forms (graphite, diamond) under conditions relevant for Mars is thus warranted.

Plate tectonics and surface weathering in an oxidizing, biogenic atmosphere are dominant geological forces on Earth, but most likely are both absent on all other planetary bodies of the solar system. In this paper those transformation processes of organic matter are discussed that are independent of plate tectonics or the presence of an oxygen-rich atmosphere. Organic matter transformed by analogous processes is present on Earth in various settings, and an investigation of these helps to guide the search for extraterrestrial organic matter of possible biogenic origin.

2. IMPACT-INDUCED TRANSFORMATION OF ORGANIC MATTER

2.1 Impact-induced formation of graphite

Graphite, the characteristic product of thermal maturation of sedimentary organic material, typically is formed during regional metamorphism under upper greenschist to amphibolite facies conditions, due to deep burial in mountain belts as result of plate-tectonic processes. On earth, the amounts of biogenic graphite formed due to sources of heat unrelated to plate tectonics, such as intra-plate volcanism, are very minor. To address the possible occurrence of graphite on other planetary bodies lacking plate tectonics but where impact cratering is an important process of energy delivery to possible near surface sedimentary rocks containing organic carbon, some terrestrial impacts provide an excellent analogue. Graphite is not generally described as a newly formed phase in impact-metamorphosed rocks. However, the Sudbury impact crater (Ontario) is an interesting case in this respect because the target rocks included sediments rich in organic matter (Becker, Poreda et al. 1996; Bunch, Becker et al. 1999; Heyman, Dressler et al. 1999; Mossman, Eigendorf et al. 2003). The origin of this organic matter has been much debated, but a powerful argument for the pre-impact origin of much of the carbon in the Onaping formation (suevitic fallback breccia) is the occurrence of carbon-rich clasts (Bunch, Becker et al. 1999), and own observations. The detection of fullerenes in the Onaping formation (Becker et al., 1996; Mossman et al., 2003) is controversial (Buseck, 2002) and may, if confirmed, indicate the presence of an additional extraterrestrial fraction of organic carbon. The basal parts of the Onaping formation are generally devoid of carbon (Ames 1999). Still, outcrop-sized areas in this basal zone contain abundant well-crystallized graphite concentrated in clasts, also supporting a pre-impact sedimentary origin (B. Hofmann, work in progress, Fig. 13-1).

Figure 13-1. Graphite flakes in thermally metamorphosed impact breccia (Onaping Formation) close to the contact with underlying impact melt sheet (Sudbury "Igneous" Complex). Joe Lake, Sudbury structure. Reflected light, oil immersion. Field of view 120 μm.

The general lack of graphite in the lower parts of the Onaping directly overlying the impact melt sheet ("Sudbury igneous complex", SIC) is therefore most likely a secondary effect, due to dissolution by high-T fluids associated with the SIC. The Sudbury example demonstrates that large impacts generating thick melt sheets provide a suitable environment to induce the formation of graphite from organic matter present in any pre-impact target rocks. In addition, fallback material containing organic material overlying a major melt sheet, as in Sudbury, will produce mobile hydrocarbons, preserved in Sudbury as veins of highly mature pyrobitumen.

2.2. Impact-induced formation of diamond

The recognition of terrestrial impacts has spurred research on impact-induced shock metamorphism of target materials. High-pressure phases of carbon, most notably diamond and lonsdaleite (hexagonal diamond) and possibly chaoite (El Goresy and Donnay 1968) and silicon carbide (not necessarily a high pressure phase) are now well-established minerals known from more than 10 terrestrial impact craters: Zapadnaya, Ilyinets, Obolon, Terny (all Ukraine) (Gurov, Gurova et al. 1995), Ries (Germany) (Hough, Gilmour et al. 1995; Schmitt, Siebenschock et al. 1999), Popigai (Siberia, Russia) (Koeberl, Masaitis et al. 1997; Masaitits 1998; Shelkov, Verchovsky et al. 1998), Sudbury (Canada) (Masaitis, Shafranivsky et al. 1999), Puchezh-Katunki (Russia), (Vishnevsky and Raitala 2000), Kara (Russia) (Vishnevsky and Raitala 2000), Gardnos, Norway (Gilmour, French et al. 2003), Lappajärvi, Finland (Pratesi, Giudice et al. 2003) and in Chicxulub ejecta (Hough, Gilmour et al. 1997). Diamond/lonsdaleite typically were formed from preexisting graphite in these craters, with the exception of Kara, where coal reportedly was the target material (Vishnevsky and Raitala 2000). The size of craters containing impact diamonds ranges from small (Zapadnaya: 3.2 km) to the largest known impacts (Sudbury: 250 km). Impact-associated diamonds can, in theory, also be derived from the impactor. Various types of meteorites are well known to contain diamonds of either presolar origin (carbonaceous chondrites) or shock-formed diamonds (ureilites, some iron meteorites).

Very fine-grained diamonds of impact origin (nanodiamonds) are thought to represent products of condensation from a vapour phase (Vishnevsky and Raitala 2000). This type of diamond is more problematic for a use as biosignature, due to possible isotope exchange processes during formation and the difficulty to resolve mixing with nanodiamonds of other (meteoritic) sources.

The carbon isotopic composition of impact-generated diamonds from terrestrial occurrences typically ranges from –7 to –22 ‰ $\delta^{13}C$ (PDB) (Eldridge, Compston et al. 1991; Hough, Gilmour et al. 1995; Hough, Gilmour et al. 1997; Koeberl, Masaitis et al. 1997; Shelkov, Verchovsky et al. 1998). These values are lighter than the majority of mantle diamonds (peak at –5‰), but heavier than typical terrestrial organic carbon (-25‰). Precursor graphite shows similar values in the Popigai structure (Koeberl, Masaitis et al. 1997), indicating that the range of carbon isotopic compositions likely is due to partial metamorphic equilibration between precursor graphite and heavier carbonates or mobile carbon phases. Published data on nitrogen isotopes in impact diamonds from Chicxulub ejecta (Gilmour, Russell et al. 1992) indicate a terrestrial origin. Popigai impact diamonds also show $\delta^{15}N$ values similar to crustal materials (Shelkov, Verchovsky et al. 1998).

The size of impact-diamonds often corresponds to the size of precursor graphite flakes, typically several 100 μm (Fig. 13-2).

The abundance of diamonds in impact rocks depends on the abundance of carbon phases in the target rocks. Graphite contents in metamorphic gneisses typically are in the range of 0.05-1wt% (corresponding to the concentration of organic carbon in nonmetamorphosed sedimentary equivalents). Since impact breccias and impact melt rocks usually are derived from a variety of different target rocks, the bulk diamond concentration in impactites can be expected to range up to a fraction of a wt% only in exceptional cases, with more typical values in the low ppm range. The recognition of diamonds in such rocks relies on total decomposition in acid (HF-HCl) or basic (NaOH) media followed by XRD, TEM or Raman spectroscopy of insoluble residues. The presently known suite of impact craters known to contain impact diamonds is largely governed by the restriction of such investigations to a few sites. Very likely diamonds are a minor, but very common product of impact in terrestrial craters. Diamond as one of the most stable minerals survives all processes of surface-related weathering and represents not only one of the most robust indicators of impacts, but also a potential biosignature provided the carbon isotopic composition is not too strongly altered by pre-impact metamorphic processes.

Figure 13-2. Impact diamonds (transformed from graphite flakes) from the Popigai crater, Siberia. The largest grains in this image measure up to 4 mm in size.

3. CARBONADO – A STRANGE FORM OF DIAMOND

One of the most enigmatic forms of diamonds is a variety called "carbonado" or "carbon". Carbonado is a microcrystalline (0.5 to 80 μm) form of diamond occurring as angular fragments up to several cm in size in placers (ancient river sediments) in Bahia (Brazil), Venezuela, Guyana and in two areas in the Ubangi area of the Republic of Central Africa (Fettke and Sturges 1933; Kerr, Graf et al. 1948; Trueb and Butterman 1969; Kagi, Takahashi et al. 1994). On both continents, carbonados are associated with normal (mantle derived) diamonds, but a common origin is unproven since both types are in secondary deposits. Before the opening of the South Atlantic Ocean both areas were in relative proximity, and a common origin for both Brazilian and African carbonados appears possible (Smith and Dawson 1985). The unusual and sometimes conflicting porperties of carbonado diamonds can be summarized as follows:

- fine-grained composite, grain size of 0.5-80 μm, with high and heterogeneously distributed porosity (Fig 13-3)
- mineral inclusions indicating crustal environment (Trueb and Butterman 1969; Trueb and De Wys 1971), other inclusions (elemental Fe, Ti, Si, SiC, Fe-Cr) indicate a strongly reducing environment (De, Heaney et al. 1998)
- trace element characteristics indicating crustal environment (Shibata, Kamioka et al. 1993), while other workers report mantle characteristics (Kagi, Takahashi et al. 1994)
- noble gases indicate presence of fissiogenic Xe and Kr, nucleogenic Ne and abundant ^{4}He, indicating a uranium-rich environment (Ozima, Zashu et al. 1991)
- absence of lonsdaleite, otherwise characteristic for impact-derived diamonds (our own XRD analyses)
- C, N isotopes: carbon isotopic composition with a narrow range between –26 and –28‰ δ^{13}C, nitrogen (1000 ppm) shows δ^{15}N 0 to +5‰ (Shelkov, Verchovsky et al. 1995). Both C and N isotopes are in the range for organic materials. An in-situ ion probe study (Vicenzi and Heaney 2001) confirmed light carbon isotopes (-24.7 to –32.0 ‰) but showed quite large variations of δ^{15}N from +7.4 to –16.1.
- The presence of N platelets indicates annealing in the diamond stability field for geological time periods (Kagi, Takahashi et al. 1994)
- U-Th-Pb systematics indicate the presence of highly radiogenic Pb, but a lack of parent radioisotopes, indicating recoil implantation. Pb model ages range from 2.6 to 3.8 Ga (Ozima and Tatsumoto 1997).

The origin of carbonado presently must be considered as unresolved. One possible problem with most analysis is the long residence in crustal rocks coupled with high porosity, facilitating contamination even by noble gases (e.g. by implantation from uranium minerals physically associated, but having a different origin). Proposed models of formation include impact transformation of massive carbon precursor (e.g. coal) (Smith and Dawson 1985), radiation-induced diamond formation in a U-Th-rich crustal environment (Ozima and Tatsumoto 1997), clathrate implosion (Haggerty 1995), mantle sources combined with a complex crustal history (Kagi, Takahashi et al. 1994), and formation in an extraterrestrial setting followed by delivery as meteorites (Haggerty 1998). Irrespective of the process, the source of carbonado most likely was organic carbon in a massive and pure form (coal or hydrocarbon). Carbonado thus represents a poorly understood example of a biosignature in diamond form.

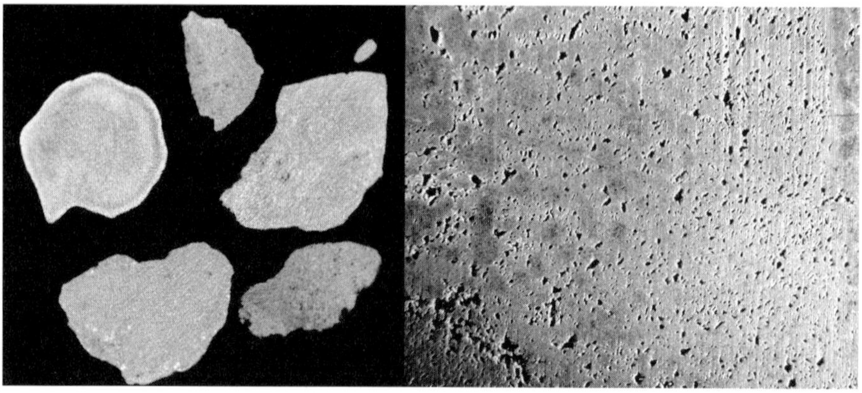

Figure 13-3. Left: X-ray tomographs of carbonado samples showing porosity, mineral inclusions and rims rich in strongly absorbing contaminations. Largest sample 25 mm in size. X-ray tomography performed at EMPA, Dübendorf, Switzerland. Right: Interior of e large carbonado fragment sectioned using a laser. Note high porosity. Field of view 10 mm. Natural History Museum Bern samples.

4. ALTERATION OF ORGANIC MATTER DUE TO IRRADIATION

Dose rates of ionizing radiation from the decay of natural radioisotopes, dominantly uranium, thorium and their daughter elements, are highly variable in crustal materials. High dose rates are locally present in and near minerals enriched in U and Th, such as uraninite UO_2, thorianite ThO_2, monazite $(REE,Th,U)PO_4$ or zircon $(Zr,U)SiO_4$, to name but a few common carriers of radioisotopes. Trace amounts of these minerals occur in dispersed form in many crustal rocks, notably granites and sediments, providing very local (on a millimeter scale) environments of high dose rates in and near the single mineral grains. More uncommon are accumulations of such radioactive minerals in small or large ore-bodies, due to mechanical separation and accumulation of minerals (e.g., placers) or local preferred precipitation. Some orebodies are massive accumulations of radioisotopes (0.1 to >10wt% U,Th) with high dose rates within 10^3 to 10^6 m^3 of rock.

In the case of dispersed radioactive minerals, high dose-rates are exclusively due to alpha-radiation because alpha-particles have a range of only tens of μm in minerals, resulting in the deposition of all of their energy in the immediate vicinity of the radioactive mineral grain, whereas β- and γ-radiation have larger ranges, resulting in a dispersion of their energy over large volumes of rock. In case of massive ore-bodies, most of the β- and γ-radiation also are retained within the volume of high dose-rate, adding to the dose.

Due to the ubiquity and mobility of organic carbon, and in many cases also because organic matter acts as a chemical reductant causing the immobilization of uranium, it is not uncommon to find organic matter tightly associated with radioactive minerals. It can be inferred, mainly by geological reasoning and by the observation of transitions to unmineralized zones containing organic material not affected by irradiation, that the type of organic matter initially present included either low-maturity plant debris (peat, coal), mobile humic substances or mobile hydrocarbons (oil).

As a consequence of irradiation over geological time periods, organic material is strongly altered and transferred into a form resembling highly mature kerogen, also known as "carburan" or "thucholite" (for the major components Th, U, C, H, O). Such radiation-altered organic carbon is known from many radioactive mineral occurrences, including the natural fission reactors of Oklo, Gabon (Nagy, Gauthier-Lafaye et al. 1991) and from radioactive heavy minerals (Parnell 2003). This carbon form is insoluble in solvents, strongly anisotropic in reflected light (indicating graphite-like ordering, Fig. 13-4), and composed dominantly of aromatic moities. Nanodiamonds have been reported from such materials from northern Russia (Daulton and Ozima 1996). Information about the biological origin of the precursor material is present only in the carbon isotopic signature, which typically reflects that of either plant debris ($\delta^{13}C$ -25±3 ‰ approx.) or mobile hydrocarbons (-30 to –40‰). The literature on organics in uranium deposits is vast, in the following just a few examples are given:

Pegmatitic uraninite transformed into "thucholite", Parry Sound, Ontario, Canada: Uraninite was formed as a high-temperature phase during solidification of a volatile-rich silicate melt to form so-called pegmatites in the Precambrian shield (Ellsworth 1928; Spence 1928; Parnell 1990). This material forms pseudomorphs after uraninite, with only 10-30 wt% of the precursor material remaining. Carbon isotope values for this "thucholite" range from –34.8 to –48.6 ‰ PDB (mean –37.9±4.8, n=7, Hofmann, unpublished data), consistent with an origin from mobile, ^{13}C-depleted hydrocarbons. This is further supported by the occurrence of still fluid oil-like hydrocarbons in the same pegmatite (Spence 1928).

Witwatersrand palaeoplacers, South Africa: The world's largest gold deposits, the palaeoplacers (fossil enrichments of heavy minerals) in the Archean Witwatersrand conglomerates, are also enriched in uranium and thorium, mainly in the form of detrital grains of uraninite. The radioactive orebodies are bedding-parallel, with a thickness of 0.1 to 0.3 m and large lateral extension. Several of the Au-U-bearing orebodies also contain high amounts of "thucholite". This material was originally interpreted as fossilized "plants" or cyanobacterial mats (Hallbauer and Van Warmelo 1974; Hallbauer 1975). Carbon isotopic values of the carbon leader reef (major occurrence of thucholite) peak at –35‰ $\delta^{13}C$, clearly indicating a biotic origin (Hoefs and Schidlowski 1967). More recent studies conclude that the uranium-associated carbonaceous matter results from short-scale intrabasinal hydrocarbon migration (Spangenberg and Frimmel 2001).

Reduction spots in red-bed sediments: Oxidized (hematite-bearing) sediments in continental and marine settings often do contain small-scale (mm to dm) spheroidal reduction phenomena, sites enriched in redox-sensitive elements (U, V, Se, Au, S and many more) and surrounded by a halo of hematite dissolution. The nature of the reductant responsible for the diagenetic formation of these features is unknown. In-situ sulfide formation (indicated by mineralogy) as well as S isotopes indicate a possible involvement of sulfur-reducing microbes (Hofmann 1990a,b; Hofmann 1991). Organic matter is present only rarely, but typically associated with uranium minerals (Parnell 1985; Hofmann 1993). These occurrences provide evidence that even in a generally oxidizing environment devoid of organic carbon, the process of radiation-induced polymerization can lead to the formation of uranium-organic matter associations that can persist over geological time periods.

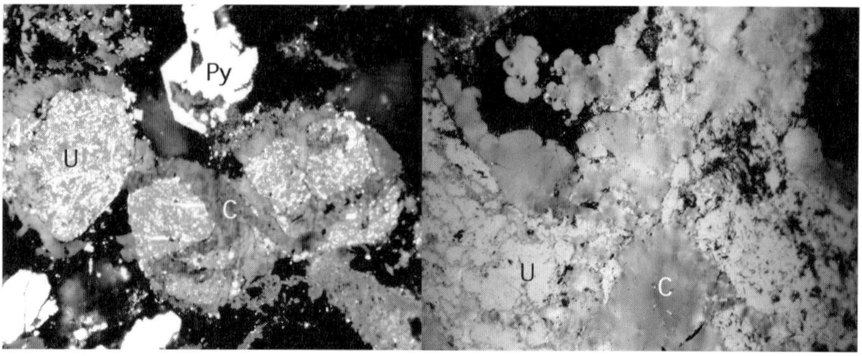

Figure 13-*4*. Left: Photomicrograph of detrital uraninite with enclosing radiation-damaged carbonaceous material. Blijvooruitzicht Mine, Witwatersrand, South Africa. Right: Radiation-damaged carbonaceous material associated with diagenetic uraninite from reduction spot in Permian red beds, Weichach well, Northern Switzerland. U=uraninite, C=carbonaceous matter, Py=pyrite. Field of view is 0.3 mm in both images.

Dispersed mineral grains in rocks are the most common form of radioactive phases in the earth's crust. Radiation-induced polymerization of organics in the vicinity of such grains has not received great attention. In a pioneering study, Rasmussen, Glover et al. (1993) described radioactive minerals coated with organics in sandstones (hydrocarbon reservoir rocks) in the Perth (Permian to Jurassic) and Bonaparte Basins in Australia. In these hydrocarbon source rocks, envelopes generally 40-50 μm thick are present around monazites and zircons. Often, these trace amount of radiation-immobilized hydrocarbons is the only trace of organics remaining in the reservoir rocks.

Based on the examples discussed above several common features of organics in high-radiation environments appear:

- Highly insoluble nature, similar to thermally mature kerogen
- Relatively light carbon isotope indicating derivation from mobile light hydrocarbons
- All occurrences are very similar, indicating that kerogen-like organics represent the end stage of radioation-induced alteration
- Even though the presence of small amounts of nanodioamonds in irradiated carbonaceous material has been reported, occurrences with the highest dose rates (e.g., the fission reactors of Oklo) do not contain carbonado-like materials.

5. OXALATE MINERAL CARBON ISOTOPE SYSTEMATICS

The calcium oxalate minerals whewellite and weddellite are the only relatively stable minerals (persisting over geological time periods) of any organic acid. These oxalate minerals are quite rare, but known to occur in weathering environments associated with lichen and plant debris, as a diagenetic formation in sedimentary rocks, and in low-T hydrothermal settings. A survey of the carbon isotopic composition of oxalate minerals from the majority of known occurrences (Hofmann and Bernasconi 1998) showed a very large range of $\delta^{13}C$ values from –31.7 to +33.7‰ PDP. In all cases isotopically "normal" ($\delta^{13}C$ -25 to –30‰) organic matter was associated with oxalates. In pairs of coexisting oxalate and organic matter, oxalate is enriched in ^{13}C by 0.7 to 6.2‰ in the case of near-surface/biological samples. Diagenetic oxalates are enriched by 7.7 to 60‰, and hydrothermal oxalates by 1.8 to 23.5‰. Fractionation must be the result of microbial and/or thermal degradation of oxalate in dissolved state prior to mineral crystallization. Oxalate mineral carbon isotopic compositions are proof of strong isotopic fractionations when oxalate is present in dissolved form in aqueous fluids at elevated temperature. A somwhat comparable situation is encountered in carbonaceous chondrites, where macromolecular bulk insoluble carbon is significantly lighter (-10 to –18‰) (Kerridge 1985) than soluble organics (-23 to +44‰) (Botta and Bada 2002, and references therein).

6. DISCUSSION

Several examples of organic materials of terrestrial origin (impact-generated graphite and diamond; carbonado diamond of unknown origin; radiation-altered organic matter; calcium oxalate minerals) demonstrate pathways by which biogenic organic material is transformed into mineral forms that may or may not retain biomarker information.

Both graphite and diamond are phases with biosignature potential on Mars. If there was any organic matter present in significant amounts in early Martian sediments, it is highly probable that at least a part of this has been transformed into graphite and/or diamond. Graphite is likely formed in thermally metamorphosed crater-fill materials overlying thick impact melt sheets in large craters. Impact diamonds probably were formed in craters of any size provided the target material did contain reduced carbon. A distinction of graphite and diamond derived from biogenic and nonbiogenic precursor materials (e.g. of meteoritic or planetary prebiotic origin) may be achieved by stable isotope analyses, if the sources are significantly different. A major advantage of graphite and especially diamond is the stability of these phases. It is likely that they persist even at the Martian surface today.

Radiation-induced polymerization is a process that has potential application in the search for biogenic organic matter. Terrestrial examples show that even single grains of radioactive minerals are able to immobilize mobile hydrocarbons, and become coated with organics by this process. Even in an environment with very low general abundances of U and Th as on Mars, rare grains of radioactive minerals (e.g., in porous sediments) may thus be interesting targets in the search for indigeneous organics.

Evidence for very strong ^{13}C enrichment in oxalate as compared with source organic of up to 60‰ indicates that the interpretation of carbon isotope data of mobile and thermally instable compounds, e.g., in carbonaceous chondrites, needs to take into account the possibility of isotopic fractionations during aqueous phase transport.

7. CONCLUSIONS

Graphite and diamond are minerals with a good potential to carry isotopic biosignatures. The importance of impact processes on Mars and the high stability of these minerals implies that if there were significant amounts of biogenic organic matter on early Mars, a fraction of this must have been transformed into these phases. Future exploration strategies should include the search for and analysis of diamond and graphite. Accessory U and Th minerals should also be investigated for carbonaceous coatings of biogenic origin, a possible setting being thermally altered impact breccias in large impact craters.

ACKNOWLEDGEMENTS

Parts of this research were supported by the Swiss NSF. Ph. Häuselmann and M. Eggimann are thanked for XRD work. Work on Sudbury is part of a project supported by the NASA Exobiology program. Constructive comments on an earlier version by Chris Koeberl improved the manuscript.

REFERENCES

Ames, D.E.: 1999, Nature of hydrothermal alteration in subaqueous impact craters: Sudbury structure, Ontario, Canada. PhD Thesis, Carlton University, Ottawa, Canada.

Becker, L., Poreda, R.J. and Bada, J.L., 1996. Extraterrestrial helium trapped in fullerenes in the Sudbury impact structure. Science, 272: 249-252.

Botta, O. and Bada, J.L., 2002. Extraterrestrial organic compounds in meteorites. Surveys in Geophysics, 23: 411-467.

Bunch, T.E. et al., 1999. Carbonaceous matter in the rocks of the Sudbury Basin, Ontario, Canada. In: B.O. Dressler and V.L. Sharpton (Editors), Large meteorite impacts and planetary evolution II, Geol. Soc. America. Spec. Paper 399, pp. 331-343.

Buseck, P.R., 2002. Geological fullerenes: review and analysis. Earth Plaet. Sci. Lett., 2003: 781-792.

Daulton, T.L. and Ozima, M., 1996. Radiation-induced diamond formation in uranium-rich carbonaceous materials. Science, 271: 1260-1263.

De, S., Heaney, P.J., Hargraves, R.B., Vicenzi, E.P. and Taylor, P.T., 1998. Microstructural observations of polycrystalline diamond: a contribution to the carbonado conundrum. Earth and Planetary Science Letters, 164: 421-433.

El Goresy, A. and Donnay, G., 1968. A new allotropic form of carbon from the Ries crater. Science, 161: 263-264.

Eldridge, C.S., Compston, W., Williams, I.S., Harris, J.W. and Bristow, J.W., 1991. Isotope evidence for the involvement of recycled sediments in diamond formation. Nature, 353: 649-653.

Ellsworth, R.V., 1928. Thucholite, a remarkable primary carbon mineral from the vicinity of Parry Sound, Ontario. American Mineralogist, 13: 419-442.

Fedo, C.M. and Whitehouse, M.J., 2002. Metasomatic Origin of Quartz-Pyroxene Rock, Akilia, Greenland, and Implications for Earth's Earliest Life. Science, 296: 1448-1452.

Fettke, C.R. and Sturges, F.C., 1933. Structure of carbonado or black diamond. American Mineralogist, 18: 172-174.

Gilmour, I. et al., 2003. Geochemistry of carbonaceous impactites from the Gardnos impact structure, Norway. Geochimica et Cosmochimica Acta, 67(20): 3889-3903.

Gilmour, I. et al., 1992. Terrestrial carbon and nitrogen isotopic ratios from Cretaceous-Tertiary boundary nanodiamonds. Science, 258: 1624-1626.

Gurov, E.P., Gurova, E.P. and Rakitskaya, R.B., 1995. Impact diamonds in the craters of the Ukrainian shield. Meteoritics, 30: 515-516.

Haggerty, S.E., 1995. Carbonado, clathrate and cavitation: a model for the acoustic induction of diamond. 6th Int. Kimberlite Conference, Extended Abstracts: 217-219.

Haggerty, S.E., 1998. Diamond-carbonado: Geological implications and research-industrial applications. Proc. 5th NIRIM Intl. Ssymposium in Advances Materials: 39-42.

Hallbauer, D.K., 1975. The plant origin of Witwatersrand "carbon". Min. Sci. Eng., 7: 111-131.

Hallbauer, D.K. and Van Warmelo, K.T., 1974. Fossilized plants in thucholite from Precambrian rocks of the Witwatersrand, South Africa. Precambrian Research, 1: 199-212.

Heyman, D. et al., 1999. Origin of carbonaceous matter, fullerenes, and elemental sulfur in rocks of the Whitewater group, Sudbury impact structure, Ontario, Canada. In: B.O. Dressler and V.L. Sharpton (Editors), Large meteorite impacts and planetary evolution II, Geol. Soc. America. Spec. Paper 399, pp. 345-360.

Hoefs, J. and Schidlowski, M., 1967. Carbon isotope composition of carbonaceous matter from the Precambrian of the Witwatersrand system. Science, 155: 1096-1097.

Hofmann, B., 1990a. Possible microbiological origin of reduction spots in red beds. In: F. C.B. and H. T.C. (Editors), First International Symposium on Microbiology of the Deep Subsurface. WSRC Information Services, Orlando, Florida, pp. 2-125 -2-133.

Hofmann, B., 1990b. Reduction spheroids from northern Switzerland: Mineralogy, geochemistry and genetic models. Chemical Geology, 81: 55-81.

Hofmann, B., 1991. Mineralogy and geochemistry of reduction spheroids in red beds. Mineralogy and Petrology, 44: 107-124.

Hofmann, B., 1993. Organic matter associated with mineralized reduction spots in red beds. In: J. Parnell, H. Kucha and P. Landais (Editors), Bitumens in ore deposits. Springer, pp. 362-378.

Hofmann, B.A. and Bernasconi, S.M., 1998. Review of occurrences and carbon isotope geochemistry of oxalate minerals: implications for the origin and fate of oxalate in diaganetic and hydrothermal fluids. Chem. Geol., 149: 127-146.

Hough, R.M. et al., 1995. Diamond and silicon carbide in impact melt rock from the Ries impact crater. Nature, 378: 41-44.

Hough, R.M., Gilmour, I., Pillinger, C.T., Langenhorst, F. and Montanari, A., 1997. Diamonds from the iridium-rich K-T boundary layer at Arroyo el Mimbral, Tamaulipas, Mexico. Geology, 25(11): 1019-1022.

Kagi, H., Takahashi, K., Hidaka, H. and Masuda, A., 1994. Chemical properties of Central African carbonado and its genetic implications. Geochimica et Cosmochimica Acta, 58(12): 2629-2638.

Kerr, P.F., Graf, D.L. and Ball, S.H., 1948. Carbonado from Venezuela. American Mineralogist, 33: 251-255.

Kerridge, J.F., 1985. Carbon, hydrogen and nitrogen in carbonaceous meteorites: Abundances and isotopic compositions in bulk samples. Geochim. Cosmochim. Acta, 49: 1707-1714.

Koeberl, C. et al., 1997. Diamonds from the Popigai impact structure, Russia. Geology, 25(11): 967-970

Masaitis, V.L. et al., 1999. Impact diamonds in the suevitic breccias of the Black Member of the Onaping formation, Sudbury structure, Ontario, Canada. Geol. Soc. America Spec. Paper, 339: 317-321.

Masaitits, V.L., 1998. Popigai crater: Origin and distribution of diamond-bearing impactites. Meteoritics and Planetary Science, 33(349-359).

Mojzsis, S.J. et al., 1996. Evidence for life on Earth before 3800 million years ago. Nature, 384: 55-59.

Mojzsis, S.J. et al., 2002. Origin and significance of Archean quartzose rocks at Akilia, Greenland. Science, 298: 917.

Mossman, D. et al., 2003. Testing for fullerenes in geologic materials: Oklo carbonaceous substances, Karelian shungites, Sudbury Black Tuff. Geology, 31(3): 255-258.

Nagy, B. et al., 1991. Organic matter and containment of uranium and fissiogenic isotopes at the Oklo natural reactors. Nature, 354: 472-475.

Ozima, M. and Tatsumoto, M., 1997. Radiation-induced diamond crystallization: origin of carbonados and its implications on meteorite nano-diamonds. Geochim. Cosmochim. Acta, 61(2): 369-376.

Ozima, M., Zashu, S., Tomura, K. and Matsuhisa, Y., 1991. Constraints from noble-gas contents on the origin of carbonado diamonds. Nature, 351: 472-474.

Parnell, J., 1985. Uranium/rare earth-enriched hydrocarbons in Devonian sandstones, northern Scotland. N. Jb. Miner. Mh., 1985, H. 3: 132-144.

Parnell, J., 1990. Mineralogy of rare-earth bearing "thucholite" from Parry Sound, Ontario. Canadian Mineralogist, 28: 357-362.

Parnell, J., 2003. Mineral radioactivity promotes organic complexity on rocky planets. Lunar and Planetary Science, 34: Abstract 1119.

Pratesi, G., Giudice, A.L., Vshnevski, S., Manfredotti, C. and Cipriani, C., 2003. Cathodoluminescence investigations on the Popigai, Ries, and Lappajärvi impact diamonds. American Mineralogist, 88: 1778-1787.

Rasmussen, B., Glover, J.E. and Foster, C.B., 1993. Polymerisation of hydrocarbons by radioactive minerals in sedimentary rocks: Diagenetic and economic significance. In: J. Parnell, H. Kucha and P. Landais (Editors), Bitumens in ore deposits. Springer, Berlin, pp. 490-509.

Schmitt, R.T., Siebenschock, M. and Stöffler, D., 1999. Distribution of impact diamonds in the Ries crater, Germany. Meteoritics and Planetery Science, 34S: A102.

Shelkov, D., Verchovsky, A.B., Milledge, H.J. and Pillinger, C.T., 1995. Carbonado: Am comparison between Brazilian and Ubangui sources based on carbon and nitrogen isotopes. 6th Int. Kimberlite Conference, Extended Abstracts: 518-520.

Shelkov, D.A., Verchovsky, A.B., Milledge, H.J., Kaminsky, F.V. and Pillinger, C.T., 1998. Carbon, nitrogen, argon and helium study of impact diamonds from Ebeliakh alluvial deposits and Popigai crater. Meteoritics & Planetary Science, 33: 985-992.

Shibata, K., Kamioka, H., Kaminsky, F.V., Koptil, V.I. and Svisero, D.P., 1993. Rare earth element patterns of carbonado and yakutite: evidence for their crustal origin. Mineralogical Magazine, 57: 607-611.

Smith, V.J. and Dawson, J.B., 1985. Carbonado: diamond aggregates from early impacts of crustal rocks? Geology, 13: 342-343.

Spangenberg, J. and Frimmel, H.E., 2001. Basin-internal derivation of hydrocarbons in the Witwatersrand basin, South Africa: evidence from bulk and molecular 13C data. Chemical Geology, 173: 339-355.

Spence, H.S., 1928. A remarkable occurrence of thucholite and oil in a pegmatite dyke, Parry Sound district, Ontario. American Mineralogist, 15(11): 499-520.

Trueb, L.F. and Butterman, W.C., 1969. Carbonado: a microstructural study. American Mineralogist, 54: 412-425.

Trueb, L.F. and De Wys, E.C., 1971. Carbon from Ubangi - a microstructural study. American Mineralogist, 56: 1252-1268.

van Zuilen, M.A., Lepland, A. and Arhenius, G., 2002. Reassessing the evidence for the earliest traces of life. Nature, 418: 627-630.

Vicenzi, E.P. and Heaney, P.J., 2001. The carbon and nitrogen isotopic composition of carbonado diamond: an in situ study. 11th V.M. Goldschmidt Conference Abstract 3886.

Vishnevsky, S. and Raitala, J., 2000. Impact diamonds as indicators of shock metamorphism in strongly-reworked Precambrian impactites. In: S. Bhattacharji, G.M. Friedman, H.J. Neugebauer and A. Seilacher (Editors), Impacts and the Early Earth, Lecture Notes in Earth Sciences, V. 91, I. Gilmour and C. Koeberl, Editors. Springer, Berlin, pp. 229-247.

Chapter 14

INSOLUBLE ORGANIC MATTER IN CARBONACEOUS CHONDRITES AND ARCHEAN CHERTS

An insight into their structure by Electron Paramagnetic Resonance

Laurent Binet[a], Didier Gourier[a], Audrey Skrzypczak[a], Sylvie Derenne[a] and François Robert[b]

[a]Ecole Nationale Supérieure de Chimie de Paris, Paris, France; [b]Museum National d'Histoire Naturelle, Paris, France

Abstract: The macromolecular carbonaceous material contained in the meteorites of Orgueil, Murchison and Tagish Lake, and in an early Archean chert from the Warrawoona Group (Australia) was analyzed by Electron Paramagnetic Resonance. The meteoritic macromolecular material distinguishes itself from its terrestrial counterpart by the presence of diradicaloids (alkyl poly-substituted aromatic moieties bearing two unpaired electrons), which appear as an extraterrestrial signature. Electron Paramagnetic Resonance (EPR) provided information about the oxygen content and the maximum size of the aromatic moieties of the macromolecular material in the Warrawoona chert. Comparison with biogenic macromolecular organic matter showed that the EPR linewidth could be an easily measurable indicator of the sample age.

Key words: Insoluble organic matter, radicals, Electron Paramagnetic Resonance, carbonaceous chondrites, cherts, Orgueil, Murchison, Tagish Lake, Warrawoona group, Gunflint Formation, Clarno Formation

1. INTRODUCTION

The issues of the origin of life on Earth and of the search for extraterrestrial life raise many important questions. One of them is the origin of the organic matter on Earth and its evolution from the abiotic precursors to living beings. This in turn raises some major questions:

P. Ehrenfreund et al. (eds.), Astrobiology: Future Perspectives, 333–357.

- What are the sources and the different types of pristine organic matter?
- How is this pristine material processed in a planetary environment to provide precursors of biological interest and ultimately living beings?
- What is the post-mortem evolution of the organic material from the first living organisms and what are the possible biomarkers remaining after several billions of years?

Carbonaceous chondrites are among the most primitive objects of the Solar System. They may contain up to 3% of organic matter and are considered as one of the possible sources of organic material on Earth [1]. The organic matter in the present carbonaceous chondrites has been in a frozen state since the Solar System formation and is reasonably considered as representative of the organic matter delivered on Earth during the meteoritic bombardment between *ca.* –4.5 and –4 Ga. A minority part (10-30%) of the organic matter in the carbonaceous chondrites is made of a complex mixture of soluble molecules, some of which are similar to those found in biochemical systems [1]. However the major fraction (70-90%) is made of an insoluble macromolecular carbonaceous material [1], the structure of which is complex and still not fully elucidated. The determination of the structure of this insoluble organic matter (IOM) is of primary importance both to elucidate its synthetic pathways and to assess its possible contribution to the production of prebiotic precursors. Stepping forward in time, the carbonaceous matter in the oldest Archean rocks is the memory of the first steps of life on Earth. However, when dealing with such old materials, usual biomarkers (isotopic, chemical or petrologic) must be considered with caution, for it is never fully excluded that they may arise from non-biological processes or by later contamination by more recent living organisms. This issue is exemplified by the recent debate about the biogenicity of the carbonaceous matter in the cherts (microcrystalline sedimentary quartz) from the Warrawoona group (Australia) dated at –3.5 Ga [2, 3, 4] and even in Aikila rocks (Greenland), age of –3.8 Ga [5]. This debate points to the necessity of a better understanding of the structure of the early Archean carbonaceous matter, and more generally the need for reliable biomarkers. When it comes to the chemical and structural analysis of the IOM in the carbonaceous chondrites and in the Archean cherts, the highly refractory and amorphous nature of these materials raises major analytical issues. Only a combination of several techniques, both chemical and physical, can lead to a reasonable structural picture of these materials, and due to the complexity of the task, no technique should be discarded. Electron Paramagnetic Resonance (EPR) has already been a successful analytical tool for coals [6], but hardly ever applied to meteorites or early Archean rocks. This contribution reports on the potential of Electron Paramagnetic

Resonance (EPR) in giving new insights into the structure of the IOM from these extraterrestrial or terrestrial rocks.

2. STRUCTURAL ANALYSIS OF THE MACROMOLECULES: BRIEF OVERVIEW OF METHODS

Table 14-1. Advantages and disadvantages of some common techniques for the analysis of the insoluble organic matter, and EPR is compared alongside them.

Technique	Potentials and advantages	Disadvantages
Thermal or chemical degradation	Production of molecular fragments	Destructive
Transmission Electron Microscopy (TEM)	Nanometer scale structure (size and stacking of the aromatic moieties)	- Difficulties with non-crystalline materials - Possible selection artifacts - Possible damage by electron beam
Spectroscopic techniques		
Infrared, Raman	- High sensitivity - Possibility of local analysis	Low resolution and selectivity
EPR	- Good resolution, selectivity and sensitivity - Many potential paramagnetic species	- Paramagnetic species not always representative of the bulk structure - Poor selectivity if interactions between species
NMR (^{1}H, ^{13}C)	- High resolution and selectivity - Distinction and quantification of different chemical groups	- Low sensitivity - Difficulties with H-poor or paramagnetic organic matter

Table 14-1 summarizes the advantages and disadvantages of a few common analytical techniques such as thermal or chemical degradation techniques, electron microscopy or spectroscopic techniques, and the situation of EPR among them. Degradation techniques, though directly providing some of the molecular fragments composing the material have the serious drawback of being destructive. If we now focus on the spectroscopic techniques, we find that when passing from high photon energies (Infrared and Raman spectroscopies) to low photon energies [Nuclear Magnetic Resonance (NMR)], the loss of sensitivity is compensated with an increase in resolution. Resolution is a key requirement when dealing with a material containing multiple functionalities. Therefore NMR is one of the most efficient technique for this kind of material. However NMR suffers from difficulties when the material is paramagnetic, for the interaction of the probed nuclei with unpaired electrons broadens the NMR signal, sometimes to the extent that the signal disappears. Precisely in the case of paramagnetic substances, EPR can be an attractive technique. This technique is based on the absorption of a microwave radiation by unpaired electrons in paramagnetic species submitted to an external magnetic field. The absorption spectrum of the unpaired electrons is influenced by the local magnetic fields internal to the material. Therefore, these electrons reveal the local chemical and electronic structure of the material around them. The analysis may be complicated by the fact that paramagnetic species are not always representative of the bulk structure. However the existence and the nature of paramagnetic species may be as such a source of information. Another difficulty is the loss of information on the individual species when strong interactions occur between them, and then information is rather derived indirectly by, for instance, changing the experimental conditions.

3. BASIC NOTIONS ABOUT EPR

3.1 Electron magnetic moment and paramagnetic substances

It is known from quantum physics that a free electron carries an intrinsic angular momentum, the so-called spin $\vec{S}$. This angular momentum was formerly, but falsely, associated to the rotation of the electron about itself. Actually, this is a pure quantum mechanical quantity, which has no classical counterpart. However, as in classical physics, a magnetic moment $\vec{\mu}$ is associated to the spin $\vec{S}$ according to:

$$\vec{\mu} = -g\beta\vec{S} \tag{1}$$

In the above equation, g is the g-factor (for a free electron, $g = g_e \approx 2.0023$), and $\beta \approx 9.26 \times 10^{-24}$ J.T^{-1} the electron Bohr magneton. Note that, for an electron, spin and magnetic moment have opposite directions; this is due to the negative electric charge of the electron.

In many substances, the electrons are paired with opposite spins, so that their magnetic moments mutually cancel out. Such substances are termed diamagnetic for an external magnetic field induces only a very small macroscopic magnetization opposite to the field and of no consequence for EPR, due to the distortion of the electronic cloud. However, some other substances contain a few unpaired electrons. Due to their intrinsic magnetic moments, the latter behave as small magnets, which preferentially align parallel to an external magnetic field. This results in a strong macroscopic magnetization along the field and the substance is termed paramagnetic. Unpaired electrons are generally found with defects or conduction electrons in solids or with transition metal or rare earth ions. Organic paramagnetic species are generally termed radicals. In organic substances, unpaired electrons often arise from bond cleavage about carbon atoms. In that case, each fragment takes up one of the two electrons ensuring the bond and a single electron remains in a dangling orbital on each fragment (Fig. 14-1a and b). Unpaired electrons can also be found in moieties made of condensed benzene rings and containing an odd number of carbon atoms (Fig. 14-1c). Radicals in which the unpaired electron is delocalized over an aromatic moiety (Fig. 14-1b and c) are particularly stable, and are thus dominant in substantially aromatic carbonaceous matter.

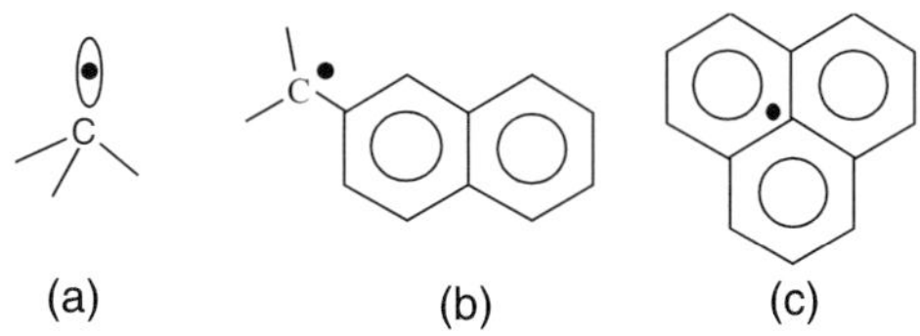

Figure 14-1. Examples of organic radicals. The black dot represents the unpaired electron, which is indeed delocalized over the aromatic moiety in the case of radicals (b) and (c).

3.2 Magnetic resonance transitions

In a magnetic resonance experiment, the paramagnetic substance is submitted to an external static magnetic field $\vec{B}_0$. The magnetic moments of the unpaired electrons interact with this magnetic field; this is the so-called Zeeman effect. Quantum physics tells us that the spin and hence the electron magnetic moment can be found only either parallel or antiparallel to the magnetic field. These two possible orientations give rise to two different electron energies, $E_- = -g\beta B_0 / 2$ and $E_+ = +g\beta B_0 / 2$, corresponding to the spin antiparallel and to the spin parallel to $\vec{B}_0$, respectively (Fig. 14-2).

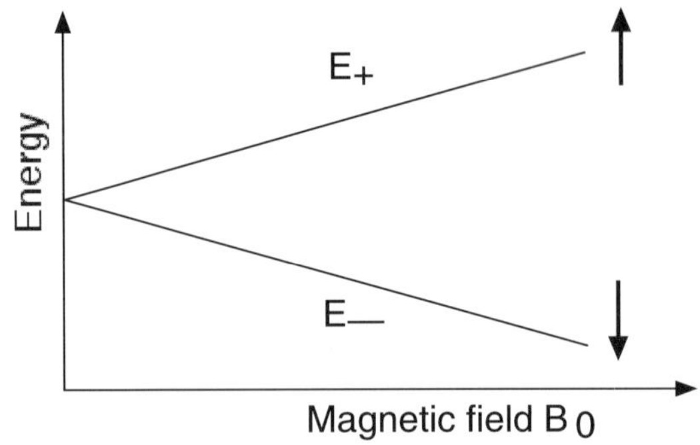

Figure 14-2. Magnetic energy levels of an electron in a magnetic field. The arrows on the right indicate the orientation of the spin angular momentum with respect to the magnetic field.

At a given temperature both energy levels are populated, with however a slight excess of unpaired electrons in the lower lying level E_-. In addition to the static field $\vec{B}_0$, the unpaired electrons are submitted to an electromagnetic field in the microwave range (frequency usually about 9.4 GHz). The electromagnetic field induces transitions between the two levels E_- and E_+ provided that the photon energy $h\nu$, with ν the microwave frequency, matches the energy splitting $\Delta E = E_+ - E_-$, *i.e.*:

$$h\nu = g\beta B_0 \qquad (2)$$

This condition is termed the resonance condition. The result of the transitions is a net absorption of the microwave energy. For technical reasons, an EPR spectrum is recorded by scanning slowly the external field B_0 at a fixed microwave frequency ν. Besides, in order to enhance the signal-to-noise ratio, a synchrone detection is used, with the application of a small and slowly oscillating magnetic field parallel to $\vec{B}_0$. Therefore an EPR spectrum appears as an absorption derivative with respect to the field B_0.

3.3 Information derived from an EPR spectrum

Here we only consider the simple case of a spectrum made of a single EPR line (Fig. 14-3). This is the situation encountered in the IOM of the carbonaceous chondrites and in some cherts.

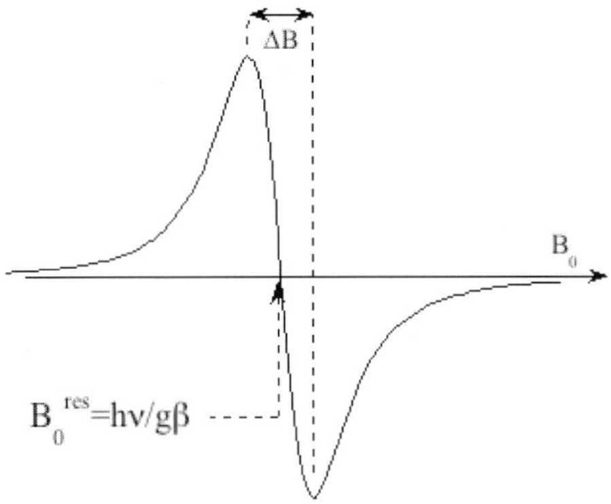

Figure 14-3. Typical EPR spectrum corresponding to an absorption derivative.

The first EPR parameter is the intensity *I*. The intensity is defined as the area under the absorption curve and thus corresponds to the double integral of the actual EPR spectrum. The intensity is proportional to the number of paramagnetic species in the sample. Thus this number can be measured by comparing the sample intensity with that of an appropriate standard sample containing a known number of spins. Some information can also be obtained from the temperature dependence of the EPR intensity. The latter is related to temperature *T* by:

$$I = aN(T)/T \tag{3}$$

In this equation, *a* is a constant depending both on the sample and on the instrumental settings, $N(T)$ the number of spins at temperature *T*, and the factor $1/T$ arises from the population difference between the two spin levels involved in the EPR transition. The number of spins $N(T)$ depends on the electronic structure of the radicals and thus bears some information about their nature. Two extreme cases can be distinguished. The first case, referred to as Curie magnetism, corresponds to the situation where the radicals have very large splittings between electron energy levels and thus remain in the same electronic state, whatever the temperature. In this case, $N(T)$ is *independent* of temperature. In the second case, referred to as Pauli magnetism, $N(T)$ varies *linearly* with temperature. This corresponds to paramagnetic species with negligibly small splitting between energy levels. This happens when the unpaired electrons are highly delocalized, such as in metals or in highly mature coals containing very large aromatic domains.

The second EPR parameter is the field at resonance B_0^{res}, which is measured at the intercept between the spectrum and the baseline. This field indeed corresponds to the maximum of absorption, and thus fulfills the condition given by Eq. (2). The value of the g-factor is thus derived from B_0^{res} according to $g = h\nu / \beta B_0^{res}$. The g-factor generally deviates from the free electron g value $g_e \approx 2.0023$, due to an additional contribution to the electron magnetic moment arising from the electron motion about atoms. Consequently, the g-value depends both on the wavefunction of the unpaired electron and on the type of atoms over which the wavefunction expands. Therefore the g-factor provides information about the electronic structure (wavefunction and energy levels) and the elemental composition of the paramagnetic species. The third parameter is the peak-to-peak linewidth ΔB. The latter results from several contributions. Some of them such as g-factor anisotropy in the case of amorphous or polycrystalline sample, magnetic dipole interactions between unpaired electrons or unresolved magnetic interactions with nearby nuclei induce a line broadening. When these contributions are dominant, the EPR line is the envelope of many unresolved overlaping individual lines. The EPR is then said inhomogeneously broadened. On the contrary, exchange interactions between unpaired electrons, electron motion or chemical exchange tend to narrow the line by averaging out the former unresolved interactions. In that case the EPR line is said homogeneously broadened.

4. INSOLUBLE ORGANIC MATTER IN THE CARBONACEOUS CHONDRITES

Reviews of the state of knowledge about the macromolecular organic matter in carbonaceous chondrites can be found in [1, 7]. Sometimes compared to type-III kerogens on the basis of its elemental composition and because it is made of interconnected aromatic moieties, this meteoritic macromolecular organic matter has indeed no terrestrial counterpart. NMR is one of the most powerful techniques to investigate the structure of such a macromolecular material and was applied to the meteorites of Orgueil, Murchison [8, 9] and Tagish Lake [10, 11]. A structural feature common to macromolecules in these three meteorites is the existence of small and highly substituted aromatic moieties. Recent studies by Electron Transmission Microscopy of the macromolecules in Orgueil and Murchison meteorites [12] showed that the aromatic moieties were made of about 10-15 condensed benzene rings and were stacked in domains of about 2-3 aromatic moieties thick. Pioneer EPR analysis of the organic matter of the Mighei [13], Cold

Bokkeveld [14] and Orgueil [15] meteorites was performed in the 1960s. Here we introduce a thorough EPR analysis of the radicals in the macromolecular organic matter from the Orgueil, Murchison and Tagish Lake meteorites. The discovery of specific radicals, never encountered in terrestrial macromolecular matter, provides additional information about the structure and the history of the chondritic macromolecular organic matter.

4.1 Organic radicals in the carbonaceous chondrites: a possible extraterrestrial signature

4.1.1 General aspects of the EPR spectra

EPR analysis was performed on the IOM isolated from the mineral matrix by an HF/HCl treatment. Three samples of terrestrial coals labelled A1, A2, A3 ranging from the less mature to the most mature were used for comparison of the EPR of terrestrial IOM with that of meteoritic IOM. Miocene coals A1 and A2 originate from the Mahakam Delta (Indonesia) and the Namurian coal A3 originates from the Solway basin (Great Britain). These coals exhibit similar elemental compositions as the meteoritic IOM. The elemental compositions of the samples are given in table 14-2.

Table 14-2. Elemental compositions of the IOM of the Orgueil, Murchison and Tagish Lake and of the coals A1, A2 and A3. [a] from [8]. [b] from [10]. [c] from [16]. n.m.: not measured.

	H/C	O/C	N/C	S/C
Orgueil[a]	0.72	0.18	0.03	0.02
Murchison[a]	0.77	0.22	0.03	0.06
Tagish Lake[b]	0.46	0.15	0.10	n.m.
A1[c]	0.82	0.20	n.m.	n.m.
A2[c]	0.75	0.09	n.m.	n.m.
A3[c]	0.49	0.04	n.m.	n.m.

A typical EPR spectrum at 9.4 GHz of the IOM is shown in Fig. 14-4, corresponding to the case of the Orgueil meteorite. Very similar spectra are observed with the IOM of the Murchison and Tagish Lake meteorites.

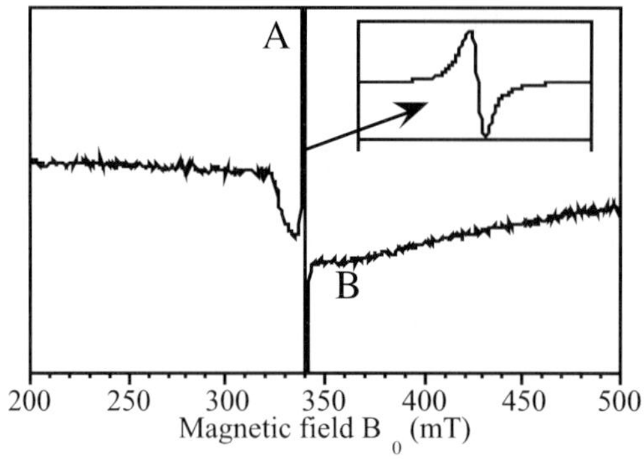

Figure 14-4. EPR spectrum at room temperature of the insoluble organic matter of the Orgueil meteorite. The inset shows an expanded view of signal A.

The spectrum is made of a sharp line, labelled A, at *g*=2.0032 for Tagish Lake and *g*=2.0031 for Orgueil and Murchison, and corresponding to the organic radicals of the IOM, and a broad feature, labelled B, arising from magnetic mineral residues mixed with the IOM. The radicals in the three coals also exhibit an EPR signal similar to signal A. The extreme simplicity of signal A may at first sight make it hopeless to get valuable information about the radicals in the IOM. Indeed the multiple line structure of the EPR spectra generally encountered with isolated organic radicals and providing most of the structural information, is missing here. This is the consequence of strong local concentrations in radicals, which favor the onset of exchange interactions (interactions resulting from electrostatic repulsion between electrons). These interactions induce a long-range spin diffusion so that the local magnetic fields experienced by the electron spins and responsible for informative structures in the EPR spectra of non-interacting radicals are here averaged to zero. The spectrum is then narrowed to a single EPR line. However, EPR offers the possibility of restoring some information by changing the experimental conditions such as the sample temperature or the incident microwave power. The analysis of the EPR signal of the radicals in the IOM of the meteorites of Orgueil, Murchison and Tagish Lake revealed two specific features:

- a heterogeneous distribution of the radicals in the IOM [16];
- the existence of particular radicals, referred to as diradicaloids [17].

As far as we are aware, these features have never been reported for the terrestrial IOM and consequently may constitute an extraterrestrial signature of the meteoritic IOM.

4.1.2 Distribution of the radicals

The spatial distribution of the radicals in the organic matter has a direct influence on the interactions (exchange or magnetic dipolar), which depend on the distance between neighboring radicals. These interactions determine the electron spin relaxation times. Two spin relaxation times, T_1 and T_2, are considered in magnetic resonance. When some microwave energy is absorbed by one spin, this energy is spread among the whole spin system in the time T_2. This energy diffusion is caused by the interactions between the spins and therefore T_2 is also termed the spin-spin relaxation time. Then this energy is released from the spin system to the surrounding matter in the time T_1. Therefore T_1 is termed the spin-lattice relaxation time. A heterogeneous distribution in a sample with local concentrations larger than the average concentration will result in enhanced interactions between radicals and thus abnormally short T_1 and T_2 values, with respect to a similar sample with a homogeneous radical concentration. The relaxation times, more precisely the product $T_1.T_2$, can be estimated from the change in the EPR response upon changing the microwave power [18]. The relaxation times of the radicals in the meteoritic IOM were compared to those of the radicals in the coals A1, A2 and A3. Radicals in coals are considered as homogeneously spread [19]. Figure 14-5 shows the quantity $T_1.T_2$ as a function of the mass averaged concentration. The latter is simply calculated by dividing the number of spins deduced from the EPR intensity, by the sample mass.

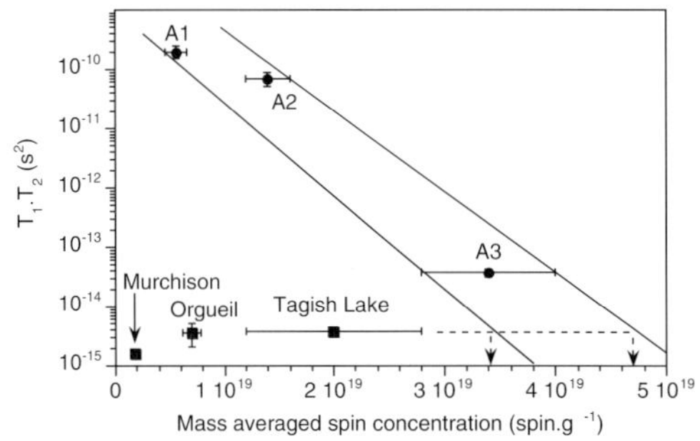

Figure 14-5. Product of the spin-lattice T_1 and spin-spin T_2 relaxation times *vs* mass averaged spin concentration for the radicals in the insoluble organic matter of the Orgueil, Murchison and Tagish Lake meteorites, and in the coals A1, A2 and A3.

The coals exhibit the expected trend, a decrease of $T_1.T_2$ with increasing concentration. The radicals in the meteoritic IOM show very close values of $T_1.T_2$, despite strong differences in average concentrations between meteorites. Besides, they have extremely low values of $T_1.T_2$ when compared to coals with similar average spin concentrations. This indicates a heterogeneous radical distribution in the meteoritic IOM, with a local concentration significantly larger than the average one. By shifting horizontally the points representing the meteorites in Fig. 14-5 to the coal domain, this local concentration is estimated to about 4×10^{19} spin.g^{-1}. It is important to notice that this value is the same for the three meteorites, despite their different nature.

4.1.3 The existence of diradicaloids

As already mentioned, the temperature dependence of the number of spins $N(T)$, especially when there is a deviation from the two cases described in § 3.3, is potentially informative about the electronic structure of the paramagnetic species, and hence about their chemical nature. Figure 14-6 shows the temperature dependence of the number of spins, normalized to the value N(100K) at 100 K, in the IOM of the three meteorites and in the coal A3 (coals A1 and A2 exhibit the same behavior as A3).

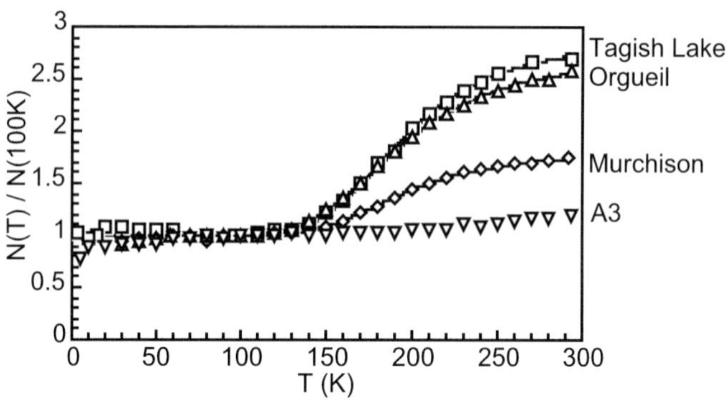

Figure 14-6. Normalized spin concentration *vs* sample temperature for the insoluble organic matter of the Orgueil, Murcison and Tagish Lakes meteorites, and coal A3. The full lines are simulations with Eq. (4).

Except for highly mature coals which may exhibit a Pauli magnetism, other types of coals are always reported to exhibit the usual Curie type magnetism [20], with $N(T)$ independent of T, as examplified by A3 in Fig. 14-6. This is due to the fact that radicals in the coals are monoradicals *i.e.* they each bear only one unpaired spin and are each characterized by a total electron spin S=1/2. The radicals in the meteorites exhibit a Curie magnetism only below 120K. A strong deviation from this type of magnetism is observed above 120K with a significant increase of the number of spins.

This specific behavior of $N(T)$ is explained by the presence of diradicaloids in the meteoritic IOM [17]. Figure 14-7a depicts the structure of a diradicaloid. The fact that during its history, the meteoritic IOM may have been submitted to irradiation or thermal stress over long periods must have caused the cleavage of a few chemical bonds. The breaking of a bond on a carbon atom directly linked to an aromatic moiety is more favorable, for the remaining unpaired electron is indeed stabilized by delocalization over the aromatic moiety. In the case of the meteoritic IOM, the high substitution level of the aromatic moieties statistically enhances the probability of bond cleavage on two carbon atoms linked to the same moiety thus leaving two single electrons in this moiety. In a diradicaloid and at low temperatures, these electrons are paired in the same molecular orbital, thus giving rise to a diamagnetic ground state for a diradicaloid, in which the magnetic moments of the two electrons cancel out, and which is then characterized by a total electron spin S=0 (Fig. 14-7b). This state is termed a singlet state. When the diradicaloids are in this state, they have no EPR response. Therefore only monoradicals, with the normal Curie magnetism, are detected in the meteoritic IOM below 120K. At higher temperatures, enough thermal energy is available to excite one of the two electrons from the low-lying molecular orbital to the upper empty orbital. This gives rise to a paramagnetic excited state for the diradicaloids, in which the magnetic moments of the two electrons add together, and which is characterized by a total spin S=1. This state is termed a triplet state. The diradicaloids in this state give an EPR signal, which adds to the signal of the monoradicals, and thus explains the increase of the number of spins in the meteoritic IOM when the temperature is raised.

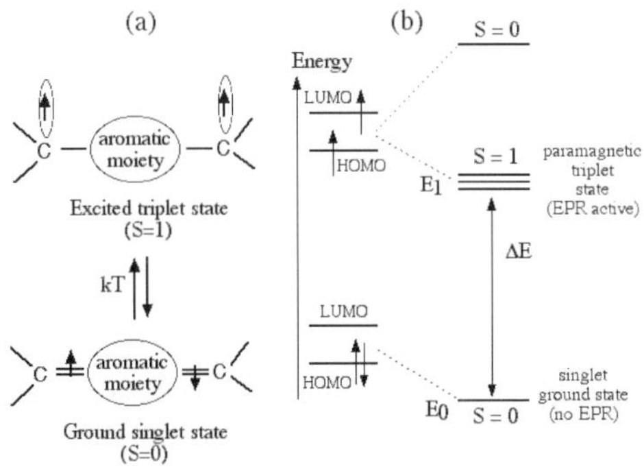

Figure 14-7. (a) Schematic structure of a diradicaloid, made of an aromatic moiety with two alkyl substituants bearing each an unpaired electron, whose spin orientation is indicated by an arrow. (b) Electron energy diagram of a diradicaloid. At left, the individual electron energy levels with their occupancy (HOMO=highest occupied molecular orbital, LUMO=lowest

unoccupied molecular orbital); at right, the energy levels of the whole system taking into account the effects of electron repulsion.

The diradicaloids are characterized by their singlet-triplet gap $\Delta E = E_1 - E_0$, and the entropy difference $\Delta\sigma = \sigma_1 - \sigma_0$, with E_1 (σ_1) and E_0 (σ_0) being the energies (entropies) of the triplet and the singlet states, respectively. The simulation of the data in Fig. 14-6 with the following equation [17]:

$$N(T) = N_{mono} + \frac{8}{3} N_{di} \frac{1}{1 + exp(-\Delta\sigma / k) exp(\Delta E / kT)} \quad (4)$$

enables to determine the values of ΔE and $\Delta\sigma$, and the proportion $N_{mono}/(N_{mono}+N_{di})$ of diradicaloids, which are given in Table 14-3.

Table 14-3. Singlet-triplet gap ΔE, entropy differences $\Delta\sigma$ for the diradicaloids, and proportion of diradicaloids among radicals in the meteorites of Orgueil, Murchison and Tagish Lake.

Meteorite	ΔE (eV)	$\Delta\sigma$(cm^{-1}.K^{-1})	proportion
Orgueil	0.101±0.003	4.2±0.1	40%
Murchison	0.104±0.004	4.3±0.1	24%
Tagish Lake	0.117±0.003	4.8±0.1	42%

The relative amount of diradicaloids varies among meteorites but is significant in all cases, ranging from 24% for Murchison to about 40% for Orgueil and Tagish Lake. In contrast, very close values are obtained for ΔE and $\Delta\sigma$ for the diradicaloids in the three meteorites. Since these parameters are intrinsically related to the chemical structure, this means that the same type of diradicaloids has to be considered in the meteorites of Orgueil, Murchison and Tagish Lake.

4.1.4 Diradicaloids and the structure of the IOM

The existence of diradicaloids specifically in the IOM of the meteorites, and not in the terrestrial IOM is consistent with the particular structure of the meteoritic IOM containing highly substituted aromatic moieties of small size. It was mentioned above that the high substitution level was a first favorable factor for the occurrence of diradicaloids. Quantum chemical calculations on model molecules indicated that the small size of the aromatic moieties was also an energetically favorable factor [17]. Calculations were performed on model molecules made of an aromatic moiety with two methyl substituants. A monoradical is formed by removing a hydrogen atom from one of the methyl groups. This corresponds to reaction path (a) in Fig. 14-8. The change in electron energy for this reaction is ΔE_1. When both methyl groups loose a hydrogen atom (reaction path (b) in Fig. 14-8), a diradicaloid is obtained, with an energy cost ΔE_2.

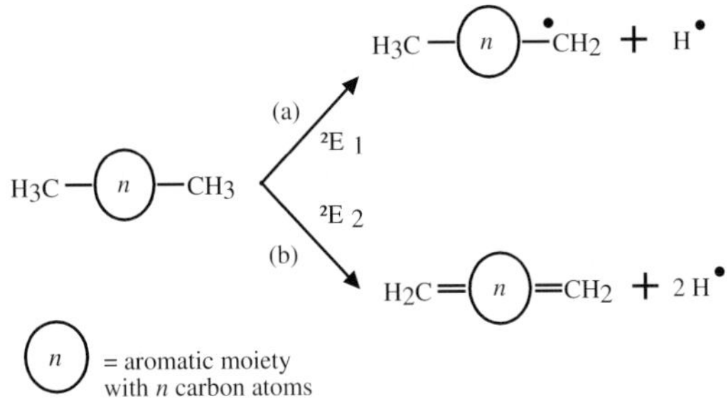

Figure 14-8. Reactions of formation of a monoradical [path (a)] and of a diradicaloid [path (b)] from a methyl doubly substituted aromatic unit. ΔE_1 and ΔE_2 represent the changes in electronic energy corresponding to reactions paths (a) and (b), respectively.

A stabilization energy of a diradicaloid with respect to two separate monoradicals can be defined as:

$$\Delta E_{stab} = 2\Delta E_1 - \Delta E_2 \quad (5)$$

When ΔE_{stab} is positive, the formation of a diradicaloid is energetically more favorable than the formation of two separate monoradicals. Quantum chemical calculations with the Extended Hückel method of ΔE_{stab} for different sizes of the aromatic moiety showed that the occurrence of diradicaloids (ΔE_{stab}>0) was favored for aromatic moieties containing less than *ca.* 30 carbon atoms (Fig. 14-9).

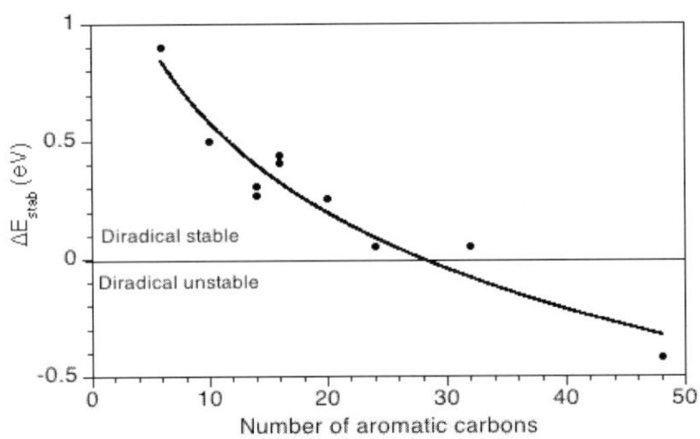

Figure 14-9. Stabilization energy of a diradicaloid with respect to two separate monoradicals *vs* the number of carbon atoms in the aromatic moiety. The full line is an interplolation.

The size of the aromatic moieties of the diradicaloids can be estimated from their singlet-triplet gap ΔE. Quantum chemical calculations of the singlet-triplet gap of the same model molecules as above, were performed by the Extended Hückel method and by a method based on the Density Functional Theory [17]. In the first method, which neglects electron repulsion, the singlet-triplet gap is approximated by the gap between the highest occupied molecular orbital (HOMO) and the lowest unoccupied molecular orbital (LUMO). The second method, which considers electron repulsion, gives in principle a more accurate value. Both methods showed that the experimental singlet-triplet gap, about 0.1 eV, was reached with diradicaloids containing *ca.* 35-40 carbon atoms (Fig. 14-10), i.e. 10-15 benzene rings, which is fully consistent with the mean size of the aromatic moieties in the meteoritic IOM determined by Transmission Electron Microscopy [12].

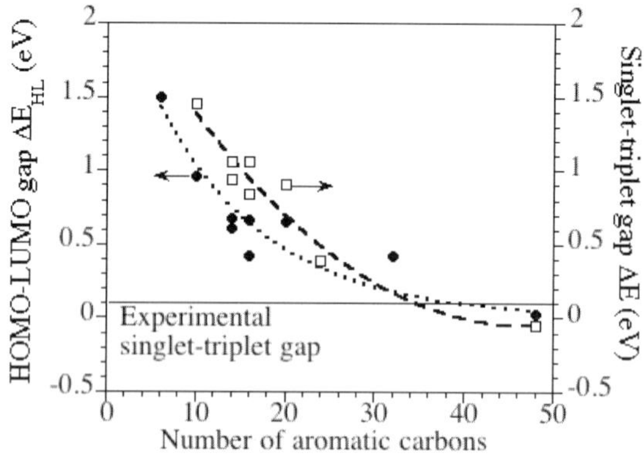

Figure 14-10. Singlet-triplet gap ΔE of a diradicaloid calculated with a DFT method (right-hand vertical axis) and gap between the highest occupied molecular orbital (HOMO) and the lowest unoccupied molecular orbital (LUMO) calculated with the Extended Hückel method (left-hand vertical axis), as a function of the number of carbon atoms in the aromatic moiety. The dashed and dotted lines represent interpolations.

4.2 Possible implications of the existence of diradicaloids

The existence of monoradicals and diradicaloids in the IOM of carbonaceous chondrites shows that this IOM was submitted to irradiation or to thermal stress, or both, in the course of its history. The presence of similar diradicaloids, based on aromatic moieties, in meteorites having different degrees of alteration on the parent body and distinguishing themselves both by the mineralogy and the organic carbon systematics, points to a common origin for the IOM in these different carbonaceous chondrites. This strongly strengthens the idea that the chondritic IOM is a very primitive material, at least partially synthesized from PAH precursors in the interstellar medium or in the proto-solar nebula. The effects on the IOM of the processes on the parent bodies must have been secondary, possibly modifying the composition in heteroelements, and as far as the radicals are concerned, their distribution and the relative abundance of diradicaloids. Models of synthesis of the meteoritic IOM were so far mainly constrained by isotope compositions and especially D/H ratios. The diradicaloids must be considered as additional markers of the synthetic process, and as such should be taken into account in the models or the experiments reproducing the IOM synthesis.

5. MACROMOLECULES IN CHERTS

5.1 Introduction

Macromolecules in cherts occur in a context different from that of the macromolecules in the carbonaceous chondrites, but they set similar issues in terms of structural analysis. The interest in the structure of the organic material embedded in cherts has been raised by the recent debate about the biogenicity of microstructures observed in the Apex chert aged of –3.5 Ga from the Warrawoona Group in Australia [2]. These microstructures contain carbonaceous material and were interpreted on a morphological basis as microbial fossils [2]. However, since the samples came from a hydrothermal vein where temperatures in the range 250-350°C may have prevailed, it was also proposed that these microstructures be only hydrothermal artifacts mimicking microbial fossils [4]. This controversy raises the necessity of determining the structure of the carbonaceous material in the cherts from the Warrawoona Group in order to elucidate its origin. It should be noted that carbonaceous matter in these cherts is found not only in the above-mentioned microstructures, but also at the quartz grain boundaries where it could have been deposited by hydrothermal fluids [21]. Only little information is known about the structure of the carbonaceous matter in the cherts from the Warrawoona Group. This is a partially aromatic material, as shown by Raman spectroscopy [3, 4]. The N/C ratio is about 0.02 [22], but the amount of the other heteroelements, especially oxygen, is still to be determined as well as the degree of aromaticity. Partial answers to some of these questions can be given by EPR.

5.2 Structural information given by EPR about the carbonaceous matter in the archean cherts

5.2.1 EPR spectrum of the Warrawoona chert

Figure 14-11 shows the EPR spectrum at room temperature of a raw sample of chert from the Warrawoona Group (sample PPRG 006 from Towers Formation). The sample weight is about a few milligrams and the amount of carbonaceous matter within is only of the order of a few micrograms. This illustrates the possibility offered by EPR to analyze *in situ* minute amounts of carbonaceous material embedded in a mineral matrix without destroying the sample.

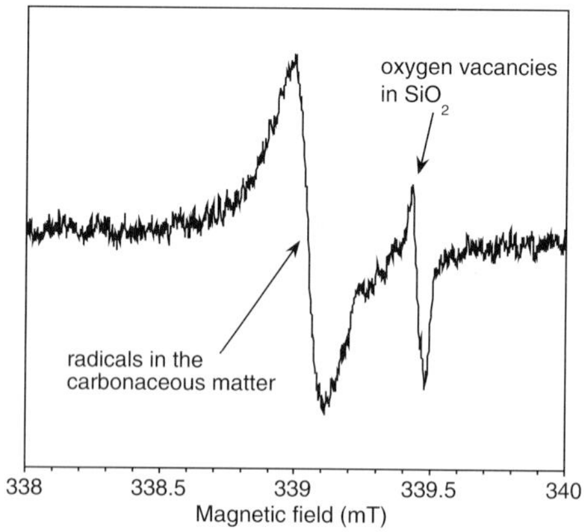

Figure 14-11. EPR spectrum of a chert from the Warrawoona Group, at room temperature and low microwave power (P=0.02 mW).

The spectrum in Fig. 14-11 exhibits a strong line due to the radicals in the carbonaceous matter and a small feature corresponding to oxygen vacancies in the quartz matrix. Similarly to the EPR signal of radicals in the meteorites, the EPR line of the radicals in the Warrawoona chert is narrowed by exchange interactions. Other paramagnetic species such as Fe^{3+} and VO^{2+} ions can also be revealed by exploring other magnetic field ranges or lowering the temperature. Here, only the information provided by the carbonaceous radicals will be discussed.

5.2.2 Indications about the chemical composition

The chemical composition can be probed by the g-factor. Due to the interactions between radicals, the g-factor gives information on the average chemical composition of the domains spanned by the wavefunctions of the unpaired electrons, rather than indications on the individual radicals. The values of the g-factor for a set of four samples are found in the range 2.0029-2.0037, with a mean value 2.0033. These values are significantly different from the g-values for pure amorphous carbon, which lie in the 2.0024-2.0029 range [23]. This demonstrates that the carbonaceous matter in the Warrawoona chert is different from pure carbon. The shift of the g-values from the free electron value 2.0023 indicates the presence of a significant amount of heteroelements. Since nitrogen is in too a weak amount to contribute significantly to the g-factor, the most probable element to be considered is oxygen. The amount of oxygen can be assessed by comparison with the g-factors in coals, which are also partially aromatic macromolecular compounds and thus the most relevant reference materials for the carbonaceous matter in the cherts. Figure 14-12 shows that the radicals in the Warrawoona chert with the highest g-values contain up to approximately 20 weight % of oxygen. The amount of oxygen in the radicals with the lowest g-values is uncertain, due to the strong dispersion of the data for the coals in this range, but could be as low as 2 weight %. The data for the coals are taken from [24, 25].

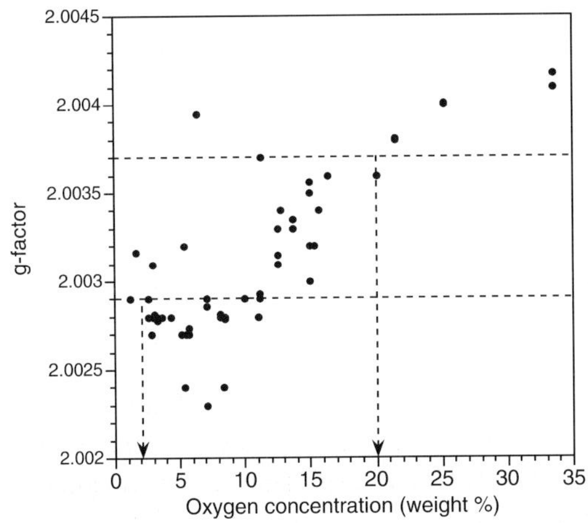

Figure 14-12. g-factors of coals (black circles) from [24, 25] *vs* oxygen concentration. The horizontal dashed lines delimitate the range of g-values of the radicals in four samples of the Warrawoona chert, and the arrows the deduced range of oxygen concentration.

5.2.3 Size of the aromatic moieties

The size of the aromatic moieties has a direct influence on the splitting between the electron energy levels of the radicals and on the extension of the wavefunction of the unpaired electrons. These electron structure features in turn determine the type of magnetism of the radicals. Consequently, the analysis of the EPR intensity as a function of the sample temperature can give some indications on the size of the aromatic moieties. Figure 14-13 shows the temperature dependence of the product $I.T$ of the EPR intensity I of the radicals by the temperature T. From Eq. (3), this product is directly proportional to the number of spin $N(T)$.

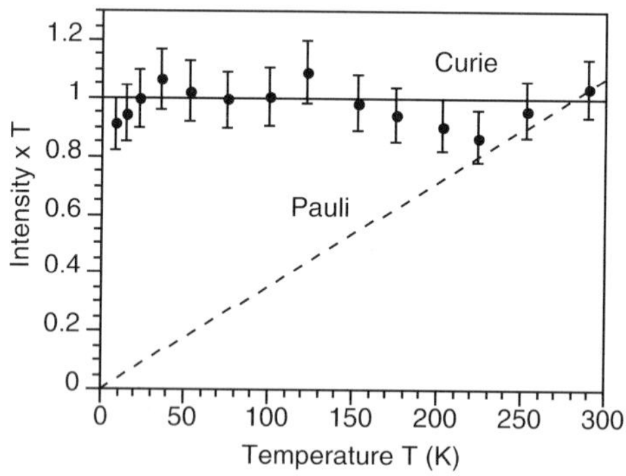

Figure 14-13. Temperature dependence of the product of the EPR intensity by temperature for the carbonaceous radicals in the Warrawoona chert, exhibiting a Curie-type magnetism. The dashed line represents the temperature dependence in case of a Pauli-type magnetism.

Figure 14-13 shows that $N(T)$ is independent of temperature, which is typical of a Curie-type magnetism. This means that the aromatic moieties covered by the unpaired electron wavefunctions are not too large, otherwise a Pauli-type magnetism, shown by the dashed line in Fig. 14-13 would have been observed. The Curie-magnetism is observed when the splitting between the electron energy levels is much larger than the thermal energy kT, with k, the Boltzman constant. From quantum mechanics, it is known that the splitting between the energy levels of an electron delocalized over a domain of size L is of the order of h^2/mL^2, with h the Planck constant and m the electron mass. Thus the size of the aromatic moieties is:

$$L \ll \sqrt{\frac{h^2}{mkT}} \tag{5}$$

If we take $T\approx 300$K, the above equation gives a maximum size $L \ll 10$ nm for the aromatic moieties. The carbonaceous matter in the Warrawoona chert is thus a nanometer scale material.

5.3 Comparison of the carbonaceous matter in the Warrawoona chert with biogenic macromolecular organic matter: toward an indicator of age

The radicals in the Warrawoona chert, with a Curie-type magnetism are clearly different from those in the carbonaceous chondrites. This feature along with their significant amount of oxygen and the limited size of their aromatic moieties make them rather comparable to radicals found in kerogens. This is not however an indication of a possible biogenic origin. Indeed, the absence of diradicaloids in the carbonaceous matter in the Warrawoona chert could also be accounted for by a loss of the memory of its extraterrestrial origin, due to abiotic terrestrial processing over geological times. As far as the origin of the organic matter in the Warrawoona chert is concerned, one has first to make sure that this organic matter is contemporary to the chert formation. Consequently, relevant EPR parameters are to be found, which could be indicative, at least of the age of the organic matter, and ultimately of its origin. In this prospect, the Warrawoona chert was compared to materials containing biogenic organic matter of different ages. These materials were cherts from the Gunflint formation (Ontario, Canada) aged of –2 Ga and from the Clarno formation (Oregon, USA) aged of –45 Ma, and the three coals mentioned in § 4.1. Figure 14-14 shows the EPR spectra at room temperature of the Clarno and Gunflint cherts. The organic matter in the Clarno chert, being at an earlier stage of evolution than that of the Gunflint and Warrawoona cherts, exhibits a greater richness in radicals. Along with kerogen-like radicals like those found in Gunflint and Warrawoona cherts, isolated molecular radicals, namely methyl radicals and another radical indicated by stars in Fig. 14-14, are observed.

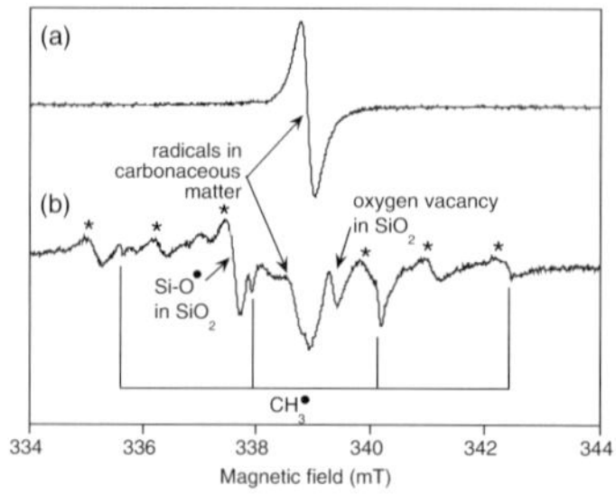

Figure 14-14. Room temperature EPR spectra of samples of cherts from the (a) Gunflint Formation and (b) the Clarno Formation.

The comparison of the spectra of the three cherts in Fig. 14-11 and 14-14 show that as the chert gets older, the isolated molecular radicals progressively disappear, while the amount of kerogen-like radicals increases. The similarity of the spectra of the radicals in the meteorites, in the Gunflint chert and in the Warrawoona chert unfortunately shows that the gross spectral shape alone is neither an indicator of the origin nor of the age of the organic matter. However, a correlation, shown in Fig. 14-15 has been observed between the linewidth of the radicals in the Clarno and Gunflint cherts and in the coals. This shows that the EPR linewidth may be an indicator of the age of a biogenic material. The fact that the points corresponding to the radicals in the Warrawoona chert are located on the correlation line cannot be considered as a clear-cut evidence of the biogenicity of the organic matter in the Warrawoona chert. However this shows that if this organic matter is biogenic, it is actually contemporary to the chert formation, and do not arise from a latter contamination by a biogenic material.

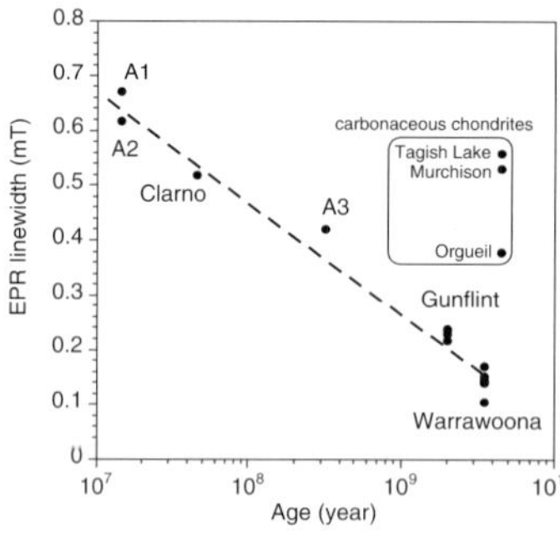

Figure 14-15. EPR linewidth of the organic radicals in coals A1, A2 and A3, in a chert from the Warrawoona Group, in the cherts of the Clarno and Gunflint Formations, and in carbonaceous chondrites, as a function of the sample age.

6. CONCLUSION

The potentialities of EPR to provide structural clues on such complex macromolecular carbonaceous materials as those found in meteorites or in terrestrial rocks were demonstrated. One major advantage of this technique is its ability to analyse *in situ* minute amounts of carbonaceous matter without any constraint of sample processing, thus preserving the intimate mineral/organic relation. One major finding is the ability to distinguish by EPR an extraterrestrial macromolecular material from a terrestrial one, thanks to the existence of extraterrestrial makers, namely diradicaloids, in the meteoritic insoluble organic matter. This could have applications in the analysis of the organic matter, if present, within samples from Mars. More precisely, the presence or not of diradicaloids could help to determine wether the Martian organic matter has a well preserved extraplanetary origin or, on the contrary, wether it was strongly processed under geological conditions which prevailed on Mars.

ACKNOWLEDGEMENTS

The authors gratefully acknowledge the persons who provided the different samples: J. William Schopf for the Clarno and the Warrawoona cherts, Stanley M. Awramik for the Gunflint chert, Sandra Pizzarello and Luan Becker for the insoluble organic matter of the Tagish Lake meteorite. The authors would also like to thank Ilaria Ciofini for performing the quantum calculations by the DFT method.

REFERENCES

1. O. Botta and J. L. Bada "Extraterrestrial organic compounds in meteorites", *Surveys in Geophysics* **23**, 411-467 (2002)
2. J. W. Schopf "Paleobiology of the Archean" in "The proterozoic biosphere", J. W. Schopf and C. Klein (Eds), Cambridge University Press (1992)
3. J. W. Schopf, A. B. Kudryatsev, D. G. Agresti, T. J. Widowiak and A. D. Czaja "Laser-Raman imagery of Earth's earliest fossils", *Nature* **416**, 73-76 (2002)
4. M. D. Brasier, O. R. Green, A. P. Jephcoat, A. K. Kleppe, M. J. Van Kranendonk, J. F. Linsdsay, A. Steel and N. V. Grassineau "Questioning the evidence for Earth's oldest fossils", *Nature* **416**, 76-81 (2002)
5. C. M. Fedo and M. J. Whitehouse "Metasomatic origin of quartz-pyroxene rock, Aikilia, Greenland, and implications for Earth's earliest life", *Science* **296**, 1448-1452 (2002)
6. R. E. Botto and Y. Sanada (Eds) "Magnetic resonance of carbonaceous solids", Advances in chemistry series 229. American Chemical Society. Washington DC, (1993)

7. P. Ehrenfreund, W. Irvine, L. Becker, J. Blank, J. R. Brucato, L. Colangeli, S. Derenne, D. Despois, A. Dutrey, H. Fraaije, A. Lazcano, T. Owen, F. Robert, an International Space Science Institue ISSI-Team "Astrophysical and astrochemical insights into the origin of Life", *Rep. Prog. Phys* **65**, 1427-1487 (2002)
8. A. Gardinier, S. Derenne, F. Robert, F. Behar, C. Largea and J. Maquet "Solid state CP/MAS ^{13}C NMR of the insoluble organic matter of the Orgueil and Murchison meteorites: a quantitative study", *Earth and Planetary Sciences Letters* **184**, 9-21 (2000)
9. G. D Cody III, C. M. O'D. Alexander and F. Tera "Solid state (^{1}H and ^{13}C) nuclear magnetic resonance spectroscopy of insoluble organic residues in the Murchison meteorite", *Geochim. Cosmochim. Acta* **66**, 1851-1865 (2002)
10. S. Pizzarello, Y. Huang, L. Becker, R. J. Poreda, R. A. Nieman, G. Cooper and M. Williams "The organic content of the Tagish Lake meteorite", *Science* **293**, 2236-2239 (2001)
11 G. D Cody III, C. M. O'D. Alexander and F. Tera "Compositional trends in chondritic organic solids within and between meteoritic groups", 34th Lunar and Planetary Science Conference, Houston, abstract # 1822 (2003)
12. S. Derenne, J. N. Rouzaud, C. Clinard, F. Robert "Direct characterization of the polyaromatic units in the insoluble organic fraction of the Orgueil and Murchison meteorites through quantitative high resolution transmission electron microscopy", submitted to *Geochim. Cosmochim. Acta*
13. A. P. Vinogradov, G. P. Vdovykin and I. N. Marov "Free radicals in the Mighei meteorite", *Geoch. Int.* **1**, 395-398 (1964)
14. J. Duchesne, J. Depireux and C. Litt "Concerning the nature of free radicals in the Cold Bokkeveld meteorite", *Geoch. Int.* **1**, 1022-1024 (1964)
15. K. F. Schulz and R. M. Elofson "Electron spin resonance of organic matter in the Orgueil meteorite", *Geochim. Cosmochim. Acta* **29**, 157-160 (1965)
16. L. Binet, D. Gourier, S. Derenne and F. Robert " Heterogeneous distribution of paramagnetic radicals in insoluble organic matter from the Orgueil and Murchison meteorites", *Geochim. Cosmochim. Acta* **66**, 4177-4186 (2002)
17. L. Binet, D. Gourier, S. Derenne, F. Robert and I. Ciofini "Occurence of Abundant Diradicaloid Moieties in the Insoluble Organic Matter from the Orgueil and Murchison Meteorites: A Fingerprint of its Extraterrestrial Origin?", *Geochim. Cosmochim. Acta* (in press)
18. J. E. Wertz and J. R. Bolton *Electron Spin Resonance : Elementary Theory and Practical Applications*, McGraw-Hill, New-York, p.456 (1972)
19. H .Thomann, B. G. Silbernagel, H. Jin , L. A. Gebhard, P. Tindall and G. R. Dyrkacz "Carbon Radical Spin Dynamics in Isolated Vitrinite Coal Macerals", *Energy & Fuels* **2**, 333-339 (1988)
20. K. S. Rothenberger, R. F. Sprecher, S. M. Castellano and H. L. Retcofsky "Temperature Dependence of the Electron Paramagnetic Resonance Intensity of Whole Coals" In *Magnetic Resonance of Carbonaceous Solids* (eds. R. E. Botto and Y. Sanada). Advances in chemistry series 229. American Chemical Society. Washington DC, (1993)
21. B. T. De Gregorio and T. G. Sharp "Determining the biogenicity of microfossils in the Apex chert, Western Australia, using Transmission Electron Microscopy", 34th Lunar and Planetary Science Conference, Houston, abstract # 1267 (2003)
22. V. Beaumont and F. Robert "Nitrogen isotope ratios of kerogens in Precambrian cherts: a record of the evolution of atmosphere chemistry?", Precambrian Research **96**, 63-82 (1999)
23. R. C. Barklie "Characterization of defects in amorphous carbon by electron paramagnetic resonance", *Diamond and Related Materials* **10**, 174-181 (2001)

24. H. L. Retcofsky, J. M. Stark and R. A. Friedel "Electron spin resonance in American coals", *Analytical Chemistry* **40**, 1699-1704 (1968)
25. T. F. Yen and S. R. Sprang "Contribution of ESR analysis toward diagenic mechanisms in bituminous deposits", *Geochim. Cosmochim. Acta* **41**, 1007-1018 (1977)

Chapter 15

THE CHEMISTRY OF THE ORIGIN OF LIFE

Oliver Botta
Astrobiology Laboratory, Leiden Institute of Chemistry, University of Leiden, The Netherlands

Astrobiology is an emerging interdisciplinary science that investigates experimentally the origin of life on the Earth and its distribution in the Universe. It encompasses scientific disciplines from astronomy to geology, chemistry, paleontology, biology and ecology with the goal to understand – to name only a few examples - the formation of solar systems (through detection of extrasolar planetary system and the exploration of our Solar system), to investigate the origin and prebiotic formation of organic compounds that could have been important for the origin of life on the Earth (including the origin of chirality), to determine the age of the oldest molecular fossils and other indicators on Earth to constrain the time for the origin of life, or to understand the complexity of the earliest organisms by phylogenetic and biochemical analysis and of modern organisms. This interdisciplinary approach is necessary to understand the implications in the different fields and to close the gaps in our knowledge, for example on the role of RNA in ancient organisms, the origin of protein biosynthesis or, in another field, the surface conditions on the early Earth. Astrobiology also has strong influences on future space missions, such as the *Rosetta* mission to investigate the composition of a comet, or future missions to Mars that very likely will carry instruments to search for traces of extinct or extant life on that planet.

1. INTRODUCTION

It was exactly 50 years ago, when a paper by a young graduate student from the University of Chicago was published in the journal *Science*. It described a simple apparatus in which a mixture of gases was subjected to a spark discharge and the products were condensed into an aqueous solution

P. Ehrenfreund et al. (eds.), Astrobiology: Future Perspectives, 359–391.

(Figure 15.1)[1].With his work, Stanley Miller, at that time in the laboratory of Harold Urey, presented experimental proof that it is possible to synthesize biologically relevant organic compounds such as amino acids from simple non-biological compounds – gases such as methane, ammonia, hydrogen and water – under conditions that could have occurred on the primitive Earth. This experiment is considered to be the starting point of experimental exobiology, an interdisciplinary field of research that investigates the origin of life on the Earth and its possible occurrence elsewhere in the universe.

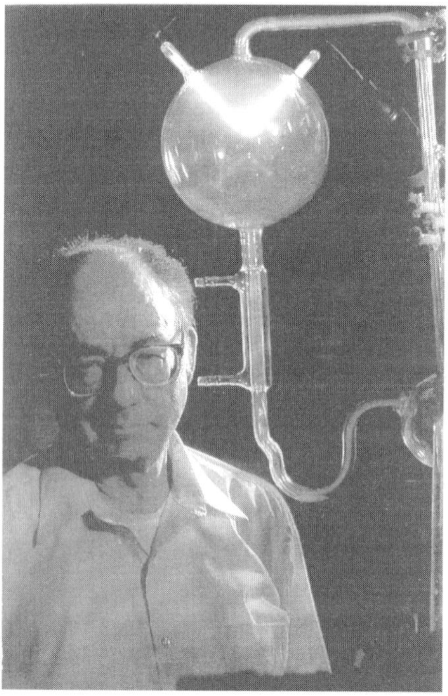

Figure 15.1. Prof. Stanley Miller in front of his apparatus that was used to produce organic compounds from hydrogen, methane, ammonia and water under potential prebiotic conditions and marked the beginning of experimental research into the origin of life.

Since that time, the field of exobiology has gone a long way [2]. In 1957, the first International Conference on the Origin of Life was held in Moscow. The International Society for the Study of the Origin of Life (ISSOL), which has been organizing subsequent meetings over the last 45 years, has been a leading institution in the efforts to bring together researchers form different scientific disciplines (chemists, geologists, planetary scientists, paleontologists, astronomers, biologists and ecologists) to discuss and coordinate their research. But it was not until the

1990's that scientific efforts on the subject of the origin of life on Earth and its distribution in the universe have been getting a big boost, both in public awareness and in financial support. Starting with two relatively small NASA Specialized Centers for Research and Training (NSCORT) in Exobiology, located at Scripps Institution of Oceanography in La Jolla, CA, and at Rensselaer Polytechnic Institution in Troy, NY, NASA began to support this interdisciplinary research in the early 1990's. Today, the NASA Astrobiology Institute (NAI), a virtual institute that currently consists of 16 lead teams distributed all over the USA as well as several international partners, is the largest operational body that provides significant funding for Astrobiology. In addition, several smaller associations have been formed, in particular in Europe, to coordinate research efforts in the origin of life. There is still much specialized terminology around which makes it difficult to understand papers and presentations for scientists that do not work in a particular field. Also, sometimes scientists are surprised that their research might contribute to the understanding of certain aspects in Astrobiology.

One of the major problems of Astrobiology is the lack of geochemical or paleontological evidence left on the Earth from the time the origin of life occurred, mainly due to fact that the Earth is an active planet, and all traces of that time have been erased by erosion or plate tectonics. Also, even the phylogenetically most ancient organisms that we find and characterize today are far more evolved than the first forms of life. We do not even have a good idea about the habitat of early life, although some hypotheses can be derived (for a review see Ref. [3]). It is generally thought that life began relatively soon after the conditions on the surface of the Earth allowed it, although there is no real evidence that this is really true [4]. From a detailed study of the mass flux to the ancient Earth-Moon system – especially during the Hadean time period between ~4.4 and ~3.8 Gyr – Ryder concluded that the accretionary flux to the Earth was relatively benign during that period, and that even the late cataclysm at ~3.9 Gyr [5] would not produce any ocean-evaporating impacts [6,7]. This is in contrast to the idea that during the first 600 Myr of Earth's history, large impactors may have annihilated any forms of life that had developed [8], making the origin of life an event that may have happened several times. Data from zoned detrital zircons found in metamorphosed sediments at Jack Hills, Western Australia, provide evidence for the existence of continental crust and even a liquid hydrosphere on the Earth, only 100 Myr after its formation [9]. The earliest evidence for life on Earth comes from carbon isotope data of graphite inclusions from the 3.8 Gyr old Isua supercustal belt in Greenland [10], although the validity of these results has been seriously challenged later by results based on the geological setting and isotope data. Geologic, petrologic and geochemical evidence strongly suggests that the rocks that were previously considered to be sedimentary precipitated chemically from seawater (banded iron formations [BIFs]) are actually metasomatized ultramafic igneous in origin, making it highly improbable that they hosted life at the time of their formation [11]. The carbon isotope data for the graphite in these rocks have also been reassessed, and it has been shown that the graphite is produced abiogenically by disproportionation of ferrous carbonate at high temperature and pressure [12]. The earliest morphologic evidence in the form of preserved microfossils was also found in Western Australia, and was dated to be 3.465 billion years old [13,14]. But also this seemingly solid evidence is under discussion, based on new data and re-interpretation of the microfossil-like structures as abiotic artefacts formed from amorphous graphite [15,16]. The earliest still unchallenged evidence for life on Earth are molecular fossils of biological lipids, such as 2α-methylhopanes, from 2.7 Gyr old shales from the Pilbara Craton, Australia. The fact that most of the "milestones" are being questioned now, and the new findings about the variety of conditions on the early Earth during the first billion years of its existence, leaves a wide "window of opportunity" for life to begin and evolve. Neither geological nor paleontologic or phylogenetic data provide currently sufficient evidence to pinpoint the origin of life to a certain physical or chemical setting. Life could have originated on the Earth anytime between 4.4 and 2.7 Gyr ago.

There are three competing theories for the chemical beginning of life. First are *gene-first* theories that suggest that the origin of life be intrinsically tied to the appearance of the ability to pass on acquired catalytic abilities. Second are *membrane-first* theories that speculate that that the appearance of encapsulating lipid bilayers made the origin of energy harvesting membranes possible. Both these theories have in common that they require an initial set of organic compounds to be present on the early Earth (heterotrophic origin) [17]. The third theory is a *metabolism-first* hypothesis, which speculates that life began with mutually catalytic chemical reactions that could also occur at elevated temperatures, and that is often associated with hydrothermal systems. In this theory, carbon (in the form of atmospheric CO_2) is fixed through chemical reactions and transformed into organic compounds (chemoautotrophic origin).

In this chapter, first ideas about the origin of organic molecules on the early Earth – required for a heterotrophic origin of life – will be discussed, followed by a section on chirality. The subsequent section will discuss the abiotic polymerization of organic compounds in solution and on mineral surfaces and their implications for the formation of the first information-carrying molecule. After reviewing the concept of the RNA world, the origin and importance of membranes will be discussed. A more biological section will describe the concept of horizontal gene transfer and its influence on the evolution of the earliest organisms, followed by an overview about the current opinion on the properties of the last universal common ancestor. Before the summary and outlook, new results on the efforts to synthesize life in the laboratory will be presented.

2. ORIGIN OF ORGANIC MOLECULES ON THE EARLY EARTH

There are three principal theories for the origin of prebiotic organic compounds on the early Earth (and perhaps on other planets): The atmospheric/oceanic synthesis via Miller-Urey-type synthesis using electric discharges, UV light or other high-energy radiation as energy sources. The second possibility is mineral catalyzed synthesis in and around hydrothermal vents, perhaps via Fischer-Tropsch-type (FTT) synthesis. And the third theory states extraterrestrial synthesis, either in the interstellar medium or inside asteroids and comets, followed by delivery to planetary surfaces. At present, since very little data are available regarding the atmospheric, oceanic or geological conditions on the early Earth, it is impossible to determine conclusively which of these sources was the most significant one, and probably all three sources contributed to the inventory (for a review see Ref. [18]).

Although the Miller-Urey (MU)-Experiment is considered the starting point of experimental exobiology [1], its relevance to the origin of organic compounds on the early Earth has been somewhat diminished. The main reason for this was the changing picture about the composition of the Earth's early atmosphere. While Miller, based on Urey's ideas, used a mixture of reduced gases in his experiment, contemporary geochemical models tend to favor a more oxidized composition of mainly CO_2, CO and N_2, with minor amounts of reduced gases such as ammonia and methane [19]. If the MU-experiment is carried out under these conditions, the yield on amino acids is drastically reduced, and the only amino acid detected in significant amounts is glycine [20]. However, there is the possibility that reduced conditions could have been present in localized environments such as volcanic plumes, where discharges may have driven prebiotic synthesis. The investigations of the synthetic mechanisms leading to the formation of the compounds found in the MU-experiment has contributed to the understanding of how these molecules could be formed abiotically [21,22].

An alternative pathway for the formation of prebiotic organic compounds on the early Earth can be identified as part of the "metabolist" theory proposed by Wächtershäuser and coworkers [23,24]. Central to this thesis is the role of iron sulfide as both catalyst and a source of energy (in the form of reductive power) for life attempting to thrive using oxidized, mantle-derived volatiles (such as CO_2, CO, H_2S, N_2 etc.). It has been proposed that carbon fixation in such a "Iron-Sulfur-World" would be accomplished via the reductive citrate cycle (RCC), in which CO_2 would be the source of carbon and reduced carbon compounds such as citric acid would be the major products. Although individual steps of the RCC have been shown experimentally to be viable under geochemical conditions [25,26], the overall concept of the emergence of self-organizing biochemical cycles has been criticized [27]. In addition, this theory is generally placed in a hydrothermal environment at temperatures of 100 °C or higher and high pressure (which has been thought to be consistent with the apparent ancient phylogeny of hyperthermophiles, see Section 8). Under these conditions, organic compounds, and particularly biologically relevant polymers such as RNA, are decomposed very rapidly [28,29,30]. In contrast to this 'hot' origin of life on the Earth, experiments based on the stability of prebiotic organic molecules have provided additional evidence that a 'moderate' or even 'cold' origin of life should be favored [31]. With respect to RNA folding, it was shown that high temperatures strongly reduce the stability of three-dimensional, and therefore functional, structures of RNA molecules [32].

The Belgian scientist Christian de Duve tried to merge these two hypotheses in his contributions to the origin of life research. Recognizing the importance of thioesters in chemical condensation reactions, as discovered by Wieland in 1951 [33], and the analogy to biochemical pathways in modern cells, De Duve placed the chemical formation of thioesters into the "metabolistic" pre-RNA-world (the "Thioester-World"), where amino acid thioesters would be formed and lead to the synthesis of simple peptides. The sulfur molecules that would be necessary to form the thioesters would come from H_2S that outgassed from the Earth's crust in volcanic regions [34]. Some of the formed peptides would possess catalytic properties, which then, in combination with energy sources such as UV light and protometabolism, would become part of a reaction network that would be able to synthesize more complicated molecules such as nucleotides and RNA, which then would lead to the RNA world (see Section 5). Although there is no experimental evidence for this hypothesis, it is a compelling way to unite two apparently contradicting ideologies of prebiotic chemistry [17].

HCN polymerization is another compelling process for the formation of organic compounds under prebiotic conditions [35]. Concentrated solutions of HCN produce nucleic acid bases and amino acids, whereas in dilute solutions hydrolysis becomes dominant. For example, one of the biological nucleobases, adenine, can be synthesized from 10 to 15 M ammonium cyanide solution at temperatures between 27 and 100 °C [36,37], while guanine is formed at yields 10 to 40 times lower than adenine [38]. Steady state concentrations of HCN in the primitive ocean were estimated to be maximal $2x10^{-6}$ M (at 0 °C) [39]. These concentrations are too low for HCN to polymerize. Because HCN is more volatile than water, evaporation in lagoons or in drying beaches can not occur if the pH is lower than the pK_a of HCN (9.2 at 25 °C) [39]. It is suggested that eutectic freezing may be a potential mechanism to concentrate HCN in an aqueous solution, for example in a lake or an ocean. The eutectic temperature of an HCN-water mixture is –21 °C, at which point the mixture contains 78 weight-% HCN, which is concentrated enough to polymerize [39]. In an experiment that demonstrated the possible formation of nucleic acid bases from HCN, a wide variety of these compounds were identified as products of a dilute frozen ammonium cyanide solution that had been held at –78 °C for 27 years [38,40]. These experiments suggest a "cold-start" for life on Earth because of the need for freezing of the water bodies.

The third source for abiotic organic compounds is extraterrestrial delivery by asteroids, comets, meteorites and interplanetary dust particles (IDPs). Amino acids have been identified in a class of meteorites called carbonaceous chondrites, in the early 1970s [41]. Since then, more than 70 different amino acids were identified in the Murchison meteorite (Figure 2) at a total concentration of approximately 60 parts per million (ppm) [42]. Nucleobases were also identified in the Murchison, Murray and Orgueil meteorites at a level of several hundred parts per billion (ppb), and based on the low terrestrial contamination, were assumed to be indigenous [43,44]. Only guanine could be identified in a set of Antarctic meteorites [45]. Finally, sugar-related compounds were also detected in Murchison and Murray [46]. For a review on the organic composition and possible extraterrestrial formation mechanisms of these compounds see Ref. [47]. Briefly, it can be said that the composition of carbonaceous chondrites with respect to their organic constituents reflects a mixture of different origins. While some of the precursor compounds that originated in the interstellar or circumstellar medium survived the conditions during solar system formation mostly unchanged, such as the polycyclic aromatic hydrocarbons (PAHs) [48], others were processed in the accretionary disk or inside growing planetesimals. It is assumed from molecular and isotopic evidence that amino acids were formed on the parent body of carbonaceous chondrites during aqueous alteration from interstellar precursors, most probably by the Strecker-cyanohydrin synthetic mechanism. Laboratory data on the deuterium retention of amino acids during Strecker synthesis indicates that, with the exception of glycine, meteoritic amino acids can be formed by this mechanism from highly-deuterated interstellar precursors [49]. Very different amino acid compositions have been detected in CM and CI carbonaceous chondrites, and the chemical reasons for these differences have not been completely understood, although it clearly points to an origin from different types of parent bodies [50,51].

Figure 15.2. . Image of a piece of the Murchison meteorite that fell in Australia on September 28,1969.

Although it is difficult to estimate the flux of extraterrestrial carbon to the early Earth from the current data [4, 5, 52], it was estimated that between 10^9 (at 4.4 Gyr) and $5*10^7$ kg (at 3.5 Gyr) of organic carbon could have been delivered to the planet per year in the first billion years of its existence, mostly in the form of interplanetary dust particles IDPs [53]. IDPs, called micrometeorites when they enter the atmosphere, have been found to be the major source of extraterrestrial organic carbon, due to their high flux rate (under the assumption of a similar organic composition as carbonaceous chondrites). Complex aromatic molecules have been identified in IDPs [54]. However, one problem associated with the delivery of organic compounds by micrometeorites is that, depending on their mass and size, they can suffer full-depth heating to temperatures (~200° to 1200°C for IDPs with a diameter of ~100 μm) during atmospheric entry deceleration [55]. This means that organic compounds present inside the IDP that are thermally unstable at these temperatures, such as amino acids, would be decomposed during this flash-heating event. It has been suggested that sublimation could be a possible mechanism by which volatile organic compounds could survive atmospheric entry heating by vaporizing off the surface of IDPs or even larger meteorites, before they are melted and destroyed (see for example Ref. [56]). Although most pure amino acids sublime with high recoveries at temperatures around 500 °C, only glycine could be detected after sublimation from Murchison meteorite powder at temperatures above 150°C, providing the only evidence that amino acids could sublime from micrometeorites and survive atmospheric entry heating [55]. None of the nucleobases present in the meteorite powder sublimed at a temperature of 450°C. Although it is not fully understood why most of the amino acids, purines and pyrimidines do not sublime from Murchison, experimental evidence suggests that divalent cations such as Ca^{2+} and Mg^{2+}, and/or the presence of kerogen-type organic polymers in Murchison may inhibit the sublimation of these compounds [55]. Obviously, organic compounds survive entry and impact in small to medium sized (few tens of meters diameter) meteorites, but the organic carbon fluxes from these objects are estimated to be about five orders of magnitude lower than for the IDPs.

Theoretical studies about impact processes on the Earth suggested that most organic compounds contained in a big impactor such as an asteroid or comet would be destroyed by the high temperatures produced in these collisions [57]. In contrast, extraterrestrial amino acids were detected in the sedimentary layers that were formed when the bolide that formed the Chicxulub crater impacted the Earth 65 Myr ago [58]. Experimental studies of impact events in the laboratory can be extended to planetary scales of hundreds of kilometers by computer simulations. New high-resolution hydrocode modeling simulations of asteroid and comet impacts, which trace the impactor's thermodynamic evolution, coupled with experimental data for amino acid pyrolysis in the solid phase suggest that amino acids would survive the shock heating of large (kilometer-radius) cometary impacts at the percent level [59]. Exogenous delivery by asteroids, comets and IDPs appears to have been an important mechanism for seeding the terrestrial planets with complex organic compounds. These compounds probably increased the molecular complexity that may have been necessary for the origin of life to occur.

3. CHIRALITY

One of the biggest enigmas in biochemistry is the emergence of the preference of one of two versions of asymmetric molecules in biosynthesis. All amino acids used in biology on the Earth are left-handed (L-enantiomer), while all ribose units in DNA and RNA are in D-conformation. Homochirality in the monomers is necessary to create secondary and tertiary structures in the polymers that are able to carry out chemical conversions (in the case of enzymes) or recognize small molecules such as hormones (in the case of receptors), for example. How this homochirality has originated has been an area of considerable discussion. Prebiotic syntheses of amino acids and hydrocarbons are expected to yield equal amounts of the isomers, therefore an enantiomeric selection had to take place at some stage in the origin or evolution of life [60]. Whether homochirality is a prerequisite or a consequence of life is still an open question. However, significant results have been obtained in the areas of chiral autocatalysis as well as in the detection of enantiomeric excesses in chiral organic compounds in meteorites, indicating that at least small enantiomeric excesses of some sort could have been present before life originated. Theories for a purely physical explanation of symmetry breaking, such as parity violation in the weak nuclear force, will not be covered in this chapter, and the reader is referred to recent reviews [61,62].

The theoretical foundations of symmetry breaking and chiral autocatalytic systems have also been reviewed [63]. In purely chemical experiments, which are not relevant for the origin of life *per se*, but dramatically demonstrate the potential of chiral autocatalysis in the amplification of homochirality, Soai and coworkers have shown that the 2-alkynylpyrimidylalkanol **2** with an initial chirality level of only 10^{-5} % ee (ee: enantiomeric excess) automultiplies with significant amplification of chirality in the addition of diisopropylzinc to 2-alkynylpyrimidine-5-carbaldehyde **1**, producing itself with the corresponding configurations in almost enantiomerically pure form (Figure 3) [64]. Without adding any chiral substances to the reaction mixture, the same reaction was shown to generate optically active pyrimidyl alkanols in a stochastic distribution, meaning that it is unpredictable which enantiomer will be in excess at the end of the reaction, but equal numbers of experiments with resulting (R) or (S) excesses were obtained [65]. This stochastic behavior is necessary for spontaneous asymmetric synthesis [63]. Finally, it was also demonstrated that a chiral catalyst that had self-improved its enantiomeric excess by autocatalysis can function as an enantioselective catalyst for another symmetric synthesis (the addition of alkylzincs to aldehydes) to yield products with enantiomeric excesses of up to 99% [66]. All these results indicate that it is in principle possible to amplify very small enantiomeric excesses by autocatalysis to form enantiomerically pure products. This underscores the importance of small enantiomeric excesses in precursor compounds that could have been available on the early Earth (or other planets).

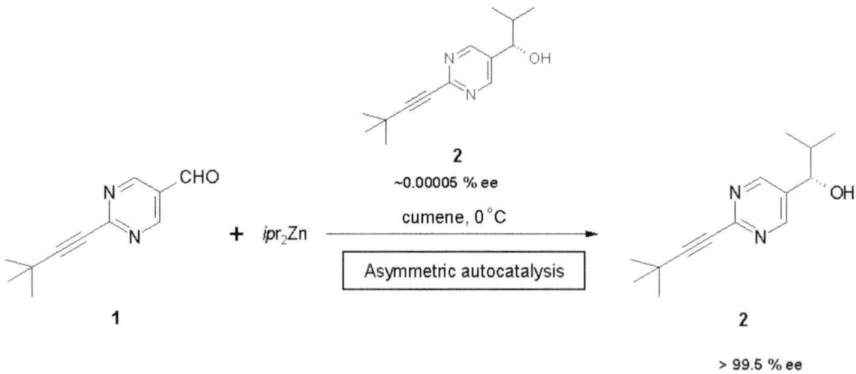

Figure 15.3. Experimental reaction system as an illustration for asymmetric autocatalysis. The alkylation of 2-alkynylpyrimidine-5-carbaldehyde **1** with diisopropylzinc leads to the formation of (S)- and (R)-(2-alkynylpyrimidyl)alkanol **2**. The enantiomeric excess of **2** in the initial mixture (in which it was present at catalytic concentration) was determined to be ~0.00005%. After three subsequent experiments, in which the resulting product **2** was used as the catalyst in the next reaction, the enantiomeric excess of **2** has increased to > 99.5%. The yields of the reactions were ≥ 90%. Adapted from ref. [64].

One of the major discoveries in the field of meteorite research has been the detection of non-racemic amino acids in the Murchison meteorite by two separate laboratories. GCMS analyses of Murchison hot-water extracts gave ee's of the L-enantiomer of two diastereoisomers of 2-amino-2,3-dimethylpentanoic acid (D,L-α-methylisoleucine and D,L-α-methylalloisoleucine) as well as isovaline [67,68]. Unlike these authors, who found excesses in amino acids that do not exist in the terrestrial biosphere and therefore cannot be contaminated by terrestrial sources, a second group reported an ee of 33 % for L-alanine, an amino acid used by all life on Earth [69]. The presence of enantiomeric excesses in α-methyl-amino acids may be due to the fact that racemization of these compounds is much slower than in amino acids that contain an α-H-atom, and that they have therefore retained their initial enantiomeric ratios during the period of aqueous alteration on the parent body. However, different formation processes for the α-methyl- and α-H-amino acids that would result in asymmetric enantiomeric distribution for the former and racemic mixtures for the latter class of compounds cannot be ruled out. More recently, these investigations have been extended to include the Murray meteorite, where lower enantiomeric excesses for 2-amino-2,3-dimethyl-pentanoic acid, isovaline, α-methylnorleucine, α-methylvaline and α-methylnorvaline were found in comparison with Murchison [70]. In the most recent study [68], it was found that the enantiomeric excesses of the amino acid isovaline in Murchison vary significantly between different samples and also within one stone. Enantiomeric excesses between 0 and 15.2 % were found, ranging much wider than previously established. This upper limit extends well beyond the theoretical boundaries set for enantiospecific amino acid decomposition by UV circular polarized light (CPL) [71].

The above mentioned results – differences between Murchison and Murray and inadequacy of UV CPL to account for enantiomeric excesses – strongly suggest that parent body processes must be responsible for the formation of the amino acids, during which these asymmetries were induced (by so far unknown mechanisms). Although α-alkyl-amino acids are not used in terrestrial biology today (only in special cases in a few fungal peptides), it can be assumed that they were delivered to the early Earth intact, where their enantiomeric excesses could have provided seeds of molecular asymmetry, which could have been amplified to lead to enantiomeric excesses in monomeric building blocks [72].

In another approach to the problem, the concept of stochastic co-oligomerization of libraries of activated, short oligomers of critical length leading to the self-assembly of large and predominantly homochiral sequence libraries as a pathway to symmetry-breaking has been discussed [73]. In the experimental set-up, tetramers of pyranosyl-RNA (p-RNA) were used as activated oligomers. The idea is as follows: One would create a mixture of all possible oligonucleotide tetramers (racemic enantiomer pairs of all eight existing diastereoisomers of all 4^4 possible sequences = 4096 tetramers), and then stochastically co-oligomerize them by chirospecific ligation to (D) and (L) libraries of higher oligomers. These libraries would consist of equal amounts of homochiral all-(D) and all-(L) oligomers, but, taken together, would no longer constitute a racemic mixture, because the number of *possible* sequences, increasing with the length of the oligomer, would quickly rise far beyond the number of *formed* sequences [72]. The two libraries will have different constitutional composition and have therefore, in principle, different chemical properties. These results solve, at least in principle, the problem of the generation of homochiral sequences.

4. POLYMERIZATION AND SURFACE REACTIONS

If one assumes that biopolymers, such as RNA and DNA, played an important role in the origin of life on the Earth, for example as the key components of the RNA world (see Section 5), then they had to be formed abiotically on the early Earth (no biopolymers have been detected in meteorites). The two big challenges for their abiotic formation are a) the abiotic synthesis of the monomeric building blocks (and their chemical activation), and b) their assembly to oligo- and polymers. There seems to be a consensus that polymer lengths in the range 30 to 100 are necessary for a self-replicating system, which corresponds to the minimum length (~40mer) for an RNA to fold into a three-dimensional functional structure. Synthesis of peptides or oligonucleotides of such length in homogeneous aqueous solution is very unfavorable, and no reports in this direction have been published. In order to form such oligomers or even polymers, reactions on surfaces of minerals have to be considered, because they provide two important advantages: 1) They increase the concentration of the reactants on a local scale by adsorptive binding and 2) they increase the reaction rate by providing local microenvironments that help to a) place the reactants in favorable conformations and b) help in the translocation of charges.

To end up with an RNA molecule, two possible chemical synthetic strategies can be envisioned. Either the activated monomers (e.g. nucleotides) were synthesized abiotically and then reacted to form short and then longer polymeric molecules, or a simpler polymeric organism precursor existed that predated RNA which was made out of simpler monomers, and from which RNA took over its genetic information carrying role.

With respect to the first idea, considerable progress has been made in the assembly of RNA oligomers from activated monomers on the surface of the mineral montmorillonite. Up to 40-mers were synthesized from unblocked monomers, using 1-methyladeninephosphate as the active leaving group at the 5'-end of the starting nucleotides [74]. The activated monomers bind more rapidly to montmorillonite than other activated monomers such as 5'-phosphorimidazolides. Detailed studies on the montmorillonite-catalyzed formation of ribonucleotide oligomers using imidazolide-activated mononucleotides have shown an overall regioselectivity for the 3',5'-phosphodiester bond of 2:1 relative to the 2',5'-bond (compared to the uncatalyzed ratio of 0.4:1) as well as sequence selectivity [75]. Longer oligomers (up to 50-mers) have been formed [76], reaching into the realm of the minimal chain length needed for active ribozymes.

The caveat with this approach is of course the synthesis of the pool of activated nucleosides [77]. Nucleotides are complicated molecules made out of an aromatic base, a sugar moiety (ribose) and a phosphate unit. The synthesis of sugars from formaldehyde (the formose reaction) gives complex mixtures, in which ribose is always a minor component. In synthesizing (D,L) ribose-2,4-biphosphate with a yield of 33% (measured in the crude reaction mixture) at room temperature in basic solution from glycoaldehyde monophosphate and formaldehyde, it was shown that there are potential pathways to generate ribose-derivatives selectively under prebiotic conditions [78]. Nevertheless, the formation of a nucleoside from a base and a sugar is also not an easy synthetic reaction (not even in the laboratory) and, at least for pyrimidine nucleosides, has not been achieved under prebiotic conditions [77]. (A ribozyme has been selected that catalyzes the formation of Uridine-5'-phosphate from uridine and activated ribose [79]. This demonstrates that, once catalytic RNAs were formed, they could catalyze the formation of their monomeric building blocks.)

The formation of a nucleotide from a nucleoside involves the addition of a phosphate to one of the hydroxy-groups of the ribose moiety (normally the 5'-OH). This process, called phosphorylation, is one of the most important biochemical reactions in organisms and is catalyzed by a large number of enzymes, for example in cellular signaling cascades. On the early Earth, however, phosphorylation, under geologically plausible conditions, is hampered by two intrinsic problems. One is that the only major source of phosphorous on the Earth, the mineral apatite, cannot be utilized for phosphorylation without the introduction of a mechanism for its protonation and/or solubilization [80]. The second problem is the efficiency of phosphorylation in aqueous solution, requiring a concentration mechanism for apatite and the activating agent [81]. Although a few experimental results have been published that demonstrated the successful phosphorylation of nucleosides with apatite and the addition of co-agents [78, 82], prebiotic phosphorylation is still considered to be a very inefficient process, adding to the problems of nucleotide synthesis and the formation of RNA.

Another possible sequence of events that can be envisaged is the idea that a genetic material made out of simpler components predated the RNA. A systematic study of properties of nucleic acid analogues was presented by Eschenmoser [83]. Not only the pyranosyl analog of RNA (p-RNA, Figure 4), but also a whole family of diastereomeric pentopyranosyl-(4'→2') oligonucleotide systems, were found to form Watson-Crick base pairing that is uniformly stronger than that of RNA itself [84]. However, although these structures could be regarded as being 'improvements' to the standard nucleic acids, their abiotic synthesis is probably not easier than that of their naturally occurring isomers. More recently, the synthesis and characterization of an oligonucleotide system containing a sugar moiety with only four carbons (called TNA) that is capable of cross-paring with RNA and DNA was reported (Figure 4) [85]. This molecule has the potential to serve as a template in nonenzymatic template-directed formation of RNA. It also stands out with regards to chemical prospects for constitutional self-assembly [85].

template-directed formation of RNA. It also stands out with regards to chemical prospects for constitutional self-assembly [85].

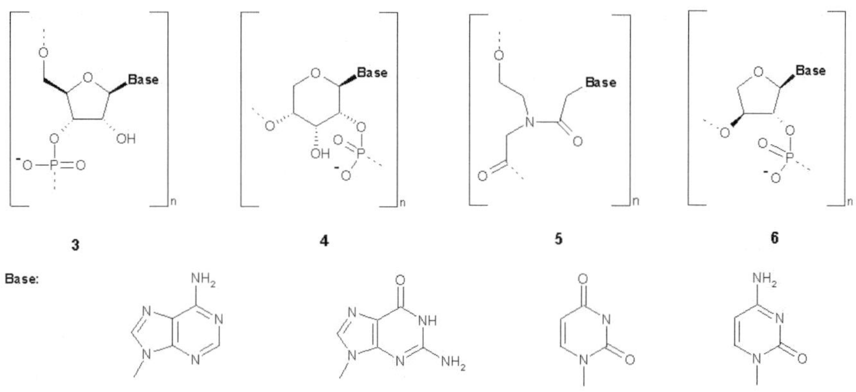

Figure 4. Constitutions of RNA (**3**) , p-RNA (**4**) , PNA (**5**), and TNA (**6**). The bases are: A = adenine; G = guanine; U = uracil; C = cytosine.]

Peptide nucleic acid (PNA, Figure 4) is another RNA analog in which the sugar-phosphate backbone has been replaced by peptide linkages; it has been studied, mainly in the context of antisense research [86]. PNA forms very stable double helices with complementary RNA and DNA, opening the possibility of information transfer from PNA to RNA and vice versa, in template-directed reactions [87]. The components of PNA are a) the nucleobases (standard or any other combination) and b) N-(2-aminoethyl)glycine and c) an acetic acid linker. All three components are potential prebiotic compounds that could have been present on the early Earth [88]. However, the assembly of the monomers from the components and a mechanism of polymerization have still to be worked out (PNA dimers tend to cyclize fairly easily, thus preventing growth beyond that stage). Also, due to its non-chiral backbone, PNA cannot induce a preferred direction in the helical coil when DNA and RNA strands would be synthesized in the template-directed fashion, assuming the activated nucleic acid monomers are present in racemic mixtures. This would theoretically result in a racemic mixture of left- and right-handed helices. However, if a racemic mixture of activated monomers were used, another problem appears, namely enantiomeric cross-inhibition. This effect was observed to occur during synthesis of poly-G from racemic mixtures of activated guanosine on a PNA C_{10} template [89]. It seems to be a very serious problem, and the authors concluded that "...it now seems unlikely that the choice of a new template, whether chiral or achiral, will overcome enantiomeric cross-inhibition so generally as to permit the template-directed replication of oligomers long enough to seed the direct emergence of the RNA world from a solution of racemic activated ribomononucleotides" [89].

5. FROM A RNA WORLD TO A DNA/PROTEIN WORLD

It was suggested in the 1960s that there was a time in the early history of life on Earth when RNA served both as the carrier of the genetic information as well as the agent of catalytic function [90,91,92,93]. Nothing is known about this postulated "RNA world", it is still a hypothesis [94,95,96]. Most of the ideas of its potential properties arise from inference from the role of RNA in contemporary organisms, in particular its role as component in the translation machinery, or by studies of RNA in the laboratory.

In today's organisms, RNA plays a major role in translation of the genetic code into specific protein sequences. For example, transfer RNAs (tRNAs), which are the specific adapter molecules that provide the link between the genetic code and the encoded protein sequence during the translation process, interact not only with their corresponding codon on the messenger RNA (mRNA), but also with the ribosome itself. This interaction is important to stabilize the binding of the tRNA to the ribosome and also involves directly functional processes such as maintaining the correct translational reading frame, translocational movement of tRNAs within the ribosome, and catalysis of peptide bond formation. As predicted by many earlier studies, and confirmed spectacularly with the high-resolution crystal structure of the ribosome [97], these interactions are mostly based on contacts between the tRNA and the ribosomal RNA (rRNA). The facts that a) the functional core of the modern ribosome, its decoding site and its peptidyl transferase center, are made up primarily of RNA, and b) the majority of its proteins are found on its outer surface, can be taken as strong evidence that the first ribosome-like biomolecules were very likely composed entirely of RNA [98].

One of the most important discoveries in biochemistry in the last 20 years is the potential of RNA to act as a chemical catalyst. After the discovery of the first catalytic functionality in the cellular RNA-splicing and processing machinery, a number of small (<1000 nucleotides [nt]) naturally occurring catalytic RNAs has subsequently been identified ([99] and references therein). Using the technique of *in vitro* selection, in which active sequences are selected against a large pool of randomized sequences, it was found that catalytic RNA molecules (so-called ribozymes) can be selected for almost every chemical reaction, including the specific aminoacylation of tRNAs, analogous to modern aminoacyl-tRNA synthetases [100]. However, it has also been shown that only ribozymes that contained functionalized nucleotides are able to catalyze certain chemical reactions, such as the Diels-Alder reaction that forms two C-C-bonds simultaneously [101]. Linking an amino acid with an RNA may therefore be another way to drastically increase the catalytic potential and specificity afforded by the resulting aminoacyl-ribozyme compared to the non-acylated one (due to the additional functionality provided by the side chain of the amino acid, for example histidine [102]), which would give the adduct an advantage under selective pressure. This may have been the first step in a process that shifted the function of bio-catalysis from polynucleotides to proteins, giving rise eventually to the formation of peptide-ribozyme complexes and, later, ribonucleoproteins (RNPs) [103]. Ribozymes that catalyze other reactions that are important in the modern translation process, namely amino acid activation [104] and peptide bond formation [105], have also been selected and characterized. The specificity of the RNA-aminoacylation reaction catalyzed by a primordial synthetase, be it RNA- or protein-based, may have been provided already very early by an operational RNA code for amino acids that is embedded in the acceptor stems (of modern tRNAs) [106]. Increasing importance of the second (anticodon) oligonucleotide domain lead to the emergence of tRNAs with a quasi-random combination of anticodons that corresponded to the amino acid. Finally, it has been shown that mRNAs, which have so far been thought of as passive carriers of genetic information, can assume a more pro-active role by serving as a metabolite-sensing "genetic switch" [107]. In this binding a distinct RNA structure is established that is likely responsible for inhibition of ribosome binding, enabling a direct negative-feedback loop to control the status of cellular metabolism. Taken together, these factors support the hypothesis that once an RNA-dominated world existed that, under selective pressure, evolved into one in which large and complex proteins took over the role of biological catalysts.

All modern cells contain large double-stranded DNA genomes. In contrast, the first RNA genomes were probably significantly smaller and possibly single-stranded. One of the critical questions is: Why did DNA appear and "take over" the function of the genetic coding from RNA? The usual argument is that DNA is more stable and can be replicated with higher fidelity. These two properties are essential for the possibility of evolution of the large genomes that encode multiple and complex proteins [108]. However, such selective pressure could only have operated in the framework of competition between populations of RNA and DNA organisms [109]. So, the emergence of proteins must have occurred in parallel with the evolution of larger and larger genomes; a role for which RNA is neither stable enough nor can be replicated accurately enough [108].

Neither chemical self-replication nor RNA-catalyzed replication offers a plausible explanation for the *spontaneous* appearance of an RNA world in the prebiotic environment: The conclusion is then that life did not start with RNA; RNA followed in the evolutionary footsteps of other replicating molecules. It might even be possible that RNA was skipped during the evolution from a pre-RNA world to the (modern) DNA/protein world [110].

6. MEMBRANES

The emergence of life on the early Earth required probably not only chemical substances that had the capability to store genetic information and perform catalytic functions, but also required the presence of defined boundaries to separate it from the environment [111,112]. These boundary structures, which in today's cells are made out of lipid bilayers, are required for speciation, energy capture and transduction. In contemporary cells, functions such as transmembrane transport of nutrients, sensor mechanisms, energy capture, and electron transport are carried out by membrane-associated proteins, which were presumably absent in the first form of cellular life [112]. Early cellular structures were probably much simpler than today.

Which compounds would have been available on the early Earth that had lipid-like properties and would form membranes under the right conditions? One possibility is that phospholipids, which are composed of fatty acids, glycol and a phosphate moiety, were synthesized during prebiotic chemical evolution. However, due to their complex structure, and the availability of phosphorous, this scenario is, in analogy to the nucleosides, unlikely. It has been shown that there are amphiphilic components in the Murchison meteorite that can be extracted using organic solvents, and that have the capacity to assemble into membranous vesicles [113,114]. These vesicles have been shown to capture and maintain concentration gradients of fluorescent dye markers, confirming the presence of a permeability barrier [115]. The composition of these amphiphiles has not been established in detail, but it seems to be clear that substantial amounts of monocarboxylic acids are present with hydrocarbon chains of C8 to C12. These relatively short fatty acids have a high critical bilayer concentration (CBC), requiring high concentrations of acids to trigger bilayer formation. This in turn raises the problem of their availability in the environment of the early Earth. A reasonable speculation may be that these fatty acids could have been concentrated in evaporating pools (see above), together with other solutes, which they may have encapsulated upon vesicle formation [112]. In a related study, Dworkin *et al.* have shown that similar vesicles are formed by components produced by UV irradiation of interstellar ice mixtures [116].

Under prebiotic conditions, three plausible entrapment processes can be envisioned: 1) simultaneous dehydration/rehydration of the vesicle and solutes, 2) production of amphiphilic molecules, from non-bilayer-forming precursors, in an environment containing solutes followed by amphiphile self-assembly, and 3) aerosol-based vesicle formation [112]. Once the catalytic species have been encapsulated in the vesicles, substrates and energy to carry out metabolic processes are required. Assuming that both (substrates and energy) are present – for example nucleotides for replication, and some sort of pigment system based on PAHs to capture radiative energy – and can be transported into the cell through the membrane, chemical reactions can occur inside the vesicle. If these reactions can be directed towards genetic information transfer (e.g. replication), then some sort of primitive cell can be constructed (see also section 9). Such precellular structures were already studied by Oparin [see Ref. [117] and references therein], and later in the laboratory of Luisi, who approached the question of self-reproducing compartments by using vesicles formed from fatty acids [117]. These authors successfully demonstrated the enzymatic polymerization of ADP to poly(A) catalyzed by polynucleotide phophorylase (PNPase) inside reverse micelles and vesicles, confirming and expanding Oparin's initial results. The results of these studies have shown for example that diffusion of molecules as large as nucleotides provided enough substrate concentration for gene amplification to take place, and further underlined the importance of the compartment in the early evolution of cellular life.

7. HORIZONTAL GENE TRANSFER

The universal tree of life displays the relationships of all living species based on the phylogeny of the small subunit (SSU) ribosomal RNA (rRNA) molecules. This molecule was chosen because it is abundant, it is coded for by organellar as well as nuclear and prokaryotic genomes, it has slow- and fast-evolving portions, and it has universally conserved structure [118]. In addition, the universal tree of life also embraces the phylogenies of ancient duplicated protein-coding genes, and the species distribution of genes and biochemical pathways. However, even before the publication of the first complete bacterial genome, many trees derived from genes and single proteins from Archaea, Bacteria and Eurkaryotes showed incongruencies with the universal tree of life. For example, only 20–50 % of protein trees supported the monophyly of domains and the sisterhood of Archaeas and Eukaryotes. These results have been taken as evidence for the exchange of genes between domains, so called horizontal (or lateral) gene transfer

(HGT), coupled with the loss of genes for specific functions, which might or might not be replaced by an acquired gene [119].

It is known that bacteria depend on gene transfer to facilitate recombination; in contrast, Eukaryotes invented sexual reproduction, tightly linking recombination between members of the same species to propagation [120]. It was therefore assumed that HGT between divergent species would be a process predominantly occurring among bacteria [121]. Interspecies gene transfer was thought to be restricted to endosymbiosis. The most famous example for this process has been the widely accepted hypothesis that mitochondria and plastids evolved from bacterial endosymbionts [122]. This hypothesis is supported by phylogenetic analyses that show that rRNA genes of mitochondria cluster as a monophyletic assemblage within the α-proteobacteria, and many protein-coding genes, which are in fact found in the nucleus, encode proteins that are transported back into the mitochondria, indicating that such genes were transferred to the nucleus during the reductive evolution of this organelle [123]. During this process, the organellar DNA replaces the corresponding nuclear gene in the host. This is an accepted form of interdomain HGT, but it was assumed that it represented a rare example. Another process of HGT could include vector-like plasmids and phages (such as the Tumor-inducing plasmid of the *Agrobacterium* genus [118]).

It has been found that many genes give different phylogenies for the same organisms [124], and the indication is that most of these genes are the result of HGT [125]. This would imply that HGT is not an exceptional process, but has occurred very frequently, particularly early in evolutionary history, and has affected all sorts of genes. If that is the case, then it would seem to have the capacity to affect the entire genome, and, given enough time, could completely erase the genealogical trace of an organism [126]. However, new analyses that are based on whole genome data from all three domains indicate the existence of an underlying phylogenetic signal that is generally in agreement with the universal tree [127]. Differences between gene trees may be caused by the compatibility of a transferred gene with the overall organization of the recipient cell [126]. For example, aminoacyl-tRNA synthetases (aaRS) are notable for relatively frequent and phylogenetically broad-range gene transfers, while on the other hand, the ribosomal proteins are notable for the virtual lack thereof. While aaRS are loosely coupled, modular components of the cell, ribosomal RNA is tightly coupled to the overall organization of the cell. This means that if the cell would be organized purely modularly, all of its components could potentially be horizontally exchanged, and the genealogical record would be eroded. This could have been the case in the early stages of cellular evolution, when cells were still very primitive (and therefore similar), which would mean that cellular evolution at its beginning would have been driven primarily by HGT [108,126]. Eventually, cells would evolve into more and more complex units, leading to a decreasing importance of HGT due to incompatibility of the exchanged genes between the "cellular environments" in which it exists and the organism, which can then be called a species in the classical sense [126]. From that point on, vertical heritage becomes the more important pathway of gene flow, with HGT being relegated to more rare events.

8. THE LAST UNIVERSAL COMMON ANCESTOR

Most contemporary theories envision the universal ancestor of all living organisms as an entity with a uniquely definable, but very primitive phenotype. Attempts to reconstruct the proteome of that last universal common ancestor (or 'LUCA') have been directed towards comparisons of the gene contents of modern genomes. Due to the apparently earliest branching of the tree separating the bacteria on the one side from the archea and eukaryotes on the other, it has been suggested that LUCA, which is placed at or close to the "root" of the tree, was a prokaryotic-like organism. In this 'classical' phylogenetic tree, hyperthermophiles emerged as the most 'primitive' organisms still existing. This, and other arguments, made these organisms the prime candidates for the most primitive forms of life present on the Earth. During this period (the Archean), the Earth was still thought to possess an oxygen-poor atmosphere and to be bombarded by impactors (see section 1). Combined with the widespread hypothesis that prokaryotes are ancestors of eukaryotes because they are simpler, it was suggested that LUCA had been a hyperthermophilic prokaryotic-like organism (e.g. Ref [128]). However, there are arguments that doubt the prokaryotic rooting of the tree [129], including the observation that many RNA molecules typical for eukaryotes could be relicts of the RNA world and thus should have been present in LUCA. The presence of many eukaryal traits in archeal proteins involved in the core of the transcription and translation systems could be explained by both extensive streamlining of these systems and further non-orthologous displacement in bacteria [128,129]. Taking this argument even further, it was proposed that hyperthermophily was one of the major streamlining forces (the 'thermoreduction' hypothesis in [130]) for the emergence of prokaryotes as the result of adaptation of a non-thermophilic LUCA to hot environments. But even if LUCA was a prokaryote, analyses of the G+C content in bacteria, which strongly correlates with their optimal growth temperature [131], as well as refined phylogenetic analyses of the rRNA tree [132], indicate that the short branches leading to thermophilic lineages in the universal tree of life do not reflect the affinity with the ancestor. Thus this would make hyperthermophily a secondary adaptation of life to very high temperatures, deconstructing the support from rRNA sequences for a hot origin of life. If these arguments were confirmed by future studies, the importance of hyperthermophilic organisms in the origin of life on Earth would be challenged [133]. This notion is consistent with thermal stability constraints from biomolecules such as RNA (see section 2).

As discussed above, newer results have also increased the importance of HGT between species, phyla or domains so that the reconstruction of the genome of LUCA seems to be doomed to failure [124]. Finally, as already mentioned above, Woese has brought up the most radical idea against the existence of a LUCA. He argues that there never was a universal ancestral cell or species, but rather a complex population of heterogeneous genetic entities, so called 'progenotes', that gave rise to the modern cellular lineages [108,134]. This population, called the last common community (LCC) by Line [135], was a "communal, loosely knit, diverse conglomerate of primitive cells that evolved as a unit, and it eventually developed to a stage where it broke into several distinct communities" (Figure 5), giving rise to the three modern domains, or lineages, without passing through a bottleneck represented by a single cellular universal ancestor [108]. The most important evolutionary driving mechanism in this progenote stage is not vertical inheritance, but HGT, leading to extensive genetic mixing. This evolution on a subcellular level, in which innovations are taken over by direct "invasion" instead of selection, can be seen as a transition between the early abiotic chemical evolution that involved organic molecules and modern Darwinian evolution.

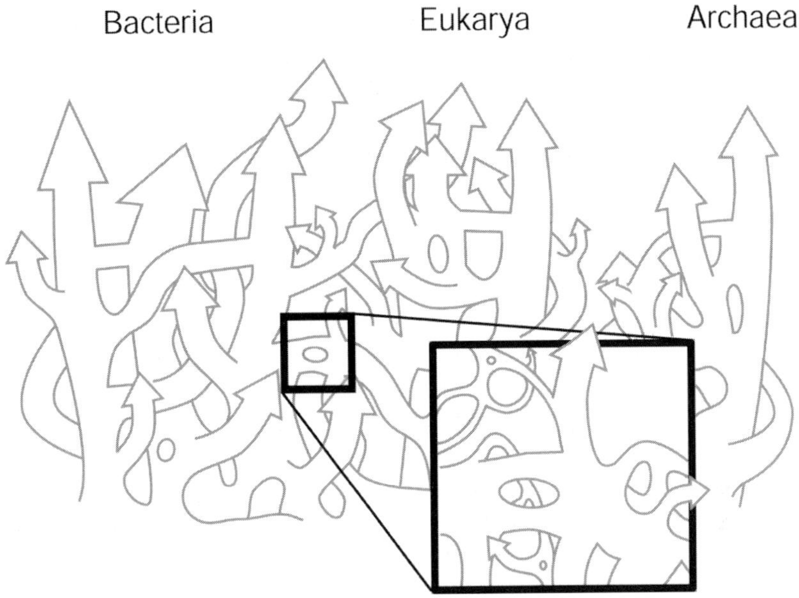

Figure 15.5. A reticulated version of tree of life, displaying at its bottom the last common community (LCC) instead of a single organism (LUCA) (adapted from Ref. [123] and [124], courtesy J. Garry).

9. SYNTHETIC LIFE

One of the biggest dreams and challenges for science has always been to understand the principle of the phenomenon 'life' to such an extent that it would become possible to create 'artificial chemical life' in the laboratory. The question is: How simple can a system be and still be considered living? First one has overcome the difficulty to define what life really is in order to clarify intellectually the experimental approach one wants to take (for a discussion on the definitions of life see [136,137]). Then, one has to understand that the objective is *not* to reconstruct the simplest form of life today – to build all their genomes and enzymes by total synthesis –, but a more fundamental one: Through experimental research one should find the minimal, as well as sufficient, set of structural and functional prerequisites of a proccss that could be agreed upon to be called 'life' [138]. From such studies, one can not expect that they result in a model that would exactly reproduce the origin of life on Earth, but several models, which probably differ widely. Some of these models will be valid in the sense that they fulfil the conditions that constrain the chemical and physical (e.g. geological) environment on the early Earth, while others will be clearly outside that realm. The most promising model, at least from an organic chemist's point of view, is that life is a property that emerges from the union of two fundamental different kinds of replicating systems: the informational genome (genotype) and the three-dimensional structure in which it is contained (phenotype) [111,138]. The best current candidates to serve as carriers for genetic information are polymeric (bio)molecules such as RNA and DNA (or perhaps simpler analogous forms such as TNA or PNA, Figure 4). But, having the genotype alone is insufficient for the system's potential to evolve. Therefore, the phenotype has to be coupled to the genotype through a coding mechanism (a task that in modern cells is achieved through the extremely complicated transcription and translation apparatus). In the simpler concept of the RNA world, the molecular structures that constitute the genotype also play the role of the phenotype and are therefore capable of evolving. In this world, 'life' uses RNA molecules to store genetic information and to act as metabolic enzymes (ribozymes). This latter property is key to this concept, because it opens the possibility for RNA to work like a polymerase that can replicate its own sequence (from a second RNA molecule). In addition, some sort of compartment is necessary to keep replicase and template together, not only to ensure frequent interaction, but also to actually give "good" replicators an evolutionary advantage over "bad" ones (see above) [111]. Consequently, a simple protocell would be composed of an RNA replicase replicating inside a replicating membrane vesicle, and both of these components would be self-assembling [111]. Finally, the two components would be coupled functionally, and the organism would then be able to evolve in vesicles with improved catalytic functionalities that would have an advantage in growth and replication. Such a concept looks relatively simple, but it holds tremendous chemical challenges to synthesize it. Efforts towards the synthesis of such artificial cells are well underway in the laboratories of Szostak, Bartel, Luisi and others. For example, the inclusion of montmorillonite grains covered with RNA oligoribonucleotides in the interior of membrane vesicles was achieved [139].

10. SUMMARY AND FUTURE PERSPECTIVES

The interdisciplinary scientific field of Astrobiology aims at the investigation of the origin of life on Earth and the search for signs of life outside the Earth. It encompasses the disciplines of astronomy, geology, paleontology, chemistry, biology and sub-disciplines such as planetary science, meteoritics, and genetics, as well as social sciences, and combines it with engineering in the fields of biochip production (for the development of "life-detection instruments") and spaceflight (for solar system exploration and extra-solar planet detection), to name only two. Astrobiology also aims to bring together scientists from different fields to combine their efforts towards the goal of investigating some of humankind's most intriguing questions: How did life begin? What is our place in the Universe? Is there life on other planets?

One of the foci of astrobiological research has been the investigations in the extremes of life on Earth. Two decades ago, the limits of life had still been thought to be restricted to near-neutral pHs, temperatures < 80 °C, and pressures of a few bars maximum. In the meantime, the limits of environmental conditions under which life can exist (there is a difference between the tolerance of an extreme environment by an organisms and the actual *requirement* of the extreme conditions) have been extended significantly. For example, an organism has been detected that is able to grow at a temperature of 121 °C [140]. Other "extremes" include resistance against high radiation (up to 20 kGy of gamma radiation by the bacterium *Deinococcus radiodurans*), pH (high pH up to 11, low pH down to 0, for *example Cyanidium caldarium*), and high oxygen levels (*Homo sapiens*) – [133]. Indisputable, this research has given incredible new insights into the mechanisms with which these organisms have adapted – *evolved* – into the niches that have been present in and on the surface of the Earth as well as in the ocean over hundreds of millions of years. It remains to be demonstrated, however, how these results will help us to understand how life *originated* on the Earth in the first place.

The future will probably see several lines of research converging towards the origin of life on Earth. It is up to the geologists and paleontologists to establish the constraints on the environmental conditions during the Archean era in more detail. As mentioned earlier, the current "gap" lies between 4.4 Gyr, where liquid water is present on the surface, to 2.7 Gyr, the undisputed presence of molecular fossils. Between these two time points 1.7 billion years apart, life must have originated and evolved, but we do not know where and when, and what the conditions were. The point is that it is crucial to establish a record on the constraints on temperature, radiation, bombardment, and presence of a solid surface on the Earth in order to be able to pinpoint the origin of life more accurately. This leads to the next point, which has to come from the biological and chemical research. These disciplines need to approach the problem by establishing constraints on the *needs* for life to originate, for example energy sources, formation of compartments and availability and stability of prebiotic organic compounds. It will be an iterative process that will eventually lead to our ability to pinpoint the time and location (in the sense of the physical and chemical environment) of the origin of life.

Last, but not least, two other important disciplines will increase our understanding of our position in the universe. Both are related to the exploration of space. One is the exploration of the Solar system. Current and future space missions to Mars, Saturn and Titan, Europa, comets, and asteroids will produce a lot of new insights into the history of our Solar system. We will try to measure *in situ* the mineral composition or presence of water ice, and later perhaps organic molecules, in the subsurface of Mars to find out how "wet" this planet really was in the past, and if life ever existed on it. (At the time of this writing, data from the Mars Exploration Rover (MER) missions to Mars were presented that provide very strong evidence for extended bodies of water on Mars in the past.) The next planned Mars Lander missions, NASA's Mars Science Laboratory (MSL) and ESA's ExoMars, are in their planning stages, and it is almost certain that they will feature instruments that should be able to search for signs of life, either past or present, on or in the surface of Mars. One candidate instrument is the Mars Organic Detector (MOD), and instrument that uses sublimation to extract organic compounds from a soil sample and fluorescence to detect organic compounds, including amino acids and PAHs [141].

Future missions to small bodies, such as the European *Rosetta* Mission to comet 67P/Churyumov-Gerasimenko, will tell us more about how the solar system formed, and will help to create a chemical "inventory" of the impactors that contributed to the chemical composition of the inner planets. The second leg in this exploration scheme is the detection of Earth-like planets around other stars and the spectroscopic detection of individual compounds on their surfaces and atmospheres. This task will require large space-based interferometers, like for example the Terrestrial Planet Finder, under study by NASA, and the European DARWIN mission. Although these missions are still more than a decade away from being launched, they will provide important insights into the frequency of planetary systems, and Earth-like planets in particular, in the galaxy and hopefully find traces of life in other worlds far away.

ACKNOWLEDGEMENT

I would like to thank Prof. Alan Schwarz and Dr. Jason Dworkin for valuable comments that improved that quality of this paper, and the European Space Agency (ESA) for financial support.

REFERENCES:

1 Miller, S. L., A Production of Amino Acids under Possible Primitive Earth Conditions, *Science* **117**, 528-529 (1953).

2 Bada, J. L., Lazcano, A., Prebiotic Soup – Revisiting the Miller Experiment, *Science* **300**, 745-746 (2003).

3 Nisbet, E. G., Sleep, N. H., The Habitat and Nature of Early Life, *Nature* **409**, 1083-1091 (2001).

4 Lazcano, A., Miller, S. L., How Long Did it Take for Life to Begin and Evolve to Cyanobacteria?, *J. Mol. Evol.* **39**, 6, 546-554 (1994).

5 Cohen, B. A., Swindle, T. D., Kring, D. A., Support for the Lunar Cataclysm Hypothesis from Lunar Meteorite Impact Melt Ages, *Science* **290**, 1754-1756 (2000).

6 Ryder, G., Mass Flux in the Ancient Earth-Moon System and benign implications for the Origin of Life on Earth, *J. Geophys. Res.* **107** (E4) Art No. 5002 (2002)

7 Ryder, G., Bombardment of the Hadean Earth: Wholesome or Deleterious?, *Astrobiology* **3**, 3-6 (2003).

8 Maher, K. A., Stevenson, D. J., Impact Frustration of the Origin of Life, *Nature* **331**, 612-614 (1988)

9 Wilde, S. A., Valley, J. W., Peck, W. H., Graham, C. M., Evidence from Detrital Zircons for the Existence of Continental Crust and Oceans on the Earth 4.4 Gyr ago, *Nature* **409**, 175-178 (2001).

10 Mojzsis, S. J., Arrhenius, G. McKeegan, K. D., Harrison, T. M., Nutman, A. P., Friend, C. R. L., Evidence for Life on Earth Before 3,800 million Years Ago, *Nature* **384**, 55-59 (1996).

11 Fedo, C. M., Whitehouse, M. J., Metasomatic Origin of Quartz-Pyroxene Rock, Akilia, Greenland, and Implications for Earth's Earliest Life, *Science* **296**, 1448-1452 (2002).

12 Van Zuilen, M. A., Lepland, A., Arrhenius, G., Reassessing the Evidence for the Earliest Traces of Life, *Nature* **418**, 627-630 (2002).

13 Schopf, J. W., Microfossils of the Early Archean Apex Chert: New Evidence of the Antiquity of Life, *Science* **260**, 640-646 (1993).

14 Schopf, J. W., Kudryavtsev, A. B., Agresti, D. G., Wdowiak, T. J., Czaja, A. D., Laser Raman Imagery of Earth's Earliest Fossils, *Nature* **416**, 73-76 (2002).

15 Brasier, M. D., Green, O. R., Jephcoat, A. P., Kleppe, A. N., Van Kranendonk, M. J., Lindsay, J. F., Steele, A., Grassineau, N. V., Questioning the Evidence for Earth's Oldest Fossils, 2002, *Nature* **416**, 76-81 (2002).

16 Brasier, M., Green, O., Lindsay, J., Steele, A., Earth's Oldest (~3.5 Ga) Fossils and the 'Early Eden Hypothesis': Questioning the Evidence, *Origins Life Evol. Biosphere* **34**, 257-269 (2004).

17 Bada, J. L., Lazcano, A., Some Like it Hot, But Not the First Biomolecules, *Science* **296**, 1982-1983 (2002).

18 Ehrenfreund, P., Irvine, W., Becker, L., Blank, J., Brucato, J. R., Colangeli, L., Derenne, S., Despois, D., Dutrey, A., Fraaje, H., Lazcano, A., Owen, T., Robert, F., Astrophysical and Astrochemical Insights into the Origin of Life, *Rep. Prog. Phys.* **65**, 1427-1487 (2002).

19 Kasting, J. F., Pavlov, A. A., Siefert, J. L., A Coupled Ecosystem-Climate Model for Predicting the Methane Concentration in the Archean Atmosphere, *Origins Life Evol. Biosphere* **31**, 271-285 (2001).

20 Schlesinger, G, Miller, S. L., Prebiotic Synthesis in Atmospheres Containing CH_4, CO and CO_2. I. Amino Acids, *J. Mol. Evol.* **19**, 376-382 (1983).

[21] Miller, S. L., The Mechanism of Synthesis of Amino Acids by Electric Discharges, *Biochim. Biophys. Acta* **23**, 480-489 (1957).

[22] Miller, S. L., Schlesinger, G., Carbon and Energy Yields in Prebiotic Syntheses Using Atmospheres Containing CH_4, CO and CO_2, *Origins Life* **14**, 83-90 (1984).

[23] Wächtershäuser, G., Groundworks for an Evolutionary Biochemistry: The Iron-Sulfur World, *Prog. Biophys. Molec. Biochem.* **58**, 85-201 (1992).

[24] Wächtershäuser, G., Life As We Don't Know It, *Science* **289**, 1307-1808 (2000).

[25] Huber, C, Wächtershäuser, G., Activated Acetic Acid by Carbon Fixation on (Fe,Ni)S Under Primordial Conditions, *Science* **276**, 245-247 (1997).

[26] Cody, G. D., Boctor, N. Z., Hazen, R. M., Brandes, J. A., Morowitz, H. J., Yoder, H. S., Jr., Geochemical Roots of Autotrophic Carbon Fixation: Hydrothermal Experiments in the System Citric Acid, H_2O-(±FeS)-(±NiS), *Geochim. Cosmochim. Acta* **65**, 3557-3576 (2001).

[27] Orgel, L. E., Self-Organizing Biochemical Cycles, *Proc. Natl. Acad. Sci. USA* **97**, 12503-12507 (2000).

[28] Miller, S. L., Bada, J. L., Submarine Hot Springs and the Origin of Life, *Nature* **334**, 609-611 (1988).

[29] Miller, S. L., Lazcano, A., The Origin of Life – Did It Occur at High Temperatures?, *J. Mol. Evol.* **41**, 689-692 (1995).

[30] Levy, M., Miller, S. L., The Stability of RNA Bases: Implications for the Origin of Life, *Proc. Natl. Acad. Sci USA* **95**, 7933-7938 (1998).

[31] Bada, J. L., Bigham, C., Miller, S. L., Impact Melting of Frozen Oceans on the Early Earth: Implications for the Origin of Life, *Proc. Natl. Acad. Sci. USA* **91**, 1248-1250 (1994).

[32] Moulton, V., Gardner, P. P., Pointon, R. F., Creamer, L. K., Jameson, G. B., Penny, D., RNA Folding Argues Against a Hot-Start Origin of Life, *J. Mol. Evol.* **51**, 416-421 (2000).

[33] Wieland, T, Schäfer, W., Bokelmann, E., Über eine bequeme Darstellungsweise von Acylthiophenolen und ihre Verwendung zu Amid- und Peptid-Synthesen, *Liebigs Ann. Chem.* **573**, 99-104 (1951).

[34] De Duve, C. (1991), *Blueprint for A Cell*, Neil Patterson Publishers, Burlington, NC.

[35] Ferris, J. P., Hagan, W. J., Jr., HCN and Chemical Evolution: The Possible Role of Cyano Compounds in Prebiotic Synthesis, *Tetrahedron* **40**, 1093-1120 (1984).

[36] Oró, J., Synthesis of Adenine from Ammonium Cyanide, *Biochem. Biophys. Res. Commun.* **2**, 407-412 (1960).

[37] Oró, J., Kimball, A. P., Synthesis of purines under possible primitive Earth conditions. I. Adenine from hydrogen cyanide. *Arch. Biochem. Biophys.* **94**, 217-27 (1961).

[38] Levy, M., Miller, S. L., Oró, J., Production of Guanine from NH_4CN Polymerization, *J. Mol. Evol.* **49**, 165-168 (1999).

[39] Miyakawa, S., Cleaves, H. J., Miller, S. L., The Cold Origin of Life: A. Implications Based on the Hydrolytic Stabilities of Hydrogen Cyanide and Formamide, *Origins Life Evol. Biosphere* **32**, 195-208 (2003).

[40] Miyakawa, S., Cleaves, H. J., Miller, S. L., The Cold Origin of Life: B. Implications Based on Pyrimidines and Purines Produced from Frozen Ammonium Cyanide Solutions, *Origins Life Evol. Biosphere* **32**, 209-218 (2003).

[41] Kvenvolden, K., Lawless, J., Pering, K., Peterson, E., Flores, J., Ponnamperuma, C., Kaplan, I. R., Moore, C., Evidence for Extraterrestrial Amino-Acids and Hydrocarbons in the Murchison Meteorite, *Nature* **228**, 923-926 (1970).

42 Cronin, J. R. and Chang, S.: 1993, Organic Matter in Meteorites: Molecular and Isotopic Analyses of the Murchison Meteorites, in J. M. Greenberg, C. X. Mendoza-Gomez & V. Pirronello (eds.), *The Chemistry of Life's Origin*, Kluwer Academic Publishers, The Netherlands, pp. 209-258.

43 Stoks, P. G., Schwartz, A. W., Uracil in Carbonaceous Chondrites, *Nature* **282**, 709-710 (1979).

44 Stoks, P. G., Schwartz, A. W., Nitrogen-Heterocyclic Compounds in Meteorites: Significance and Mechanisms of Formation, *Geochim. Cosmochim. Acta* **45**, 563-569 (1981).

45 Shimoyama, A., Hagishita, S., Harada, K., Search for Nucleic Acid Bases in Carbonaceous Chondrites from Antarctica, *Geochemical Journal* **24**, 343-348 (1990).

46 Cooper, G., Kimmich, N., Belisle, W., Sarinana, J., Brabham, K., Garrel, L., Carbonaceous Meteorites as a Source of Sugar-Related Organic Compounds for the Early Earth, *Nature* **414**, 879-883 (2001).

47 Botta, O., Bada, J. L., Extraterrestrial Organic Compounds in Meteorites, *Surv. Geophys.* **23**, 411-467 (2002).

48 Messenger, S., Amari, S., Gao, X., Walker, R. M., Clemett, S. J., Chillier, X. D. F., Zare, R. N., Lewis, R. S., Indigenous Polycyclic Aromatic Hydrocarbons in Circumstellar Graphite Grains from Primitive Meteorites, *Astrophys. J.* **502**, 284-295 (1998).

49 Lerner, N. R., Peterson, E., Chang, S., The Strecker Synthesis as a Source of Amino Acids in Carbonaceous Chondrites: Deuterium Retention During Synthesis, *Geochim. Cosmochim. Acta* **57**, 4713-4723 (1993).

50 Ehrenfreund, P., Glavin, D. P., Botta, O., Cooper, G., Bada, J. L., Extraterrestrial Amino Acids in Orgueil and Ivuna: Tracing the Parent Body of CI Type Carbonaceous Chondrties, *Proc. Natl. Acad. Sci. USA* **98**, 2138-2141 (2001).

51 Botta, O., Glavin, D. P., Kminek, G., Bada, J. L., Relative Amino Acid Concentrations as a Signature for Parent Body Processes of Carbonaceous Chondrites, *Origins Life Evol. Biosphere* **32**, 143-163 (2002).

52 Love, S. G., Brownlee, D. E., A Direct Measurement of the Terrestrial Mass Accretion Rate of Cosmic Dust, *Science* **262**, 550-553 (1993).

53 Chyba, C. F., Sagan, C., Endogenous Production, Exogenous Delivery and Impact-Shock Synthesis of Organic Molecules: An Inventory for the Origins of Life, *Nature* **255**, 125-132 (1992).

54 Clemett, S. J., Maechling, C. R., Zare, R. N., Swan, P. D., Walker, R. M., Identification of Complex Aromatic Molecules in Individual Interplanetary Dust Particles, *Science* **262**, 721-725 (1993).

55 Glavin, D. P. Survival of Prebiotic Organic Compounds During Exogenous Delivery. Implications for the Origin of Life on Earth and Potentially on Mars. Ph.D. Thesis, University of California, San Diego, 2001.

56 Glavin, D. P. and Bada, J. L, Survival of Amino Acids in Micrometeorites During Atmospheric Entry, *Astrobiology* **1**, 259-269 (2001).

57 Chyba, C. F., Thomas, P. J., Brookshaw, L. and Sagan, C. Cometary Delivery of Organic Molecules to the Early Earth. *Science* **249**, 366-373 (1990).

58 Zhao, M. and Bada, J. L. Extraterrestrial Amino Acids in Cretaceous/Tertiary Boundary Sediments at Stevns Klint, Denmark. *Nature* **339**, 463-465 (1989).

59 Pierazzo, E. and Chyba, C. F. Amino Acid Survival in Large Cometary Impacts. *Meteorit. Planet. Sci.* **34**, 909-918 (1999).

60 Bada, J. L., Extraterrestrial handedness?, *Science* **275**, 942-943 (1997).

[61] Feringa, B. L., van Delden, R. A., Absolute Symmetric Synthesis: The Origin, Control, and Amplification of Chirality, *Angew. Chem. Int. Ed.* **38**, 3418-3438 (1999).

[62] Podlech, J., Origin of Organic Molecules and Biomolecular Homochirality, *Cell. Mol. Life. Sci.* **58**, 44-60 (2001).

[63] Kondepudi, D. K., Asakura, K., Chiral Autocatalysis, Spontaneous symmetry Breaking, and Stochastic Behavior, *Acc. Chem. Res.* **34**, 946-954 (2001).

[64] Sato, I., Urabe, H., Ishiguro, S., Shibata, T., Soai, K., Amplification of Chirality from Extremely Low to Greater than 99.5% *ee* by Asymmetric autocatalysis, *Angew. Chem. Int. Ed.* **42**, 315-317 (2003).

[65] Soai, K., Sato, I., Shibata, T., Monyia, S., Hayashi, M., Matsueda, Y., Imamura, H., Hayase, T., Morioka, H., Tabira, H., Yamamoto, J., Kowata, Asymmetric Synthesis of Pyrimidyl Alkanol without Adding Chiral Substances by the Addition of Diisopropylzinc to Pyrimidine-5-carbaldehyde in Conjunction with Asymmetric Autocatalysis, *Tetrahedr. Asymm.* **14**, 185-188 (2003).

[66] Sato, I., Urabe, H., Ishii, S., Tanji, S., Soai, K., Asymmetric Synthesis with a Chiral Catalyst Generated from Asymmetric Autocatalysis, *Org. Lett.* **3**, 3851-3854 (2001).

[67] Cronin, J. R., Pizzarello, S., Enantiomeric Excesses in Meteoritic Amino Acids, *Science* **275**, 951-955 (1997).

[68] Pizzarello, S., Zolensky, M., Turk, K. A., Nonracemic Isovaline in the Murchison Meteorite: Chiral Distribution and Mineral Association, *Geochim. Cosmochim. Acta* **67**, 1589-1595 (2003).

[69] Engel, M. H., Macko, S. A., Isotopic Evidence for Extraterrestrial Non-Racemic Amino Acids in the Murchison Meteorite, *Nature* **389**, 265-268 (1997).

[70] Pizzarello, S., Cronin, J. R., Non-Racemic Amino Acids in the Murray and Murchison Meteorites, *Geochim. Cosmochim. Acta* **64**, 329-338 (2000).

[71] Bonner, W. A., Rubenstein, E., Supernovae, Neutron Stars and Biomolecular Chirality, *BioSystems* **20**, 99-111 (1987).

[72] Pizzarello, S., Weber, A. L., Prebiotic Amino Acids as Asymmetric Catalysts, Science 303, 1151 (2004).

[73] Bolli, M., Micura, R., Eschenmoser, A., Pyranosyl-RNA: Chiroselective Self-assembly of Base Sequences by Ligative Oligomerization of Tetranucleotide-2',3'-cyclophosphates (with a Commentary Concerning the Origin of Biomolecular Homochirality), *Chem. Biol.* **4**, 309-320 (1997).

[74] Huang, W., Ferris, J. P., Synthesis of 35–40-mers of RNA Oligomers from Unblocked Monomers. A Simple Approach to the RNA World, *Chem. Comm.* 1458-1459 (2003).

[75] Miyakawa, S., Ferris, J. P., Sequence- and Regioselectivity in the Montmorillonite-Catalyzed Synthesis of RNA, *J. Am. Chem. Soc.* **125**, 8202-8208.

[76] Ferris, J. P., Montmorillonite Catalysis of 30–50-mer Oligonucleotides: Laboratory Demonstration of Potential Steps in the Origin of the RNA World, *Origins Life Evol. Biosphere* **32**, 311-332 (2002).

[77] Orgel, L. E., The Origin of Life – a Review of Facts and Speculations, *Trends Biochem. Sci.* **23**, 491-495 (1998).

[78] Müller, D., Pitsch, S., Kittaka, A., Wagner, E., Wintner, C. E., Eschenmoser, A., Aldomerisierung von Glycoaldehy-phophaten und (in Gegenwart von Formaldehyd) Racemischen Pentose-2,4-diphosphaten: *rac*-Allose-2,4,6-triphosphat und *rac*-Ribose-2,4-diphosphat sind die Reaktionshauptprodukte, *Helv. Chim. Acta* **73**, 1410-1468 (1990)

[79] Unrau, P. J., Bartel, D. P. RNA-Catalyzed Nucleotide Synthesis, *Nature* **395**, 260-263 (1998)

80 Schwartz, A. W., Prebiotic Phosphorylation – Nucleotide Synthesis with Apatite, *Biochim. Biophys. Acta* **281**, 477-480 (1972).

81 Lohrmann, R., Orgel, L. E., Prebiotic Synthesis: Phosphorylation in Aqueous Solution, *Science* **161**, 64-66 (1968).

82 Schwartz, A. W., van der Veen, M., Bisseling, T., Chittenden, G. J. F., Prebiotic Nucleotide Synthesis – Demonstration of a Geologically Plausible Pathway, *Origins Life* **6**, 163-168 (1975).

83 Eschenmoser, A. Towards a Chemical Etiology of Nucleic Acid Structure, *Origins Life Evol. Biosphere* **27**, 535-553 (1997).

84 Beier, M., Reck, T., Wagner, R., Krishnamurthy, R., Eschenmoser, A., Chemical Etiology of Nucleic Acid Structure: Comparing Pentopyranosyl-(2'→4') oligonucleotides with RNA, *Science* **283**, 699-703 (1999).

85 Schöning, K.-U., Scholz, P., Guntha, S., Wu, X., Krishnamurthy, R., Eschenmoser, A., Chemical Etiology of Nucleic Acid Structure: The α-threofuranosyl-(3'→2') oligonucleotide system, *Science* **290**, 1347-1351 (2000).

86 Egholm, M., Buchardt, O., Nielsen, P. E., Berg, R. H., Peptide Nucleic Acid (PNA). Oligonucleotide Analogues with an Achiral Peptide Backbone, *J. Am. Chem. Soc.* **114**, 1895-1897 (1992).

87 Schmidt, J. G., Nielsen, P. E., Orgel, L. E., Information Transfer from Peptide Nucleic Acid to RNA by Template-Directed Syntheses, *Nucleic Acids Res.* **25**, 4797-4802 (1997).

88 Nelson, K. E., Levy, M., Miller, S. L., Peptide Nucleic Acids Rather Than RNA May Have Been the First Genetic Molecules, *Proc. Natl. Acad. Sci. USA* **97**, 3868-3871 (2000).

89 Schmidt, J. G., Nielsen, P. E., Orgel, L. E., Enantiomeric Cross-Inhibition in the Synthesis of Oligonucleotides on a Nonchiral Template, *J. Am. Chem. Soc.* **119**, 1494-1495 (1997).

90 Buchanan J. M. (1965), Chairman's remarks. The Origin of Prebiological Systems and of Their Molecular Matrices (Fox SW, ed.). Academic Press, New York, pp. 101-104.

91 Haldane J. B. S. (1965), Data needed for a blueprint of the first organism. The Origin of Prebiological Systems and of Their Molecular Matrices (Fox SW, ed.). Academic Press, New York, pp. 11-18.

92 Crick, F. H. C., The Origin of the Genetic Code, *J. Mol. Biol.* **38**, 367-379 (1967).

93 Orgel, L. E., Evolution of the Genetic Apparatus, *J. Mol. Biol.* **38**, 381-393 (1967).

94 Joyce, G. F., The Rise and Fall of the RNA World, *The New Biologist* **3**, 399-407 (1991).

95 Gilbert, W., The RNA World, *Nature* **319**, 618 (1986).

96 Gesteland, R. F. & Atkins, J. F. (eds.), *The RNA World.* New York: Cold Spring Harbor Press, 1993.

97 Yusupov, M. M., Yusupova, G. Zh., Baucom, A., Lieberman, K., Earnest, T. N., Cate, J. H. D., Noller, H. F., Crystal Structure of the Ribosome at 5.5 Å Resolution, *Science* **292**, 883-896 (2001).

98 Moore, P. B. & Steitz, T. A., The Involvement of RNA in Ribose Function, *Nature* **418**, 229-235 (2002).

99 Scott, W. G. & Klug, A., Ribozymes: Structure and Mechanism in RNA Catalysis, *Trends Biochem. Sci.* **21**, 220-224 (1996).

100 Lee, N., Bessho, Y., Wei, K., Szostak, J. W., Suga, H., Ribozyme-Catalyzed tRNA Aminoacylation, *Nature Struct. Biol.* **7**, 28-33 (2000).

101 Tarasow, T. M, Tarasow, S. L., Eaton, B. E., RNA-Catalyzed Carbon-Carbon Bond Formation, *Nature* **389**, 54-57 (1997).

[102] Roth, A. & Breaker, R. R., An Amino Acid as a Cofactor for a Catalytic Polynucleotide, *Proc. Natl. Acad. Sci. USA* **95**, 6027-6032 (1998).

[103] Schimmel, P. & Kelley, S. O., Exiting an RNA World, *Nature Struct. Biol.* **7**, 5-7 (2000).

[104] Kumar, R. K. & Yarus, M., RNA-Catalyzed Amino Acid Activation, *Biochemistry* **40**, 6998-7004 (2001).

[105] Zhang, B. & Cech, T. R. , Peptide Bond Formation by *in vitro* Selected Ribozymes, *Nature* **390**, 96-100 (1997).

[106] Schimmel, P., Giegé, R., Moras, D., Yokoyama, S., An Operational RNA Code for Amino Acids and Possible Relationship to the Genetic Code, *Proc. Natl. Acad. Sci USA* **90**, 8763-8768 (1993).

[107] Nahvi, A., Sudarsan, N., Ebert, M. S., Zou, X., Brown, K. L., Breaker, R. R., Genetic Control by a Metabolite Binding mRNA, *Chem. Biol.* **9**, 1043-1049 (2002).

[108] Woese, C., The Universal Ancestor, *Proc. Natl. Acad. Sci. USA* **95**, 6854-6859 (1998).

[109] Forterre, P., The Origin of DNA Genomes and DNA Replication Proteins, *Curr. Opinion Microbiol.* **5**, 525-532 (2002).

[110] Dworkin, J. P., Lazcano, A., Miller, S. L., The Roads to and from the RNA world, *J. Theor. Bio.* **222**, 127-134 (2003).

[111] Szostak, J. W., Bartel, D. P., Luisi, P. L., Synthesizing Life, *Nature* **409**, 387-390 (2001).

[112] Monnard, P.-A., Deamer, D., Membrane Self-Assembly Processes: Steps Toward the First Cellular Life, *Anat. Rec.* **268**, 196-207 (2002).

[113] Deamer, D. W., Boundary Structures Are Formed by Organic Components of the Murchison Carbonaceous Chondrite, *Nature* **317**, 792-794 (1985).

[114] Mautner, M. N., Leonard, R. L., Deamer, D. W., Meteorite Organics in Planetary Environments: Hydrothermal Release, Surface Activity, and Microbial Utilization, *Planet. Space Sci.* **43**, 139-147 (1995).

[115] Apel, C. L., Mautner, M. N., Deamer, D. W., Self-Assembled Vesicles of Monocarboxylic acids and Alcohols: Conditions for Stability and for Encapsulation of Biopolymers, *Biochim. Biophys. Acta* **1559**, 1-9 (2002).

[116] Dworkin J. P., Deamer D. W., Sandford S. A., Allamandola L. J., Self-assembling Amphiphilic Molecules: Synthesis in Simulated Interstellar/Precometary Ices, *Proc. Natl. Acad. Sci. USA* **98**, 815-819 (2001).

[117] Walde, P., Goto, A., Monnard, P.-A., Wessicken, M., Luisi, P. L., Oparin's Reaction revisited: Enzymatic Synthesis of Poly(adenylic acid) in Micelles and Self-Reproducing Vesicles, *J. Am. Chem. Soc.***116**, 7541-7547 (1994).

[118] Woese, C. R., Bacterial Evolution, *Microbiol. Rev.* **51**, 221-271 (1987).

[119] Brown, J. R., Ancient Horizontal Gene Transfer, *Nature Rev: Genetics* **4**, 121-132 (2003).

[120] Gogarten, J. P., Gene Transfer: Gene Swapping Craze Reaches Eukaryotes, *Curr. Biol.* **13**, R53-R54 (2003).

[121] Gogarten, J. P., Doolittle, W. F., Lawrence, J. G., Prokaryotic Evolution in Light of Gene Transfer, *Mol. Biol. Evol.* **19**, 226-2238 (2002).

[122] Margulis, L., (1970), *Origin of Eucaryotic Cells*, Yale Univ. Press, New Haven, CT.

[123] Doolittle, W. F., You Are What You Eat: A Gene Transfer Ratchet Could Account for Bacterial Genes in Eukaryotic Nuclear Genomes, *Trends Genet.* **14**, 307-311 (1998).

[124] Martin, W., Mosaic Bacterial Chromosomes: A Challenge En Route to a Tree of Genomes, *BioEssays* **21**, 99-104 (1999).

[125] Doolittle, W. F., Phylogenetic Classification and the Universal Tree, *Science* **284**, 2124-2128 (1999).

126 Woese, C. R., On the Evolution of Cells, *Proc. Natl. Acad. Sci. USA* **99**, 8742-8747 (2002).

127 Daubin, V., Moran, N. A., Ochman, H., Phylogenetics and the Cohesion of Bacterial Genomes, *Science* **301**, 829-832 (2003).

128 Xu, Y., Glansdorff, N, Was Our Ancestor a Hyperthermophilic Prokaryote?, *Comp. Biochem. Phys. A* **133**, 677-688 (2002).

129 Forterre, P., Philippe, H., Where Is the Root of the Universal Tree of Life?, *BioEssays* **21**, 871-879 (1999).

130 Forterre, P., Thermoreduction, a Hypothesis for the Origin of Prokaryotes, *CR Acad. Sci. Paris* **318**, 415-422 (1995).

131 Galtier, N., Tourasse, N., Gouy, M., A Nonhyperthermophilic Common Ancestor to Extant Life Forms, *Science* **283**, 220-221 (1999).

132 Brochier, C., Philippe, H., A Non-Hyperthermophilic Ancestor for Bacteria, *Nature* **417**, 244 (2002).

133 Rothschild, L. J., Mancinelli, R. L., Life in Extreme Environments, *Nature* **409**, 1092-1101 (2001).

134 Doolittle, W. F., The Nature of the Universal Ancestor and the Evolution of the Proteome, *Curr. Opinion Struct. Biol.* **10**, 355-358 (2000).

135 Line, M. A., The Enigma of the Origin of Life and its Timing, *Microbiol.* **148**, 21-27 (2002).

136 Luisi, P. L., About Various Definitions of Life, *Origins Life Evol. Biosphere* **28**, 613-622 (1998).

137 Cleland, C. E., Chyba, C. F., Defining Life, *Origins Life Evol. Biosphere* **32**, 387-393 (2002).

138 Eschenmoser, A., Kisakürek, M. V., Chemistry and the Origin of Life, *Helv. Chim. Acta* **79**, 1249-1259 (1996).

139 Hanczyc, M. M., Kujikawa, S. M. Szostak, J. W., Experimental Models of Primitive Cellular Compartments: Encapsulation, Growth, and Division, *Science* **302**, 618-622 (2003).

140 Kashefi, K., Lovley, D. R., Extending the Upper Temperature Limit for Life, *Science* **301**, 934 (2003).

141 Kminek, G., Bada, J. L., Botta, O., Glavin, D., Grunthaner, F., MOD: An Organic Detector for Future Robotic Exploration of Mars, *Planet. Space Sci.* **48**, 1087-1091 (2000).

Chapter 16

A NOVEL SYNTHESIS OF BIOMOLECULAR PRECURSORS

Raffaele Saladino[1], Claudia Crestini[2], Fabiana Ciciriello[3], Giovanna Costanzo[3], Rodolfo Negri[4], and Ernesto Di Mauro[5]

[1]INFM, Dipartimento A.B.A.C., Università della Tuscia, 01100 Viterbo, [2]Dipartimento di Scienze e Tecnologie Chimiche Università "Tor Vergata", 00133 Rome, [3]Istituto di Biologia e Patologia Moleculari, CNR, 00185 Rome, [4]Dipartimento di Biologia Cellulare e dello Sviluppo Università "La Sapienza", 00185 Rome, [5]Fondazione "Istituto Pasteur, Fondazione Cenci Bolognetti", Università "La Sapienza", 00185 Rome

Abstract: We discuss the role of formamide, a product of hydrolysis of hydrogen cyanide (HCN), as precursor of relevant components of nucleic acids in prebiotic conditions and describe the efficient synthesis of purine, adenine, cytosine, thymine, and 5-hydroxymethyluracil. The remarkable formation of some purine acyclonucleosides is also reported, providing a possible solution to the problem of the elusive origin of nucleosides under prebiotic conditions. The role of catalysts as $CaCO_3$, silica, alumina, TiO_2 and others in enhancing and variegating the yields of these compounds is described. In addition, formamide acts as a selective agent in the degradation of bases, nucleosides and DNA oligomers. Taken together, these observations reveal a formamide-based synthesis/degradation cycle whose properties provide an equilibration mechanism for the pool of prebiotic precursors.

Key words: formamide, pyrimidines, acyclonucleosides, nucleic acids precursors, synthesis-degradation equilibrium

1. OUR OPERATIVE DEFINITIONS

A definition of life is, at best, elusive. We have conservatively adopted as a reference frame for our studies the simplest and broadest of the definitions: *life is the perpetuation of chemical information.*

P. Ehrenfreund et al. (eds.), Astrobiology: Future Perspectives, 393–413.

Several corollaries are implicit:

(i) Perpetuation does not imply eternity. Life may have ended and started several times.

(ii) Chemical information must be endowed with a minimal level of complexity. Operationally, we (*a*) do not to call *life* the condensation of NH_2COH (formamide) into a nucleobase and its successive recovery in pure form (in a cycle similar to those that will be described below). However, we consider (*b*) *life* to be a chemical process that transmits a templating molecular information. Life starts somewhere between the limits set by *a* and *b*. From *(ii)* two corollaries follow:

(iii) the template is the product of previous reactions. Implicit in this is:

an evolutive process in a pure neo-Darwinian sense. The perpetuation of chemical information has occurred in a changing environment, thus being intrinsically and necessarily an evolutive process. The *emergence* of new properties is an inborn attribute of the system. Given that the prebiotic emergence of self-replicating molecules *a priori* did not take advantage of supporting enzymatic systems, emergence of life as defined here must necessarily have been a robust phenomenon.

2. LIFE IS A ROBUST PHENOMENON

For the sake of clarity in describing our experimental approach (and the results obtained) we report in Figure 16.1 the general scheme for the evolutionary sequence pioneeringly suggested by Fox and Dose (1972).

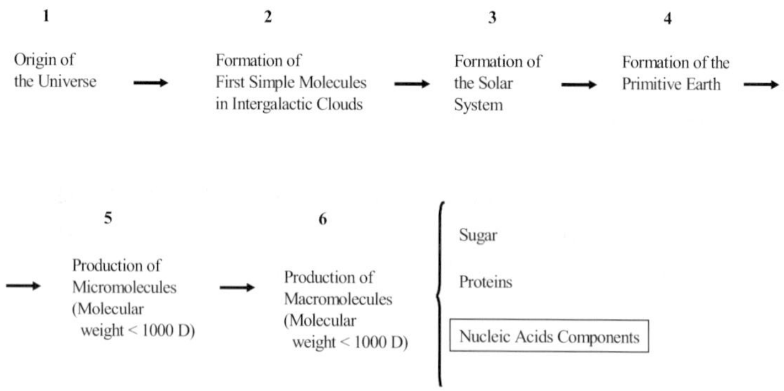

Figure 16.1. The ideal evolutionary sequence (from Fox and Dose, 1972, modified).

Since the time of its formulation, this scheme has not been controversial. In addition, we now know that both micro- and macro-molecules are synthesized (and are stable enough to survive) in the interstellar environment. Macromolecular material has been found in carbonaceous chondrites, probably made out of aromatic cores with aliphatic chains [for an up-dated treatment of this matter the reader is referred to Ehrenfreund et al. (2002), chapter 3]. The molecules on which we have focused our attention (hydrogen cyanide HCN and formamide $HCONH_2$) are also present in the interstellar medium (*ibidem*). Our experimental approach is an effort to explore the mechanisms (and the physico-chemical conditions allowing their embodiment) for the passage from simple to macro-molecules. In order to keep the experimental approach into a maneuverable dimension, we further limit (arbitrarily and operationally) our goal by assuming that:

i. The macromolecules of interest are those involved in life "as we know it" on Earth. Even if generated and present in interstellar media, the precursor chemical species considered require the appropriate environment for their evolution. This can only be provided by planetary systems and by Earth-like planets. Naturally, one should keep in mind that our planet is special and unique but that it is just one aggregate of chemical compounds which are present elsewhere, in different amounts and combinations.

ii. The operative definition according to which *life is the perpetuation of chemical information,* entails a replication process. Two general mechanisms are conceivable: (1) each information-containing molecule (ICM) replicates as a single unit, exactly and completely. (2) Each ICM replicates through a template-driven process, each molecular generation being the chemical mirror image of both the previous and the following ones. The sentence by Watson and Crick: "*it has not escaped our notice that the specific pairing we have postulated immediately suggests a possible copying mechanism for the genetic material*" (Watson and Crick, 1953) has clarified both this mechanism and its prevalence in templated life (again: as we know it). Nucleic acids are found in living systems as both RNA or DNA, and both can be in single- or double-stranded form. In the physico-chemical conditions tested chemical stability is higher for DNA than for RNA and for the double- than for the single-stranded forms (Kochetkov and Budovskii, 1972). When trying to identify precursor molecules and mechanisms that could have acted in prebiotic conditions (see below), the property of higher-stability of the DNA double strand makes it in principle a favored candidate. On the other hand the intrinsic ability of RNA to act as a catalyst (Pace & Marsh, 1985; Buzayan et al., 1986; Cech and Bass, 1986; Cech, 1986; Nissen et al., 2000; Joyce, 2002) has been described. Catalytic properties of DNA have also been reported (Santoro *et al.* 2000; Sheppard *et al.*, 2000). The bias "stability versus reactivity" is far from being solved. In the present study we focus on *stability*. One of our starting considerations is that in a prebiotic enzyme-free environment, DNA or RNA started their ICM role directly as an alternating series of (short) mirror-template strands; that is: essentially as a double strand, for sake of stability.

iii. The information-containing ur-molecule which is endowed with the set of chemical properties allowing it to replicate if and when sufficient materials, energy and time are provided, is a genotype and a phenotype at the same time. Such distinction will only be justified in later, more complex specialized structures. The catalysts that were likely to be relevant in the pristine syntheses of nucleic bases (see below) and that were possibly involved in the first polymerization steps, should not be considered as a component of the process endowed with genetic properties, neither genotypic nor phenotypic.

iv. Genetic systems are characterized by evolution. The probability of the existence of an organism as genetically complex as is each one of us, is statistically marginal. Given the very nature of the closed system into which we exist and reproduce, the possibility of the occurrence of another genome identical to mine is not given. Each genome can only be accounted for by the serial occurrence of a number of selection events, each exerted on a large ensemble of possible alternative combinations, and each exerted on the results of previous choices. Each combination descends from stochasticity and selection. Focusing on simple self-replicating systems we adopt a neo-Darwinian approach: selection provides a powerful tool to solve the major problem set by the relatively low *stability* of the precursors.

v. In any physico-chemical scenario dealing with the origin of life, *stability* of the precursor molecules is a major concern. The problem is usually solved by invoking conditions in which the rate of synthesis is higher than that of degradation. The degradation rates of precursor nucleic bases has been analyzed and discussed (Levy and Miller, 1998) and found to be relatively rapid. These observations have lead some to hypothesize that the temperature of the environment in which prebiotic molecules formed and evolved was from moderate to freezing (of water) [as discussed in Levy and Miller (1998); Bada and Lazcano (2002)], thus presumably favoring synthesis over degradation. Stability of the products in prebiotic scenarios is a serious concern. Our working hypothesis is that the problem set by the instability of nucleic bases as precursors is difficult to solve if considered only as a negative attribute of the system. If on the contrary instability could be functionally incorporated into a genetically evolving chemical scheme (see point *iv*), it could play a positive role in the equilibration of the pool of precursors. A balanced composition at equilibrium of the mixture of precursors is necessary for the formation of a genotypically defined (see point *ii*) information-bearing molecule of the type of DNA "as we know it" (see point *i*). Therefore an important consequence of considering an *instability process* to be instrumental for the equilibration of prebiotic pools of precursors is that the *emergence* of productive pools would not have had to wait for the random accumulation of all the four bases (adenine, guanine, cytosine, thymine) in the equilibrated mixture. It would on the contrary be a self-sustaining and catalyst-controlled event, making polymerization a much more likely possibility. In the case that the required catalysts were not rare trace components but were among the most commonly occurring minerals, the chances of an *emergence* would be even more robust.

3. ASPECTS OF THE CHEMISTRY OF FORMAMIDE IN RELATION TO PREBIOTIC SYNTHESES

Our focus on formamide as a relevant compound in prebiotic matters depends on the fact that it can react either at the carbonyl group or at the amide nitrogen to give a large number of nitrogen-containing heterocyclic compounds. Examples include the general synthesis of relevant prebiotic intermediates in the synthesis of nucleic bases, such as imidazoles (Bredereck et al., 1959), oxazoles, and pyrimidine (Bredereck et al., 1957). In addition, the physico-chemical properties of formamide described below are largely compatible with its possible prebiotic role on early Earth.

Herein we report some results on the use of formamide in the synthesis of several purine and pyrimidine derivatives and in the selective degradation of DNA as prebiotic processes.

Formamide is a clear liquid, soluble in water, alcohols, and glycols, and insoluble in hydrocarbons, chlorinated solvents, and ethers. Formamide is a good solvent for proteins and nucleic acids owing to its high dielectric constant. The boiling point of formamide is at 210 °C and its freezing point is at 2.55 °C. Azeotropic effects for a 5-95% formamide solution are limited (ECT Kirk-Othmer, 1978).

The lowest temperature at which formamide undergoes appreciable decomposition is 180-190°C (Kirkpatrick, 1944; ECT Kirk-Othmer, 1978). Formamide decomposes at 210°C to NH_3, CO, HCN, and H_2O at a value of about 0.5%/min. HCN may be obtained in high yields by heating formamide at above 350 °C in the presence of inorganic catalysts. Formamide is hydrolyzed slowly by water at room temperature to give formic acid + ammonia. Recently, the revisitation of the hydrolysis of formamide in water by studying of the solvent deuterium kinetic isotope effect was performed, giving a value of the constant k_{hyd} of $1.1x10^{-10}$ s^{-1} that corresponds at $t_{1/2}$ = 200 yr at 25 °C, pH 7.0 (Slebocka-Tilk et al., 2002). Formamide is formed by hydrolysis of HCN, one of the most studied prebiotic precursors for the synthesis of amino acid and nucleic bases. Calculation of the steady-state concentration of HCN in the primitive ocean was performed assuming the hydrolysis to formamide (and further to formic acid) as the only pathway for the loss of HCN (Miller, 1987). Results reported in the literature for several pH values and temperatures showed that low temperature and pH favour relatively high concentration of HCN, but even at pH 7 and 0 °C, low concentrations of HCN are expected. At high temperature (100 °C) and at high pH value (pH 8) the steady state concentrations of HCN and formamide were calculated to be $7x10^{-13}$ M and 1x 10^{-15} M, respectively (Miyakawa et al., 2002). Thus, concentration mechanisms are required for the synthesis of adenine from HCN (Sanchez et al., 1967). Pioneering studies have shown that purines can be synthesized from neat formamide when treated at high temperature (Brederek et al., 1959; Yamada and Okamoto, 1972; Yamada et al., 1978a). Evidence was obtained for the incorporation of both HCN and formamide molecules in the adenine ring (Yamada et al., 1978b).

4. THE BOTTOM-UP APPROACH

We have focused on HCN chemistry and obtained further evidence of its relevance as a main route for the prebiotic synthesis of purines and, more interestingly, of pyrimidines. As previously reported, in addition to the condensation into nucleobases, HCN hydrolyses to formamide (then to formic acid), a possible additional down-the-line candidate for the synthesis of nucleobases. Our approach consists of the following:

The standard Linear Model (the bottom-up approach) entails the study of the possible prebiotic precursors and their function in the syntheses of purine or pyrimidine bases, their transformation into nucleosides, then into nucleotides, and then eventually into oligomers and polymers, perhaps of DNA. Major obstacles have been encountered: low yields of pyrimidine bases, low reactivity and selectivity in the formation of the glycosidic bond with sugars to yield nucleosides, few efficient prebiotic reactions to phosphorylate the nucleosides, and rapid rates of degradation of the bases before any chance of their incorporation into a polymer.

We have asked the question: do compounds exist which react with one or more components of the double strand and which can thus be hinted to as chemically plausible candidates for synthetic pathways of prebiotic interest?

4.1 Constructing DNA. Syntheses of Nucleic Bases and Nucleosides by Formamide

Pioneering studies on the synthesis of purines from formamide (Brederek et al., 1959; Yamada and Okamoto, 1972; Yamada et al., 1978a) led us to consider the possibility to perform the reaction under catalytic conditions as a more plausible and efficient prebiotic route. Silica, alumina, and more complex oxides such as perovskites, spinels, clays and zeolites, were presumably present on the early Earth. These inorganic oxides have already shown to be of value in a large range of prebiotic catalytic transformations due to acidic or basic properties, to cation-exchange capacity, and to their ability to accommodate copious quantities of water or other polar molecules (Thomas and Thomas, 1997; Strasak et al., 1991; Ristori et al., 1991). To investigate a possible role of metallic oxides on the prebiotic synthesis of nucleic bases from formamide we reacted neat formamide in the presence of $CaCO_3$, silica, alumina, kaolin, zeolite (Y type) as representative prebiotic models of heterogeneous catalysts.

The results of the condensation of formamide **1** in the presence of metallic oxides (Saladino et al., 2001) reveal that:

- the main reaction products are purine **2**, adenine **3**, cytosine **4**, and 4(3H)-pyrimidinone **5** (Figure 16.2).
- The yield of the various products markedly changes depending on the catalyst used. In the presence of a general base catalysis with $CaCO_3$, the simple purine ring was obtained as the only recovered product in a yield that is very high for a prebiotic synthesis (Saladino et al., 2001). HCN, necessary for the synthesis of more complex purine derivatives (e.g., adenine), is not formed from formamide under highly basic conditions. It is noteworthy that when the reaction was performed with an acidic catalyst such as SiO_2 (and to a lesser extent Al_2O_3), a different regioselectivity was observed and pyrimidine derivatives (i.e., compounds **4** and **5**) were recovered in appreciable amounts, in addition to adenine and purine. Using zeolite (which is characterized by a microporous structure providing a favourable microenvironment and a dehydrating system) the yield of purine was second only to that produced by $CaCO_3$. Regardless of the specific reaction conditions, these results show that formamide is a possible prebiotic substrate for the synthesis of purine and pyrimidine derivatives. They also suggest that inorganic catalysts could have played a major role in the distribution of reaction products.

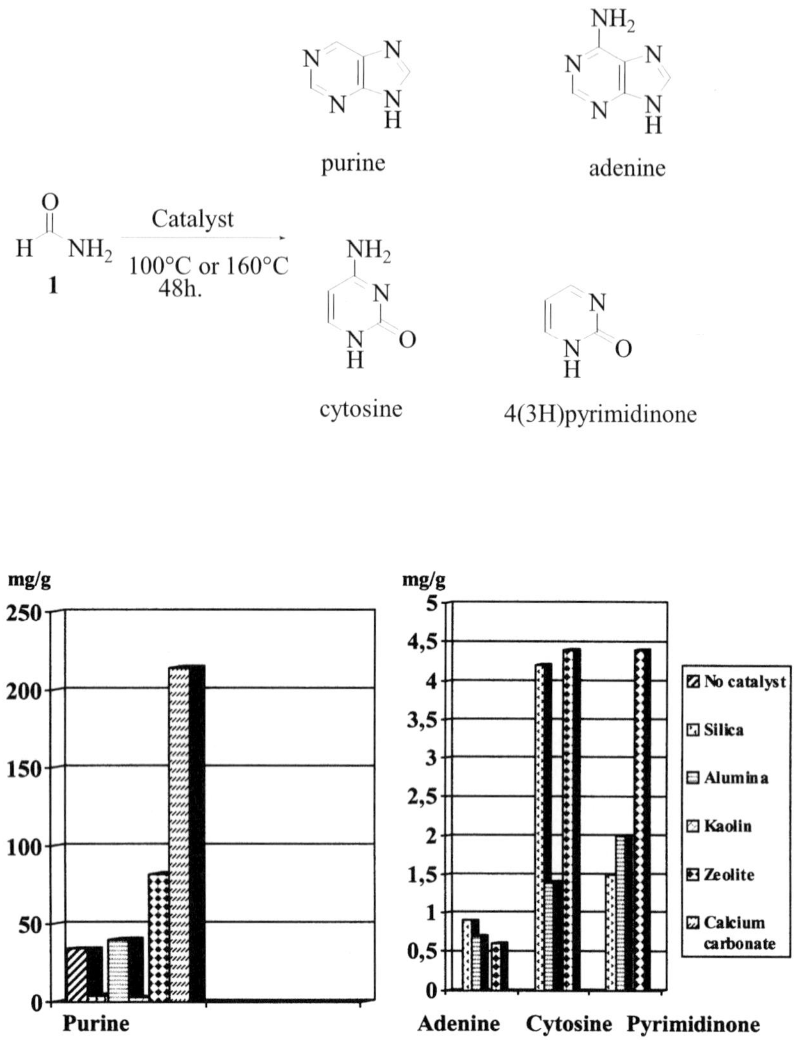

Figure 16.2. Ring formation of purines and pyrimidines from formamide in the presence of catalysts. Panel A: Only purine was formed at 100°C. All other compounds were formed at 160°C. Panel B: ordinate = mg of product for g of formamide.

4.2 The Distinct Effect of TiO_2

In the presence of TiO_2 , the condensation of formamide resulted in the formation of nucleic acid components not previously observed: thymine, 5-hydroxymethyluracil, and N^9-formylpurine, N^6, N^9-diformyladenine acyclo-nucleosides (Saladino et al., 2003) (Figure 3). This latter class of compounds is the step towards next level of complexity, potentially leading to informational macromolecules. Formaldehyde, formic acid and traces of uracil were also recovered from the reaction mixture, as observed by gas chromatography-mass spectroscopy. TiO_2-photocatalysed degradation of amides has been reported and can be accompanied by the release of formaldehyde useful for the formose condensation (Friesen et al., 1999 and Refs therein).

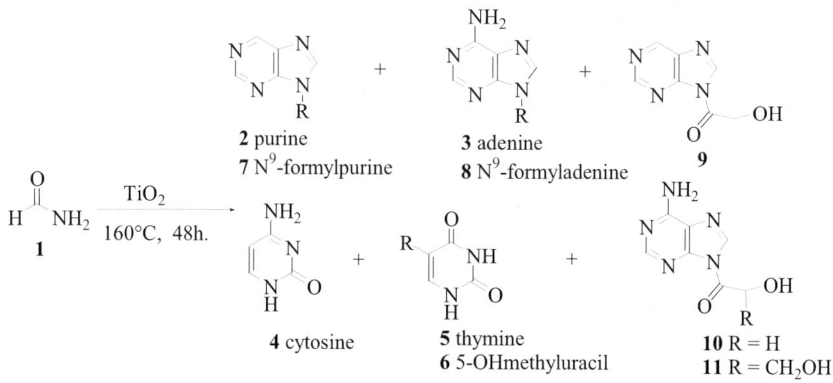

2, 3: R= H. **5**: R= CH_3. **6**: R= CH_2OH. **7,8**: R= CHO. **9, 10, 11**: purine acyclonucleosides

Figure 16.3. Synthesis of purines and pyrimidines bases, N9-formylpurines and acyclonucleosides from formamide in the presence of TiO_2.

N^9-formylpurine, and N^6, N^9-diformyladenine acyclonucleosides may form from N^9-formylpurine or N^6, N^9-diformyladenine by formose condensation (as detailed in Saladino et al., 2003). In this reaction formaldehyde is converted into a mixture of monosaccharides by an aldol-like reaction [Butlerow, 1861; Decker et al., 1982 and Refs therein]. Alumina, and naturally occurring aluminosilicates, kaolinite, illite, and hydroxyapatite may catalyse this transformation and several sugars, including ribose and 2'-deoxyribose, may be formed by the action of ultraviolet light on a dilute solution of formaldehyde (Ponnamperuma, 1965;

Reid and Orgel, 1967). TiO_2 may catalyse the formose condensation by stabilizing the anion resulting from the ionisation of formaldehyde hydrate, rendering the N^9-formyl moiety much more subject to nucleophilic attack (Schengchen and Falconer, 1999). Examples of carbon-carbon bond formation *via* aldolization of aldehydes on TiO_2 have been described (Rekoske and Barteau, 1999 and Refs therein).

The relevance of the condensation of formamide into acyclonucleosides resides in the fact that it was long assumed that nucleotides are produced by the condensation of a sugar with preformed purine and pyrimidine bases. However, major problems arose in demonstrating this condensation in prebiotic experimental conditions (Orgel, 1998). Apart from the synthesis of cytosine arabinonucleoside, no prebiotic condensation of uracil or cytosine with ribose is known (Sanchez and Orgel, 1970).

Thus, the possibility to build stepwise the sugar moiety starting from formylpurine derivatives, in which a masked glycosidic bond is already formed, is a potential alternative route to nucleosides.

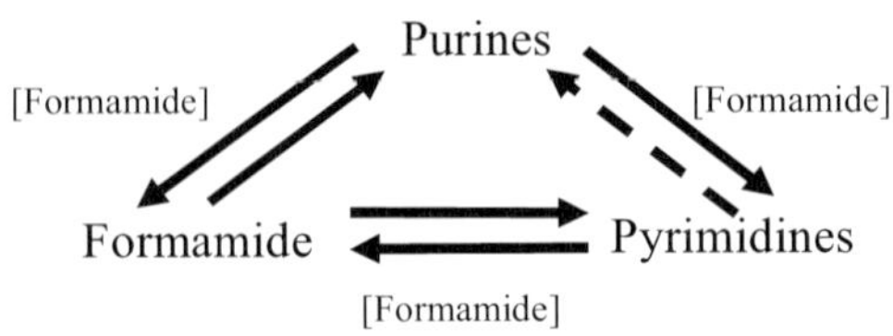

Figure 16.4. In the presence of formamide, additional formamide is recovered from the degradation of purine and pyrimidine bases (solid outer arrows). Synthetic reactions also occur (inner arrows) leading from formamide to purines and pyrimidines. The arrows do not refer to specific chemical equilibria, but merely indicate the possibility that one compound is transformed into another.

A pre-RNA world in which side-chain different from ribose might be embedded into informational molecules cannot be completely ruled out, even if the structure of such substituents is actually unknown. Acyclonucleosides are a well-known class of nucleoside analogues characterized by relevant biological activities, mainly exerted upon incorporation in DNA or RNA molecules (Saladino et al., 2000). Acyclonucleosides often retain the ability to pair with natural purine and pyrimidine bases, even if in some cases the possibility to elongate the oligonucleotide polymeric chain may be lost (Saladino et al., 2001). A number of acyclonucleosides have been proposed as open-chain, flexible, and prochiral prebiotic alternatives to natural sugars (Joyce et al., 1987). Among them, adenine and guanine-N^9-acetic acids, and cytosine and uracil-N1-acetic acids have been recently reported as possible precursors for the synthesis of peptide nucleic acid derivatives as a possible pre-RNA world alternative (Nelson et al., 2000).

5. DEGRADING DNA BY FORMAMIDE (THE TOP-DOWN APPROACH)

We have experimentally identified formamide as a compound that at temperatures around 100°C selectively degrades the DNA bases in the following order of decreasing reactivity: G>A>C>T (Saladino et al., 1996, 1997). When the reaction is performed on polynucleotides, the degradation of nucleic bases is accompanied by scission of the glycosidic linkages through β-elimination (Di Mauro et al., 1994; Negri et al., 1996). The overall reaction of formamide with DNA follows the mechanism: (i) degradation of the purine and pyrimidine bases and (ii) β-elimination of the acidic sugar protons with subsequent β-elimination of the phosphate residue. The 3' β-elimination occurs first, causing an increase in acidity of the reactive proton Hγ in the α,β unsaturated sugar residue, thus allowing 5' residue elimination. The chemical mechanisms leading to the degradation were determined for both purine (Saladino et al., 1996) and pyrimidine (Saladino et al., 1997) nucleosides and unambiguously proven by isolation and spectral characterization (^{1}H-NMR, ^{13}C-NMR, mass spectroscopy) of all main intermediates involved in the transformation (*ibidem*). For purines, degradation occurs by nucleophilic attack in C(8) position of the heterocyclic ring, leading to ring opening of the imidazole ring and formation of formyl-pyridine nucleoside derivatives; for pyrimidines, degradation occurs by nucleophilic attack in C(6) and C(4) positions of the heterocyclic ring, leading to different ring opening products, such as formamide, ammonia, β–dicarbonile derivatives, and urea. Alterations of the chemical structure of the natural nucleic bases [i.e., C(7) *deaza* modifications for purine residues and C(5) substitution with an electron-donating or electron-withdrawing group for pyrimidines] markedly change the sensitivity of the nucleic bases to formamide attack (Saladino et al., 1996). The degradation and removal of the heterocyclic purine or pyrimidine bases by hydrolysis of the β-nucleosidic bond induce in the 2'-deoxy ribose moiety an equilibrium between the hemiacetalic and the aldehydic form. The two reactive protons (Hα and Hγ, indicated in Figure 16.5) present in the aldehydic form are available for reaction and are eliminated.

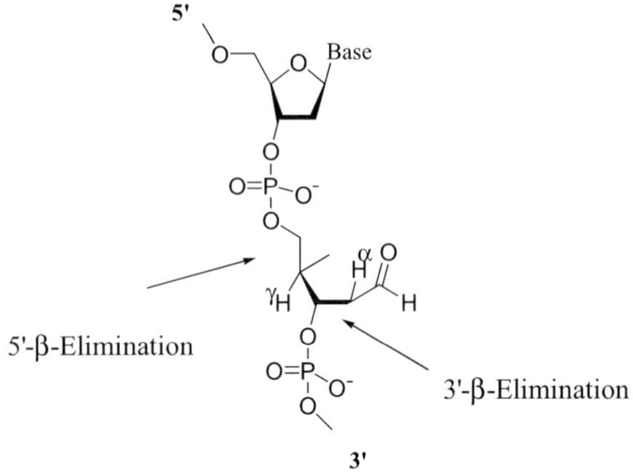

Figure 16.5. Cleavage of the 3' and 5' phosphodiester bonds by amides. Schematic representation of the degradation of the sugar moiety indicating the two different β-eliminations (Negri et al., 1996).

This leads to cleavage of 3' and 5' phosphodiester bonds, respectively. In the case of formamide (a weakly basic compound), cleavage of the 5' bond occurs less efficiently than cleavage of the 3' bond because of the lower acidity of proton Hγ. If more strongly basic compounds (i.e., *N*-methylformamide or *N/N*-di-methylformamide) are tested, the base selective cleavage is maintained and the increased efficiency of β-elimination at the 5'-bond is observed (Negri *et al.*, 1996; Costanzo et al., 1997).

Figure 16.6 reports three degradative pathways and shows that in every instance formamide adds to the nucleic base then leaves, causing its degradation. Specifically, we describe (Panels A and B) the reaction of formamide with 2'-deoxyadenosine and 2'-deoxyguanosine (Saladino et al., 1996), and (panel C) with 2'-deoxycytidine (Saladino et al., 1997).

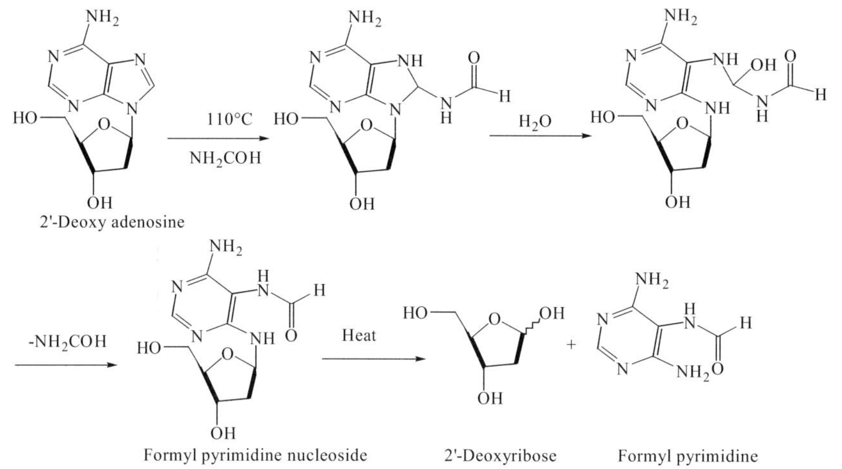

panel A

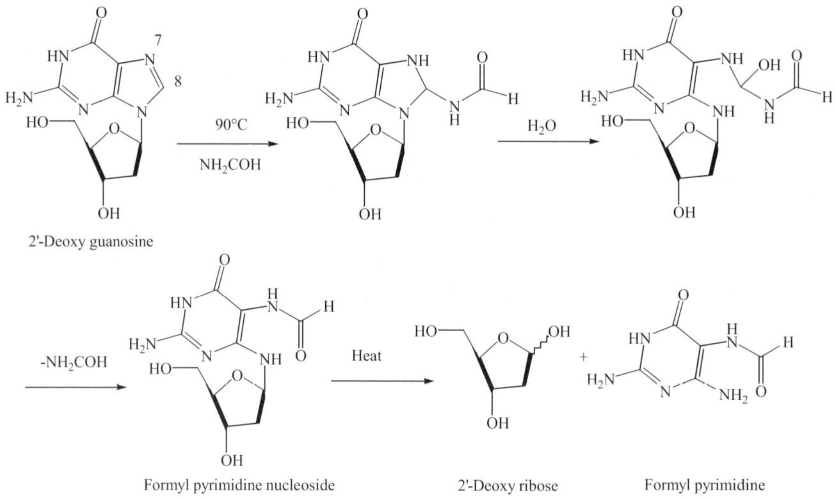

panel B

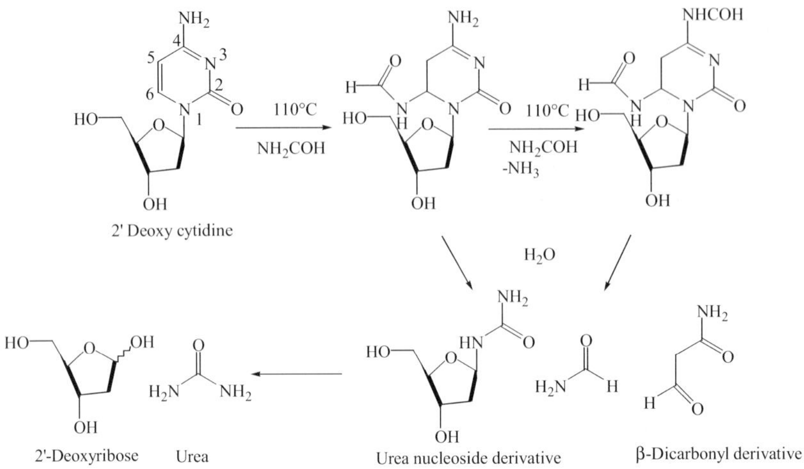

panel C

Figure 16.6. Degradation of 2'-deoxyadenosine (Panel A), 2'-deoxyguanosine (Panel B) and 2'-deoxycytidine (Panel C) by nucleophilic addition of formamide (from Saladino et al., 1996; 1997, respectively).

In conclusion, we observe that formamide:

- Selectively degraded DNA polynucleotides, as well as all of the major components of DNA: including the bases, the β-glycosidic bond, the 2'-deoxyribose moiety, and the phosphodiester bonds.
- Formamide is newly formed after its reaction with the nucleic bases, and pyrimidine derivatives can be obtained by degradation of purine nucleosides (Figure 16.4). Is formamide also a valuable product for the reconstruction of DNA? In Figure 4 the outer arrows refer to the process of degradation of DNA by formamide, while the inner arrows refer to recently reported constructive reactions. Evidence for the occurrence of these latter reactions was obtained, as described in the previous sections.

6. EQUILIBRATING THE POOL

Based on the results reported above, on the chemical attributes of formamide and on physico-chemical considerations described in (Saladino et al., 2001), the most favorable set of conditions using this precursor for the synthesis of nucleic bases would be high formamide concentration, anhydrous conditions, the presence of a catalytic system, a local stable microenvironment, and a temperature higher than 100°C and lower than 180°C. We have described that in these conditions a full set of nucleic bases forms in variable yields and compositions, depending on the specific catalyst used (Saladino et al., 2001; 2003). Such dependence of the prebiotic events on the environment is clearly exemplified in the instance reported in Figure 2 and was observed for every catalyst tested (Saladino et al., 2001; 2003), the TiO_2 system included. The resulting salient feature is a pool of potential DNA precursors.

Imbalance of the precursors is a problem as serious as their instability. The two properties are connected. This is a consequence of the fact that DNA in solution is chemically stabilized by its double-stranded structure and that a double strand requires for optimal pairing equal amounts of purines and pyrimidines. The possibility of non-canonical Watson-Crick base-pairing [i.e., Hoogsteen and reverse-Hoogsteen base pairing; (Pulleyblank et al., 1985; Lyamichev et al., 1986; Wells, 1988)] increases the number of possible strand pairings, but does not change the basic need for an equilibrated starting pool of precursors. By *equilibration* we refer to the attainment of a pool of precursors whose composition is compatible with the possibility of giving rise to base-sequence complementary polymers, as typically occurring in DNA.

Pool equilibration is greatly favored by the existence of a cycle of the type described in Figure 16.4. Formamide condenses into a base; then, in the continued presence of formamide, the base is degraded to release the precursor back into the medium. This process prevents the precursor formamide from being locked into the base that happens to be produced more efficiently and that, for the very reason of its relative abundance, would remain unutilized in base pairings. Thus, cycling may provide a solution to the instability problem caused by the very property of formamide of being at the same time an operator of the construction and of the degradation of nucleic bases. We have reported that relative abundances change depending on the catalyst present (Figure 16.2). It is appropriate to note that we are not dealing here with polymers and that we are considering only various pools of mixed nucleic bases. Polymerization remains a major challenge to the scenario, as the eventual way out from these otherwise futile interconversion cycles.

Catalysts could also alleviate the imbalance of the resulting products. Thus, focusing on TiO_2, we have analyzed the effects on the degradation of nucleosides and of polynucleotides under the same experimental conditions that it serves as a catalyst for formamide condensation into purines, pyrimidines and acyclonucleosides (Saladino et al., 2003). The differential effects of TiO_2 on the degradation of polynucleotides were studied for homogeneous or mixed-sequence polymers. Starting from the highly differential sensitivity of the DNA-bound bases (G>A>C>T, as in Negri et al., 1996) in its absence, the presence of a threshold concentration (3 mg/ml) of TiO_2 brought the sensitivity of the four bases to similar levels. Together with the observed effects on the degradation of nucleosides (Saladino et al., 2003), these results reveal a role of this specific catalyst in equilibrating the pool.

Figure 16.7 describes the modification of the sensitivity of the four bases to formamide: decreased degradation of DNA-embedded guanine (G) and adenine (A) as a function of the increasing concentration of TiO_2 , and the opposite effect on thymine (T) and cytosine (C). The plot shows the relative sensitivity (ordinate) of the four bases as a function of the increasing concentration of TiO_2 (abscissa). The equilibration effect is evident (from Saladino et al., 2003).

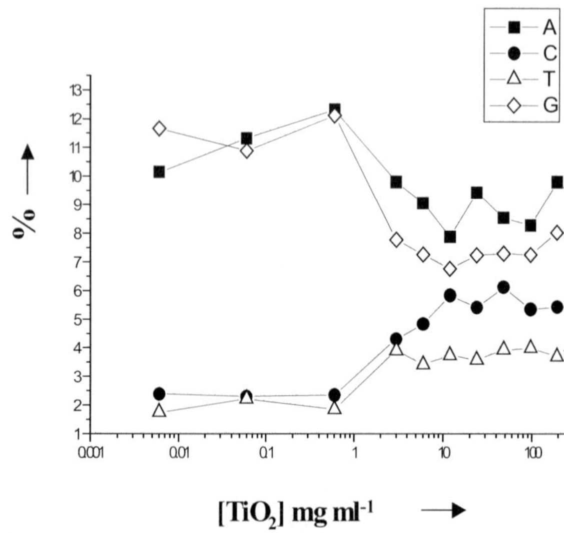

Figure 16.7. The relative sensitivity (ordinate) of the four bases as a function of increasing concentration of TiO_2 (abscissa). The values reported in ordinate represent the average sensitivity (%) to cleavage of the each indicated base when embedded in DNA and cleaved by formamide in a condition of "less than one cleavage" per DNA molecule. Experimental details in Saladino et al., 2003.

Again, one should remark that we are not dealing with polymerization reactions and that we have described, in the bottom-up approach:

- the preferential synthesis from formamide of purine, adenine, guanine, cytosine, thymine, 4(3H)-pyrimidinone and 5-OH-Me-uracil.
- the unexpectedly high difference in yields exerted by prebiotic compounds such as silica, $CaCo_3$, alumina and TiO_2.

In the top-down approach:

- the preferential degradation by formamide of guanine, adenine, cytosine and thymine, in this order. Degradation in the same order of preference also occurs when the compounds are embedded in DNA as when they are in the nucleoside form.
- Degradation is selectively affected by metal oxides.
- The equilibration role exerted by TiO_2 on the pool of bases.

No matter how effective the pool equilibration process ruled by the cycles described above, these compounds would be bound, in the absence of a way-out towards more stable forms, to repeated degradations and to an idle cycling. Our working hypothesis is that such a way-out can only be provided by polymerization, with its induced self-stabilization effect. A first indication that the formamide condensation offers a plausible solution is provided by the observed presence in the same pools of acyclonuclesides (Figure 16.8), compounds which are chemically close to nucleosides.

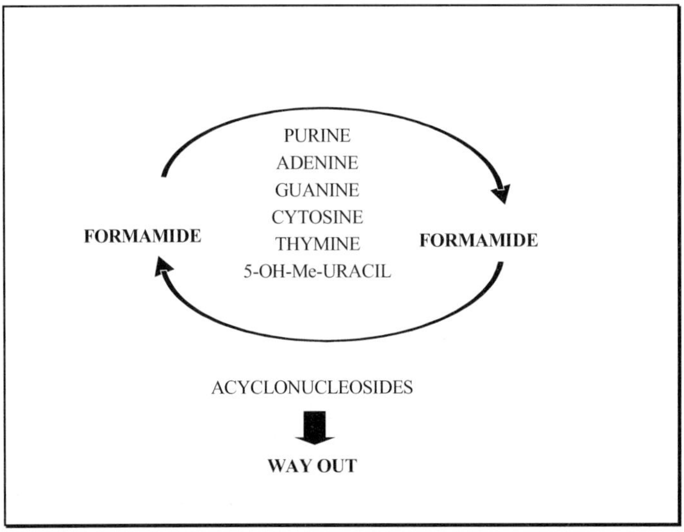

Figure 16.8. Acyclonucleosides are a plausible step towards next level of chemical complexity.

7. UR-GENOTYPES AND PHENOTYPES

An initial polymer does not need to be very complex and genetically meaningful to win the game of evolution. It only has to be better that the other players. The important quality involved is the capability of a base to recycle its sub-components in order to facilitate the production of other less abundant bases, thus increasing its own chances of base-pairing and stabilization.

We have described that compounds like TiO_2 are quite efficient in stimulating an effective equilibration of the pool of precursors. Once polymerization of nucleic bases has occurred (following a mechanism for which we have no hints, yet), positive selection of information-bearing molecules relative to their fellow competitors can take place based on the actual composition of the pool and depending on the catalysts present. This positive selection is exerted not on simple molecules (as single bases) but on the products of (short) polymerizations, and is a process that may bring as an extra bonus increased stability. It is at this moment that the precursor bases escape from the futile cycle of informationless interconversions and become part of a molecule with genetic potential.

Thus, an initial polymerization creates out of a pool of precursors both an initial genotype (the information that will replicate itself) and, at the same time, a phenotype consisting of physico-chemical properties of the molecule and of its local fitness. At this stage evolution may start, based on chemical selection. We would like to point out that:

- the description of the syntheses taking place from reactive prebiotic materials of the chemical system $HCN/HCONH_2/HCOOH$, leading to all of the biologically relevant nucleic bases,
- the fact that this chemistry offers a possible solution, through the synthesis of acyclonucleosides, to the long-standing problem of the poor reaction between bases and sugars,
- the activation and variegation of the syntheses of genetic precursors by abundant prebiotic catalysts,
- the possibility to equilibrate the pool of the candidate precursors that is offered by the chemistry of the system, favoring in perspective the nucleic acids strand-pairing,

are obviously far from offering a satisfactory scenario for the complete series of steps necessary to bridge a small compound to full-fledged genetic material. Among the points remaining completely open:

- which were the steady-state concentrations of $HCN/HCONH_2/HCOOH$ in water solutions? how were the K_{eq}s of the various reactions involved affected by these concentrations and by the varying physico-chemical conditions?

- what was the range of temperatures allowing both the polymerization and the survival of the polymers?
- where did the phosphate of the DNA backbone come from?

The very fact that the energy conditions required for the synthesis of precursors may well have been different from those of their survival and of their successive interactions keeps open a wide array of possibilities.

To conclude, our preferred prebiotic scenario is an internally interconverting pool of compounds made from HCN (which hydrolyzes to formamide), contained in a basin of TiO_2, alumina, silica, or $CaCO_3$, or of all of them; and laying on the slopes of an active volcano as a heat source. The scenario is still largely hypothetical.

ACKNOWLEDGEMENTS

This work was supported by ASI contract CNR/ASI I/R/315/02, HFSP grant RGP0207/2001, CNR Genomica Funzionale, COFIN n. 2001057198.001, FIRB n. RBNE01KMT9.001, BEMM "La Sapienza".

REFERENCES

Bada, J.L. and Lazcano, A.: 2002, Origin of life. Some like it hot, but not the first biomolecules, *Science* **296,** 1982-1983.

Bredereck, H., Gompper, R., and Klemm, K.: 1959, Acid amide reaction. XVI. Self-condensation of N,N-disubstituted acid amides, *Chemische Berichte* **92** , 1456-1460.

Bredereck, H., Gompper, R., Morlok, G.: Formamide reactions. VIII. A new pyrimidine synthesis, *Chemische Berichte* 1957, **90**, 942

Butlerow, A.: 1861, Formation of Monosaccharides from Formaldehyde, *Compt. Rend.* **53**, 145-167.

Buzayan, J.M., Gerlach W.L., and Bruening, G.: 1986, Non enzymatic cleavage and ligation of RNAs Complementary to a plant virus satellite RNA, *Nature* **323**, 349-353.

Cech, T.R., Bass, B.L.: 1986, Biological catalysis by RNA. *Annu. Rev. Biochem.* **55**, 599-629.

Cech, T.R.: 1985, Self-splicing RNA: implications for evolution, *Int. Rev. Cytol.*, **93**, 3-22.

Costanzo, G., Negri, R., Di Nicola-Negri. E., De Leo, R., and Di Mauro, E.: 1997, Detection of human genomic mutations by chemical single-reaction DNA sequencing, *Trends in Genetics* T40030.

Decker, H., Schweer, H. and Pohlmann R.: 1982, Identification of formose sugars, presumable prebiotic metabolites, using capillary gas chromatography/gas chromatography mass spectrometry of n-butoxime trifluoroacetates on OV-225, *J. Chromat.* **244,** 281-291.

Di Mauro, E., Costanzo, G., and Negri, R.: 1994, One-lane chemical sequencing of PCR amplified DNA: the use of terminal transferase and of the base analogue inosine, *Nucleic Acids Res.* **22,** 3811-3812.

Ehrenfreund, P., Irvine, W., Becker, L., Blank, J. Brucato, J., Colangeli, L., Derenne, S., Despois, D., Dutrey, A., Fraaije, H., Lazcano, A., Owen, T. and Robert F.: 2002,

Astrophysical and astrochemical insights into the origin of life, *Reports on Progress in Physics* **65**, 1427-1487.

ECT Kirk-Othmer Encyclopedia of Chemical Technology, Formic acid and derivatives (Formamide), Wiley Interscience, Ed.: 1978, **11**, 258-262.

Friesen, D.A., Haedely, J.V., Langford, C.H.,: 1999, The Photooxidative Degradation of *N*-Methylpyrrolidinone in the Presence of $Cs_3PW_{12}O_{40}$ and TiO_2 Colloid Photocatalysts, *Environ. Sci. Technol.* **33,** 3193-3198

Fox, S.W., Dose K.: 1972, "Molecular Evolution and Origin Of the Life", W.H. Freeman Ed., USA.

Kirkpatrick, E.C.: 1944 U.S. Patent (to E.I. Dupont de Nemours & Co. Inc), **2**, 346-425.

Kochetkov, N.K., and Budovskii, E.I.: 1972, *Organic Chemistry of the Nucleic Acids*, Plenum Press (London, New York).

Joyce, G.F., Schwartz, A.W., Miller, S.L., and Orgel L.E.: 1987, The case of an ancestral genetic system involving simple analogues of the nucleotides, *Proc. Natl. Acad. Sci. USA*, **84**, 4398-4402.

Joyce, G.F.: 2002, The antiquity of RNA-based evolution, *Nature*, **418,** 214-221.

Levy, M. and Miller, S.L.: 1998, The stability of the RNA bases: implications for the origin of life, *Proc. Natl. Acad. Sci. U S A.* **95,** 7933-7938.

Lyamichev, V.I., Mirkin, S.M. and Frank-Kamenetskii, M.D.: 1986, Structures of homopurine-homopyrimidine tract in superhelical DNA, *J. Biomol. Struct. Dyn.* **3**, 667-669.

Miller, S.L.: 1987, Which Organic Compounds Could Have Occurred on the Prebiotic Earth? *CSH Symposia on Quantitative Biology*, **LII**, 24-25.

Miyakawa, S., Cleaves, H.J., and Miller, S.L.: 2002, The cold origin of life: B. Implications based on pyrimidines and purines produced from frozen ammonium cyanide solutions, *Orig. Life Evol. Biosph.*, **32**, 195-208.

Negri, R., Costanzo, G., Saladino, R. and Di Mauro, E.: 1996, One-step, one-lane chemical DNA sequencing by N-methylformamide in the presence of metal ions, *BioTechniques* **21**, 910-917.

Nelson, K.E., Levy, M., and Miller, S.M.: 2000, Peptide nucleic acids rather than RNA may have been the first genetic molecule, *Proc. Natl. Acad. Sci. USA* **97,** 3868-3871.

Nissen, P., Hansen, J., Ban., N., Moore, P.B. and Steitz, T.A.: 2000, The structural basis of ribosome activity in peptide bond synthesis, *Science* **289**, 920-930.

Orgel, L.E.: 1998, The origin of life, a review of facts and speculations, *TIBS* **23,** 492-495.

Pace, N.R. and Marsh, T.L.: 1985, RNA catalysis and the origin of life, *Orig. Life Evol. Biosphere* **16,** 97-116.

Ponnamperuma, C.: 1965, in The origins of prebiological systems and of their molecular matrices, (Ed.: S.W. Fox), Academic Press, New York, 221-236.

Pulleyblank, D.A., Haniford, D.B. and Morgan A.R.: 1985, A structural basis for S1 nuclease sensitivity of double-stranded DNA, *Cell* **42**, 271-280.

Reid, C. and Orgel, L.E.: 1967, Synthesis in sugars in potentially prebiotic conditions, *Nature* **216**, 455-457.

Rekoske. J. E. and Barteau, M. A.: 1999, Competition between acetaldehyde and crotonaldehyde during adsorption and reaction on Anatase and Rutile titanium dioxide, *Langmuir* **15**, 2061-2070.

Ristori, G. G., De Nobili, M., and Fusi, P. 1991: Non-biological degradation of oxamide adsorbed on homoionic montmorillonite, *Applied Clay Science* **6**, 171-179.

Saladino, R., Ciambecchini, U., Crestini, C., Costanzo, G., Negri, R. and Di Mauro, E.: 2003, One-pot TiO_2-Catalyzed Synthesis of Nucleic Bases and Acyclonucleosides from Formamide: Implications for the Origin of Life, *ChemBioChem* **4**, 514-521.

Saladino, R., Crestini, C., Costanzo, G., Negri, R. and Di Mauro, E.: 2001, A possible prebiotic synthesis of purine, adenine, cytosine, and 4(3H)-pyrimidone from formamide: implications for the origin of life, *Bioorg. Med. Chem.* **9**, 1249-1253.

Saladino, R., Crestini, C., Palamara, A.T., Danti, M.C., Corelli, F., and Botta, M.: 2001, Synthesis, biological evaluation, and pharmacophore generation of uracil, 4(3H)-pyrimidinone, and Uridine derivatives as potent and selective inhibitors of parainfluenza 1 (sendai) virus, *J. Med. Chem.* **44**, 4554-4562.

Saladino, R.: 2000, Recent advances in the oxidative functionalization of purine and pyrimidine nucleic acids components, *Targets in Heterocyclic Systems, Chemistry and Properties* (Eds.: P.A. Attanasi, D. Spinelli) **3**, 4-36.

Saladino, R., Crestini, C., Mincione, E., Costanzo, G., Di Mauro, E., and Negri, R.: 1997, Mechanism of degradation of 2'-deoxycytidine by formamide: implications for chemical DNA sequencing procedures, *Bioorg. Med. Chem.* **11,** 2041-2048.

Saladino, R., Mincione, E., Crestini, C., Negri. R., Di Mauro, E. and Costanzo, G.: 1996, Mechanism of degradation of purine nucleosides by formamide. Implications for chemical DNA sequencing procedures, *J. Am. Chem. Soc.* **118**, 5615- 5619.

Sanchez, R. A., Ferris, J.P., and Orgel, L.E.: 1967, Studies in prebiotic synthesis. II. Synthesis of purine precursors and amino acids from aqueous hydrogen cyanide, *J. Mol. Biol.* **30**, 223-253.

Sanchez, R.A. and Orgel, L.E.: 1970, Studies in prebiotic synthesis. V. Synthesis and photoanomerization of pyrimidine nucleosides, *J. Mol. Biol.* **47**, 531-543.

Santoro S. W., Joyce G.F., Sakthivel, K., Gramatikova, S. and Barbas, C. F. III: 2000, RNA Cleavage by a DNA Enzyme with Extended Chemical Functionality *J. Am. Chem. Soc.,* **122**, 2433 -2439

Sheppard, T.L., Ordoukhanian, P. and Joyce, G. F.: 2000, A DNA enzyme with *N*-glycosylase activity, *Proc. Natl. Acad. Sci.* **97**, 7802-7807

Schengchen, L. and Falconer, J.L.: 1999, Acetone and acetaldehyde oligomerization on TiO_2 surfaces, *Catalysis Lett.* **57**, 89-93.

Slebocka-Tilk, H., Sauriol, F., Monette, M., and Brown, R.S.,: 2002, Aspects of the hydrolysis of formamide: revisitation of the water reaction and determination of the solvent deuterium kinetic isotope effect in base, *Canadian J. Chem.* **80**, 1343-1350.

Strasak, M., Sitek, J., Sersen, F., Janik, R., and Slosiarikova, H.: 1991, Prebiotic and biomimetic reactions on montmorillonites**,** *Conf. Coord. Chem.* 279-284.

Thomas, J.M. and Thomas, W.J.: 1997, In Principles and Practice of Heterogeneous Catalysis, *VCH: Weinheim* 339-365 and 626-627.

Watson, J.D. and Crick, F.H.C.: 1953, A structure for Deoxyribose Nucleic Acid, *Nature* **171**, 737-738.

Wells, R.D.: 1988, Unusual DNA structures, *J. Biol. Chem.* **263,**1095-1098.

Yamada, H. and Okamoto, T.: 1972, Polycyclic N-hetero compounds XIV. Reactions of methylpyridines with formamide, *Chem. Pharm. Bull.* **20**, 623.

Yamada, H., Hirobe, M., Higashiyama, K., Takahashi, H. and Suzuki, K.: 1978a, Reaction mechanism for purine ring formation as studied by 13C-15N coupling, *Tetrahedron Lett.* **42**, 4039.

Yamada, H., Hirobe, M., Higashiyama, K., Takahashi, H. and Suzuki, K.: 1978b, Thermal fission and reformation of the carbon-nitrogen bond in formamide as studied by carbon-13-nitrogen-15 coupling, *J. Am. Chem. Soc.* 100.

Chapter 17

MARS, EUROPA, AND BEYOND

Understanding the Potential for Life in the Solar System

John D. Rummel
Office of Space Science, NASA Headquarters, Washington, D.C. 20546, USA

Abstract: The two other solar system bodies thought to be most compatible with "life as we know it" are the planet Mars and Europa, a natural satellite of the planet Jupiter. These worlds appear to harbor the potential for past and/or present-day liquid water, biologically useful energy sources, and a significant and rich organic chemistry. Such traits are under active investigation both through ongoing, targeted, solar system exploration missions and from the extensive analysis of data from previous missions and astronomical observations. And both bodies are the subject of astrobiologically inspired future missions. The nature of both Mars and Europa fuels speculation about the prospects for life, and the established facts about each of them, added to more recent observations, can explain their astrobiological interest. Nonetheless, such data can only form a circumstantial case for that interest, and further investigations of water (in all of its forms), energy, and organic chemistry are sure to be required before astrobiological investigations can be further targeted—and data from any biological observations can be properly interpreted. Most important will be a dedication to understanding both Mars and Europa for the environments that they possess—and the nature and distribution of those environments in space and time—rather than trying to understand these worlds by simple analogy to the modern Earth. It is clear that both Mars and Europa have characteristics that may be similar to those of Earth when studied over its entire history, but it is equally true that each of them have characteristics that are unlike anything presented by the Earth system at any single time in its past. The same can be said of Saturn's moon, Titan, which presents a compelling mix of organic chemistry, water ice, and atmosphere—but must represent a significant departure from any historical Earth. This is not necessarily a disadvantage. In fact, through the study of Mars, Europa, and Titan we may come to recognize circumstances on the early Earth that are presently unknown, or at least unappreciated. Such is the promise of astrobiology—not simply a "search for life," but a change in perspective that can shed light on established disciplines, while it allows for the pursuit of answers to some of science's most compelling questions.

Key words: Mars, Europa, Titan, astrobiology, exobiology, life

P. Ehrenfreund et al. (eds.), Astrobiology: Future Perspectives, 415–443.

1. INTRODUCTION

"Now my own suspicion is that the Universe is not only queerer than we suppose, but queerer than we *can* suppose."
—J. B. S. Haldane

When envisioning opportunities for life in this solar system outside of the Earth, one is limited by the lack of perspective gained through the study of Earth-life itself. Because all known life on this planet appears to be related [1]—making use of the same, basic, biochemistry throughout—there is no data available on alternatives which may prove to be feasible. Accordingly, any study of extraterrestrial life is limited by what we know about the myriad, but entirely related, forms of life on Earth—"life as we know it."

That being said, when applying the template provided by Earth life to this solar system, three bodies appear to be the most promising targets for an exploration of extraterrestrial biology—or at least provide locations on which to pursue studies of prebiotic chemistry. On two of these bodies, the planet Mars and Europa, a moon of the planet Jupiter, analogues to life-supporting Earth environments may exist today or may have existed in the past. Both appear to possess the fundamental characteristics of such terrestrial environments—the prospect of at least small quantities of liquid water, multiple sources of biologically useful energy, and the potential to harbor organic compounds and an organic chemistry fueled either by materials formed in space or by compounds produced by their own planetary processes. Farther out, Titan, Saturn's largest satellite, has a thick, nitrogen-methane atmosphere in which photochemistry is continuously producing hydrocarbons, nitriles, and a thick layer of organic smog. For sheer richness of material, Titan may well be the best place in the solar system to study prebiotic organic chemistry—especially, but not only, atmospheric chemistry—and these investigations will be pursued by the Cassini-Huygens mission beginning with Saturn-orbital-insertion in July 2004 and probe entry into Titan's atmosphere in January 2005.

While we don't expect to find life on Titan, because of its low (94°K) surface temperature, we must hold open the possibility of ecological niches in the deep interior, where liquid water could very well be available. Time, and data, will tell. In the interim, Mars and Europa clearly represent the best possibility for finding evidence of past or present extraterrestrial life—despite the fact that, so far, neither liquid water nor organic molecules have been demonstrated to exist on either Mars or Europa. The suggestion that even "life as we know it" could exist on either, or both, seems intriguing, though the actual data lag far behind.

Why then, are these bodies of so much interest to astrobiology?

1.1 Astrobiology is . . .

As possessor of a recently revisited name and as heir to one of the nascent concepts of the space age [2], astrobiology may be considered to contain all of the interdisciplinary perspectives of exobiology, plus extensions into biospheric research and evolutionary biology. Accordingly, astrobiology may be defined as "the study of the origins, evolution, distribution, and future of life in the universe" [3], with three fundamental questions:

- How does life begin and evolve?
- Does life exist elsewhere in the universe? and
- What is the future of life on Earth and beyond?

When attempting to provide a "roadmap" for astrobiological inquiry [3], these three questions are seen as the drivers for a number of specific goals. And all three questions can be related to missions now planned or underway to the planet Mars, but here we will address only the first two questions, as well as two of the astrobiology goals that are derived from them in the roadmap [3], namely:

- Understand how life emerges from cosmic and planetary precursors; and
- Explore for past or present habitable environments, prebiotic chemistry and signs of life elsewhere in our Solar System;

In fact, it may be that the study of other worlds will be critical to our understanding of how life emerged on Earth—the critical perspective from space first identified by Lederberg in 1960 [2]—and how it may emerge elsewhere. Of the planetary scale bodies that have been discovered in our Solar System, Europa and Mars now appear most-likely to provide that perspective.

1.2 An Origin of Life on Other Worlds?

Elsewhere in this volume [4] there is a discussion of some of the concepts presently put forward to explain the origin of life on this planet. For now, of course, there are no competing hypotheses explaining the origin of life on Europa or Mars. Rather than being guided by concepts focused on the unique aspects of nonterrestrial worlds, at this stage in the history of astrobiology, the study of Mars and Europa is more likely to be driven by considerations that are analogous to those found on the Earth, where it can be hypothesized that life was originated *de novo*. Such may or may not be the case—but even the *deus ex machina* explanation of panspermia (however posed, cf., [5-7]) can only push the origin of life into some other realm, to some other planet or moon at some other time. So in setting the stage for the astrobiological interest in Mars and Europa, perhaps it would be instructive to look at the environments and circumstances thought to be important to the origin of life here—or perhaps wherever it may have taken place—recognizing that in the final analysis the actual circumstances for both of those bodies may not correlate to those on Earth.

But what are those important circumstances? Others (e.g., [3]) have suggested that four elements must have converged to admit the origin of life. These include:

- A source of organic compounds;
- The origin and evolution of functional biomolecules;
- The origin of energy transduction; and
- The origins of cellularity.

A number of schemes have been outlined that might explain the convergence of these components on Earth. Perhaps most applicable are those recently discussed by Ehrenfreund et al. [8], whereby organic compounds delivered to the Earth by asteroids, comets, and smaller fragments, such as meteorites and interplanetary dust particles, form the source material for the functional biomolecules, energy transduction components, and cellular systems to follow. These schemes are attractive particularly because they should be applicable, in some degree, to all solar system bodies—and because complex chemistry is clearly possible in dense interstellar clouds and in the forming protostellar disk and planets. Alternatively, the concept of an endogenous formation and buildup of organic molecules, either through natural energy discharge in a planetary atmosphere (e.g., [9-11]) or by processing of material within hydrothermal systems [12-14] has a relatively long history as a generic concept. For example, over 100 years ago Darwin [15] espoused the buildup of such organics "in some warm little pond" prior to the origin of life because they would not be "absorbed or devoured...before living creatures were formed."

Nonetheless, the energetics of the situation make it unlikely that the study of the origin of life will be advanced greatly if it is confined to a search for the next gooey warm pond, even an extraterrestrial one. Instead, leading concepts of organic-buildup-leading-to-the-origin-of-life have invoked a dynamic energetic environment—whether it be freeze-thaw cycles on an ice-covered world [16], or the natural dynamics of temperature and pH that accompany the dissipation of energy in deep-sea hydrothermal systems [17]—to explain the disequilibrium conditions that are thought to have allowed for self-replicating chemical systems to be sustained, and eventually evolved, on Earth. For those interested in the once volcanically active Mars, and the ever-so-frozen Europa, the attraction of those ideas is obvious. Coupled with concepts that allow for the development of significant prebiotic chemistry, while spanning the range of energy sources, pH conditions, and mineral assemblages likely to be encountered in an aqueous medium (e.g., [18-24]), the study of a terrestrial origin of life is very encouraging to astrobiologists who are concentrating on the extraterrestrial possibilities of Europa and Mars.

In any event, the sequence of events that leads from interstellar cloud to forming solar system to planetary chemistry and physics seems to allow for a great deal of participation by organic compounds—and to different degrees—along the way. And different aspects of the origin-of-life puzzle may also be influenced by inorganic compounds, and those may become important anywhere along the way. For example, Ricardo et al. [24] recently reported the effects, in the laboratory, of borates on the formation of simple sugars critical to life on Earth. To them, the story seems clear—"These experiments suggest that the formation of pentoses appears to be the natural outcome of the chemical transformation of organic molecules present in the nebula that form stars and planets in the presence of borate minerals." Consider the other potentially important chemical products and the processes that may have formed them, and the picture becomes amazingly complex, but similar statements, with perhaps different chemical midwives, may be true when explaining the availability of other compounds important to the origin of life.

If, indeed, nebular chemistry is the key, then there is every reason to suspect that those important compounds should turn up elsewhere in the Solar System. Accordingly, whether for their abundant water ice, the dynamical nature apparent in their geology, or for their past or even present volcanism—and the potential contributions of organic material made during their formation and later from space—bodies such as Mars and Europa continue to promise astrobiological insights of a fundamental nature.

1.3 Life at the Extremes

Earth was once a fairly luxurious place for living organisms. A surplus of liquid water provided a planet-spanning test-tube for investigations into the interactions between oceans and other planetary processes. Volcanism was prevalent as the Earth cooled after its formation and things had calmed down a bit since the really *big* impact that launched the Moon. Sure, there were still some really large impactors, but those kept things warm despite the relatively faint young Sun (which still provided some high-quality photons), and the effects of having the Moon as close as it was stoked the tides into truly gigantic proportions and kept the ocean cycling and churning. Such an active circumstance continually exposed new portions of the cooling planet to the ocean, and provided the ocean's new life with access to varying, new sources of chemical and physical energy. Moreover, some of that life was exposed to even more novel environments as the Earth's heat-engine started the conveyor belt of planetary renewal that we now know as plate tectonics. Most life forms were unlikely to survive subduction and burial processes, but they were slowly and surely introduced to every livable portion of the planet.

And perhaps the best thing about the early Earth was that there was almost no unbound oxygen—anywhere. Earth life was still free of this most dangerous of molecules.

The late Robert D. MacElroy was the first to christen organisms which thrive in extreme environments as "extremophiles" [25]. While it may pay to appreciate the fact that humans, too, can be considered as extremophiles (for our overt dependence on oxygen), the extreme environments of interest with respect to planetary exploration are decidedly *not* those favorable to humans. Recently, Rothschild and Mancinelli reviewed some characteristics of the Earth organisms found in extreme environments [26], some of which may be more pertinent to Mars and Europa. Most importantly, it should be noted that extremophiles of interest to extraterrestrial environments (all microbial, at present) need not be super-bugs. As outlined below, both Mars and Europa can be anticipated to possess environments that are, if not warm and nurturing to humans, well within the tolerance of quite a number of Earth organisms. Whether there are martian or europan organisms is quite another question—it may be possible to identify circumstantial evidence that life could exist on one or both of these bodies, but only direct investigation can tell whether or not it already exists.

Table 17-1 provides an update on some of the more relevant characteristics of known extremophilic Earth organisms.

Table 17-1. Some Characteristics of Earth Extremophiles of Interest to the Astrobiology of Mars and Europa. Claims are not definitive, and may refer to restricitive and non-natural situations (table derived chiefly, but not entirely, from [26, 43] and their references).

Characteristic	Maximum	Minimum
Temperature for Growth	120°C	-10°C
Temperature for Survival	≤ 160°C	-196°C or lower
Pressure	1,680 MPa	0 MPa
Radiation	≤ 20 kGy	0
pH	≥ 11-13	≤ 0
Dormancy	≤ 250,000,000 years	0

2. MARS

2.1 Astrobiology on Mars: Mysteries of the Red Planet

With the arrival of space flight, it has been possible to envision answers to some of the most compelling questions about Mars—and all of those questions relate to its astrobiological promise. Since the first successful Mars mission was flown in 1964 (see Table 17-2.), those answers have been replacing past suppositions at an increasing rate, and since the mid-1990s our ability to address them has, literally, skyrocketed. Furthermore, there is an explicit acceptance that astrobiology, and the underlying promise of Mars exobiology, is a central driver for agency missions.

For instance, NASA's Mars strategy is "follow the water," to understand Mars and its hydrology as they interact with martian geology, climatology, and the prospects for life. Not entirely by coincidence, NASA's Mars program is essentially enacting a strategy for the exobiological study of Mars published by its Exobiology Program in 1995 [27], which recommended global reconnaissance for the first explorational phase, focusing on the role of water, past or present, and on the identification of potentially fruitful sites for landed missions, followed by the landed missions themselves. These would provide in situ descriptions of promising sites identified during the first phase, especially concentrating on broad-based geochemical and mineralogical characterization, and later on the elemental, molecular and isotopic analysis of the biogenic elements in a variety of microenvironments,

Table 17-2. Significant Mars Missions to Date
(modified from Frommert, H., <http://www.seds.org/~spider/mars/mars-l.html> 2004, and <http://marsweb.jpl.nasa.gov/>)

Launch Year	Mission Name (Responsibility)	Mission Characteristics
1964	Mariner 4 (NASA):	First successful Mars mission. Returned 22 TV pictures of its surface, discovering its cratered nature.
1969	Mariner 6 (NASA)	Successful fly-by. Returned data and 75 photos, mainly from the equatorial region. Found that most of Mars' atmosphere was made of carbon dioxide.
1969	Mariner 7 (NASA)	Successful fly-by. Returned data and 126 photos flying over the south polar region.
1971	Mariner 9 (NASA)	Mars' first artificial satellite—successfully achieved Mars orbit in November 1971. Returned 7,329 TV pictures covering the entire surface of Mars, providing the first full photographic atlas, or photo globe, of a celestial body. Discovered volcanoes, flow channels, and more surface structures.
1975	Viking 1 (NASA)	Orbiter and lander mission. Reached Mars orbit on June 19, 1976, the lander softlanded on Mars on July 20, 1976, in Chryse Planitia at 22.48° North areographic latitude, 48.01° Western longitude. Viking Lander 1 operated until November 13, 1982.
1975	Viking 2 (NASA)	Orbiter and lander mission. Reached Mars orbit on August 7, 1976, lander softlanded on September 3, 1976, in Utopia Planitia 47.97° N, 225.74° W, 7,420 km North-East of Viking 1. Both the Viking 2 orbiter and lander were equally successful as the sister craft to Viking 1; Viking Orbiter 2 was active until July 25, 1978, and Viking Lander 2 returned data up to August 7, 1980, when Viking Orbiter 1, which had served as a communications relay, was shut down.
1988	Phobos 2 (USSR)	Successfully reached Mars orbit on January 29, 1989, and returned data and photos of Mars and Phobos. Transmissions were interrupted shortly thereafter, and not reacquired.

Table 17-2 (cont.). Significant Mars Missions to Date

Launch Year	Mission Name (Responsibility)	Mission Characteristics
1996	Mars Global Surveyor (NASA)	Mars orbiter. Reached Mars and successfully entered Mars orbit on September 11, 1997. Conducting orbital investigations in the visual and IR, plus laser altimetry and UHF communications relay.
1996	Mars Pathfinder (NASA)	Soft-landed on Mars on July 4, 1997, in Ares Vallis, at 19.5° N, 32.8° W. Sojourner was released to the Martian surface on July 6, and performed investigations of Martian soil and rocks. Both spacecraft operated until the last data transmission on September 27, 1997.
2001	Mars Odyssey (NASA)	Mars orbiter. After interplanetary cruise, the spacecraft arrived at Mars and was successfully inserted into Mars orbit on October 24, 2001. This mission is studying Mars from orbit, using visual, IR, gamma-ray, proton, and neutron detectors, and is serving as a UHF communications relay.
2003	Mars Express (ESA) Beagle-2 lander	Orbiter and Lander. Mars Express orbiter was successfully inserted into martian orbit on December 25, 2003. The Beagle 2 lander was successfully separated on December 19, and reached Mars in direct approach on December 25, 2003. The lander could not immediately be contacted after landing.
2003	Spirit / MER-A (NASA)	First of two sister spacecraft. A large (174 kg) rover, based on the Athena payload, landed using an airbag system without an active stationary lander. The successful soft landing occurred on January 4, 2004, 0435 UT (January 3, 8:35 p.m. PST) in Gusev crater. The mission is investigating the landing site region, and is intended for at least 90 days of science operations, including rover operations away from the lander.
2003	Opportunity / MER-B (NASA)	Twin payload to MER-A, with a similar mission and duration. Successful soft landing occurred on January 25, 2004, 0005 UT (January 24, 9:05 p.m. PST) in a small crater on Meridiani Planum, the so-called "hematite" site.

including analysis of volatile species. Only after those tasks had been accomplished did the strategy call for deployment of exobiologically focused experiments, which would comprise a detailed characterization of organic compounds, search for biomarkers and for morphological evidence of formerly living organisms, and, in the event that extant life seemed plausible, experiments to test for metabolism in living systems, based on a knowledge of conditions and resources at specific sites.

A fourth phase, also under consideration by NASA (and now by ESA under the Aurora program), will involve the robotic return of martian samples to Earth—greatly improving characterization of the organic inventory and essential for verification of any in situ evidence for extinct or extant life obtained on Mars itself. Finally, a fifth phase of the exobiology strategy could involve human missions and the establishment of a detailed geological context for any exobiologically significant observations made previously, while aiding in the detection of "oases"—on or under the surface of Mars—which may be capable of promoting or supporting life.

Scientifically, top-level astrobiological investigations of Mars seek to answer the questions, "Was Mars ever warm and wet?" and "If so, when and where?" If the answer to the first question appears to be "yes" (even if Mars may not have been very warm, or very wet—and the meteorite ALH84001 suggests that this must have been so, somewhere [28]), then the "when and where" question is essential to an understanding of whether life ever could have originated on Mars, and what may have happened to it. It could very well be that life never originated on Mars, even if the conditions for life were pervasive. Such a circumstance would provide us with an opportunity to understand the origin of life from a null perspective—something impossible to achieve on Earth. On the other hand, if life did originate on Mars—or was imported to Mars from Earth or elsewhere [cf., 7]—there are profound lessons to be learned from either its extinction or persistence in the martian environment.

2.2 Mars Today and in the Past

Some of the basic characteristics of today's Mars are given in Table 17-3. Many of these characteristics were known in whole or in part prior to the operation of the missions listed in Table 17-2, but nearly all of details about the surface and atmosphere are recent—a product of the space age. Perhaps equally as important as the observations themselves have been the new insights that exploration missions have provided. Increasingly, puzzles

Table 17-3. Characteristics of Mars and Europa

Characteristic	**Mars**	**Europa**
Primary	Sun	Jupiter
Equatorial Diameter (Earth = 12,756 km)	6,780 km	3,138 km
Approximate Mass (Earth = 5.97×10^{24} kg)	64.19×10^{22} kg	4.8×10^{22} kg
Relative Mass (Earth = 1)	0.107	0.008
Surface Gravity (Earth = 1)	0.379	0.135
Average Density (Earth = 5.5 g/cm^3)	3.9 g/cm^3	3.01 g/cm^3
Mean Distance from Primary	227,700,000 km	670,900 km
Mean Distance from Sun (Earth = 1 AU)	~2.4 AU	~5.2 AU
Orbital Eccentricity (present-day)	0.094	0.009
Orbital Inclination (Earth = 0.0 degrees to plane of the ecliptic)	1.850 degrees to plane of the ecliptic	0.470 degrees to Jupiter's equator
Orbital Period	687 days	3.55 days
Rotational Period	24 hrs, 39.6 minutes	3.55 days
Surface Temperature	Mean: -53°C; Range: -133°C to +25°C	Mean: -170°C; Range: -145°C day to ~ -188°C night
Surface Composition	Basalt minerals, CO_2 & water ice, dust, etc.	Water ice, SO_2, CO_2, and hydrated minerals
Atmospheric Composition	95.3% CO_2; 2.7% N_2; 1.6% Ar; 0.13% O_2; 0.08% CO; trace H_2O	Trace oxygen and Na
Atmospheric Pressure	Average 6.36 mb at mean radius (6.9 mb - 9 mb at Viking 1 site)	Nil

about the true nature of Mars are being raised, and while some answers will need to await future data, there are new theoretical insights that suggest the nature of what can be discovered about Mars. While it has been productive to address initial findings by analogy to the Earth, the limitations of that approach are gradually coming to light. Some aspects may be similar, but Mars is *different,* and we don't yet know the why and the how of it.

Perhaps the most critical finding is the effect that the changes in the obliquity and eccentricity of the martian orbit [29, 30] can have on both the overall and local insolation received at Mars, and the potential that these can drive the cyclical accumulation and degradation of frost, snow, and ice—building up both polar caps and glaciers—on the martian surface [31, 32] (Figure 17.1). Over the last 10 Myr these swings in obliquity for Mars ranged from 14° to 48° (compared to 22° to 24.5° for the Earth), while the range of the martian orbital eccentricity was between 0 and 0.06—twice that of the Earth. The net effect is summer insolation values at the poles that can vary greatly from that of today (~220 W/m^{-2}), with a low of just under 150 W/m^{-2}, and a high of over 500 W/m^{-2} (over 400 W/m^{-2} within the last 1 Myr) [31]. The net effect on insolation at the poles is periodic, although recently asymmetrical [32] with a period of about 51,000 yr [31].

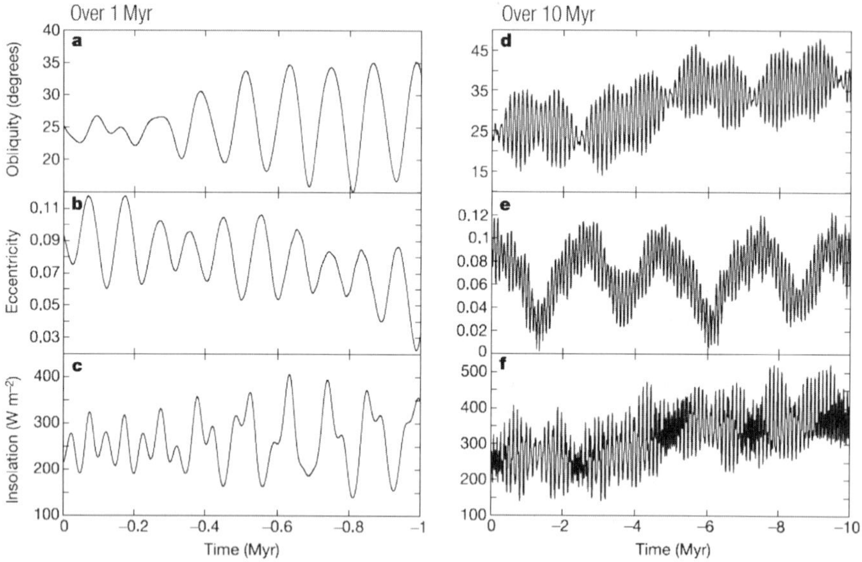

Figure 17-1. Obliquity, eccentricity and insolation of Mars at the north pole surface at the summer equinox. a–c, Over 1 Myr; d–f, over 10 Myr. In the first 0.5 Myr of c, as the obliquity variations are small, the insolation is dominated by precession, while in the remaining part, the insolation variations becomes dominated by the obliquity cycle (Taken from Laskar et al. 2002 [31], used by permission.)

Today, and for perhaps billions of years, it is these swings in obliquity, eccentricity, and insolation that are thought to drive a cold hydrological cycle on Mars that may span thousands of years. And that cycle, complexed with ice/water interactions with the bedrock of the planet and the omnipresent dust, in turn may provide an explanation of the layering of the rock that is seen in many places on Mars [33], and the potential buildup of ice in the middle and upper latitudes of the martian surface [34] (where it may still be abundant—as suggested by Mars Odyssey data [35-37]). This ice in places may make up 50% (by mass) of the material in the upper 1 m of the martian surface. The martian obliquity cycle can also explain the buildup of snowpack and glacial ice that, also complexed with the abundant dust as an insulating layer, has been invoked to explain at least some of the widespread gully features [cf. 38., 39]. A Mars Orbiter Camera image depicting one such gully system is given in Figure 17-2. It could also be that, in some places, this ground ice could form a layer of subsurface water that might later be manifested through an artesian spring, another explanation that has been given for the gullies [40].

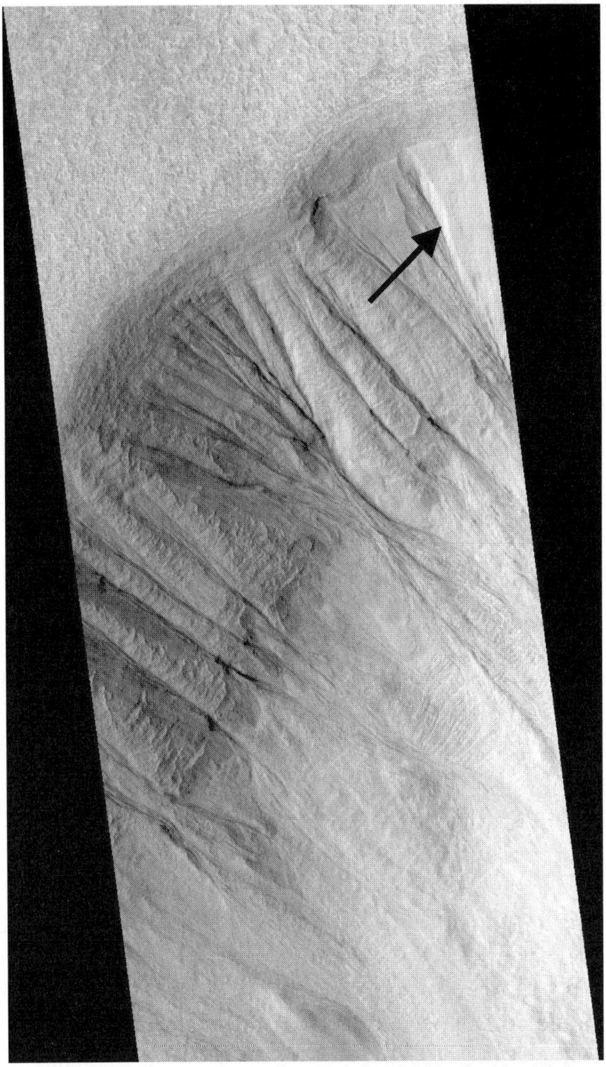

Figure 17-2. A crater wall on Mars, located at 33.3 S, 92.9 E, cited by Christensen [39] as an example of snowmantle-derived gullies, with a portion of the mantle still in place (see arrow). This image (M09-02875) was obtained by the Mars Orbiter Camera system aboard the Mars Global Surveyor spacecraft, launched in 1996 (credit NASA/JPL/Malin Space Systems).

Clearly, in periods of enhanced polar sunlight there is a strong propensity for melting in the martian icecaps [31, 32]. But there may be other ways that such melting could be forced, either at the poles or elsewhere that water ice may exist. For example, the Tharsis volcanoes and Olympus Mons attest to intense, long-lived volcanism in the martian past. Can it exist today? There is an absence of information about martian heat flow from which to derive first-order estimates of subsurface heat on Mars, and to date no extant volcanism has been detected. Nonetheless, Yoder and his colleagues [41] recently reported that by analyzing radio tracking data from the Mars Global Surveyor spacecraft, they were able to monitor solar tidal deformation of Mars, and from those observations have inferred a martian core with a radius of between 1520 and 1840 km, and that is at least partially liquid—thus leaving the door open to the present-day existence of volcanism on Mars.

Another way that ice on Mars may be liberated as liquid water involves impact events. For example, the morphology of cratering on Mars—and particularly flow features associated with liberated groundwater—has been used to identify areas on Mars that can contain ground ice (e.g., [34]), and it may be that certain impact features are correlated with massive outflow features on the martian surface [42]. These features cannot typically be relegated to the distant past, and some of them [42] may be geologically recent—formed within the last 100 million years or even more recently.

So it appears that liquid water, or water in the form of buried ice (and potentially, liquid water beneath that ice), can exist on or under the surface of Mars today. But does that make it likely that martian life exists there? Perhaps not, but a comparison with the capabilities of Earth life may be instructive. Recently, Jakosky et al. [43] have shown that water in liquid form can exist in contact with ice and/or dust grains when the temperature is at or above about -20°C. Such temperatures can exist at the martian poles when obliquities are greater than about 40°, while under some conditions Mars polar temperatures may rise above 0°C. In contrast, when warmed to -20°C, an active metabolism can be measured in organisms isolated from Siberian permafrost—and it has been shown that those organisms will grow at temperatures as low as -10°C. Hence, it is likely that the polar environment on Mars has possessed habitable conditions within relatively recent epochs—and Mars appears on the edge of being habitable at present—at least at or near the poles, or at subsurface locations where even small amounts of liquid water could very well exist.

In other locations, however, the results of a search for life on Mars have at best been equivocal. The Viking missions of the mid-1970s tried a number of strategies to detect life [cf., 44], but despite some encouraging results [45], the lack of organics detectable by the Viking Gas Chromatograph-Mass Spectrometer [46] in the martian regolith at the parts-per-billion level was seen by the majority of the Viking Biology Team as a strong indication that the martian soil material was greatly affected by a strongly oxidizing substance or substances it contained, and that these had mimicked life during some aspects of the mission. In particular, the Team believed that the martian soil material had even destroyed organics that may have been brought to Mars via meteorite falls, etc., and might destroy any other organics as well.

Recently, Benner and his colleagues [47], reexamined the assumptions of the GC-MS, and concluded that nonvolatile salts of benzenecarboxylic acids, and perhaps oxalic and acetic acid, could be metastable intermediates of meteoritic organics under oxidizing conditions. In particular, they found that the salts of these organic acids would have been largely invisible to the Viking GC-MS, though one of them, benzenehexacarboxylic acid (mellitic acid), is generated by oxidation of organic matter of the type known to come to Mars. Mellitic acid is rather stable to further oxidation, and would not have been easily detected by the Viking experiments. If that is the case, then altered organic compounds may have existed at the Viking sites—and it would seem that issues associated with the detection of life on the martian surface will have to be approached carefully as exploration progresses to more benign places than Chryse or Utopia.

While the issues associated with extant life are perhaps more contentious, whether Mars may have been more habitable in the past is another question that has received extensive debate. In fact, the concept of a warmer, wetter Mars in the past has a long history within the scientific exploration of the planet (see summary in [48]). And that concept has recently received some fairly persuasive support. While the most striking results so far from the Mars Odyssey mission have been the discovery of abundant subsurface hydrogen on Mars, interpreted as ice, the Mars Global Surveyor (MGS) mission has revolutionized our understanding of the planet through both the high resolution imaging of the Mars Orbiter Camera and through the fantastic accuracy of the Mars Orbiting Laser Altimeter (MOLA). MOLA has characterized the elevations of the planet with absolute accuracy measured in meters, and we now know the elevations on Mars better than we do those on the continents of the Earth. Through MOLA data [49] three distinct basins have been identified on Mars (northern plains, Hellas, Argyre/Solis), and drainage features leading into them have been mapped extensively, along with almost all of the remainder of the planet.

In the MOLA mapping comes some of the first support for a warm, wet, early Mars that isn't strictly an interpretation of the channel morphologies known since the 1960s. By calculating the mass of the Tharsis bulge and Olympus Mons, the MOLA team [50] speculated that the release of carbon dioxide and water that accompanied the emplacement of Tharsis's ~3 x 10^8 cubic kilometers of magma may have resulted in Mars having a warmer climate that would have enabled the formation of ancient valley networks during the Late Noachian time, beginning ~4.3 to 3.85 billion years ago (Ga) and ending ~3.8 to 3.5 Ga. Because many of these valley network orientations are controlled by the slopes that comprise Tharsis, that would make the bulk of Tharsis, and its accompanying CO_2 and water, largely Noachian in age. Coupled with the potential that early Mars had a significant magnetic field [51], the stage is set for early Mars to have been a warm, wet, and protected niche for life.

How that niche was arrayed is still an unknown. With MOLA data in hand, Head et al. [52] have suggested that the northern plains of Mars may have once formed an ocean—one that would have included a drainage basin approximately 3/4 of the surface of Mars [49]. Seemingly opposed to that concept is the interpretation of the MGS-TES data from Bandfield et al., [52], which suggests that a major component of the martian surface under the proposed ocean is andesite, a mineral that is common on Earth in areas affected by hot-spot, flood volcanism, and not particularly characteristic of standing water. That interpretation, however, has been challenged by Wyatt and McSween [54], who suggest a weathered basaltic model to explain the TES observations. Wyatt and McSween's interpretation is supported by the analysis of Clifford and Parker [55], who speculate that at least 1/3 of Mars would have supported standing bodies of water during those times. Nonetheless, as outlined in Jakosky and Phillips [56], a thick martian atmosphere was not to be—rather it was lost to space by hydrogen escape or impact erosion, and the Mars of today emerged from a wetter, and perhaps warmer, past. But signatures of that time may still appear on the planet. Recently, the Mar Orbiter Camera has uncovered [57] what appears to be an ancient river delta on the planet, providing further evidence of a wetter past. And the other signature of interest—and certainly the most important to astrobiologists—could be life itself. Only upcoming missions of exploration will tell for sure.

2.3 Questions, Process, and Promise

> "In some ways the debate has really moved little since the days of Flammarion and Lowell. The most interesting information remains right at the limit of resolution, be it meters in satellite images of gullies, or nanometers in microscopic images of magnetite crystals. Always life on Mars seems just beyond the fields that we know."
> —Kevin Zahnle [58]

If we establish that Mars once had, or now has, environments where Earth life could grow and thrive, then we will either be in a position to investigate a habitable world without life—or to learn about another life form that we find there. On the other hand, if we find that no such niche ever existed we will be poised to better understand life's limits, and those of the martian environment. Either way, we will learn more about our solar system, and our place in it.

Following up on the initial resurgence of robotic missions that took place in the late 1990s and in 2001 will be a suite of science missions that will look to extend geological and climatic insights about Mars. The twin Mars exploration rovers, Spirit and Opportunity (see Figure 17-3), are visiting Gusev Crater and Meridiani Planum, respectively, to conduct ground truthing of the orbital data from Mars Global Surveyor and Odyssey—seeking to understand the history of water on Mars through mineralogical and elemental analysis of martian rocks in the context of two compelling field sites. Simultaneously, Mars Express will seek to build on the orbital database with a stereo camera and to extend new sensors such as an orbital radar to sound the subsurface for water and ice, a Fourier transform spectrometer, Vis/IR mineralogical mapping spectrometer, and UV/IR atmospheric spectrometer to better characterize martian materials and their abundances on the surface and in the atmosphere, and to field an energetic neutral atom analyzer to better understand the interaction of the martian atmosphere with the solar wind. Beyond these missions, all launched in 2003, there will be further orbital reconnaissance at high resolution, followed by an attempt to visit the most ice-rich terrains of the northern plains and characterize the relationship between ice and soil at that site.

And eventually there will be an emplacement of additional supporting orbiters, more capable roving vehicles, and there are plans for sample return missions—offering to make available the most capable instruments on this planet to complete the study of Mars. Once a sample return is completed successfully, a fully informed decision can be made about human exploration, its costs and liabilities. As we proceed to learn more, and to ask better questions, we can address them on Mars. When we move beyond the fields that we know we can unlock those questions, and push back on the fundamental questions of life.

Figure 17-3. The MER-B / "Opportunity" rover on Meridiani Planum, Mars, showing the first rock-outcrop seen by a lander on the martian surface (image credit: NASA/JPL).

3. EUROPA

> "I think, therefore, that I shall not have done amiss if the First is called by me Io, the Second Europa, the Third, on account of its majesty of light, Ganymede, the Fourth Callisto . . . This fancy, and the particular names given, were suggested to me by Kepler, Imperial Astronomer . . . I hail him as joint father of these four stars."
> —Simon Marius, 1614

3.1 Discovery and Realization

History credits Galileo, in 1610, as the discoverer of the satellites that as a group now bear his name, but Marius was the one who applied the individual names that we now use to describe these fascinating worlds. For centuries after, however, these bodies were only obscurely known to science due to their distance and relatively small size. Nonetheless, ground-based telescopes were capable of studying them to some extent, and as time progressed through the mid-1900s their true nature began to be appreciated. At approximately the same time that Lewis [59] was constructing the first conjectural model of Europa as an icy world with the potential for an ammonia-rich ocean under its surface, Pilcher et al. [60] detected water frost over most or all of Europa.

Subsequently, Europa, though not the largest of the Galilean satellites, became increasingly a target-of-interest to passing spacecraft (e.g., Pioneers 10 and 11), and especially Voyager 2. Voyager 1 had taken only distant, low-resolution images of Europa, and though those displayed a large number of intersecting linear features, they were not understood. When Voyager 2 was targeted to obtain closer, high- resolution images [61], the moon was so lacking in topographic relief that the nature of the surface was not immediately discerned. Later, Voyager scientists acknowledged the possibility that Europa could be internally active due to tidal heating (based on its orbital eccentricity around Jupiter and the influence of the other Galilean satellites). While early models of ocean formation (cf., [62]) neglected tidal heating and showed that solid-state convection in a thick ice cover could lead to significant heat loss and complete freezing of the water layer, more complete models have developed since that time, and have included tidal heating (e.g., [63-66]). These models have shown that beneath a crust of brittle water ice—and depending on the actual heat flow and tidal parameters chosen, a thick convective ice layer—Europa might have oceans as deep as 50 kilometers, or deeper. The implications of a possible ocean under the ice were quickly grasped.

3.2 The Surface of Europa

Europa is the smallest of the four Galilean satellites of Jupiter, though it is still the 6th largest satellite in the Solar System. Europa is slightly smaller than our own Moon, with a diameter of 3,138 km, but its very bright surface is approximately 5 times brighter. The surface of Europa is extremely cold, with the temperature ranging from about -220°C at the poles to a balmy

-150°C near the equator. Based on observations taken to date, Europa is the smoothest object in the solar system, with no feature exceeding 1 km in height. Some additional characteristics are given in Table 17-3.

Although Voyager 2 was able to observe Europa, the best images and measurements were to be one of the prime achievements of the Galileo spacecraft, launched from the US Space Shuttle in 1989 and arriving at Jupiter in 1995, only after completing a flyby of Venus, once, Earth, twice, and two asteroids. It was more than a full year later, in December 1996, that Galileo made its first close flyby of Europa. Nonetheless, even prior to that first flyby, images such as that shown in Figure 17-4 had suggested that Europa would be a particularly interesting world. Subsequent flybys during the original mission, and during the extended mission known as the "Galileo Europa Mission" provided even more images, as well as data from the other spacecraft instruments—in particular the Near-Infrared Mapping Spectrometer. Europa, though seemingly smooth from a long distance, proved to be covered by an incredible and complex array of cracks, ridges, pits, and various terrains and textures (Figures 17-5, 17-6). While many portions of the satellite were covered with ridges—often formed atop other ridges in a convoluted pattern—many portions of the satellite appeared to have been taken apart and only partially put back together. This jig-saw puzzle quality appeared at different scales—from the largest to almost the smallest discernable.

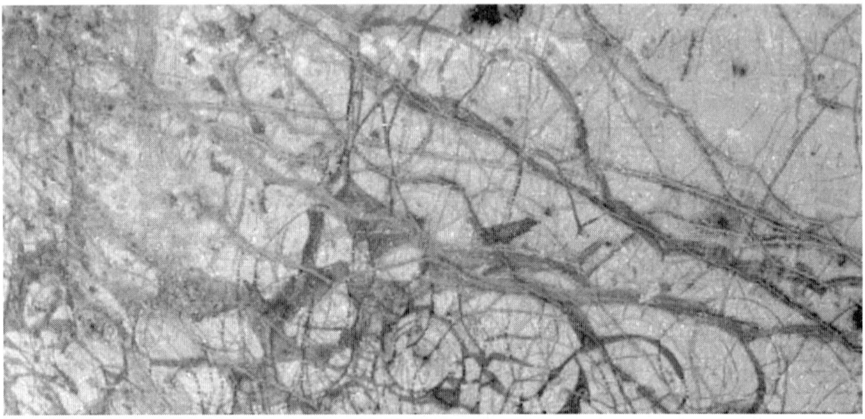

Figure 17-4. The surface of Europa, as seen in August 1996 from the Galileo spacecraft--then 155,000 kilometers away. The large-scale puzzle-like qualities of the surface and the filled-in cracks were early indications that the surface ice might conceal something more fluid below (image credit: NASA/JPL).

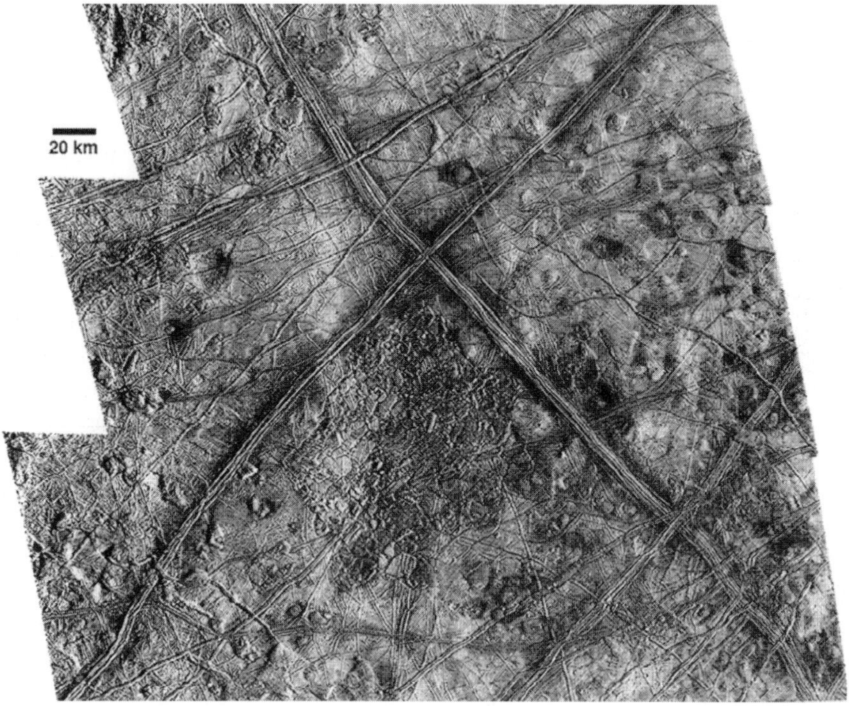

Figure 17-5. Europa's pits, spots, domes and ridges surround the so-called Conamara Chaos region. This composite image derives from Galileo's E6 encounter with Europa in February 1997, with a closest approach of 586 km (image credit: NASA/JPL).

Data from Galileo's Near-Infrared Mapping Spectrometer (NIMS, [67]) suggested that the ice on Europa's surface was not pure water, but rather that it likely contained hydrated magnesium and/or sodium salts, which were positively correlated with the presence of cracks in the surface of Europa. In addition, NIMS detected H_2O_2 on the surface (it had already been detected [68] in the Europan atmosphere), suggesting that the surface is highly

Figure 17-6. Closest pictures of Europa taken by the Galileo mission—a very high resolution view of the Conamara Chaos region. Scale bar, applicable to both images, is 1 km (image credit: NASA/JPL).

influenced by radiolysis that is caused by Europa's immersion in Jupiter's strong radiation belts. The same features that have been interpreted as magnesium and sodium salts, meanwhile, have also been interpreted as deriving from sulfur compounds (including sulfuric acid, sulfur dioxide, and various sulfur polymers) that may participate in a radiolytic sulfur cycle within and on the ices of Europa. Figure 17-7 provides an idea of the intense radiation environment at the surface of Europa, and the significant levels of protection provided by relatively small amounts of water ice.

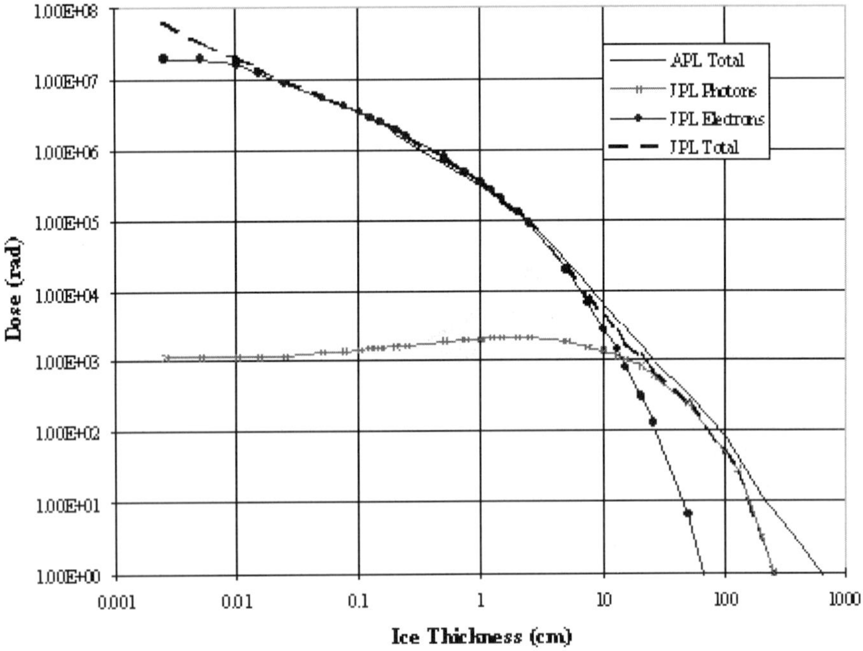

Figure 17-7. Radiation dose models for Europa, in rads per month (30.4 days) of exposure below varying thicknesses of ice. The results of two independent evaluations are given, "JPL Total" and "APL Total." In the JPL Total model, the separate contributions of electrons and photons (bremsstrahlung) are shown. The APL Total model has higher proton fluxes at very high energies. Theoretical differences aside, natural variations of up to an order of magnitude have been observed in Jupiter's trapped-particle intensities. (From the Space Studies Board, National Research Council report, *Preventing the Forward Contamination of Europa*, NRC, 2000, used by permission).

3.3 What Lies Beneath?

Based on the encouragement of theoretical models, initial results from the Galileo encounters were seen to provide confirmation of a Europan ocean [70]. Moreover, supporting evidence from gravity data [71, 72] and eventually from the detection of an induced magnetic field (induced by

Jupiter's field on an otherwise non-magnetic Europa) [73-75] have greatly constrained the possibilities of what is going on beneath Europa's icy exterior. Pappalardo et al. [76] summarized much of the evidence for an ocean, and suggested future observations to confirm one to the satisfaction of all, but there are few other hypotheses that explain so much of the evidence from Galileo's observations.

The specifics of how an ocean and the Europan surface might interact, however, are far from agreed-upon. The depth of the ice-layer, in particular, has been subject to different hypotheses, while the nature of the ridge systems on Europa [e.g., 77-78] and the role of faulting in crustal cycling [e.g., 79-80] have been debated in a variety of ways. In particular, there is a keen interest [e.g., 81] in discerning the present-day geological activity of Europa, and the rate of recycling of the crust—both to constrain the crustal thickness and depth of any ocean, as well as to characterize the amount of energy that might be available for life, if any is to be found on Europa.

As to that possibility, early concepts such as those of Reynolds et al. [82] provided some concepts of life as it might exist in and around a Europan ocean, although those under consideration early-on derived chiefly from the perspective of photosynthetic life within cracks and crevices, and therefore protected from the extremes of cold and radiation, somehow groveling for photons in the dim Europan sunshine. Later concepts [84, 85] developed around the basis of an autotrophic population of organisms that would make use of chemical disequilibria driven by a hypothetical hydrothermal circulation under the Europan ocean, though the potential limitations of these situations were explored in Gaidos et al. [84]. Later on, Chyba [86] took into consideration the potential contribution of radiolytic materials from the surface, and particularly H_2O_2, that could be delivered into a Europan ocean by crustal turnover processes. Such oxidized compounds, if regularly introduced into an otherwise reduced ocean chemistry, could provide chemical energy to support a surprisingly large amount of biomass. In an even more optimistic case, Chyba and Hand [88] considered additional sources of molecular oxygen and hydrogen that might be formed on Europa by simultaneous production in Europa's bulk ice and in the ocean itself, due to the decay of radioactive ^{40}K. Such processes could allow a biomass production of ~10^{10}-10^{12} g yr^{-1}, a considerable potential contribution to the biological universe.

Future studies of Europa, *in situ,* are the provenance of a mission currently under study by NASA [89], the Jupiter Icy Moons Orbiter (JIMO). While the mission has yet to complete its definition phase, in early studies it has been formulated as a highly capable mission that would use nuclear electric propulsion to achieve successive orbits around Callisto, Ganymede, and Europa, conducting remote sensing investigations of those bodies, and possibly even landing probes on Europa. JIMO's constraints are mostly a function of launch mass—it is desirable to use a single launch vehicle for the mission—but the benefits of a lander are particularly clear in the areas of seismology [90] and astrobiology [91]. Currently, the mission is being studied for a $\geq$ 2015 launch date.

4. AND BEYOND

> "All these worlds are yours except Europa, attempt no landings there."
> —A. C. Clarke, *2010: Odyssey II,* 1982

Contemplating the potential lessons of Mars and Europa, which may have life, and may not, is made much easier in the face of Titan. While it might be forgivable to be disappointed if it turns out that no life exists on either of the former bodies, when Titan is added to the set the array of opportunities to understand prebiotic chemical evolution and planetary processes is stunning. Not just the "Goldilocks" paradigm of Venus-Earth-Mars (where one is too hot, one is too cold, and one is just right), but a suite of conditions that spans the gamut of those that may have been faced by the early Earth. And despite the likelihood that each of them has specific characteristics that are different than anything on the Earth at any single time in its past—they are their own worlds—by the study of Mars, Europa, and Titan we may come to recognize characteristics of the early Earth that are presently obscure, and to appreciate the contributions of those characteristics to the origin, or at least presence, of life here. Alternatively, on Titan we may also find out the implications of a chemistry that may be complex enough to support life, but in the absence of liquid water [92]. As we pursue the studies of those worlds, paradigms will shift, and understanding can grow. This is the promise of an astrobiological perspective on the universe—not a simple "search for life," but a different perspective on the whole consideration of life, and its role in our world and worlds beyond.

And if, along the way, we also can shed light on established disciplines, such as understanding what circumstances may really allow for multicellularity in living systems (maybe we do find life elsewhere), or what conditions may lead to the establishment of a functioning biosphere (by studying other biospheres, or even by studying the null condition), then we will be making the best use of the opportunities presented by space flight and astrobiology, together.

ACKNOWLEDGEMENTS

The author acknowledges the use of Mars Orbiter Camera images processed by Malin Space Science Systems that are available at http://www.msss.com/moc_gallery/ permission from Jacques Laskar to use one of his figures, and the long-standing patience of the Editor of this volume. Thanks, also, to Michael Meyer, Steve Saunders, Luann Becker, Toby Owen, and the rest of the ISSI Team on "Prebiotic matter..." for their comments and reading of the manuscript.

REFERENCES

1. Woese, C.R., Kindler, O., Wheelies, M.L. Towards a natural system of organisms: Proposals for the domains Archaea, Bacteria, and Eucarya. *Proc. Nat. Acad. Sci. USA*, 87, 4576-4579, 1990.
2. Lederberg, J. Exobiology: Approaches to Life Beyond the Earth, *Science* 132, 393-400, 1960.
3. Des Marais, D. J., Allamandola, L. J., Benner, S. A., Boss, A. P., Deamer, et al. The NASA astrobiology roadmap. *Astrobiology* 3, 219-235, 2003.
4. Botta, O. The chemistry of the origin of life. *This volume*, 2004.
5. Thomson, W. [Lord Kelvin], At the British Association for the Advancement of Science, Edinburgh meeting—Inaugural address of Sir William Thomson, LL. D., F. R. S., President. *Nature* 4, 262-270, 1871.
6. Arrhenius, S. Panspermy: The Transmission of Life from Star to Star, *Scientific American*, 196, 196, 1907.
7. Mileikowsky, C., Cucinotta, F., Wilson, J.W., Gladman, B., Horneck, G., Lindgren, L., Melosh, H.J., Rickman, H., Valtonen, M.J., Zheng, J.Q. Natural transfer of viable microbes in space. Part 1: From Mars to Earth and Earth to Mars. *Icarus*, 145, 391–427, 2000.
8. Ehrenfreund, P. et al. Astrophysical and astrochemical insights into the origin of life. *Rep. Prog. Phys*. 65, 1427-1487, 2002.
9. Miller, S. L. A production of amino acids under possible primitive earth conditions. *Science* 117, 528-529, 1953.
10. Ferris J. P. and Chen, C. T. Chemical evolution. XXVI. Photochemistry of methane, nitrogen, and water mixture as a model for the atmosphere of the primitive earth. *J. Am. Chem. Soc.* 97, 2962-7, 1975.
11. Sagan, C. and Chyba, C. The early faint Sun paradox: organic shielding of ultraviolet-labile greenhouse gases. *Science* 276, 1217-1221, 1997.
12. Shock, E. L. Chemical environments in submarine hydrothermal systems, in Marine Hydrothermal Systems and the Origin of Life (N. Holm, ed.), *Orig. Life Evol. Biosphere,* 22, 67-107, 1992.
13. Shulte, M. and Shock, E. L. Thermodynamics of Strecker synthesis in hydrothermal solutions. *Orig. Life Evol. Biosphere*, 25, 161-173, 1995.

14. Shock, E.L. 1996. Hydrothermal systems as environments for the emergence of life. Evolution of Hydrothermal Ecosystems on Earth (and Mars?), CIBA Foundation Symposium No. 202, 40-52.
15. Darwin, C. R. Letter to J. D. Hooker. 1 February 1871.
16. Bada, J. L., Bigham, C., and Miller, S. L. Impact Melting of Frozen Oceans on the Early Earth: Implications for the Origin of Life. *Proc. Nat. Acad. Sci. USA,* 91, 1248-1250, 1994.
17. Corliss, J.B. On the creation of living cells in submarine hot spring flow reactors: attractors and bifurcations in the natural hierarchy dissipative systems. *Orig. Life Evol. Biosphere*, 19, 381-382, 1986.
18. Wächtershauser, G. Before enzymes and templates: theory of surface metabolism. *Microbiol. Rev.,* 52, 452–484, 1988.
19. Huber, C. and Wächtershauser, G. Peptides by activation of amino acids with CO on (Ni,Fe)S surfaces: implications for the origin of life. *Science* 281, 670-672, 1998.
20. Huber, C., Eisenreich, W., Hecht, S., Wächtershauser, G. A possible primordial peptide cycle. *Science* 301, 938-940, 2003.
21. Braterman, P. S., Cairns-Smith, A. G. Sloper, R. A. Photo-oxidation of hydrated Fe2+ - significance for banded iron formations, *Nature*, 303, 163-164, 1983.
22. Arrhenius, G. Dysoxic environments as models for primordial mineralisation. In Cairns-Smith, A.G. Hartman, H.(eds) *Clay and Clay Minerals and the Origin of Life.* Cambridge University Press, 97-104, 1986.
23. Hanczyc, M. M., Fujikawa, S. M., Szostak, J. W. Experimental models of primitive cellular compartments: encapsulation, growth, and division. *Science* 302, 618-622, 2003.
24. Ricardo, A., Carrigan, M. A., Olcott, A. N., Benner, S. A. Borate minerals stabilize ribose. *Science* 303, 196, 2004.
25. MacElroy, R. D. Some comments on the evolution of extremophiles. *Biosystems* 6, 74–75, 1974.
26. Rothschild, L. J., Mancinelli, R. L. Life in extreme environments. *Nature* 409, 1092-1101, 2001.
27. Kerridge, J. F. (Chair), et al. An Exobiological Strategy for Mars Exploration. NASA SP-530, NASAHeadquarters, Washington, D.C., USA, 1995.
28. McKay, D. S., Gibson, E. K., Thomas-Krepta, K. L., Vali, H., Romanek, C. S., Clemett, S. J., Chillier, X.D.F., Maechling, C. R., Zare, R.N. Search for past life on Mars: possible relic biogenic activity in martian meteorite ALH84001, *Science* 273, 924-930, 1996.
29. Ward, W.R. Long-Term Orbital and Spin Dynamics of Mars, in *Mars,* (H.H. Kieffer, B.M. Jakosky, C.W. Snyder, and M.S. Matthews, eds.), University of Arizona Press, Tucson, 298–320, 1992.
30. Touma, J. & Wisdom, J. The chaotic obliquity of Mars. *Science* 259, 1294–1297, 1993.
31. Laskar, J., Levrard, B., Mustard, J. F. Orbital forcing of the martian polar layered deposits, *Nature* 419: 375-377, 2002.
32. Head, J. W., Mustard, J. F., Kreslavsky, M. A., Milliken, R. E., Marchant, D. R. Recent ice ages on Mars, *Nature* 426, 797-802, 2003.
33. Malin, M. C., Edgett, K. S. Sedimentary rocks of early Mars," *Science* 290: 1927-1937, 2000.
34. Squyres, S. W., Carr, M. H. Geomorphic evidence for the distribution of ground ice on Mars," *Science* 231, 249-252, 1986.
35. Boynton, W. V., Feldman, W. C., Squyres, S. W., Prettyman, T. H., Brückner, J. et al. Distribution of hydrogen in the near-surface of Mars: evidence for subsurface ice deposits," *Science* 297, p. 81-85, 2002.

36. Feldman, W. C., Boynton, W. V., Tokar, R. L., Prettyman, T. H., Gasnault, O., et al., Global distribution of neutrons from Mars: results from Mars Odyssey," *Science* 297, 75-78, 2002.
37. Mitrofanov, I., Anfimov, D., Kozyrev, A., Litvak, M., Sanin, A., et al., Maps of subsurface hydrogen from the High Energy Neutron Detector, Mars Odyssey, *Science* 297, 78-81, 2002.
38. Costard, F., Forget, F., Mangold, N., Peulvast, J. P. Formation of recent martian debris flows by melting of near-surface ground ice at high obliquity. *Science* 295, 110–113, 2002.
39. Christensen, P. Formation of recent martian gullies through melting of extensive water-rich snow deposits, *Nature* 422, 45-48, 2003
40. Malin, M. C., Edgett, K. S. Evidence for recent groundwater seepage and surface runoff on Mars," *Science* 288: 2330-2335, 2000.
41. Yoder, C. F., Konopliv, A. S., Yuan, D. N., Standish, E. M, Folkner, W. M. Fluid core size of Mars from detection of the solar tide, *Science*, 300, 299-303, 2003.
42. Carr, M. H. *Water on Mars* (Oxford Univ. Press, New York, 1996).
43. Jakosky, B. M., Nealson, K. H., Bakermans, C., Ley, R. E., Mellon, M. T. Subfreezing activity of microorganisms and the potential habitability of Mars' polar regions. *Astrobiology* 3, 343-350, 2003.
44. Klein, H. P. The Viking mission and the search for life on Mars, *Reviews of Geophysics and Space Physics* 17, 1655-1662, 1979.
45. Levin, G. V., Straat, P. A. Viking Labeled Release biology experiment: interim results, *Science* 194, 1322-1329, 1976.
46. Biemann, K., Oró, J., Toulmin III, P., Orgel, L.E., Nier, A.O. et al., The search for organic substances and inorganic volatile compounds in the surface of Mars," *Journal of Geophysical Research* 82, 4641-4658, 1977.
47. Benner, S. A., Devine, K. G., Matveeva, L. N., Powell, D. H. The missing organic molecules on Mars, *Proc. Nat. Acad. Sci*, 97, 2425-2430, 2000.
48. McKay, C. P., Mancinelli, R. L., Stoker, C. R., Wharton, R. A. The possibility of life on Mars during a water-rich past," in *Mars,* (H.H. Kieffer, B.M. Jakosky, C.W. Snyder, and M.S. Matthews, eds.), University of Arizona Press, Tucson, 1234–1245, 1992.
49. Smith, D. E., Zuber, M. T., Solomon, S. C., Phillips, R. J., Head, J. W., et al. The global topography of Mars and implications for surface evolution," *Science* 284, 1495-1503, 1999.
50. Phillips, R. J., Zuber, M. T., Solomon, S. C., Golombek, M. P., Jakosky, B. M., et al. Ancient geodynamics and global-scale hydrology on Mars," *Science* 291, 2587-2591, 2001.
51. Acuña, M. H., Connerney, J.E.P., Ness, N.F., Lin, R.P., Mitchell, D., et al. Global distribution of crustal magnetization discovered by the Mars Global Surveyor MAG/ER experiment, *Science* 284, 790-793, 1999.
52. Head III, J. W., Hiesinger, H., Ivanov, M. A., Kreslavsky, M., Pratt, S., et al. "Possible Ancient oceans on Mars: evidence from Mars Orbiter Laser Altimeter data, *Science* 286, 2134-2137, 1999.
53. Bandfield, J. L., Hamilton, V. E., Christensen, P. R.. A global view of martian surface compositions from MGS-TES," *Science* 287, 1626-1630, 2000.
54. Wyatt, M. B., Mc Sween, H. Y. Spectral evidence for weathered basalt as an alternative to andesite in the northern lowlands of Mars, *Nature* 417, 263-266, 2002.
55. Clifford, S. M., Parker, T. J. The evolution of the martian hydrosphere: implications for the fate of a primordial ocean and the current state of the northern plains, *Icarus* 154, 40-79, 2001.

56. Jakosky, B. M., Phillips, R. J. Mars' volatile and climate history, *Nature* 412, 237-244, 2001.
57. Malin, M. C., Edgett, K. S.. Evidence for persistent flow and aqueous sedimentation on early Mars, *Science* 302, 1931-1934, 2003.
58. Zahnle, K. Decline and fall of the martian empire. *Nature* 412, 209-213, 2001.
59. Lewis, J. S. Satellites of the outer planets: their physical and chemical nature, *Icarus* 15, 174-185, 1971.
60. Pilcher, C. B., Ridgeway, S. T., McCord, T. B. Galilean satellites: identification of water frost, *Science* 178, 1087-1089, 1972.
61. Smith, B. A., Soderblom, L. A., Beebe, R., Boyce, J., Briggs, G., et al. The Galilean satellites and Jupiter—Voyager 2 imaging science results, *Science* 206, 927-950, 1979.
62. Reynolds, R. T., Cassen, P. M. On the internal structure of the major satellites of the outer planets, Geophys. Res. Lett., 25, 4157-4160, 1979.
63. Squyres, S. W., Reynolds, R. T., Cassen, P. M., Peale, S. J. Liquid water and active resurfacing on Europa, *Nature* 301, 225-226, 1983.
64. Ojakangas, G. W., Stevenson, D. J. Thermal state of an ice shell on Europa, *Icarus*, 81, 220-241, 1989.
65. Fanale, F. P., Salvail, J. R., Matson, D. L., Brown, R. H. The effect of volume phase changes, mass transport, sunlight penetration, and densification on the thermal regime of icy regoliths, *Icarus*, 88, 193-204, 1990.
66. McKinnon, W. B. Convective instability in Europa's floating ice shell, *Geophys. Res. Lett.,* 26, 951-954, 1999.
67. McCord, T. B., Hansen, G., Fanale, F. P., Carlson, R. W., Matson, D., et al. Salts on Europa's surface detected by Galileo's Near-Infrared Mapping Spectrometer, *Science*, 280, 1242-1245, 1998.
68. Hall, D. T., Strobel, D. F., Feldman, P. D., McGrath, M. A., Weaver, H. A., Detection of an oxygen atmosphere on Jupiter's moon Europa, *Nature,* 373, 677–679, 1995.
69. R.W. Carlson, R.E. Johnson, and M.S. Anderson, "Sulfuric Acid on Europa and the Radiolytic Sulfur Cycle," *Science* 286: 97-99, 1999.
70. Carr, M. H., Belton, M. J. S, Chapman, C. R., Davies, M. E., Geissler, P., et al. Evidence for a subsurface ocean on Europa, Nature 391, 363–365, 1998.
71. Anderson, J. D., Schubert, G., Jacobsen, R. A., Lau, E. L., Moore, W. B. Europa's differentiated internal structure: inferences from four Galileo encounters, *Science*, 281, 2019-2022, 1998.
72. Moore, W. B., Schubert, G. The tidal response of Europa, *Icarus* 147, 317-319, 2000.
73. Khurana, K. K., Kivelson, M. G., Stevenson, D. J., Schubert, G., Russell, C. T., et al. Induced magnetic fields as evidence for subsurface oceans in Europa and Callisto, *Nature* 395, 777-780, 1998.
74. C. Zimmer, K.K. Khurana, and M.G. Kivelson, "Subsurface Oceans on Europa and Callisto: Constraints from Galileo Magnetometer Observations," *Icarus* 147: 329-347, 2000.
75. Kivelson, M. G., Khurana, K. K., Russell, C. T., Volwerk, M., Walker, R. J., Zimmer, C. Galileo magnetometer measurements: a stronger case for a subsurface ocean at Europa, Science 289, 1340–1343, 2000.
76. Pappalardo, R. T., Belton, M.J.S., Breneman, H.H., Carr, M.H., Chapman, C. R., et al. Does Europa have a subsurface ocean? Evaluation of the geological evidence, *Journal of Geophysical Research* 104, 24015-24055, 1999.
77. Head, J. W., Pappalardo, R. T., Sullivan, R. Europa: morphological characteristics of ridges and triple bands from Galileo data (E4 and E6) and assessment of a linear diapirism model, *Journal of Geophysical Research* 104, 24223-24236, 1999.

78. Hoppa, G. V., Tufts, B. R., Greenberg, R., Geissler, P. G. Formation of cycloidal features on Europa, *Science* 285, 1899-1902, 1999.
79. Hoppa, G.V., Tufts, B. R., Greenberg, R., Geissler, P., Strike-slip faults on Europa: global shear patterns driven by tidal stress, *Icarus* 141, 287-298, 1999.
80. Prockter, L.M. and Pappalardo, R. T. Folds on Europa: implications for crustal cycling and accommodation of extension, *Science* 289: 941-943, 2000.
81. Phillips, C. B., McEwen, A.S., Hoppa, G. V., Fagents, S. A., Greeley, R., et al. The search for current geologic activity on Europa, *Journal of Geophysical Research* 105: 22559-22578, 2000.
82. Reynolds, R. T., Squyres, Colburn, D. S., McKay, C. P. On the Habitability of Europa, *Icarus* 56, 246-254, 1983.
83. Jakosky, B. M., Shock, E. L. The biological potential of Mars and Europa, *Journal of Geophysical Research* 103: 19359-19364, 1998.
84. Gaidos, E. J., Nealson, K. H., Kirschvink, J. L., Life in ice-covered oceans, *Science* 284, 1631–1633, 1999.
85. McCollom, T. M. Methanogenesis as a potential source of chemical energy for primary biomass production by autotrophic organisms in hydrothermal systems on Europa', *Journal of Geophysical Research* 104, 30729–30742, 1999.
86. C.F. Chyba, "Energy for Microbial Life on Europa," *Nature* 403: 381-382, 2000.
87. Chyba, C. and Phillips, C. B. Europa as an abode of Life. *Origins of Life and Evolution of the Biosphere* 32, 47–68, 2002.
88. Chyba, C. F., Hand, K. P. Life without photosynthesis, *Science* 292, 2026–2027, 2001.
89. Greeley, R., Khurana, K., McKay, C., Moore, W., Prockter, L., et al. Report of the NASA Science Definition Team for the Jupiter Icy Moons Orbiter (JIMO), *in preparation,* 2004.
90. Lee, S., Zanolin, M., Thode, A. M., Pappalardo, R. T., Makris, N. Probing Europa's interior with natural sound sources, *Icarus* 165, 144-167, 2003.
91. Chyba, C. F. Searching for life on Europa from a spacecraft lander, in *Signs of Life,* (Committee on the Origins and Evolution of Life, National Research Council, eds.), National Academies Press, Washington, 86-90, 2002.
92. Ball, P. Water, water, everywhere? *Nature* 427, 19-20, 2004.

Chapter 18

ASTROBIOLOGY IN THE UNITED STATES

A Policy Perspective

David H. Smith
Space Studies Board, National Research Council, The Keck Center, Washington, D.C.

Abstract: The first part of this chapter contains a brief history of some of the key events leading to initiation of NASA's Astrobiology program in the late 1990s. The perspective taken is than of a non-specialist and emphasis is placed on policy issues rather than key scientific discoveries. A basic conclusion that can be drawn from the circumstances surrounding the initiation of the program is that a combination of scientific results in diverse fields, combined with advanced planning by scientific community and NASA, enabled prompt exploitation of a political opportunity. The second part of this chapter concerns the findings and recommendations contained in a recent scientific and programmatic assessment of NASA's activities in the general area of astrobiology. The basic conclusion of this study is guarded optimism that the program is on track.

Key words: astrobiology, exobiology, space policy

1. THE ORIGINS OF NASA'S ASTROBIOLOGY PROGRAM

1.1 The Three Eras of Astrobiology

Are we alone in the universe? While speculation that we are not has waxed and waned over the centuries, it was not until thc lattcr half of the 20th Century that the question became directly amenable to concerted scientific study. Statistics on the usage of the terms astrobiology and/or exobiology[1] in publications contained in three commonly used databases (Figure 18-1), suggests that the history of this field of study can be roughly divided into three eras.

P. Ehrenfreund et al. (eds.), Astrobiology: Future Perspectives, 445–465.

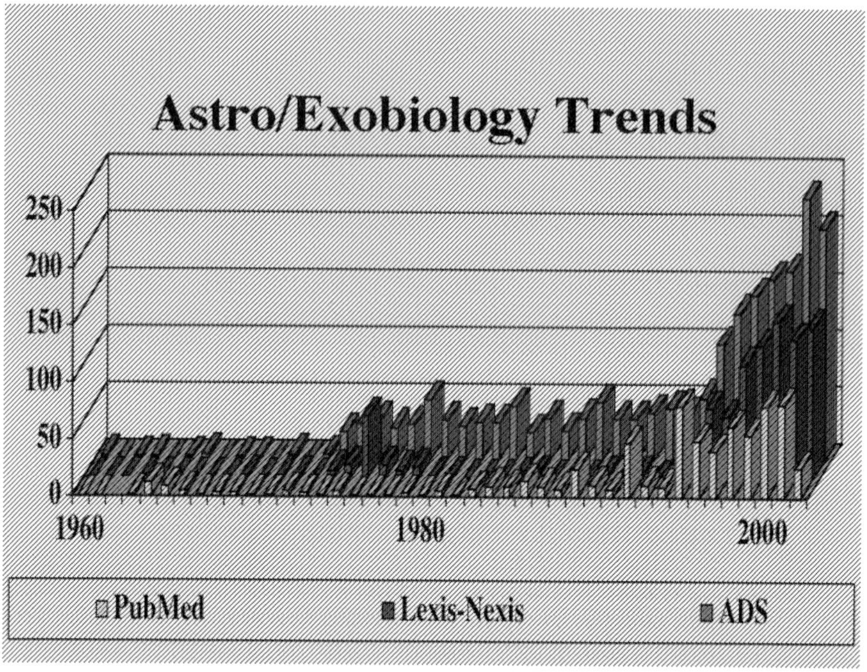

Figure 18-1. The results for searches for the use of the term astrobiology and/or exobiology in three commonly used databases: PubMed, Lexis-Nexis, and the Astrophysics Data System (ADS) which, roughly speaking, catalogue the biomedical literature, newspaper and wire-service articles, and publications in the fields of astronomy, astrophysics, and the planetary sciences, respectively. The trend in usage suggests that the history of astro/exobiology can be divided into three eras: the period up to the mid-1970s; the mid-1970s to the mid-1990s; and the mid-1990s to the present.

These eras are as follows:

- The Pre-Viking Era—The period up to 1976;
- The Post Viking Era—The years 1977 to mid-1996; and
- The Post-ALH84001 Era—From mid-1996 to the present day.

We will briefly examine a few of the key events in the development of astrobiology during the first two eras and then look at the third era in more detail. It must be emphasized that perspective used in assessing the pre- and post-Viking era is not that of a scientist. Rather, the viewpoint adopted is that which might be taken by a non specialist—e.g., a program examiner in the Office of Management and Budget[2] or a legislative aid in the office of a member of the U.S. Senate or House of Representatives—seeking an answer to the question: Is astrobiology, as practiced by NASA, a legitimate field of scientific enquiry? To such an individual, key scientific papers and major discoveries will be of less importance than answers to two subsidiary questions:

1. What is astrobiology's scientific heritage? That is, is it something new or is there a long history of activity in this area.
2. Have NASA's astrobiology activities been validated by independent groups? That is, does the general scientific community share and endorse NASA's goals in this area.

1.2 The Pre-Viking Era

The period up to 1976 was marked by a steady increase in scientific interest in questions related to the origin, evolution, and distribution of life in the universe. Although the term astrobiology, in its current incarnation was coined in the mid 1990s, references to astrobiology—defined as a field intermediate between astronomy and biology concerned with problems of life on other celestial bodies—can be found as early as 1953.[3] A gathering—Problems Common to the Fields of Astronomy and Biology—described as "the first American symposium in astrobiology" was held at Lowell Observatory in Flagstaff, Arizona, on 18 June, 1957.[4] The symposium's participants discussed, amongst other things, the nature of the martian surface and the survival of terrestrial organisms on Mars. The papers presented were later published by the symposium's sponsor, the Astronomical Society of the Pacific.[5-10]

On 4 October 1957, a few months after the Flagstaff meeting, the Soviet Union launched Sputnik 1. The event attracted worldwide attention and added impetus to all space-related studies and, indeed, to many other scientific activities. The U.S. response to Sputnik was to greatly increase the pace of its space activities. And, this, in turn, created a demand for scientific input on the conduct of these research activities. Therefore, in response to a joint request from the National Science Foundation, the National Advisory Committee for Aeronautics, and the Advanced Research Projects Agency for independent scientific guidance on burgeoning U.S. space activities, the National Academy of Sciences created the Space Science Board (SSB) on 26 June, 1958.[11] A few months later, the National Aeronautics and Space Administration (NASA) is created from the preexisting National Advisory Committee for Aeronautics.

Although use of the term astrobiology appears to have died out in the late 1950s, interest in the interface between biology and the space sciences grew. Writing in their influential 1961 book, *Science in Space*, Lloyd Berkner and Hugh Odishaw—the first chair and executive director, respectively, of the SSB—commented, "Unquestionably the possibility that some form of living matter exists on other planets is the most exciting prospect: the origin of life under radically different conditions of environment and ecology is a subject of unprecedented significance to fundamental biology.[12] Not surprisingly,

Berkner and Odishaw's book included a chapter on exobiology[13]—defined as the study of the origin of life and its progress in independent evolutionary systems.[14]

As the 1960s progressed, scientific interest in exobiology increased. In 1966, for example, the SSB published *Biology and the Exploration of Mars* and, its companion volume, *Extraterrestrial Life: An Anthology and Bibliography*,[15-16] its first major studies devoted specifically to the search for life in space. In the former document, the SSB strongly recommends that NASA put major emphasis on the biological exploration of Mars and, in particular, the launch of an "Automated Biological Laboratory." Six years later, the SSB established the Committee on Planetary Biology and Chemical Evolution to provide continuing advice to NASA in the general area of exobiology.

Interest in exobiology and, in particular, the search for life on Mars was not confined to the scientific community. Program managers at NASA and their political masters were equally captivated by the quest for life in space. This alignment of interest between scientific, programmatic, and political communities resulted in the launch of the Viking missions to Mars in 1976. Costing some $4.1 billion (in 2003 dollars),[17] the Viking program was one of the most expensive space-science programs ever undertake.

1.3 The Post Viking Era

Unfortunately, the twin Viking landers did not find evidence for life on Mars. For this and a variety of other reasons, the period from 1977 to 1995 was a period of retrenchment for astrobiology. Scientific interest in issues relating to the origin and evolution of life remained as strong as in the preceding decades, but the failure of the Viking instruments to deliver clear-cut evidence of life on Mars undermined the political support for further major astrobiological activities at NASA. The lack of an alignment of interest between the scientific and political communities, compounded by a series of budgetary crises at NASA, resulted not only in the deferment of additional Mars missions but also in the near-abandonment of solar system exploration. Indeed, the SSB, long a champion of astrobiological research, went so far as to disestablished its Committee on Planetary Biology and Chemical Evolution in 1988.

1.3.1 Advanced Planning Continues

Although political interest in astrobiology did not began to pick up again until near the end of this era, the scientific community did not cease planning for better times. Throughout the period from the mid-1970s to the mid-

1990s, internal NASA advisory committees as well as those convened by independent groups, including the SSB, continued the scientific planning for future astrobiological studies. Increasingly, these plans focused on the biological exploration of Mars and the search for and characterization of planets around other stars. Examples of some of these important advisory committee reports are shown in Table 18-1.

Table 18-1. Important Science Strategy Reports Relevant to Astrobiology, 1977-1996

Title	Authoring Group	Year Published
Post-Viking Biological Investigations of Mars	SSB	1977
Project Orion: A Design Study of a System for Detecting Extrasolar Planets	NASA	1980
Origin and Evolution of Life–Implications for the Planets	SSB	1981
Other Worlds from Earth: The Future of Planetary Astronomy	NASA	1989
Strategy for the Detection and Study of Other Planetary Systems and Extrasolar Planetary Materials	SSB	1990
The Search for Life's Origins	SSB	1990
Toward Other Planetary Systems	NASA	1992
An Exobiological Strategy for Mars Exploration	NASA	1995
A Road Map for the Exploration of Neighboring Planetary Systems	NASA	1995
Exploration and the Search for Origins: A Vision for Ultraviolet-Optical-Infrared Space Astronomy	AURA	1996

1.3.2 Turmoil at NASA

The early 1990s were a particularly difficult period for NASA. A series of mishaps with high-profile missions, including the loss of Mars Observer and the failure of the Galileo spacecraft to deploy its high-gain antenna, generated yet more negative publicity for an agency still recovering from earlier problems, such as the catastrophic loss of the Space Shuttle Challenger and the embarrassment concerning the Hubble Space Telescope's primary mirror.

In March 1993, NASA headquarters was reorganized. Control of research activities, previously the responsibility of the Office of Space Science and Applications (OSSA), was divested to three new offices with the following responsibilities:

• Office of Space Science (Code S), under the direction of Associate Administrator Wesley T. Huntress, was given managerial responsibility for astronomy and astrophysics, solar and space physics, and solar system exploration. Significantly, responsibility for exobiology, originally a component of OSSA's life science program, is retained by Code S's Solar System Exploration Division.

• Office of Life and Microgravity Science and Applications (Code U),[18] under the direction of Associate Administrator Harry C. Holloway, would henceforth manage biomedical research and the study of fluids, combustions and other investigations into the effects of the microgravity environment on materials.

• Office of Mission to Planet Earth (Code Y),[19] under the direction of Associate Administrator (acting) Shelby Tilford (Charles Kennel is later confirmed in this position), will manage the former-OSSA's Earth remote-sensing activities.

These and other organizational changes within NASA headquarters took place against at background of a mounting budget crisis. The agency's 5-year budget projections (Figure 18-2) signaled that significant cuts were likely in the funding for many programs and, in particular, those of Code S. Compounding NASA's problems during the period from approximately 1993 to 1995 were a series of internally and externally mandated reviews of the agency's entire way of doing business. These reviews included the following:

• ***The Roles and Missions Activity*** was undertaken to "clean up the organization, remove duplication, and improve effectiveness." Significantly, it was not part of a government-wide cost-cutting activity.

• ***The NASA Federal Laboratory Review*** was chartered to "evaluate and develop recommendations which improve efficiency and effectiveness of R&D investments in NASA Laboratories." It was also tasked to "assess and define options to eliminate unnecessary redundancies and overlaps in the existing laboratory system."

• ***The Zero Base Review*** was a nation-wide initiative to streamline all aspects of the federal government. In February 1995, a component of this activity, the NASA Headquarters Red Team issued a white paper designed to stimulate radical thinking about the future of the agency.

1.3.3 NASA Science Institutes

A major issue arising in many of the studies of NASA's organization was the role of, and need for, scientists doing basic research at NASA centers. Another issue was the apparently anomalous position of NASA's Ames Research Center. In particular, Ames's diverse, interdisciplinary space research and development activities appeared, to the casual observer, duplicative of undertakings at other NASA centers. Moreover, Ames' space-science activities were a small component of a center whose main focus was aeronautical research. In other words, the chain of command within which space scientists at Ames were embedded and to whose direction they responded, led to the aeronautics leadership rather than the scientific leadership at NASA headquarters.

While, a more careful, less superficial assessment of Ames' research portfolio revealed that much, if not all, of the apparently redundant activities actually complemented other NASA undertakings, this did not prevent several of the review panels calling for radical restructuring of research activities at Ames. Indeed, the Headquarters Red Team went so far as to recommends that all Earth, planetary, astronomical, and biological science activities at Ames Research Center be either transferred to other NASA facilities or terminated.

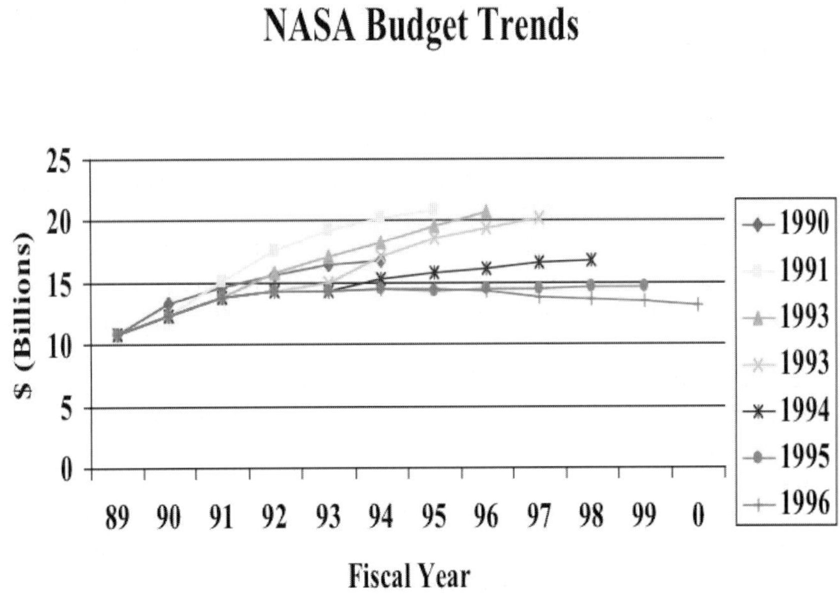

Figure 18-2. NASA's 5-year budget projections for fiscal years 1990 to 1996. At the beginning of this period, federal budget planning assumed that the agency's budget would increase annually. By the end of this period, however, the expectation was that NASA's budget would be flat if not actually declining. As a result, projects initiated in the heady days of the late-1980s/early-1990s were under significant financial stress just a few years later.

In response to the Red Team's sweeping proposals, NASA's chief scientist, France Cordova, and the associate administrators of Codes S, Y, and U issued their own statement supporting the important role of scientists at NASA centers.[20] In addition, these four leaders of NASA's space science activities complained that the Red Team was is too inward looking, too top down in its approach, and too concerned with apparent redundancies in terms of topical areas (e.g., solar system exploration, atmospheric science, and the like) rather than what is actually being done. Cordova et al. concluded by describing a plan to preserve and enhance a thriving multidisciplinary research environment at NASA's research centers. Their basic suggestion was, in essence, to transfer scientists from the centers to a series of new science institute that would be managed as private, nonprofit entity.[21] The science-institutes plan would take the existing model provided by the Jet Propulsion Laboratory—a federally funded, nonprofit research facility managed for NASA by the California Institute of Technology—and attempt to replicated it elsewhere.

The proposal had a key flaw which was recognized by its authors. NASA scientists are government employees and transferring them out of the federal employment to a private entity would have required that NASA persuade Congress to pass special legislation guaranteeing the scientists' civil-service benefits.[22] This did not prevent scientists at Ames Research Center from taking the institute approach as a model for ensuring their own survival. Ames' researchers formulated a plan for a science institute focusing on studies at the intersection between astronomy and astrophysics, planetary science, and biology.[23] Wesley Huntress dubbed the Ames' initiative "astrobiology," independently reinventing a term first used some 40 years earlier. Later, in September 1996, Ames' researchers held a workshop to further refine their plans for astrobiology and to begin work on the drafting of a manifesto, the Astrobiology Roadmap.

The turmoil within NASA and the resulting threats to the agency's science program did not go unnoticed with the Congress. Various members of the House of Representatives and the Senate began to call for a so-called space summit to discuss NASA issues and make concrete proposals about the agency's future.

1.3.4 Major Scientific Discoveries

Although 1996 saw NASA in a state of some disarray, the agency's science programs and space-science and space-science related discoveries and announcements were in the news on an almost weekly basis. Some of the science stories that made headlines in the first half of 1996 include the following:

- The identification of the first extrasolar planets;
- Observations from the Galileo spacecraft suggesting that liquid water exists below Europa's surface;
- Increasing interest in Antarctica's Lake Vostok as a terrestrial analog of an extraterrestrial environment;
- Hubble Space Telescope observations of protoplanetary disks; and
- An increasing realization that life exists in extreme terrestrial environments.

Discussion about the future of NASA's science programs reached a critical mass on 7 August 1996 with the announcement of claims of evidence of fossils in the martian meteorite ALH84001.

1.4 The Post-ALH84001 Era

A fortuitous combination of scientific discoveries in 1996, culminating in the announcement that the martian meteorite ALH84001 showed evidence that life might have existed on Mars in the distant past, brought astrobiological issues once more to the attention of politicians. The rekindling of the alignment between the scientific, programmatic, and political communities, that last existed in the pre-Viking era, was eventually to lead to the initiation of a major astrobiology program at NASA.

Galvanized into action by the publicity surrounding the martian meteorite announcement, President Clinton took up the call for a bipartisan summit on the future of the space program. The summit would "allow the Administration to work with the Congressional leadership to develop a broad consensus on a balanced NASA program for the future....[and] provide an opportunity to discuss the recent evidence that life may have existed on Mars, as well as other significant advances in space science and technology."[24]

1.4.1 The Origins Initiative

In preparation for the space summit, the White House's Office of Science and Technology Policy (OSTP) and NASA asked the Space Studies Board to hold a workshop to discuss the implications of the ALH84001 announcement and other recent scientific advances, and also to prepare for a subsequent briefing to Vice President Gore. The resulting workshop, held on 28-30 October, concluded that the study of "origins"—be it of life, planetary systems, stars, galaxies, or the universe—is a powerful organizing theme for NASA's space-science activities.

On 11 December, key participants in the SSB workshop briefed Vice President Gore on recent advances in origins. The summary message presented at the Vice President's Space Science Symposium was that recent "breakthroughs are astonishing returns being reaped from years of investment in many scientific disciplines. Now is the time to leverage that investment and to pursue the quest for origins into the 21st Century."[25]

Although the claims concerning ALH84001 were quickly questioned, the events of 1996 had a profound long-term impact on both scientists and policy makers. The dividend came on 6 February 1997 when the Clinton Administration's proposed NASA budget for fiscal year 1998 was announced. The proposal includes funds for a major new initiative in the area of origins. The so-called Origins Initiative included significant funding increases for Mars, Europa, and astrophysical missions and the initiation of a major program in astrobiology. It must be emphasized that these were new funds, not the reallocation of existing NASA resources.

Although the major beneficiary of the Origins Initiative was NASA, the National Science Foundation (NSF), the principal sponsor of the U.S. Antarctic Meteorite Collection Program, also announced some new, astrobiologically relevant activities in early 1997. These included what was to prove to be the extremely successful and popular, but now discontinued Life in Extreme Environments (LexEn) program.

1.4.2 The NASA Astrobiology Institute

The most visible element of the NASA's new Astrobiology program was to be the NASA Astrobiology Institute (NAI), to be based at Ames Research Center. NAI was intended not only to pioneer a new scientific endeavor—astrobiology—but also to pioneer new ways of doing science. Rather that being a traditional bricks-and-mortar facility, NAI was envisaged by NASA as a virtual entity. That is, NAI would be composed of geographically dispersed teams of individual scientists who would cooperate and collaborate on research projects within and between teams with the aid of advanced communications tools.

To these ends, on 31 October 1997 NASA issued a Cooperative Agreement Notice (CAN) for participation as NAI lead teams. Some 53 consortia of researchers from universities, public and private research institutions, and NASA centers subsequently submitted proposals to participate. Eleven of them were selected on 19 May 1998 and NAI officially open for business on 1 July of that same year. A second CAN was issued in June 2000 and four additional lead teams were added the following March.

1.5 Conclusions

We are now in a position to provide answers to the two questions posed by our hypothetical congressional staffer or budget examiner at the beginning of this discussion. First, irrespective of whether it is called astrobiology, exobiology, or something else, studies relating to the origin, evolution, and detection of life in the universe have a scientific heritage that predates the creation of NASA. Second, the scientific importance of studies relating to the search for life in the universe have been consistently singled out as something NASA should be concerned about by numerous internal and, more importantly, external advisory groups for a period of some 40 years.

The initiation of NASA's astrobiology program and, in particular, the events of 1996-1997 raise a number of important questions. Prime among these is why did the claims concerning ALH84001 have such a wide-reaching impact and generate hundreds of millions of dollars of new funding for the space sciences? Especially when the initial claim was met with widespread skepticism.

There appear to be at least four contributing factors. These are as follows:

1. ***Massive Publicity***—Members of the public, scientists and the press, already sensitized to life-in-space issues by a series of discoveries and announcements earlier in 1996, was electrified by the claims made about ALH84001. This attention did not escape the notice of policy makers and politicians.

2. ***Rapid Response of Scientific Community***—The ability of the scientific community to organize the Origins Workshop and Vice President Gore's subsequent Space Science Symposium so soon after the initial announcement concerning ALH84001 meant that the claims could be placed in context. In other words, scientists had a chance to explain that this was not an isolated discovery. Rather, they were able to demonstrate in a timely manner where the claims concerning ALH84001 fit in a pattern of discoveries easily explainable to non-specialists using straightforward thematic concepts.

3. ***Timing Relative to Budge-Planning Cycle***—The ALH84001 claims, the resultant publicity, and the subsequent contextual explanations came at a time when NASA and the Office of Management and Budget were formulating budget proposals and counterproposals for the subsequent fiscal year. Thus, if policy makers wanted to follow up on the discoveries, the federal purse strings were already loose and public monies could more easily be found for new activities than they would have been at a later stages in the budgetary process when funds had already been allocated.

4. ***Implementation Plans Already Formulated***—Major elements of the plans concerning what to do next to exploit the window of opportunity pried opened by ALH84001 already existed thanks to prior work by scientific advisory groups to define important science questions and priorities in areas relevant to origins. These plans, together with activities such as those of researchers at the Ames Research Center to define astrobiology and to formulate the concept of a science institute dedicated to astrobiological studies, meant that the foundations already existed for the uses to which enhanced funding would be used should it become available.

These factors therefore suggest that the space-science community should conduct regular planning activities for new initiatives even if the current budget climate does not permit their immediate implementation. Strategies should be formulated that outline the current state of scientific knowledge in particular fields, define the important open questions for which answers are needed, and the means by which answers to these questions might be obtained. These strategies need to be revisited on a regular basis so that they are current and available for implementation should the opportunity arise.

2. COEL'S LIFE IN THE UNIVERSE STUDY

2.1 Origin of Study

Although Congress approved the Clinton Administration's Origins Initiative and authorized funding for its various elements, including NASA's Astrobiology Program, it was only a matter of time before some Representatives began to wonder whether or not NASA was making the best use of its enhanced funding. In particular, they asked, how did NASA's programs complement or duplicate activities being undertaken by other federal, private, and foreign research groups?

As a result of these concerns, the NASA Authorization Act of 2001 included language calling for a review by the National Academy of Sciences of "international efforts to determine the extent of life in the universe." In response to this congressional direction, Edward Weiler, NASA's associate administrator for the Office of Space Sciences, commissioned the National Research Council's[26] Space Studies Board to undertake a study to addressing the following main topics:

- Assess the direction of the Astrobiology Program;
- Survey initiatives for seeking life in the universe conducted by other U.S. Federal and non-governmental groups, and by foreign space agencies;
- Identify enhancements to the U.S. program; and
- Recommend areas for coordination of NASA efforts with those of other parties.

Dr. Weiler's letter suggested that the review might be undertaken by the SSB's Committee on the Origin and Evolution of Life (COEL). This committee, a joint undertaking between the SSB and the NRC's Board on Life Sciences, was established in 1999 to be provide scientific advice to, and maintain oversight of, NASA's Astrobiology Program and related activities supported by other federal agencies.

COEL's review of NASA's Astrobiology program was formally initiated at a meeting held in Irvine, California, in February 2001. Here the committee's members (Table 18-2) heard presentations by officials from NASA headquarters and NAI principal investigators. The committee began drafted the report at its July 2001 meeting in Washington, D.C., during which it heard from officials of other federal agencies involved in astrobiology. Work continued in November 2001 at a meeting at the SETI Institute in Mountain View, California, where programs to search for signals from intelligent extraterrestrial life were examined. The report was completed in February 2002 at the University of Arizona in Tucson, where technologies relevant to the search for and characterization of extrasolar planets were discussed. The report was revised and updated in response to reviewer comments during a meeting in Woods Hole, Massachusetts, in May 2002 and released in an unedited, prepublication format the following July. The final, printed version of the report, *Life in the Universe,* was released in April, 2003.[27]

Table 18-2. The Committee on the Origins and Evolution of Life

Name	Affiliation
John Baross	University of Washington, *Co-chair*
Jonathan I. Lunine	University of Arizona, *Co-chair*
Luann Becker	University of California, Santa Barbara
Steven A. Benner	University of Florida
Joseph A. Berry	Carnegie Institution/Stanford University
Wendy M. Calvin	University of Nevada, Reno
David Deamer	University of California, Santa Cruz
Marilyn Fogel	Carnegie Institution of Washington
Katherine H. Freeman	Pennsylvania State University
J. Peter Gogarten	University of Connecticut
Norman Pace	University of Colorado
Sandra Pizzarello	Arizona State University
David A. Stahl	University of Washington
Lucy M. Ziurys	University of Arizona

In the course of the study, COEL received presentations from a diverse group of experts including the following:

• Officials from NASA, the National Science Foundation, the Department of Energy and the National Institutes of Health;

• The principal investigators and/or representatives from 13 of the NASA Astrobiology Institute's 15 lead teams; and

• Representatives of astrobiologically relevant research groups at the Space Telescope Science Institute, SETI Institute, Monterey Bay Aquarium Research Institute, Smithsonian Astrophysical Observatory, Salk Institute and Centro De Astrobiología.

2.2 Structure of the Astrobiology Program

NASA's Astrobiology program has the following main component:

• The NASA Astrobiology Institute, a consortium of geographically dispersed research teams conducting collaborative research facilitated by advanced videoconferencing and next-generation internet links.

• The Exobiology and Evolutionary Biology research and analysis (R&A) programs which disperse grants to individual researchers via a process of peer review; and

• The Astrobiology Science and Technology Instrument Development (ASTID) and Astrobiology Science and Technology for Exploring Planets (ASTEP) programs designed to provide individual researchers with the funds necessary to development instruments relevant to astrobiological studies and to field test advanced instrumentation in unique terrestrial environments, respectively.

2.3 The Findings of COEL's Study

The findings and recommendations of COEL's study can be organized as follows:

- Overall assessment of the Astrobiology program;
- Enhancements to the Astrobiology program
- Astrobiology outside of NASA

2.3.1 Overall Assessment

COEL commended NASA for the following:

• The remarkable progress made in the past 5 years organizing the Astrobiology program;

• Recognizing the high value of research and analysis programs related to astrobiology;

• The present level of involvement of the Astrobiology program in flight missions; and

• Developing a well-balanced Solar System Exploration program as a foundation for the central endeavor of astrobiology.

2.3.1.1 The Astrobiology Roadmap

COEL found that the 1999 Astrobiology roadmap—NASA's astrobiology manifesto—is too broad and insufficiently selective in defining the following:

- The central research goals of astrobiology;
- Peripheral goals that still may contribute; and

• Goals genuinely outside of astrobiology as an intellectually coherent endeavor.

COEL proposed that astrobiology be defined as a discipline whose central focus concerns a selected set of issues directly linked to the origin, evolution, and ubiquity of life in the cosmos. And, that an important operational goal of astrobiology is to inform NASA spaceflight missions with respect to the techniques and targets for the search for life elsewhere, and the search for clues to the steps leading to the origin of life on Earth.

The core scientific areas within astrobiology should, therefore, be specifically and selectively defined as those that deal with the origin, evolution, and occurrence of life in the cosmos as embraced in NASA's R&A programs in the general areas of exobiology, evolutionary biology, planetary origin and evolution, cosmochemistry, and astronomical studies relating to the search for, and characterization of, extrasolar planets. Finally, COEL suggested that the drafting of a revised Astrobiology roadmap should be closely coordinated with that of the Office of Space Sciences' Astronomical Search for Origins program to identify overlaps in the two activities, common areas of study, and approaches to enhance the level of interaction in research.

2.3.1.2 The Virtual-Institute Concept

Given that the NAI was founded as an experiment in collaborative research facilitated by advanced telecommunications tools, COEL was concerned that NASA was not implementing communications solutions to bring the institute in line with the original concept. As a result, NASA should critically review the communication needs and costs required to make the NASA Astrobiology Institute a virtual institute along the lines of the original vision established by NASA's Ames Research Center.

2.3.1.3 Astrobiology Missions

Although the Astrobiology program's present level of involvement in NASA spaceflight missions is appropriate, COEL cautioned NASA against attempting to force NAI or other elements of the Astrobiology program into an artificially focused role of trying to design specific astrobiology missions. While individual NAI investigators are encouraged to propose instrument concepts or whole Discovery-class (or equivalent) missions, NASA must be careful not to bias the usual peer-review selection process for instruments and missions by specially labeling proposals developed by NAI investigators.

2.3.2 Enhancements to the Current Program

2.3.2.1 Balancing Astrobiology's Programmatic Elements

While COEL believes that the various elements of the astrobiology program—consortium science, R&A grants programs, and technology-development and field-testing activities—it was concerned that the relative funding for each is not balanced. That is, the resources devoted to NAI are approximately equal to the total resources devoted to the other two programmatic elements. In COEL's view, a well-balanced triad of science and technology efforts will better ensure the long-term vigor of astrobiological research.

2.3.2.2 Balancing the Participants in Astrobiology

NASA has done a good job of highlighting the research performed by the members of NAI's lead teams. It has not, however, done as much to draw attention to the achievements of the larger community of astrobiologists with no NAI affiliation. The failure to recognize that astrobiology encompasses a much larger volume of intellectual and capital resources than NAI could, potentially, damage the long-term development of astrobiology as a new scientific discipline. To this end, COEL suggested that NAI actively encourage extramural collaborations. A particularly important mechanism for promoting interactions between NAI and non-NAI researchers are the focus groups established by NAI Director Baruch Blumberg to examine specific topical areas relevant to astrobiology. Participation in these semiformal discussion should be open to all interested researchers irrespective of their institutional affiliation.

2.3.2.3 Alternative Approaches to Consortium Science

The NAI is not NASA's only experiment in collaborative research in the general area of astrobiology. Two so-called NASA Specialized Center of Research and Training (NSCORT)—the New York Center for Studies of the Origin of Life at Rensselaer Polytechnic Institute and the Exobiology Center at the University of California, San Diego—have been supported for approximately 5 and 10 years, respectively, to perform studies relevant to the origin and evolution of life. Like NAI, the NSCORTs are consortia of individual research teams. But unlike NAI, an NSCORT's research activities are co-located at each institution (or confined to a small number of geographically adjacent institutions) rather than involving a larger number of geographically dispersed groups as is the case with NAI. Another important difference between NAI and the NSCORTs is that the latter have a particularly strong educational emphasis. Indeed, the success of the older of the two NSCORTs in producing talented and accomplished graduates recommends this program, and COEL saw them as important alternative models for doing consortium science and recommended that they should continue to run in parallel with the NAI.

2.3.2.4 Balancing Scientific Disciplines

Astrobiology research efforts are, according to COEL assessment, dominated by the biological and geological sciences. This is inappropriate since the intellectual objectives of astrobiology includes much of the planetary sciences and selected areas of astronomy and astrophysics. Involvement of planetary scientists and astronomers has been hampered in those communities by concerns about the scientific value of astrobiology as an intellectual endeavor. Even absent such concerns, there remains the difficulties inherent in collaboration between researchers whose research techniques, technical language, and experimental approaches are very different. Nevertheless, addressing long-term astrobiological objectives will require a deeper and more extensive exchange of ideas between biologists and geoscientists and practitioners of the more traditional space sciences.

2.3.3 Astrobiology Outside of NASA

2.3.3.1 Other Federal Agencies and Departments

In addition to NASA's support, a significant amount of research relevant to astrobiological goals is sponsored by a variety of other public and private institutions in the United States. Activities funded by the National Science Foundation (NSF), the National Institutes of Health (NIH), the Department of Energy (DOE) contribute, or have the potential to contribute, to astrobiology. Even some of the activities of the U.S. Department of Agriculture (USDA) are relevant to astrobiology. Closer coordination and cooperation between these organizations is key to making the best use of limited research funds.

NSF provides U.S. researchers with access to unique scientific facilities—e.g., ground-based optical and radio telescopes—and locations—e.g., Antarctica (meteorites, ice cores, Dry Valleys, Lake Vostok)—relevant to astrobiology. Moreover, NSF's pioneering Life in Extreme Environments program—unfortunately now discontinued—was important for bringing scientists in the biological and physical scientists together to explore important problems relevant to astrobiology. NASA and NSF operate is very different ways. NSF's basic philosophy is to support hypothesis-driven research and, as a result, discovery-based proposals fare poorly. Whereas, NASA's basic philosophy is mission oriented. The space agency's scientific culture favors an exploratory approach and, as a consequence, discovery-based proposals are encouraged. These complementary personalities are important for the advancement of astrobiological research.

NIH, like NSF, has a scientific culture which discourages discovery-based research. As a result, COEL believes that NIH has overlooked many insights into astrobiology that might have been gained by NASA's non-reductionist approach. Nevertheless, NIH commands financial resources far beyond anything NASA currently possesses or is likely to possess in the future. And, it is difficult to conceive of a comprehensive astrobiology strategy that does not, in some way or other take advantage of, and build upon, the diverse biomedical research activities conducted by NIH.

The contributions DOE and USDA could potentially make to astrobiology fall within the general area of genomics. DOE has a uniquely productive gene-sequencing program which NASA should find creative ways to leverage. Similarly, USDA is interested in the genomes of agriculturally important organisms. COEL pointed out that these genomes conceivably reflect the response of ancestral organisms to historical changes in physical and biological environment—including mass extinctions and major climactic variations—and suggested that NASA cooperate with USDA to enable astrobiologists to use and interpret these genetic records.

2.3.3.2 Private Institutions

Private organizations also have a role to play in astrobiology. NASA's life-detection efforts are focused almost exclusively on the detection of primitive organisms. This was not always the case, at one time NASA sponsored attempts to detect the stray radio emissions of advanced extraterrestrial civilizations. COEL commented that the search for extraterrestrial intelligence (SETI) is both the most romantic and the most contentions aspect of astrobiology. NASA's program proved too controversial and was terminated in 1993 following a heated debate in Congress. Since then, leadership in this activity has passed to the California-based SETI Institute. In COEL's view, this organization has forged a unique endeavor by the judicious combination of private and public funds. In doing so, it had maintained a high standard of scientific research and had articulated clearly and authoritatively the rationale for approaches to a comprehensive SETI program.

2.3.3.3 International Institutions

Astrobiology is not a purely U.S. activity. While European scientists have long pursued vigorous research efforts concerning the origin and evolution of life, the creation of NAI spurred the development of astrobiological institutes and consortia overseas. NASA's initiative have catalyzed the creation of similar programs in Australia, France, Spain, and the United Kingdom. COEL singles out for particular praise the efforts of Spanish astrobiologists to create a world-class astrobiology research facility, the Centro de Astrobiología (CAB). NASA and the NAI, in turn, deserve credit for doing their part in encouraging and fostering the growth of the CAB. Astrobiology is, in COEL's view, a prime example of the United States leading the rest of the world into a new discipline area and new forms of research.

2.4 Conclusions of COEL's Study

NASA's Astrobiology program is, in COEL's view, well poised to catalyze fundamentally important discoveries concerning the origin of life, its distribution in the cosmos, and the long-term fate of life on Earth. NASA scientists and managers deserve credit for implementing a broad program in an era of tight budgets. These positive conclusions were not without some reservations. COEL identified five key issues that the Astrobiology program should address in the near future to ensure its own health. These issues can be summarized as follows:

• The definition of astrobiology and its goals needs to be clarified to counter the widespread perception that astrobiology—as a nascent scientific discipline and a NASA program—is ill-defined.

• The impact NAI is having on the development astrobiology as a distinctive scientific discipline above and beyond that of a standard research and analysis program will eventually need to be assessed in a systematic manner.

• Expanding the field of astrobiology necessarily means that NAI regularly incorporate new researchers and new institutions and this requires the periodic review and retirement of the existing lead teams.

• If NAI is to be a catalyst for interdisciplinary research in astrobiology it must actively involve a much larger set of researchers than those who are members of the lead teams.

• The Astrobiology program needs to better integrate astronomical research within the context of its overall research portfolio.

ACKNOWLEDGEMENTS

The author thanks Wesley Huntress for several useful discussions, the International Space Science Institute for travel support, and Pascale Ehrenfreund for her patience.

NOTES AND REFERENCES

1. When in lower case, these terms are used interchangeably to indicate the scientific study of issues broadly relating to the origin, evolution and distribution to life in the universe. When in upper case, these terms refer to specific NASA programs.
2. That part of the Executive Office of the President responsible for preparing the federal budget submitted to Congress for approval each year.

3. Hubertus Sturghold, *The Green and Red Planet*, University of New Mexico Press, Albuquerque, 1953.
4. A.G. Wilson, "Problems Common to the Fields of Astronomy and Biology: A Symposium—Introduction," *Publications of the Astronomical Society of the Pacific* 70: pp.41-43, 1958.
5. H. Sturghold, "General Review of Problems Common to the Fields of Astronomy and Biology," *Publications of the Astronomical Society of the Pacific* 70: pp.43-50, 1958.
6. W.M. Sinton, "Spectroscopic Evidence for Vegetation on Mars," *Publications of the Astronomical Society of the Pacific* 70: pp.50-56, 1958
7. A. Dollfus, "The Nature of the Surface of Mars," *Publications of the Astronomical Society of the Pacific* 70: pp.56-64, 1958
8. J.A. Kooistra, Jr., R.B. Mitchell, and H. Strughold, "The Behavior of Microorganisms Under Simulated Martian Environmental Conditions," *Publications of the Astronomical Society of the Pacific* 70: pp.64-69, 1958
9. D.G. Simons, "Manned Balloon Capabilities for Astronomical Observations," *Publications of the Astronomical Society of the Pacific* 70: pp.69-74, 1958
10. I. Schmidt, "Application of Recent Advances in Physiological Optics to Astronomy," *Publications of the Astronomical Society of the Pacific* 70: pp.74-78, 1958

11. The Space Science Board was renamed the Space Studies Board in 1988.

12. L.V. Berkner and H. Odishaw , "Dimensions and Problems," pp.3-47 in *Science in Space*, L.V. Berkner and H. Odishaw eds., McGraw-Hill Book Company, New York, 1961.
13. J. Lederberg , "Exobiology: Experimental Approaches to Life Beyond the Earth," pp.407-425 in *Science in Space*, L.V. Berkner and H. Odishaw eds., McGraw-Hill Book Company, New York, 1961.
14. J. Lederberg and H.K. Hartline, "The Biological Sciences and Space Research," pp.403-406 in *Science in Space*, L.V. Berkner and H. Odishaw eds., McGraw-Hill Book Company, New York, 1961.
15. Space Science Board, National Research Council, *Biology and the Exploration of Mars*, National Academy of Sciences, Washington, D.C., 1966.
16. Space Science Board, National Research Council, *Extraterrestrial Life: An Anthology and Bibliography*, National Academy of Sciences, Washington, D.C., 1966.

17. Viking cost some $3.2826 billion in 1994 dollars according to Table 1.1 of *The Role of Small Missions in Planetary and Lunar Exploration*, National Academy Press, Washington, D.C., 1995. Cost converted to 2003 dollars using the inflation calculator at the U.S. Bureau of Labor Statistics website <http://www.bls.gov>.
18. Now called the Office of Biological and Physical Research.
19. Now called the Office of Earth Science.
20. Memo from NASA Chief Scientist and Associate Administrators for OSS, OLMSA, and OMTPE, "Response to February 2, 1995 White Paper."

21. NASA Science Institutes Team, *NASA Science Institutes Plan: Report of the NASA Science Institutes Team—Final Publication (Incorporating Public Comments and Revisions)*, National Aeronautics and Space Administration, Washington, D.C., February, 1996.

22. Postscript: The necessary legislative relief was never granted and the only independent science institute of the type proposed to be actually established was the National Space Biomedical Research Institute in Houston, Texas.
23. Fax from Ad Hoc Team at Ames Research Institute to T. Luewcock, City Manager, Sunnyvale, California, and others, "Reinvention and Innovation at the Intersections: Value of NASA/Ames Space and Life Sciences Community to the Nation and the Region," 15 March, 1995.
24. Letter from J.H. Gibbons, Assistant to the President for Science and Technology, to D.S. Goldin, Administrator, NASA, 25 September, 1996.
25. C.R. Canizares, A.I. Sargent, et al., *The Search for Origins: Findings of a Space Science Workshop*, National Aeronautics and Space Administration, Washington, D.C., 1996.
26. The National Research Council is the principal operating arm of the National Academy of Sciences and the National Academy of Engineering.
27. Space Studies Board and Board on Life Sciences, National Research Council, *Life in the Universe: An Assessment of U.S. and International Programs in Astrobiology*, National Academies Press, Washington, D.C., 2003

Chapter 19

ASTROBIOLOGY IN EUROPE

A. Brack[a], G. Horneck[b], the late D. Wynn-Williams[c]
[a]Centre de Biophysique Moléculaire, CNRS, Rue Charles Sadron, 45071 Orléans cedex 2, France, [b]DLR, Institute of Aerospace Medicine, Linder Höhe, 51147 Köln, Germany, [c]British Antarctic Survey, High Cross, Madingley Road, Cambridge CB3 0ET, U.K.

1. INTRODUCTION

Although the research programme designated as *Astrobiology*, the study of the origin, evolution, distribution and destiny of life, was recently launched as a NASA initiative, the question of the chemical origins of life is engraved in the European scientific patrimony, as it can be traced back to the pioneer ideas of Charles Darwin, Louis Pasteur, and more recently to Alexander Oparin. During the last decades, the European community of origin of life scientists has organized seven out of the twelve International Conferences on the Origins of Life held since 1957. This community contributed also to the enlargement of the field of research to the study of life in extreme environments and to the search for extraterrestrial life, i.e. astrobiology in the NASA-inspired terminology. The present paper aims to describe the European science background in astrobiology as well as the European networking in astrobiology (see also Brack et al., 2001).

2. THE SCIENCE BACKGROUND

Astrobiology includes the study of the origin, evolution and distribution of life in the Universe. Although it is difficult to define what is meant by the word "primitive life", one generally considers as living any chemical system able, *a minima*, to transfer its molecular information via self-reproduction and to evolve. The concept of evolution implies that the system normally

P. Ehrenfreund et al. (eds.), Astrobiology: Future Perspectives, 467–475.

transfers its information more or less faithfully but makes a few random errors, leading potentially to a higher complexity and possibly to a better adaptation to the environment.

Because of evolution, primitive life was different from life today and only hypothetical descriptions of primitive life can be proposed. Because of the limitations of time, prebiotic chemistry can never be repeated in the laboratory. Therefore, simulations may only represent possible supports for plausible hypotheses from the point of view of historical legitimacy. The only way around this difficulty is to collect clues from different disciplines.

Today, much information related to the history of terrestrial life as well as to the detection of possible extraterrestrial life has been collected by European scientists in astronomy, planetology, geology, paleontology, biology and chemistry. The data are like pieces of a puzzle which can now be put together to give a clearer image of the possible distribution of life in the Universe.

3. ASTROBIOLOGY NETWORKING ACTIVITIES IN EUROPE

Astrobiology is a multidisciplinary science. It encompasses the disciplines of chemistry, biology, palaeontology, geology, atmospheric physics, planetary physics and stellar physics. Most aspects of this research require substantial resources, in analytical or simulation laboratory facilities, ground instrumentation, spacecraft and staffing. Several European nations, e.g., Finland, France, Germany, Russia, Spain, Sweden, Switzerland and United Kingdom, have already established national networks. Collaborative links are already producing real results in Europe, such as the ROSE, Response of Organisms to Space Environment, consortium of ESA-funded experiments for the International Space Station, and the ESA Topical Team ROME on Responses of Organisms to Martian conditions. Within the framework of the European Commission COST Actions to foster cooperation in a specific research area, COST D27 *"Origin of life and early evolution"* has been approved in June 2001 for a period of 5 years.

The main objective of this action is to develop the chemistry of the origins and early evolution of life, with special attention to cosmochemistry, prebiotic chemistry of small molecules, directed evolution and origin of the genetic code. Hosted by the International Space Science Institute in Bern, the International Space Science Team "Prebiotic matter in space" is a consortium of 12 scientists, each representing a specific research field crucial to revealing the origin of life as a consequence of the evolving Universe.

The European Exo/Astrobiology Network Association, EANA, was created to co-ordinate the different European centres of excellence in exo/astrobiology or related fields. The specific objectives of EANA are :

- to bring together European researchers interested in exo/astrobiology programmes and to foster their cooperation
- to attract young scientists to this quickly evolving, interdisciplinary field of research
- to create a website establishing a database of expertise in different aspects of exo/astrobiology
- to interface the Network with European bodies such as ESA, ESF, the European Commission and with non European institutions active in the field
- to popularise exo/astrobiology to the public and to students.

Collaborative research has been developed in the different fields covered by exo/astrobiology:

- terrestrial life as a reference (Origins of life, environmental context, ingredients for primitive life, life in a test tube, diversity of bacterial life, life in the extremes, panspermia)
- exploring the Solar System (Mars, Europa, Titan, comets)
- search for life beyond the Solar System (exoplanets, Corot and Darwin missions).

The value of sharing resources on an international scale was highlighted at the Inaugural Meeting of the European Exo/Astrobiology Steering Group at the British National Space Centre, London in October 1999 and at the Strategy-oriented Meeting at CNES, Paris in October 2000. At these two meetings, senior representatives of the European Science Foundation and the European Space Agency acknowledged that exobiology in Europe should be strengthened, formalized as a Network and supported.

The First European Exo/Astrobiology Workshop, held in Frascati, Italy, May 2001, was attended by 200 scientists. The Second European Workshop was organised in Graz, Austria, in September 2002. The workshop, attended by 320 participants, was oriented particularly to the planetology aspects of astrobiology, in acknowledgement of the expertise of the local organisers.

The Third European Workshop was hosted in November 2003 by the Centro de Astrobiologia in Madrid and dedicated to the search for life on Mars.

EANA is run by an Executive Council consisting of national members presently representing 17 European nations active in the field, e.g. Austria, Belgium, Denmark, Finland, France, Germany, Hungary, Italy, Poland, Portugal, Romania, Russia, Spain, Sweden, Switzerland, The Netherlands, United Kingdom, on the basis of one representative per nation, and elected members in a number equal to the number of active nations.

EANA is affiliated to the NASA Astrobiology Institute. The formal affiliation was signed in 2002 at the Graz Workshop by Rosalind Grymes, Deputy Director of NAI, during a reception hosted by the Governor of Styria in the historical Eggenberg Castle. EANA is member of the Federation of Astrobiology Organizations, FAO, including the Australian Centre for Astrobiology (ACA), the Astrobiology Society of Britain (ASB), the Spanish Centro de Astrobiologia (CAB), the French Groupement de Recherche en Exobiologie (GDR Exobio), the American NASA Astrobiology Institute (NAI) and the Swedish Astrobiology Center (SWAN). The FAO has been created to facilitate international exchange between students and to harmonise the planning of joint astrobiology meetings.

The EANA Web Page, *http://www.spaceflight.esa.int/exobio*, is hosted as part of the ESA Virtual Institute at ESA/ESTEC in Noordwijk, The Netherlands.

4. EUROPEAN SPACE MISSIONS RELEVANT TO ASTROBIOLOGY

4.1 The Aurora Programme

In November 2001, the Ministerial Conference in Edinburgh approved a new ESA programme called Aurora that is dedicated to planetary exploration. This programme has the goal of implementing a long-term plan for the robotic and human exploration of Solar System bodies holding promise for traces of life. The primary planetary target is Mars with the Moon considered as possible target for technological developments.

Right from the beginning, this was an interdirectorate initiative in which the Directorate for Strategy, the Directorate for Manned Spaceflight and

Microgravity, the Directorate for Space Science and the Directorate for Technical Operations and Support worked together. As a basis for the Aurora long-term plan, a human Mars mission is scheduled 2030.

Between 1997 and 1999, an Exobiology Science Team performed an overview of the state of exobiology and developed recommendations specifically for the search for life on Mars (Brack et. Al, 1999; Westall et al., 2000). An integrated instrument package, which included subsurface drilling capabilities, was recommended.

This instrument package is the core of what is now called the Pasteur Payload intended to equip the rover of the ExoMars exobiology mission in 2009. The latest developments in instrumentation have been taken into account through a call for ideas. Three scientific working groups established through this call will interact with the industrial contractor for the phase A/B development of the package, to ensure readiness and best performance when the flight opportunity with the ExoMars mission occurs. In 2020, a Mars sample return mission will again provide clues in the search for life, and will include on a small scale, all the technology elements of a human Mars mission, ranging from launch, transfer to Mars, atmospheric entry and descent, landing, ascent and finally return to Earth.

4.2 The Cassini-Huygens Mission (launched in October 1997)

The NASA/ESA Cassini-Huygens spacecraft, launched in October 1997, will arrive in the vicinity of Saturn in early 2005 and perform several flybys of Titan making spectroscopic, imaging, radar and other measurements. A descent probe managed by European scientists will penetrate the atmosphere and will systematically study the organic chemistry in Titan's geofluid. During 150 minutes, *in situ* measurements will provide detailed analysis of the organics present in the air, in the aerosols and at the surface (Lebreton, 1997; Raulin, 1998).

4.3 The Mars Express Mission (launched in June 2003)

The European Mars Express mission has been launched on the 2nd of June 2003 from the Baikonur launch pad in Kazakhstan on board a Soyuz Fregat Launcher. Its main objectives are the search for water and life. The spacecraft carries seven instruments. The instruments will make observations from the main spacecraft as its polar orbit gradually moves round to give global coverage over the mission's expected lifetime of nearly one Martian (two Earth) year. The High/Super Resolution Stereo Colour Imager HRSC will take detailed images of the surface and atmosphere of Mars and will image the entire planet in full colour, 3D and at high resolution. The ground penetrating radar Subsurface Sounding Radar/Altimeter MARSIS will search for water up to 5 km below ground. Three complementary spectrometers will analyse the atmosphere and the surface : the IR Mapping Spectrometer OMEGA will map the mineralogical surface composition of Mars, the Planetary Fourier Spectrometer PFS will study the composition of the atmosphere and the UV and IR Atmospheric Spectrometer SPICAM will address the question of why is the Martian atmosphere so oxidising. The Energetic Neutral Atoms Analyser ASPERA will detect energetic neutral atoms and address the question of how the solar wind erodes the Martian atmosphere. Finally, the radio science experiment on board Mars Express MaRS will use the radio signals that convey data and instructions between the spacecraft and Earth to probe the planet's ionosphere, atmosphere, surface and even interior.

An exobiology lander, Beagle-2, was carried by Mars Express. The lander comprised an integrated suite of instruments to optimize the search for traces of life on Mars in protected subsurface and rock interior samples. The package included instruments to study sample mineralogy (composition, macroscopic and microscopic structural and textural features), organics (elemental, isotopic, molecular), oxidation state and petrology. The lander successfully separated from the orbiter on 19 December 2003, and was programmed to land on Isidis Planitia on December 25. All the attempts to contact Beagle 2 from Mars Express, Mars Odyssey and from the Jodrell Bank radio-telescope were unsuccessful, thus compelling the scientists to consider Beagle 2 as lost.

4.4 The Rosetta Mission (launched in March 2004)

Rosetta is designed to rendez-vous with Comet 67P/Churyumov-Gerasimenko, orbiting around it and making observations as it journeys towards the Sun. En route to the comet it will flyby at least one asteroid. It has been launched in March 2004 and the mission will end in 2015. The objective of Rosetta is to study the origin of comets, the relationship between cometary and interstellar material and its implications with regard to the origin of the Solar System. The measurements relevant to astrobiology will consist in the determination of the chemical, mineralogical and isotopic compositions of volatiles and refractories in the cometary nucleus.

4.5 The COROT Mission (launch expected early 2006)

The French led mission COROT, dedicated to wide field stellar photometry, aims to detect and to study stellar oscillations - asteroseismology - and to search for extrasolar planets, especially telluric ones. These terrestrial-like rocky planets will be searched for by the transit (or occultation) method which allows the precise determination of the orbital period and the size of the planet since COROT is devoted to stellar photometry, aiming both a high precision and a long observation time. Although COROT is not optimized for the detection of telluric exo-planets, mainly because of its small telescope diameter, it should nevertheless be able to achieve some significant small planet identification.

4.6 The Darwin Mission (launch expected in 2014)

Darwin will use a flotilla of six space telescopes, each of which will be at least 1.5 metres in diameter. They will work together to scan the nearby Universe, looking in the mid-infrared for signs of life on Earth-like planets. To see planets around nearby stars would require a telescope of about 30 metres in size and this is far beyond the current limits of technology.

To overcome this limitation, Darwin will use a technique known as interferometry, which consists of the use of a number of smaller telescopes that combine their individual signals to mimic a much larger telescope. Darwin must also cut out the blinding light from the central star. This will be achieved by a so-called "nulling interferometer" whereby the signals from some of the telescopes are slightly delayed.

By precisely adjusting this delay, the central bright object is 'cancelled' out, allowing the faint, nearby planet to be visible. The American space agency, NASA, is also considering a mission similar to Darwin called the Terrestrial Planet Finder. Given the ambitious nature of both projects, it is possible that NASA and ESA will collaborate on the final mission, building a joint Darwin/Terrestrial Planet Finder, which they will launch and operate together.

4.7 Astrobiology Experiments in Earth Orbit

ESA has developed two facilities to expose organic and biological specimens to the conditions of outer space and to study their responses after retrieval: (i) the BIOPAN facility on board of the Russian satellite Foton which stays 2 weeks in Earth orbit and (ii) the EXPOSE facility to be mounted at an external platform of the International Space Station. BIOPAN will resume its flights in 2005 and EXPOSE is scheduled for 2005 and probably again for 2008 and will stay up to 3 years in space. Main fields of research are directed towards understanding prebiotic chemical evolution processes and limits of life in extreme environments.

5. EUROPEAN TRAIL GUIDE IN ASTROBIOLOGY

Life - defined as a chemical system capable of self-reproduction and of evolution - originated from the reaction of reduced carbon-based organic matter in liquid water. The organic molecules required for the appearance of terrestrial life might have been formed in Earth's atmosphere, in submarine hydrothermal systems or delivered to the Earth via interplanetary dust. Schematically, the prebiotic resources needed for the emergence of life can be compared to parts of a chemical robot. By chance, some parts assembled to form a robot capable of assembling other parts to form a second identical robot. From time to time, a minor error in the assembly process generated more efficient robots. Since the oldest fossils of microorganisms discovered so far have been identified in geological horizons associated with hydrothermal systems, either subaerial or submarine, it is reasonable to consider that the robots were driven by thermal energy, around 85°C. The robots had therefore to protect themselves against hydrolysis that was also boosted by the high temperatures. They probably faced this difficulty by growing on, or within, mineral surfaces such as sulfides and/or clays.

The number of parts required for that first robot is still not known. The problem is that, on Earth, those earliest parts have been erased by geological processes such as plate tectonics. The chances of understanding the emergence of life on Earth by recreating a similar process in a test tube will obviously depend upon the simplicity of the chemical reactions leading to life. The discovery of a second genesis of life on another celestial body would strongly support the idea of a rather simple genesis of terrestrial life. The possibility that life may have evolved on Mars during an early period when water existed on the surface, and the possibility that life may still exist on that planet deep below the surface, marks it as a prime candidate in the search for life beyond the Earth. Cryovolcanic flows at the surface of Europa point to a possible subsurface ocean of water and of hydrothermal vents, which might harbour a basic life form. The discovery of the exoplanets opens the search for extraterrestrial life to the whole Universe.

The trail guide for future European research in astrobiology should target:

- the environment of the primitive Earth and the supply of organics
- the search for autocatalytic systems active on mineral surfaces in water at temperatures ranging from 50 °C to 85 °C
- the bacterial world as a reference
- the search for a second genesis of life on Mars, Europa, exoplanets with space missions and SETI.

REFERENCES

Brack, A., Fitton, B. and Raulin, F. (1999) Exobiology in the Solar System & the search for life on Mars. ESA-SP 1231, ESA Pub. Noordwijk, The Netherlands.

Brack, A., Horneck, G., Wynn-Williams, D. (2001) Exo/Astrobiology in Europe. Origins Life Evol. Biosphere 31,459-480.

Lebreton, J.P. ed. (1997) Huygens: Science Payload and Mission. ESA-SP 1177, ESA Pub. Noordwijk, The Netherlands.

Raulin, F. (1998) Titan. *in* The molecular origins of life: assembling pieces of the puzzle. ed. A. Brack, pp. 365-385, Cambridge University Press, Cambridge.

Westall, F., Brack, A., Hofmann, B., Horneck, G., Kurat, G., Maxwell, J., Ori, G.G., Pillinger, C., Raulin, F., Thomas, N., Fitton, B., Clancy, P., Prieur, D. and Vassaux, D. (2000) An ESA study for the search for life on Mars. Planet. Space Sci. 48, 181-202.

Chapter 20

FUTURE PERSPECTIVES AND STRATEGIES IN ASTROBIOLOGY

The ISSI Team
Pascale Ehrenfreund[a], Bill Irvine[b], Toby Owen[c], Luann Becker[d], Jennifer Blank[e], John Brucato[f], Luigi Colangeli[g], Sylvie Derenne[h], Anne Dutrey[i], Didier Despois[j], Antonio Lazcano[k], Francois Robert[l],
[a]Leiden Observatory, Holland, [b]University of Massachusetts, USA, [c]University of Hawaii, USA, [d]University of California, Santa Barbara, USA, [e]Lawrence Livermore National Laboratory, USA, [f]Observatory Napoli, Italy, [g]Observatory Napoli, Italy, [h]ENS de Chimie de Paris, France, [i]Observatory Bordeaux, France, [j]Observatory Bordeaux, France, [k]UNAM, Mexico, [l]Laboratoire de Mineralogie, France

The subject of Astrobiology can be approached from many different perspectives, as reflected in numerous textbooks, journals, conference proceedings, web documentation and popular literature. This book is focused on abiotic organic matter from the viewpoint of astronomy and planetary science and considers its potential relevance to the origins of life on Earth and elsewhere. Guided by the review papers in this book, this concluding chapter aims to identify key questions to motivate future research and stimulate astrobiological applications of current and future research facilities and space missions.

1. ORIGINS

In recent years outstanding progress has been made in our understanding of the structure of the Universe, its age and important epochs during its evolution. The Wilkinson Microwave Anisotropy Probe (*WMAP*) has made the first detailed full-sky map of the oldest light in the Universe. *WMAP* polarization measurements indicated that the first stars ignited as early as 200 million years after the Big Bang. It is thought that these objects produced the first heavy elements, although with a different nucleosynthetic

P. Ehrenfreund et al. (eds.), Astrobiology: Future Perspectives, 477–512.

signature than later generations of stars (Spaans 2004). The age of the Universe is found to be ~ 13.7 billion years old, with an apparent margin of error of close to 1%. In the currently popular "concordance model" of the Universe, 70% of the energy density is thought to be dark energy, 25% dark matter and 5% normal matter (http://map.gsfc.nasa.gov/). However, neither cold dark matter nor dark energy are well understood. ESA's X-ray observatory, *XMM-Newton*, surveyed distant clusters of galaxies and found puzzling differences between today's clusters of galaxies and those present in the Universe around seven thousand million years ago. Some scientists claim that this can be interpreted to mean that the 'dark energy' which most astronomers now believe dominates the Universe simply does not exist (Vauclair et al. 2003).

Measurements of targets at high redshift allow us a glimpse into the early history of our Universe. CO gas from the most distant quasar at z = 6.42 has been recently measured (Walter et al. 2003), showing directly that heavy elements were already abundant ~ 850 million years after the Big Bang. Furthermore it has been reported that galaxy clusters began forming when the Universe was only ten per cent of its present age (Miley et al. 2004). Among the fundamental cosmological questions of direct relevance to Astrobiology are:

- What are the conditions for primordial star formation ?
- When and where did the first heavy elements form ?
- How are high and low-mass protostars formed ?
- When did the first planetary systems form ?

Many parameters of current cosmological models have to be confirmed and refined by theoretical work, observations of distant objects, and space missions such as *WMAP, Planck, SST, Herschel* and *JWST*, see Appendix.

2. INTERSTELLAR MEDIUM

In our own Milky Way galaxy, which seems to be typical, dense interstellar clouds contain more than 100 identified molecular species in the gas phase (Millar 2004). In the interstellar medium the distribution of carbon is still an unsolved question (Cataldo 2004). In dense molecular cloud material a reasonable fraction of C is incorporated into CO gas (20%) and a small percentage (~ 5 %) is present in C-bearing ice species. In contrast, in diffuse interstellar clouds which are exposed to UV radiation very low levels of CO gas are observed. In such environments about 15 % of the cosmic carbon is attributed to polycyclic aromatic hydrocarbons (PAHs). PAHs seem to be prevalent in the diffuse interstellar medium but may or may not be present within molecular clouds. This leaves a large fraction (> 50 %) of the cosmic carbon unaccounted for in the interstellar medium.

Laboratory simulations in combination with interstellar observations argue that this missing carbon is incorporated into solid state macromolecular material (Pendleton and Allamandola, 2002) such as amorphous and hydrogenated amorphous carbon (Ehrenfreund and Charnley, 2000; Mennella et al. 1998). Though many different forms of carbon have been discussed, hydrogenated (and dehydrogenated) amorphous carbon provide currently the best fit to interstellar observations (UV bump at 220 nm in the interstellar extinction curve) and simultaneously the best quantitative solution for current dust models (Mennella et al. 1998).

Major questions concerning the chemistry of organic matter in the interstellar medium that will be investigated as new ground-based and space-borne instrumentation becomes available include

- What are the principal reservoirs of carbon in different interstellar environments?
- How complex are the organics that are present in molecular clouds?
- What is the detailed nature of the chemical reactions and processing on and in interstellar grain mantles and how are the molecules so produced returned to the gas phase?
- What are the similarities and differences in chemical evolution between low-mass star-forming regions relative to regions where high-mass stars form, and can such comparisons tell us in what type of region the Sun formed ?
- How is organic matter preserved and/or processed as stars form in molecular clouds and interstellar material falls into protoplanetary nebulae, and which isotopic ratios are most informative of the processes involved (cf. Ehrenfreund et al. 2002; Millar 2004) ?
- Can we determine the precise nature of the PAHs present in the interstellar medium, as well as the carriers of the diffuse interstellar bands (DIBs)?

Most carbon in the ISM seems to be in the solid grains. Given that the bulk of the organic material in carbonaceous chondrites, which can be subjected to detailed laboratory analysis, is still considered an "intractable polymer", the precise characterization of the corresponding interstellar grain component may well depend upon the collection and analysis of interstellar dust by space missions such as *Stardust* and even then significant questions may remain.

Furthermore, to the degree to which the composition and structure of the grains remain uncertain, their precise role in chemical reactions will also be (at least partially) unclear. Progress on such questions will be aided by the higher sensitivity and spatial resolution provided by the new facilities that are anticipated to come online in the relatively near future, as listed in the Appendix.

Undoubtedly major advances in our understanding of such questions as those raised above will come from future theoretical and laboratory work. Low temperature ice simulations, gas phase reaction kinetics and stochastic simulations of gas and grain surface chemistries will provide insightful knowledge on lacking parameters. Theoretical research will certainly be aided by faster and more capable computers. One might hope that advances in laboratory techniques will allow the spectra of large, complex, gas phase organic molecules such as PAHs to be precisely measured, perhaps leading to the identification of the carriers of the DIBs. There remain many reactions of importance for interstellar chemistry whose rates have not yet been measured, particularly over the temperature range appropriate to molecular clouds. Likewise, the processes relevant to isotopic fractionation and ortho/para conversion as molecules freeze out onto grains or sublimate from ice to gas phase need study for many species (e.g., Irvine et al. 2000). To the extent that they contain unmodified interstellar material, comets may provide at least partial answers to these questions in a remarkably straightforward and elegant way (see section 6).

3. PROTOPLANETARY DISKS

In circumstellar disks encountered around precursors of solar-type stars, the gas and the dust orbiting the young stars are remnant material from the parent molecular cloud, even if some chemical re-processing may have already occurred during the early phases of the stellar formation, as mentioned by Markwick and Charnley (2004). During recent years there have been important improvements in modeling of protoplanetary disks, but the following topics are still outstanding:

- How does the angular momentum of disks evolve ?
- What are the physical and chemical conditions within the disk?
- What are the timescales for disk dissipation ?

In these objects H_2 clearly remains the dominant species, but observations of chemistry with large radio telescopes and mm interferometers allow observers to detect other molecules in the gas phase in a few objects: ^{12}CO and its main isotopomers ^{13}CO and $C^{18}O$, HCO^+ (and $H^{13}CO^+$), C_2H, CS,

CN, HCN, HNC and H_2CO (Dutrey et al. 1997). Observations reveal that, as in molecular clouds, CO is the most abundant detectable species containing carbon. In general, the simple organic molecules observed so far appear under-abundant compared to nearby dense cores. Indeed, molecular surveys are sensitivity limited to the detection of the more abundant molecules observed in molecular clouds and mainly trace material located in the "cold" outer disk (radius > 50 AU). For most of the species, the under-abundance can be explained by the fact that since the outer disk is cold (with typically a mid-plane kinetic temperature around $T_k = 10$ K at radius > 100-200 AU), molecules should condense on grains. The presence of some radicals such as CN or C_2H is also indicative of a photon-dominated chemistry. Molecules can be photo-dissociated in the upper layers of the disk (Van Zadelhoff et al. 2003). The disk, heated by the radiation from the central star, can be significantly warmer at the surface than at the mid-plane even at large distances, with temperatures of 30-100 K at radius=100 AU and at 3-5 scale heights above the mid-plane (Dartois et al. 2003). Closer to the star, where the disk temperature can be significantly higher (in particular, due to viscous heating) the chemistry would be different but this region is currently out of the range of telescopes. Recent models of disks irradiated by stellar X-ray emission also show that X-rays can significantly affect the disk surface and penetrate inside the disk.

However, chemistry cannot be investigated without an accurate knowledge of the physical conditions of a protoplanetary disk (such as temperature, density or turbulence). The lack of observations is a major limiting factor of current chemical models. The forthcoming generation of optical, infrared (such as *SST* and *JWST*) and submillimeter telescopes such as *ALMA* or *Herschel* should provide in the next decade the first quantitative and complementary studies of both the physics and the chemistry for the gas and the dust. *ALMA*, for example, will have the sensitivity and angular resolution needed to determine the radial and vertical dependences of physical parameters such as the kinetic temperature or surface density at scales of a few AU in nearby protoplanetary disks. *ALMA* will also allow deep molecular surveys for the detection of relatively complex prebiotic molecules. This, together with a better determination of the physical properties, would significantly improve our knowledge of the chemistry in regions of disks relevant to planetary formation.

4. PLANET FORMATION AND EXTRASOLAR PLANETS

Observational constraints on planetary formation are based on observations of our own Solar System, circumstellar disks around young stars which are thought to be analogs of our Solar Nebula, and more recently detected exoplanets (Wuchterl 2004). As of March, 2004, some 105 extrasolar planetary systems have been identified, of which 13 contain multiple planets, for a total of 120 planets (http://www.obspm.fr/encycl/cat1.html). Almost all of these have been found by the Doppler effect technique (e.g., Marcy et al. 2002; Queloz 2000), which searches for periodic variations in the stellar radial velocity induced by the gravitational effect of the planet(s). Because this technique is not currently sufficiently sensitive to detect terrestrial-type planets, all those found to date have masses (more exactly, lower limits on their masses) characteristic of giant (jovian) planets. The majority of the detected exoplanets have semi-major axes less than 1 AU, making their systems quite unlike the Solar System. The currently favored explanation for this unexpected scenario is that the planets formed at significantly larger distances from their host stars (beyond the "snow line") and subsequently migrated inward as a result of gravitational interaction with the analog of the solar nebula (Lin and Papaloizou 1986). The effect of such migration on the orbits of pre-existing planets is a subject of active research. Indeed physical conditions in the inner disk (radius < 50 AU, about the radius of the Kuiper Belt in our Solar System), where planetary formation is thought to occur, are actually poorly known and there are no direct observational constraints on models of planet formation.

But there are some indirect constraints. At least a few percent of solar-type stars have giant planets. There is a strong correlation between the proportion of heavy elements that a star contains (its metallicity) and the probability that it will have planets (Fischer and Valenti 2003). Within this correlation, the Sun has a rather low metallicity, suggesting that in this attribute, at least, our Solar System is unremarkable. In the case of Jupiter and the Sun, the giant planet itself has a higher proportion of heavy elements than its star (Owen et al. 1999). Taken together, these two observations strongly favor core-accretion models for giant planet formation, rather than models invoking disk instabilities.

Given that terrestrial planets are not yet observable and that giant planets with Solar-System-like orbits are just becoming detectable (as the time over which radial velocity measurements have been made approaches the orbital period of such planets), it remains unclear just how common/unusual our own planetary system may be. Obviously, this is a topic of the utmost importance for future astrobiological research.

In a couple of cases transits of planets have been observed. Observations of such events are very important, since they remove the uncertainty in the inclination of the planetary orbit, allowing actual planetary masses and radii to be determined. Moreover, absorption by atmospheric gases may be measurable, providing information on atmospheric composition (e.g., Hubbard et al. 2001).

Major questions for future investigations include:

- What are the physical conditions in the inner disk leading to initial particle aggregation and subsequent growth of planetesimals ?
- How do terrestrial and giant protoplanets grow within the disk ?
- What are the implications of evolving disk structure and subsequent feedback, including migration, for the formation of protoplanets?
- Do extrasolar planetary systems like ours exist, with terrestrial planets and life?

Thanks to high angular resolution, *ALMA* will provide images of gaps created by proto-planets in inner dust disks (Wolf et al. 2002). Optical and infrared interferometry performed with instruments such as the *VLTI* will also investigate the dust distribution in very inner disks. These observations will really open a new domain since they will provide invaluable information on interactions between disks and their young planets leading to observational constraints on hydrodynamical models of planet formation and migration.

It seems likely that the detection and ultimate characterization of extrasolar Earth-like planets will require observations from space. The first systematic search for Earth-like planets will be carried out by NASA's *Kepler* mission, an orbiting telescope to be launched in 2007. *Kepler* will simultaneously observe 100,000 stars, continuously monitoring their brightnesses with a precision of 0.01% for 4 years. The goal is to detect transits of small planets in front of their stars.

The precision is adequate to detect planets as small as Mars, and the four-year lifetime allows the detection of at least three transits of a planet in a Sun-like star's habitable zone. The French-led *COROT* mission similarly observes tiny changes in brightness from nearby stars and will be launched in 2006 from Russia.

Coronographs and interferometric arrays such as those proposed for NASA's *Terrestrial Planet Finder (TPF)* mission and ESA's *IRSI-DARWIN* mission (see Appendix) are the adventures for the next decade in the search for terrestrial planets and their biospheres.

5. PLANETS AND SMALL BODIES IN OUR SOLAR SYSTEM

Our Solar System is the obvious place to begin the search for extraterrestrial life. Recent years have seen enormous progress in the exploration of planetary bodies of the Solar System thanks to ambitious space missions using innovative technologies. Actually, on-going missions and future programs of the major international space agencies trace a clear route along which our knowledge of Solar System bodies will evolve in the next several years (Rummel 2004). NASA successfully deployed two landers on the Martian surface in January of 2004, both equipped with instrumented rovers capable of exploring traverses of a kilometer or more (http://marsrovers.jpl.nasa.gov/home/). One of them, *Spirit*, is in the crater Gusev, which is probably the site of an ancient lake, judging from the fact that a number of dry river channels empty into this crater. The ancient existence of liquid water on the surface has been established by Spirit's studies of rocks and minerals at the site. Opportunity, the second rover, landed inside a small crater in a region thought to contain the mineral hematite. Studies of the surroundings are underway as we write, including the use of a grinding tool to expose a fresh surface on Martian bedrock.

5.1 Space Research and Remote Sensing

The ESA *Mars Express* spacecraft has started to return new sets of data based on observations from orbit; as examples, high spectral resolution data are returned on the composition of the planet's surface and atmosphere and a powerful radar searches for sub-surface ice, while a stereo-colour camera with a resolution of 10m will examine the surface (http://www.esa.int/export/SPECIALS/Mars_Express/).

The IR spectrometer on *Mars Express* is able to detect and identify trace gases such as methane in the Martian atmosphere, that could be signs of active life. Other than that, however, these current missions are not directly looking for evidence of life itself. Rather, they are trying to unravel the planet's geological history to determine when and for how long, conditions at the surface of Mars might have been suitable for the origin and early evolution of life.

The *Cassini-Huygens* spacecraft is presently on its way to the Saturn system and will deploy a probe into the atmosphere of Titan in January of 2005. The probe will use a gas-chromatograph-mass spectrometer (GC-MS) to analyze the composition of both the gases and the organic aerosols in the atmosphere during its 2.5 hour descent to Titan's surface. It will also carry a number of other experiments, including a near IR spectrometer and an imager. If the probe survives the landing, it will analyze whatever surface materials it encounters, which could range from bare ice through drifts of aerosol deposits to lakes of liquid hydrocarbons. Hence we expect to learn a great deal about organic chemical synthesis driven by solar UV and electron impact in a low-temperature and thus non-aqueous environment.

The *Cassini* orbiter will carry out its own investigations of Titan over the full length of the mission, which will be at least 4 years long and may last possibly two to four more. The orbiter has a radar that can penetrate Titan's haze and map the surface, showing the locations of liquid hydrocarbon lakes and rivers. It also carries a powerful far-IR spectrometer that will assay the composition of Titan's atmosphere, including evaluations of isotope ratios. A near-IR spectrometer will map surface features at several wavelengths, thus enabling identification of their compositions. Further studies of Titan will be based on the results from *Cassini-Huygens* (http://saturn.jpl.nasa.gov/index.cfm).

The key questions we are trying to answer with planetary exploration are:

- What were the mechanisms responsible for Solar System formation?
- What are the evolutionary processes that have brought the Solar System to the configuration we observe today?
- What were the circumstances that have favoured the appearance of life on Earth and, possibly, on other bodies?

The future exploration of planetary bodies requires a combination of *remote* and *in situ* analyses (Roush and Cruikshank 2004; Roos-Serote 2004). This is particularly true for the planet Mars, surely the target with the most major implications for Astrobiology. The major space agencies aim at a progressive approach to investigate this planet, seeking signs of extant or extinct life in the near future, and ultimately planning to include human exploration. Remote observations have the task to give a better and more precise characterisation of the overall context of the planet in terms of chemical, physical and geo-morphological features.

The first goal of in situ exploration is to provide a deep knowledge of local properties by means of direct analyses of surface and sub-surface samples with the aim of identifying the material properties at macro- and micro-scales and solving questions related to the possible past appearance of life on Mars and related boundary conditions (e.g., history of water). The other important goal of in situ analyses is to study on short, seasonal and annual time scales the evolution of the atmosphere and its components (i.e., gas, aerosol, dust), as well as the surface-atmosphere interactions, with the aim to describe the present climatic conditions and to trace back the climate evolution on the planet (Roos-Serote 2004). These elements are essential both to understand the past history of the planet and to pose the boundary conditions and requirements (e.g., hazard conditions) for future exploration of the planet, which may include piloted missions. In this respect, mobility of automatic devices on the surface is essential to explore large areas of the surface and properly place the local information in the context provided by remote observations. The deployment of elaborate laboratories should include drilling below the surface to acquire samples from depths of a meter or more. The idea is to reach levels that have not been exposed to the highly oxidizing environment on the surface, in order to search for ancient organic material that may be preserved from early wet periods on Mars. These samples would then be pyrolized and analyzed by gas chromatography/mass spectrometry (GC-MS) to determine the nature of any organics they may contain.

Within the Mars exploration program, a fundamental step is represented by the *return of samples* for laboratory detailed analyses. This approach will offer a very powerful tool to validate remote and in situ data with precise reference results for their proper interpretation. More generally, analyses of Martian meteorites (e.g. Bishop et al. 1998) and terrestrial analogs of Mars materials (Roush and Bell 1995; Esposito et al. 2000) performed in the laboratory have already demonstrated their importance and continue to be a key component of the process aimed at correctly interpreting observations. The Mars example illustrates a standard approach for exploration of the Solar System that can (should!) be applied to several other bodies. For example, NASA and ESA missions to Mercury will explore an "extreme" planet, to characterize its evolution and present environmental conditions. In this case, the implications for Astrobiology are not linked to the search for life markers. However, we must consider that a proper achievement to answer key question 3 (listed above), cannot be obtained without a proper clarification of key questions 1 and 2.

Observations of Europa will be the centerpiece of the Jupiter Icy Moons Orbiter (*JIMO*) mission, using a nuclear-electric spacecraft that is being studied for a possible launch in 2012-2015 (http://www.jpl.nasa.gov/jimo/). The trip time to Jupiter will be about 6 years. Once on station, the spacecraft will be able to change its orbit at will using its ion engine. The payload is still under review, but a central objective of the mission will surely be an effort to establish the existence of Europa's putative sub-crustal ocean and to characterize it as much as possible. One aspect of this study will be an investigation of dark material around spatter cones and other features on Europa's surface to see if any organic material was brought up from below. Exploration of the ocean itself, if it indeed exists, will have to wait for a future mission.

5.2 Supporting Laboratory Studies

Laboratory research has played and will continue to play a fundamental role in supporting the interpretation of data returned by missions exploring various environments of the Solar System. The laboratory is the right place to:

- perform analyses on extra-terrestrial samples (e.g., meteorites, interplanetary dust particles, samples collected from Solar System bodies) (Ehrenfreund et al. 2002, Binet 2004, Molster 2004);
- study terrestrial analogs of compounds expected in various Solar System bodies (based on observations) from refractory to organic species (see, e.g, Cataldo 2004; Roush and Cruikshank 2004, Hofmann 2004).

In particular, the interpretation of remote (e.g., spectroscopic) data concerning planetary surfaces is aimed at a proper disentangling of the contribution of different species, including the identification of minor species. This task requires the determination of the optical properties of a large variety of different species pertaining to different classes of refractory (e.g., silicates) and organic materials, depending on the body under investigation (Roush and Cruikshank 2004). To deal with absorption and scattering properties from complex mixtures of different grain species and with different morphological properties (size distribution and shape) is not an easy task. The iterative process between theoretical models and laboratory experiments is necessary to allow us to correctly interpret observations of planetary surfaces.

In future perspective, a new generation of laboratory experiments will be required to simulate processes affecting materials at different stages of Solar System evolution. This approach requires knowledge and a careful simulation of environmental conditions and a systematic control of the effects induced by agents such as UV and ion irradiation, thermal processing and impacts. This program can be achieved by preparing simulation chambers where either macro or micro-scale effects on target materials can be monitored in detail. The ultimate goal is to build a framework of laboratory data in which observations can have a proper interpretation. The availability of simulation chambers where space conditions can be reproduced is also essential to design and test new generation instrumentation for space application.

6. COMETS

Comets are important potential sources of water, carbon, nitrogen and complex organic molecules for the early Earth. The recent inventory of comets shows a large number of organic molecules (Crovisier 2004). The major questions in cometary science related to Exo/Astrobiology seem thus to be the quantitative estimation of this delivery process across time, the precise nature of the chemicals involved, their survival when reaching the Earth and their true availability for prebiotic chemistry. Answering them involves theoretical studies, laboratory experiments, ground and space-borne observations and challenging in situ analysis and sample return.

For the future missions, the main questions to be answered include:

- How were comets formed ?
- How different are comets ?
- What is the composition of comets, including the nature of refractory organics ?
- How much cometary material was delivered to Earth across time ?
- What is the influence on prebiotic chemistry ?

In order to solve these problems, comets will be explored by several different spacecraft during the next decade. *Stardust* has just successfully collected grains of dust from the coma of comet Wild 2 and will deliver them to Earth in 2006. This collection will miss cometary volatiles but will provide essential information on silicates and carbon and perhaps some refractory organics. Similarly, the Japanese mission *Hayabusa* (ex. MUSES-C) will bring back to Earth rocky material collected from an asteroid surface for laboratory analyses. Meanwhile, *Deep Impact* will send a copper projectile into (or through!) the nucleus of Comet Tempel-2 in 2005, and then examine the resulting crater and the newly liberated dust and gas.

A greater level of sophistication will occur with *Rosetta,* successfully launched in March 2004 which will couple long-term close observations to in situ measurements on the nucleus of comet 67P/Churyumov-Gerasimenko in 2013 (http://www.esa.int/export/SPECIALS/Rosetta/). The goals are a deep investigation of nucleus composition, structure and evolution and a careful characterization of the mechanisms and processes involving nucleus, coma and tail, with their solid and volatile components. Equipped with a variety of instruments, *Rosetta* will provide our closest look at a comet nucleus, both literally and figuratively. In situ spectroscopy of the coma may allow the detection of hundreds of species. Another very relevant point is the study of isotopic ratios in organics. *Rosetta* will also investigate two asteroids on its journey; suggested targets are Lutetia and Steins.

In the next years new large ground based instruments will appear, and should devote some fraction of time to cometary science (e.g., *ALMA, LMT*). As far as space telescopes are concerned, *Herschel/HIFI* will allow the systematic study of D/H ratio in comets (and hence of their contribution to ocean water). *Corot* (and possibly *Darwin or TPF*) missions will search for exocomets. Extrasolar interplanetary dust particle disks (Exozodi) will be studied by *SST*.

A comet sample return mission is the ultimate goal, since for quite a long time the extreme sensitivity of laboratory equipment will not be available for in situ studies. As in the case of meteorites, considerable information is contained in the structure of the materials. NASA has just invited proposals for a comet sample return mission. If selected, the earliest such mission would fly just after *Rosetta* reaches its goal.

Theoretical work, mainly related to dynamical evolution of the Solar System, is needed to better constrain the amount of cometary matter that was delivered to Earth across time. Constraints may also come from observations of extrasolar planetary systems. The delivery process is not a soft one, even in the case of small particles. Impacts may be studied to some extent by numerical simulations and laboratory experiments. The entry of small particles has been observed for the Leonid meteors, but no satisfactory model exists yet, and progress will require the study of the individual faint meteors of more ordinary showers. The near-Earth collection and laboratory study of particles unaltered by interaction with the terrestrial environment is the necessary input to the models, which should aim to explain the micrometeorite population and any enrichment/modification of the atmosphere by meteors. In order to assess the real importance of the delivered material for prebiotic chemistry, laboratory experiments and modeling are needed to study its influence on the early prebiotic processes on Earth. In this respect, the form under which molecules are delivered (e.g., individual compounds, or embedded in macromolecular material, or in micrometeorites) may influence the final outcome (Ehrenfreund et al. 2002).

Progress along the previous lines requires a better knowledge about similarities and differences among comets and their formation processes. To this aim future studies should include comparisons of the different comet populations (Fink 1992, A'Hearn et al. 1995). The deep analysis of a single object must be followed by more systematic studies that allow the acquisition of statistical information for a given cometary population, and variations among populations. In-space exploration, in particularly by multi-wavelength observations, of the pristine Kuiper Belt objects in the 10 - 1000 km class (including Pluto) will be extremely interesting for the study of the origin of the Solar System (de Bergh 2004). This is the goal of the recently approved NASA New Horizons Pluto-Kuiper-Belt mission. The largest of these relatives of comets may have had liquid water for some time, favoring complex organic chemistry. Because of the possible role of comets in the appearance of life on Earth, the future search for exo-comets may also represent a key observational program.

7. ORGANIC MATTER IN METEORITES AND IDPs

In carbonaceous chondrites, organic matter is distributed in two fractions: the soluble component (i.e. soluble in water and in organic solvents) and the insoluble fraction recovered as a solid residue after thorough solvent extractions and mineral dissolution through acid treatments. Although the soluble fraction represents less than 1% of the organic carbon, it has been extensively studied and its molecular and isotopic compositions are well-known, at least in the Murchison meteorite.

In contrast, due to limitations in analytical techniques, the chemical structure of the insoluble fraction (usually referred to as IOM, Insoluble Organic Matter or macromolecular carbon), which accounts for most of the organic carbon of meteorites, is far from being established (Binet et al. 2004). In IDPs, due to their small size and hence analytical problems, knowledge of the nature and distribution of the organic matter is much less than that for chondrites (Molster 2004).

Several key questions about the origin of organic matter in the chondritic meteorites and IDPs can be identified:

- What is the relationship between soluble and insoluble meteoritic organic fractions ?
- What is the molecular structure of meteoritic IOM ?
- How do the organosynthesis pathways of soluble and insoluble meteoritic organic fractions proceed ?
- What is the possible relationship with organic matter in interstellar space ?
- What is the chemical composition of IDPs at a molecular level ?

Several models have been proposed to explain the origin of the soluble fraction but at this time none of them is able to account for its molecular diversity and isotopic composition. It has been argued that this soluble fraction could be derived from the hydrolysis of an organic precursor, presumably formed in interstellar conditions. Such hydrolyses would have taken place at (or near) the surface of the meteorite parent body during the extensive hydrothermal activity, as recorded by the mineralogy and the geochemistry of the carbonaceous meteorites (Bunch and Chang 1980) and to a lesser extent of IDPs. If correct, this interpretation has profound implications for the origin of life on Earth.

Indeed among the numerous soluble species, some are the building blocks of living organisms. For example, amino acids and presugars are present in the Murchison carbonaceous meteorite (Cooper et al. 2001). As a consequence of this new possible interpretation, solid organic precursors originating in interstellar conditions would have liberated, in warm water, organic compounds that represent the potential building blocks for a primitive metabolism. It is therefore important to document the relation between the soluble and the insoluble fraction in order to judge to what extent the insoluble fraction can be regarded as the leftover product of a hydrolyzed precursor.

The use of the new high resolution in situ analysis methods (Synchrotron spectroscopy, mass spectroscopy at the nanometer scale by ion-microprobes and 3D high resolution microscopy) should permit scientists to obtain chemical, isotopic and molecular maps with resolutions at the nanometer scale in the insoluble organic matter extracted from the carbonaceous meteorites.

Among the marked improvements of these new techniques is the unique ability of ion-microprobes to record in situ and quantitatively the chemical elements constituting the organic matter (i.e. : H,C,N,O,S). Correlations among these maps should permit the reconstruction of the various molecular moieties constituting the insoluble organic matter along with their respective isotopic compositions (mostly the D/H and $^{15}N/^{14}N$ ratio whose natural variations are huge in extraterrestrial organic matter and therefore easily accessible to ion-microprobe determination). Based on the chemical compositions and the mass spectra obtained through ion-microprobe analyses, it should be possible to calculate an average molecular composition in accordance with spectroscopic data. It should then be possible to ascribe to each molecular super-structure a given isotopic composition. This approach will yield - for the first time - the possibility of obtaining experimentally the isotopic composition of individual molecular entities constituting the bulk organic insoluble matter. Of course the isotopic composition of the pure organic end-members (will not be determined in situ, but mixing models coupling molecular, chemical and isotopic compositions should allow us to extrapolate the composition of each organic end-members. The same set of problems developed in this section apply for IDPs but will involve technical challenges which are beyond anything that can be reached at the present time (Molster 2004). As stated above, investigations have always focused on the volatile fraction of meteoritic material, such as amino acids and simple subunits of nucleic acids. However, the volatile fraction is far less abundant compared to the bulk of macromolecular matter (IOM) which will have been transported to Earth. This material which will have been delivered via small bodies to the early planets is in the form of macromolecular networks, similar to terrestrial kerogens. There are no studies concerning how such material could be used in the context of the origin of life. Does it act as a useless inert solid body, could it be acting as a catalyst, could it be reformed by radiation and thermal processing, or by catalysis with iron, could aliphatic bridges within this networks be broken and release fragments which could be more handy to form functional precursors for life?

8. STUDIES OF THE EARLY EARTH

Our knowledge of the early Earth is extremely limited and interpretations are highly controversial. The successful development of life on Earth as we know it resulted from a special set of circumstances. Among these, surface temperature, the presence of a hydrosphere, and the presence of suitable chemical building blocks appear to have been crucial.

Hafnium isotope measurements constrain the Earth's core formation to within 30 million years of the formation of the Solar System. Radiometric dating using isotopes of Lu and Hf suggests that the Moon-forming impact also occurred within this temporal window (Halliday 2000). The geothermal gradient at this time was much warmer, and differentiation of the planet's interior occurred rapidly, segregating most of the iron into the Earth's core and leaving behind a more oxidized mantle. Rampant volcanic degassing of this mantle material would have produced a relatively neutral (non-reducing) atmosphere containing water, carbon dioxide, and nitrogen (Selsis 2004).

Current models suggest that our Solar System segregated rapidly 4.55 Ga (1 Ga = 1 billion years) ago to form, with increasing distance from the Sun, terrestrial, fluid, and icy planets. The majority of planetesimals that formed Earth were situated within a few AU, and these may have been relatively water-poor. Larger, water-rich planetesimals located further from the Sun (>2 AU) may have also been incorporated into the agglomerating Earth (up to 15 %, based on D/H). While Earth's oceans today have a volume of roughly $1.3x10^9$ km^3, large impacts in early Earth's history may have repeatedly vaporized primordial water oceans and atmospheres and produced temporary magma oceans at the surface and reduced atmospheres.

Neon and argon in the atmosphere today are grossly depleted in comparison with solar and cosmic abundances; this observation is consistent with loss of a large fraction of the volatile component of the planet. The Sun's luminosity increases with time, affecting the surface temperature of the planets. 4.6 Ga ago, our Sun was approximately 30 % less bright; consequently, the Earth's surface temperature might have been colder. In the absence of greenhouse gases (such as CO_2, CH_4) in the Earth atmosphere water oceans would have frozen during this time (Kasting and Catling, 2003).

On the basis of oxygen isotope ratios in detrital zircons from Australia (Valley et al., 2003), Earth may have had a stable water ocean ~ 4 Ga years ago. It is not known that the emergence of life required a marine environment, but the oldest direct evidence for life on Earth comes from fossils preserved in marine deposits that postdate this age. Microfossils in stromatolites from coetaneous sedimentary deposits in Pilbara (Australia) and Barberton (S. Africa) date to 3.5-3.3 Ga. Indirect evidence based on stable isotope ratios has been used to postulate the existence of life even earlier in Earth's record (Westall 2004). For example, low $^{13}C/^{12}C$ ratios in banded iron formations (BIFs) from Greenland might be an indication of life from 3.8 Ga (Rosing 1999).

Similarly, sulfur isotope signatures in early sediments might be a record of biological activity from a similar time (~ 3.5 Ga, Shen et al. 2001). While these new biosignature results are intriguing, they inspire as many questions as they do answers.

Open questions that are the topic of on-going research address:

- What are the timescales of Earth formation and differentiation ?
- What is the composition of the early atmosphere, as deduced from the rock record ?
- What was the influence of the impact record on potential surface habitats ?
- When did liquid water exist on the early Earth and where did it come from ?
- What are the sources of carbon and nitrogen ?
- Was there glaciation during the first billion years, and if, what was its distribution ?
- When did the first organisms arise and what was their geochemical environment ?

The answers are all dependent on the completeness of the rock record and the accuracy of our analytical methods (Hofmann 2004). However, it may be possible to augment the meager ancient rock record on Earth with rocks from Mars, which has large areas of ancient crust and may have had an early atmosphere similar to ours.

Future studies relevant to the characterization of the early Earth and the origin of life should focus on a combination of analyses of Archean samples, experiments designed to help constrain the interpretation of the rock record, collection and characterization of extraterrestrial materials, and laboratory synthesis of prebiotic organic compounds. A suite of complementary isotope systems such as Re/Os, W/Hf, and Pb/Pb applied to the oldest rock samples give us better constraints on the timing of early geochemical events in the Hadean Earth. As our ability to measure rapidly increasingly smaller samples develops, drilled rock cores through Archaean terrain and impact sites, among others, will provide new insight and higher resolution of the geochronology. On-going laboratory experiments that address element partitioning, isotope fractionation, biomarker preservation, among other processes, will increase our ability to interpret the rock record and hone our search for records of the earliest life. Upcoming sample return missions from Mars, asteroids, comets, the solar windstream (and our Moon and other moons) will provide new samples and complementary information about early Solar System history, atmospheres, and the Earth's impact record. Likewise, information about all these objects gleaned through in-situ measurements (spectroscopy, photography) made by unmanned spacecraft will add to our body of knowledge. It is clear that we cannot understand the Earth without knowing more about our sister planets, the comets, and the meteorites.

9. BIOCHEMISTRY AND GENETICS

The attributes of the first living organisms are unknown. They were probably simpler than any cell now alive, and may have lacked not only protein-based catalysis, but perhaps even the familiar genetic macromolecules, with their ribose-phosphate backbones. Such ancestral polymers may have not even involved nucleotides. Although several polymers capable of encoding genetic information have been synthesized in the laboratory, as of today there are no reports of catalytic activity in these analogues of nucleic acid molecules (Botta 2004, Saladino et al. 2004).

Questions which need to be addressed include

- Are TNA, PNA and other such analogues of nucleic acids (Botta 2004) also endowed with catalytic activity?
- Are there any additional models of pre-RNA worlds whose components can be synthesized and accumulated under plausible prebiotic conditions?
- Can the synthesis of membrane-forming compounds and other components of cells under primitive terrestrial conditions be truly sustained?
- What are the minimal requirements of a laboratory model of a minimal living system?
- Does the list of requirements needed for model systems include membranes, catalytic agents and genetic polymers only?
- How did the evolutionary transition from such hypothetical pre-RNA worlds to the current DNA/RNA/protein-based cells take place?
- How did the complex structure of extant enzymes originate?

Over the past several decades, it has been abundantly demonstrated that life does not necessarily need the benign conditions that humans require, but can even exist in quite extreme environments. These include environments of extreme pH (0.6 - 12.5), temperature (-2 - 130° C), pressure (to 110 MPa) and salinity (to 37.5% NaCl) (Cavicchioli 2002). Research on extremophiles is very important for the determination of habitats beyond the Earth and the Solar System. However, all the available evidence suggests that the most basic questions pertaining the origin of life relate to much simpler replicating entities, predating by a long (but not necessarily slow) series of evolutionary events the oldest recognizable lineages (presently hyperthermophiles and other extremophiles) represented in molecular phylogenies. The principal appeal of the idea that extremophiles are closer to the last common ancestor of all extant terrestrial life is based on molecular biological evidence, which suggests that the Archaea have evolved less rapidly than the Eucarya and the non-hyperthermophilic Eubacteria.

However, this may simply reflect the stability of certain environments over geological time, the lack of competition for these niches, and functional constraints on the structure of cellular components that allow extremophiles to flourish under such unfavorable conditions (Cleaves et al. 2004). It is worth noting that comparative studies of the rRNA genes of prokaryotes strongly suggest that their putative cenancestor, the last common ancestor shared by all extant lifeforms, was not hyperthermophilic (Galtier et al. 1999; Brochier and Philippe 2002). Because of this, it is important to distinguish between ancient and primitive organisms. Species which are located near the root of universal rRNA-based trees are ancient, but they are not endowed with primitive molecular genetic apparatus, nor do they appear to be more rudimentary in their metabolic abilities than their modern counterparts.

Thus, in spite of their tremendous significance in understanding the environmental limits within which life can thrive, these centrally rooted extremophiles, including various heat-loving prokaryotes, may tell us little about the origin of life itself. But why has no microbial species older than hyperthermophiles been identified until now? Molecular phylogenetic studies may provide clues to some very early stages of biological evolution. However, it is difficult to see how the applicability of this approach can be extended beyond a threshold that corresponds to a period of cellular evolution in which protein biosynthesis was already in operation, during the RNA/Protein world which is thought to have preceded the extant DNA/RNA/protein/cell worlds (Freeland et al. 1999, Dworkin et al. 2003).

Our understanding of the early stages of cell evolution that took place after the evolution of DNA genomes but prior to the divergence of the Bacteria, Archaea and Eucarya is hindered by several major problems, including a proper assessment of the extent of lateral gene transfer, and the fact that a significant fraction of the sequences derived from complete genome sequencing projects remain unidentified (30 to 50% depending on the organism). To what extent has horizontal transfer of genetic information limited the possibility of reconstructing microbial evolution (Doolittle 1999)? Our efforts are also complicated by the inadequate biodiversity sampling that has shaped our current databases, which represent an extremely biased set of sequenced genes and genomes. Nonetheless, although clearly incomplete, the adequacy of fully sequenced genome data bases for the reconstruction of ancestral states is probably greater than is generally realized. In spite of the extraordinary diversity of habitats and lifestyles, organisms share a surprising amount of enzymatic activities, metabolic routes, and basic biological functions, as reflected in genome replication, gene expression, and metabolic pathways.

The possibility that some of the enzymes of archaic pathways may have survived in modern organisms suggest that considerable prudence should be exerted when attempting to describe the physiology of ancestral organisms (Delaye et al. 2004). It is of course possible that extant enzymes participated in alternative pathways which no longer exist or remain to be discovered. This possibility has begun to be explored by computer searches for alternative reaction pathways, and considerable attention should be given to the possibility that significant variations of the basic pathways may have existed in the past (Zubay 1993; Delaye et al. 2004).

10. ASTROBIOLOGY WORLDWIDE

The NASA Astrobiology Institute (NAI) represents a partnership between NASA and a number of academic or other research organizations to promote, conduct, and lead integrated multidisciplinary Astrobiology research and train young researchers (http://nai.arc.nasa.gov/). In 1998 NASA established the NASA Astrobiology Institute as one element of its research program in Astrobiology. The NAI is currently composed of 16 Lead Teams, which together represent over 700 investigators across the United States, and it has international partnerships with Astrobiology research organizations around the world. NAI Lead Teams involve researchers from multiple disciplines, and they usually include team members from a variety of geographically distributed institutions. Members collaborate with one another, both in person and virtually, using communication and collaboration tools supplied by NAI. Lead Teams are supported by NASA through 5-year Cooperative Agreements with Ames Research Center. NAI members are also encouraged to obtain funding from NASA, NIH, NSF, or other sources for their research and educational activities (Smith 2004).

The European Astrobiology Network Association (EANA) was created in 2001 to coordinate the different national research centers and to promote research in Astrobiology in Europe (Brack et al. 2004) (http://www.spaceflight.esa.int/exobio).

EANA's Executive Council consists of national members presently representing 17 European nations active in the field: Austria, Belgium, Denmark, Finland, France, Germany, Hungary, Italy, Poland, Portugal, Spain, Sweden, Romania, Russia, Switzerland, The Netherlands and the United Kingdom.

EANA is affiliated to the NASA Astrobiology Institute. EANA is also a member of the International Astrobiology Circle (IAC), including the Astrobiology Society of Britain, the Australian Centre for Astrobiology (ACA), the Spanish Centro de Astrobiologia (CAB), the French Groupement de Recherche en Exobiologie (GDR-Exobio), the NASA Astrobiology Institute, NAI, in the US, and the Swedish Astrobiology Network (SWAN).

Europe hosts the largest Astrobiology institute, headed by Dr. Juan Perez-Mercader. The Centro de Astrobiologia is situated near Madrid and is focused on interdisciplinary research within Astrobiology (http://www.cab.inta.es/). Europe does not receive any financial support for Astrobiology from European organizations, with the exception of small grants in order to sponsor young researchers attending Astrobiology conferences. Astrobiology research is currently organized and financed exclusively by national organizations and then mostly on an individual basis. Astrobiology activities and available funding strongly vary from country to country. Presently, Astrobiology in Europe is not tied to space research or to the European Space Agency (ESA).

THE END:

Astronomer Frank Drake has remarked that the search for extraterrestrial life is "the greatest adventure left to humankind." ASTROBIOLOGY (http://www.astrobiology.com) envelops that adventure in a scientific discipline that includes a search for our own beginnings on our own planet. We are fortunate to be living in an era where centuries of speculation about these two ancient and fundamental problems: the origin of life and its prevalence in the Universe, are being replaced by experimental science. Today's rich array of new spacecraft, telescopes and dedicated scientists promises a steady flow of discoveries and insights that will ultimately lead us to the answers we seek, see Space Roadmap (Appendix).

We look forward to the continuing thrill of the chase! A quick glance at the history of science reveals that the most enriching views are the outcome of a multidisciplinary approach - an established aspect of Astrobiology that is both a virtue and a necessity. Thus, by its very nature and goals, Astrobiology lends itself to the development of novel perspectives.........

Acknowledgement: We are grateful to our students, postdocs and collaborators and most of all the inspiring authors which made this book possible ....

APPENDIX

Main observational facilities relevant to Astrobiology or connected areas

Large optical and radio telescopes:

The ESO Very Large Telescope consists of an array of four 8-meter telescopes which can work independently or in combined mode. In this latter mode the VLT provides the total light collecting power of a 16 m single telescope. The telescopes may also be used in interferometric mode providing high resolution imaging. The useful wavelength range extends from the near UV up to 25 μm in the infrared.

The twin KECK Telescopes (KECK 1+2), located at Mauna Kea (Hawaii) are the world's largest optical and infrared telescopes. Each KECK telescope has a revolutionary primary mirror, ten meters in diameter and is composed of 36 hexagonal segments that work in concert as a single piece of reflective glass. Together, the twin KECK telescopes have the resolving power of a single telescope 90m in diameter, able to discern sources just milliarcseconds apart.

The potential of new, large ground-based optical telescopes (such as Gemini or the Large Binocular Telescope) to contribute to the studies discussed in this book is illustrated by, e.g., recent VLT observations of icy grain mantles in star formation regions (Pontoppidan et al., 2003) and Keck measurements of cometary composition, particularly of molecular species lacking permanent dipole moments and hence inaccessible by radio techniques (Gibb et al., 2003; Mumma et al., 2002). New optical to Mid-IR interferometers, such as the combined KECK telescopes or the VLTI (VLT Interferometer) will be very uselful to probe the dense material in inner disks where planetary formation should occur.

The Atacama Large Millimeter Array (ALMA) is a joint project of the European Southern Observatory (ESO) and the US National Radio Astronomy Observatory (NRAO), with the possible participation of Japan. ALMA will consist of 64 antennas, each 12m in diameter, with baselines up to 12 km, located in the very high and dry Atacama Desert in Chile. The interferometer is expected to start operation in 2008 and to be fully operational in 2011. Frequency coverage will eventually cover the atmospheric windows between 40 and 900 GHz, with angular resolutions up to about 0.01''. The broad frequency range is important for astrochemistry, since lines from (e.g.) simple hydrides and atomic carbon occur at the higher frequencies, while those from complex organic molecules will be at the lower end of the range. The combination of unprecedented sensitivity (provided by the large collecting area), exceptional angular resolution, and the very high spectral resolution achieved by heterodyne techniques will allow detailed study of the chemistry and physics of organic matter that is collapsing to form pre-stellar cores and then protoplanetary disks. Moreover, it is anticipated that ALMA will be able to extend the searches for new prebiotic molecules more than one order magnitude lower in abundance than is currently possible (van Dishoeck 2002).

The largest filled aperture telescope operating at short millimeter wavelengths in the next decade will be the 50m-diameter Large Millimeter Telescope (LMT) at 19° North latitude in Mexico (Irvine et al. 2003). Equipped with both heterodyne and bolometric focal plane arrays for rapid mapping of extended sources in the 0.8-3 mm bands, the LMT will be the instrument of choice for sensitive studies of dust emission at moderate resolution (few arcsec), and will provide invaluable "wide-angle" views of molecular emission that can be followed up at higher spatial resolution by ALMA. Operated as a university-type facility, it will be available for "target of opportunity" observations of time-variable phenomena such as comets.

The IRAM telescopes will be still in operation. The single 30m radiotelescope in the Sierra Nevada (Spain) is operated in the bands at 3, 2 and 1.3mm and is also equipped with a bolometer array working at 1mm. The Plateau de Bure Interferometer (PdBI) is an array consisting of 6 antennas of 15m diameter and will continue to have the largest collecting area among the existing mm arrays prior to ALMA. The array will be soon upgraded to operate in all four of the bands at 3, 2, 1.3 and 0.8mm and is currently able to reach a resolution of 0.6'' at 1.3mm.

The other major new telescopes operating at (sub)millimeter wavelengths in the coming decade will be the Combined Array for Research in Millimeter-wave Astronomy (CARMA) and the Submillimeter Array (SMA). CARMA is an amalgamation of the previous OVRO and BIMA arrays, at a new higher elevation site in California (Beasley and Vogel 2003).

It will consist of 23 antennas and will operate at wavelengths of 12, 3, 1.3, and possible 0.8 mm. A resolution of 0.1" will be achieved at 1.3mm, providing a powerful instrument for the astrochemistry of comets, interstellar clouds, and star formation. The SMA has recently completed construction, and formal dedication of the telescope took place in November, 2003 (Qi 2003). It consists of 8 x 6m antennas, with a planned frequency range of 180-900 GHZ (0.3-1.7 mm), and angular resolutions of 0.1-0.5" will be obtainable.

Space missions with Infrared facilities

New space-borne telescopes are, of course, critically important for observations at wavelengths where the terrestrial atmosphere is opaque. For studies of organic molecules, this applies to significant portions of the infrared and (sub)millimeter bands. At the longer wavelengths the premier new facility will be the Herschel Space Observatory (HSO), previously known as FIRST. Herschel is "cornerstone" mission number 4 in the European Space Agency's "Horizon 2000" science plan and will be launched in 2008. It will be capable of high resolution spectroscopy as well as photometry between 60 and 670 μm wavelength (450-5000 GHz), and will thus provide important astrochemical data complementary to that at mid-infrared wavelengths from the James Webb Space Telescope (JWST) and at submillimeter wavelengths by ALMA. In particular, Herschel will study the processes by which stars, their surrounding protoplanetary disks and planets themselves are made and investigate the degree of molecular complexity in different regions and the relationships between grain surface and gas phase chemistry.

SST (Spitzer Space Telescope, former SIRTF), successfully launched in August 2003, is a cryogenically-cooled infrared observatory, with a 85 cm telescope, capable of studying objects ranging from our Solar System to the distant reaches of the Universe. Its estimated lifetime is 5 years and it will cover the wavelength range 3-180 μm. The airborne observatory SOFIA (to be operational in 2005) will observe the infrared range 5-300 μm. SOFIA is a Boeing 747SP aircraft which is currently being rebuilt by NASA and DLR to accomodate a 2.5m reflecting telescope. SOFIA is the successor of the previous KAO (Kuiper Airborne Observatory) and will be the largest airborne observatory in the world.

The James Webb Space Telescope (JWST, formerly called the NGST) will be optimized for infrared observations and will be able to penetrate the dusty envelopes around new-born stars and take a closer look at the stars themselves. The largest gain in sensitivity and spatial resolution at mid-infrared wavelengths will be provided by a camera and spectrometer on-board JWST. The JWST should be three orders of magnitude more sensitive than any ground-based telescope in the 5-30 μm region. Thus, various interstellar compounds will certainly be an important subject for study, and in particular observations in quiescent clouds and regions of low-mass star formation, where very luminous background sources are absent.

How planetary systems form and evolve, and whether habitable or life-bearing planets exist around nearby stars, are major objectives to be studied with several missions, including COROT, Kepler, TPF, and IRSI-DARWIN. With the help of nulling interferometers in the thermal infrared, to remove the parent star light, IRSI-DARWIN will search for the spectral signature of gases such as water, CO_2, CH_4 and O_3 in the atmosphere of extra-solar planets in order to identify Earth-like planets capable of sustaining life.

SPACE ROADMAP FOR ASTROBIOLOGY (updated from Foing 2002)

SPACE OBSERVATORIES 2000 2005 2010 2015 2020 2030

IR, sub-mm — 2003-SST
2005-SOFIA
2008-Herschel
JWST, IRSI-Darwin, TPF

LIFE IN SOLAR SYSTEM 2000 2005 2010 2015 2020 2030

Mars

Water — 1996-MGS (mapping 1999)
2001-Odyssey
2003-MEX,MER
2005-MRO
2007-Phoenix

Exobiology — 2009 MSL, Exomars

Rovers — 2003-MER

Aurora & US Program — Robotic outpost
2011-2013 — Sample return planned 2030

Human assisted research

Europa

Ice Ocean Surface — Galileo

Ocean and liquid niches — 2012-2015-JIMO

Comets, Asteroids

Encounter rendezvous — 1999-Stardust (rendezvous 2004, return 2006)
2003-Hayabusa (sample return 2007)
2005-Deep Impact
2004-Rosetta (rendezvous 2014)

PLANETS, EXOPLANETS 2000 2005 2010 2015 2020 2030

Exoplanets

Exo-Jupiters — Ground-Based — 2012-GAIA

Terrestrial Super Earths — 2006-COROT

Earth-like habitable — 2008-Kepler

Atmospheric imprint of life — TPF, IRSI-Darwin

Habitability

Water (Mars orbit) — 1996-MGS, 2001-Odyssey
2003-MEX,MER
2005-MRO

Organic Chemistry — 1997-Cassini-Huygens (arrival Titan 2005)

Frustration of life by impacts — 2003-SMART-1
2012-BepiColombo
2004-Messenger

SIGNATURES OF LIFE	2000	2005	2010	2015	2020	2030
Biomarkers in Solar System						
Instrumentation for life characterisation			2009-MSL, Exomars			
Biomarkers in exoplanets						
Atmospheric gases				IRSI-Darwin		
Radio bio-signals	SETI…….					

REFERENCES

A'Hearn M.F., Millis R.L., Schleicher D.G., Osip D.J., Birch P.V.: 1995, The Ensemble properties of Comets: Results from Narrowband Photometry of 85 Comets, 1976--1992. *Icarus*, **118**, 223-270.

Beasley, A.J., and Vogel, S.N.: 2003, CARMA: specifications and status, *Proc. SPIE*, **4855**, 254-264.

Binet, L. et al.: 2004, in Astrobiology: Future Perspectives (this volume)

Bishop, J. L., Pieters, C.M., Hiroi, T., Mustard, J.F.: 1998, Spectroscopic analysis of Martian meteorite Allan Hills 84001 powder and applications for spectral identification of minerals and other soil components on Mars, *Meteorit Planet. Sci.*, **33**, 668-707.

Botta, O.: 2004, in Astrobiology: Future Perspectives (this volume)

Brack, A., Horneck, G., Wynn-William, D.: 2004, in Astrobiology: Future Perspectives (this volume)

Brochier, C., H. Philippe.: 2002, A non-hyperthermophilic ancestor for Bacteria. *Nature* **417**: 244-246.

Bunch, T., E., and Chang, S.: 1980, Carbonaceous chondrites. II - Carbonaceous chondrite phyllosilicates and light element geochemistry as indicators of parent body processes and surface conditions, *Geochim. Cosmochim. Acta* **44,** 1543-77.

Cataldo, F: 2004: in Astrobiology: Future Perspectives (this volume)

Cavicchioli, R.: 2002, Extremophiles and the search for extraterrestrial life, *Astrobiology* **2**, 281-292.

Cleaves, J.H., Chalmers, J.H.: 2004, Are Extremophiles Relevant to the Origin of Life? *Astrobiology* 4, 1-9

Cooper, G. W., Kimmich, N., Belisle, W., Sarinana, J., Brabham, K., Garrel L.: 2001, Carbonaceous meteorites as a source of sugar-related organic compounds for the early Earth, *Nature*, **414** 879-83.

Crovisier, J.: 2004, in Astrobiology: Future Perspectives (this volume)

Dartois, E., Dutrey, A., Guilloteau, S. : 2003, Structure of the DM Tau Outer Disk: Probing the vertical kinetic temperature gradient *Astron. and Astrophys.* **399**, 773

De Bergh, C: 2004, in Astrobiology: Future Perspectives (this volume)

Delaye, L., Becerra, A., and Lazcano, A.: 2004, The nature of the last common ancestor. In The Genetic Code and the Origin of Life, Lluis Ribas de Pouplana (ed), Landes Bioscience, Georgetown, in press

Doolittle, W. F.: 1999, Phylogenetic classification and the universal tree, *Science* **284**, 2124-2128.

Dutrey, A., Guilloteau, S., Guelin, M. : 1997, Chemistry of protosolar-like nebulae: The molecular content of the DM Tau and GG Tau disks. *Astron. and Astrophys.* **317**, L55-L58.

Dworkin, J. P., Lazcano, A., and Miller, S. L.: 2003, The roads to and from the RNA world. *Journal of Theoretical Biology* **222**, 127-134.

Ehrenfreund, P., and Charnley, S. B.: 2000, Organic Molecules in the Interstellar Medium, Comets, and Meteorites: A Voyage from Dark Clouds to the Early Earth, *Ann. Rev. Astron. Astrophys.* **38,** 427-83.

Ehrenfreund, P., Irvine, W.M., Becker, L., Blank, J., Colangeli, L., Derenne, S., Despois, D., Dutrey, A., Fraaije, H., Lazcano, A., Owen, T., and Robert, F.: 2002, Astrophysical and astrochemical insights into the origin of life, *Repts. Prog. Physics,* **65**, 1427-1487.

Esposito, F., Colangeli, L., Palomba, E. 2000, Infrared reflectance spectroscopy of Martian analogues, *Journal of Geophysical Research*, **105**, E7, 17643-17654.

Fink, U.: 1992, Comet Yanaka (1988r) - A new class of carbon-poor comet, *Science* **257**, 1926-1929.

Fischer, D., and Valenti, J.A., Metallicities of Stars with Extrasolar Planets, in Scientific Frontiers in Research on Extrasolar Planets, ASP conference Series, Vol. 000, Eds. D. Deming and S. Seager (2003)

Foing, B.: 2002, Space activities in exo-astrobiology, in: Astrobiology. The quest for the conditions of life. Gerda Horneck, Christa Baumstark-Khan (eds.). Berlin: Springer, p. 389–398.

Freeland SJ, Knight RD, Landweber LF.: 1999, Do proteins predate DNA? Science **286**, 690-692.

Galtier, N, Tourasse, N., Gouy, M. A: 1999, Nonhyperthermophilic Common Ancestor to Extant Life Forms. *Science* **283**, 220-221.

Gibb, E.L., Mumma, M.J., dello Russo, N., Disanti, M.A., and Magee-Sauer, K.: 2003, Methane in Oort cloud comets, *Icarus*, **165**, 391-406.

Halliday, A.: 2000, Terrestrial accretion rates and the origin of the Moon, *Earth and Planetary Science Letters* **176,** 17-30.

Hofmann, B.: 2004, in Astrobiology: Future Perspectives (this volume).

Hubbard, W. B., Fortney, J. F., Lunine, J. I., Burrows, A., Sudarsky, D., and Pinto, P. A.: 2001, Theory of Extrasolar Giant Planet Transits, *Astrophys. J.,* **560**, 413-419.

Irvine, W.M., Carramiñana, A., Carrasco, L., and Schloerb, F.P.: 2003, The Large Millimeter Telescope/El Gran Telescopio Milimétrico: A New Instrument for Astrobiology, *Origins Life Evol. Biosphere,* **33**, 597-607.

Irvine, W.M., Schloerb, F.P., Crovisier, J., Fegley, B., Jr., and Mumma, M.J.: 2000, Comets: a Link Between Interstellar and Nebular Chemistry, in *Protostars and Planets IV*, V. Mannings, A. Boss and S. Russell (eds.), Tucson: Univ. Arizona Press, pp. 1159- 1200.

Kastling, J. and Catling, D.: 2003, Evolution of a Habitable Planet, *Ann. Rev. Astron. Astrophys.*, **41**, 429-463.

Lin, D. N. C., and Papaloizou, J. B. C.: 1986, On the tidal interaction between protoplanets and the protoplanetary disk. III - Orbital migration of protoplanets, *Astrophys. J.,* **309**, 846-857.

Marcy, G. W., Butler, R. P., Fischer, D. A., Laughlin, G., Vogt, S. S., Henry, G. W., and Pourbaix, D.: 2002, A Planet at 5 AU around 55 Cancri, *Astrophys. J.,* **581**, 1375-1388.

Markwick, A., Charnley S.: 2004, in Astrobiology: Future Perspectives (this volume).

Mennella, V., Colangeli, L., Bussoletti, E., Palumbo, P. and Rotundi, A.: 1998, A New Approach to the Puzzle of the Ultraviolet Interstellar Extinction Bump, *Astrophys. J.* **507**, L177–180.

Miley, G. et al.: 2004, A large population of `Lyman-break' galaxies in a protocluster at redshift z ~ 4.1, *Nature* **427**, 47-50.

Millar, T.J.: 2004, in Astrobiology: Future Perspectives (this volume).

Molster, F.: 2004, in Astrobiology: Future Perspectives (this volume)

Mumma, M.J., Disanti, M.A., dello Russo, N., Magee-Sauer, K., Gibb, E., and Novak, R.: 2002, The organic volatile composition of Oort cloud comets: evidence for chemical diversity in the giant-planets' nebular region, in B. Warmbein (ed.), ESA SP-500, pp. 753 – 762.

Owen, T., Mahaffy, P, Niemann, H.B., Atreya, S., Donahue, T., Bar-Nun, A., and De Pater, I., A Low temperature Origin for the Planetesimals that formed Jupiter 1999 *Nature*, **402**, 269-270.

Pendleton, Y. J., Allamandola, L. J.: 2002, The Organic Refractory Material in the Diffuse Interstellar Medium: Mid-Infrared Spectroscopic Constraints, *Astrophys. J. Suppl.* **138,** 75-98.

Pontoppidan, K.M., Fraser, H.J., Dartois, E., Thi, W.-F., van Dishoeck, E.F., Boogert, A.C.A., d'Hendecourt, L., Tielens, A.G.G.M., and Bisschop, S.E.:2003, A 3-5 μ m VLT spectroscopic survey of embedded young low mass stars I. Structure of the CO ice, *Astron. Astrophys.,* **408**, 981-1007.

Queloz, D., Benz, W., Delfosse, X., and Mayor, M.: 2000, Results of extrasolar planet search and their impact on future astrometric surveys, *Proc. SPIE* , **4006**, 359-364.

Qi, C.: 2004, in C.L. Curry and M. Fich (eds.), SFChem 2002: Chemistry as a Diagnostic of Star Formation, Ottawa: National Research Council Press, in press.

Roos-Serote, M.: 2004, in Astrobiology: Future Perspectives (this volume).

Rosing M.T.: 1999, 13C depleted carbon microparticles in >3700 Ma seafloor sedimentary rocks from West Greenland, *Science*, **283**, 674-676.

Roush, T.L., Bell III, J.F.: 1995, Thermal emission measurements 2000-400/cm (5-25 micrometers) of Hawaiian palagonitic soils and their implications for Mars, *Journal of Geophysical Research*, **100**, E3, 5309-5317.

Roush, T., Cruikshank, D.: 2004, in Astrobiology: Future Perspectives (this volume).

Rummel, J.: 2004, in Astrobiology: Future Perspectives (this volume).

Saladino, et al.: 2004, in Astrobiology: Future Perspectives (this volume).

Selsis, F.: 2004, in Astrobiology: Future Perspectives (this volume).

Shen, Y., Buick, R., Canfield, D.E.: 2001, Isotopic evidence for microbial sulphate reduction in the early Archaean era, *Nature* **410**, 77-81.

Smith, D.: 2004, in Astrobiology: Future Perspectives (this volume).

Spaans, M.: 2004, in Astrobiology: Future Perspectives (this volume).

Valley, J.W.; Bindeman, I.N.; Peck, W.H.: 2003, Empirical calibration of oxygen isotope fractionation in zircon, *Geochimica et Cosmochimica Acta*, **67/17**, 3257-3266.

Van Dishoeck, E.F.: 2002, in J. Alves and M. McCaughrean (eds.), The Origin of Stars and Planets -- the VLT View, ESO Astrophysics Ser. (Springer), pp. 67-76.

Van Zadelhoff, G.-J., Aikawa, Y., Hogerheijde, M. R., van Dishoeck, E.F. 2003, Axi-symmetric models of ultraviolet radiative transfer with applications to circumstellar disk chemistry, *Astron. Astrophys.,* **397**, 789.

Vauclair, S.C., Blanchard, A., Sadat, R., Bartlett, J.G., Bernard, J.-P., Boer, M., Giard, M., Lumb, D.H., Marty, P., Nevalainen, J.: 2003, The XMM-Omega project. II. Cosmological implications from the high redshift L - T relation of X-ray clusters, *Astron. Astrophys.,* **412**, p.L37-L41.

Walter, F., Bertoldi, F., Carilli, C., Cox, P., Lo, K.Y., Roberto, Xiaohui, F., Omont, A., Strauss, M.A., Menten, K.: 2003, Molecular gas in the host galaxy of a quasar at redshift z = 6.42, *Nature*, **424**, 406-408.

Westall, F.: 2004, in Astrobiology: Future Perspectives (this volume).
Wolf, S., Gueth, F., Henning, Th., Kley, W.: 2002, Detecting Planets in Protoplanetary Disks: A Prospective Study, *Astrophys. J.* **566**, 97.
Wuchterl, G. 2004, in Astrobiology: Future Perspectives (this volume).
Zubay, G.: 1993, To what extent do biochemical pathways mimic prebiotic pathways? Chemtracts-Biochem

GLOSSARY

ALMA:

Atacama Large Millimeter Array: International collaboration between Europe, the US, and probably Japan to build a synthesis radio telescope that will operate at millimeter and submillimeter wavelengths (64 x 12m antennas) http://www.eso.org/projects/alma/.

AURORA:

Aurora is part of Europe's strategy for space exploration, endorsed by the European Union Council of Research and the ESA Council in 2001. This strategy calls for Europe to explore the Solar System and the Universe, to stimulate new technology and to inspire the young people of Europe to take a greater interest in science and technology.
http://www.esa.int/SPECIALS/Aurora/index.html

BEPI COLOMBO:

Consisting of two orbiters, BepiColombo will provide the most complete exploration yet of Mercury, the innermost planet. One component of BepiColombo will map the planet and a second will investigate its magnetosphere. http://www.esa.int/esaSC/index.html

CASSINI-HUYGENS:

The *Cassini-Huygens* spacecraft is presently on its way to the Saturn system and will deploy a probe into the atmosphere of Titan in January of 2005. Huygens will be the first probe to land on a world in the outer Solar System. http://saturn.jpl.nasa.gov/index.cfm

COROT:

COnvection ROtation and planetary Transits is a mission led by the French National Space Agency, CNES. It is a 30cm diameter space telescope designed to detect tiny changes in brightness from nearby stars and to search for exoplanet transits. Launch is scheduled in 2006.
http://www.obspm.fr/encycl/corot.html

DEEP IMPACT:

Deep Impact, to be launched in 2005, will be the first mission to make a spectacular, football-stadium-sized crater, into the speeding comet. This encounter with comet Tempel-1 in 2005 will provide clues to the formation of the Solar System.
http://deepimpact.jpl.nasa.gov/

EXOMARS:

ExoMars is the first Aurora Flagship mission to be assessed. Its aim is to further characterise the biological environment on Mars in preparation for robotic missions and then human exploration. This mission calls for the development of a Mars orbiter, a descent module and a Mars rover. The Mars orbiter will have to be capable of reaching Mars and putting itself into orbit around the planet. On board will be a Mars rover within a descent module. http://www.esa.int/SPECIALS/Aurora/index.html

GAIA:

Gaia is a global space astrometry mission. Its goal is to make the largest, most precise map of our Galaxy by surveying an unprecedented number of stars - more than a thousand million with an accuracy of 10 μarcsec.
http://www.esa.int/esaSC/index.html

GALILEO:

The Galileo spacecraft began its long journey to the realm of the outer planets in 1989. Its mission was to study Jupiter and its moons in more detail than any previous spacecraft. The Galileo spacecraft's 14-year odyssey came to an end on Sunday, Sept. 21, 2003 when the spacecraft passed into Jupiter's shadow then disintegrated in the planet's dense atmosphere.
http://galileo.jpl.nasa.gov/

HAYABUSA:

The primary scientific objective of the Hayabusa (formerly Muses-C) mission, launched in 2003, is to collect a surface sample of material from an asteroid (25143 Itokawa/1998 SF36) and return the sample to Earth for analysis. http://www.isas.ac.jp/e/enterp/missions/muses-c/index.shtml

HERSCHEL:

Herschel will be the only space facility ever developed covering the far infrared to submillimetre range of the spectrum (from 80 to 670 microns). It will open up a virtually unexplored part of the spectrum which cannot be observed well from the ground. http://www.esa.int/esaSC/index.html

ISO:

Infrared Space Observatory, European infrared satellite with 4 instruments onboard, operational between 1995-1998, covering the wavelength range between 2.5 and 200 μm.
http://www.esa.int/esaSC/index.html

IRSI-DARWIN:

An array of up to 6 space telescopes, each of which will be at least 1.5 m in diameter. They will work together to scan the nearby Universe, looking for signs of life on Earth-like planets.
http://www.esa.int/esaSC/index.html

JIMO:

The Jupiter Icy Moons Orbiter is an ambitious proposed mission to orbit three planet-sized moons of Jupiter -- Callisto, Ganymede and Europa -- which may harbor vast oceans beneath their icy surfaces. The mission would launch in 2012 or later. http://www.jpl.nasa.gov/jimo/

JWST:

The James Webb Space Telescope (JWST, formerly called the NGST) is a large, infrared-optimized space telescope. It will have an 18-segment, 6.5-meter primary mirror and will reside in an L2 Lissajous orbit. JWST is scheduled for launch in 2011. http://ngst.gsfc.nasa.gov/

KEPLER:

The Kepler Mission is a space mission designed using the transit technique to detect and characterize hundreds of Earth-size planets in the habitable zone of a wide variety of stars.
http://planetquest.jpl.nasa.gov/Kepler/kepler_index.html

Large millimeter telescope (LMT):

A joint project of the University of Massachussetts (U.S.A.) and the Instituto Nacional de Astrofisica, Optica & Electronica (Mexico) to build the world's largest radiotelescope for use at millimeter wavelengths, completion expected in 2004/2005. http://www.lmtgtm.org/

MER:

NASA's twin *Mars Exploration Rovers*, now named *Spirit* and *Opportunity*, are designed to study the history of water on Mars. MER A and B are equipped with a robotic arm, a drilling tool, three spectrometers, and four pairs of cameras that allow them to have a human-like, 3D view of the terrain. Each rover can travel as far as 100 meters in one day. Both rovers have landed successfully in January 2004.
http://marsrovers.jpl.nasa.gov/home/

MEX:

Mars Express is Europe's first spacecraft to the red planet. It carries seven instruments. The orbiter instruments are remotely investigating the Martian atmosphere, surface and subsurface.
http://www.esa.int/esaSC/index.html

MGS:

The Mars Global Surveyor (MGS) mission performs high resolution imaging of the surface and studies of the topography and gravity, the role of water and dust on the surface and in the atmosphere of Mars, the weather and climate of Mars, the composition of the surface and atmosphere, and the existence and evolution of the Martian magnetic field.
http://mars.jpl.nasa.gov/

MRO:

Mars Reconnaissance Orbiter (2006) will characterize the surface, subsurface, and atmosphere of Mars, and will identify potential landing sites for future missions. http://marsprogram.jpl.nasa.gov/mro/

MSL:

The Mars Science Laboratory (2009) will analyze dozens of samples scooped up from the soil and cored from rocks as it explores with greater range than any previous Mars rover. It will investigate the past or present ability of Mars to support life. http://centauri.larc.nasa.gov/msl/

ODYSSEY:

Mars Odyssey's maps chemical elements and minerals on the surface of Mars, looks for water in the shallow subsurface, and analyzes the radiation environment to determine its potential effects on human health.
http://marsprogram.jpl.nasa.gov/

PHOENIX:

A robot explorer named Phoenix will be launched by NASA in 2007 to land and scout the high northern latitudes of Mars in 2008 (Scout mission). The lander would touch down on terrain suspected of covering a vast reservoir of water ice just a foot beneath the surface.
http://www.nasa.gov/missions/solarsystem/phoenix_water.html

ROSETTA:

Rosetta's main objective is to rendezvous with Comet 67P/Churyumov-Gerasimenko. Rosetta was launched in March 2004. The mission will study the nucleus of the comet and its environment in great detail for a period of nearly two years, and land a probe on its surface.
http://www.esa.int/esaSC/index.html

SST-SPITZER:

The Spitzer Space Telescope (former SIRTF), launched in August 2003 completes NASA's great observatories that study the Universe at wavelengths from visible light to x-rays. http://www.spitzer.caltech.edu/

SMA:

Sub-millimeter array: An array of radiotelescopes on Mauna Kea (Hawaii, U.S.A.) operated by the Harvard-Smithsonian Center for Astrophysics. http://sma-www.harvard.edu/

SMART-1:

First of ESA's small missions to test new technology, SMART-1 is also the first European spacecraft to go to the Moon. Its main objective is to test solar-electric primary propulsion for future deep-space missions. http://www.esa.int/esaSC/index.html

STARDUST:

NASA's Stardust spacecraft is currently on a mission to collect and return the first samples from the vicinity of comet Wild-2. Samples have been captured in January 2004 in a low-density aerogel and will be parachuted to Earth in 2006. http://stardust.jpl.nasa.gov/

Terrestrial Planet Finder (TPF):

A proposed NASA mission that will study all aspects of planets: from their formation and development in disks of dust and gas around newly forming stars to the presence and features of those planets orbiting the nearest stars. http://planetquest.jpl.nasa.gov/TPF/tpf_index.html

WMAP:

The Wilkinson Microwave Anisotropy Probe (WMAP) is a NASA Explorer mission measuring the temperature of the cosmic background radiation over the full sky with unprecedented accuracy. http://map.gsfc.nasa.gov/

Astrophysics and Space Science Library

Volume 302: ***Stellar Collapse,*** edited by Chris L. Fryer
Hardbound, ISBN 1-4020-1992-0, April 2004

Volume 301: ***Multiwavelength Cosmology,*** edited by Manolis Plionis
Hardbound, ISBN 1-4020-1971-8, March 2004

Volume 300: ***Scientific Detectors for Astronomy,*** edited by Paola Amico, James W. Beletic, Jenna E. Beletic
Hardbound, ISBN 1-4020-1788-X, February 2004

Volume 299: ***Open Issues in Local Star Fomation,*** edited by Jacques Lépine, Jane Gregorio-Hetem
Hardbound, ISBN 1-4020-1755-3, December 2003

Volume 298: ***Stellar Astrophysics - A Tribute to Helmut A. Abt,*** edited by K.S. Cheng, Kam Ching Leung, T.P. Li
Hardbound, ISBN 1-4020-1683-2, November 2003

Volume 297: ***Radiation Hazard in Space,*** by Leonty I. Miroshnichenko
Hardbound, ISBN 1-4020-1538-0, September 2003

Volume 296: ***Organizations and Strategies in Astronomy, volume 4,*** edited by André Heck
Hardbound, ISBN 1-4020-1526-7, October 2003

Volume 295: ***Integrable Problems of Celestial Mechanics in Spaces of Constant Curvature,*** by T.G. Vozmischeva
Hardbound, ISBN 1-4020-1521-6, October 2003

Volume 294: ***An Introduction to Plasma Astrophysics and Magnetohydrodynamics,*** by Marcel Goossens
Hardbound, ISBN 1-4020-1429-5, August 2003
Paperback, ISBN 1-4020-1433-3, August 2003

Volume 293: ***Physics of the Solar System,*** by Bruno Bertotti, Paolo Farinella, David Vokrouhlický
Hardbound, ISBN 1-4020-1428-7, August 2003
Paperback, ISBN 1-4020-1509-7, August 2003

Volume 292: ***Whatever Shines Should Be Observed,*** by Susan M.P. McKenna-Lawlor
Hardbound, ISBN 1-4020-1424-4, September 2003

Volume 291: ***Dynamical Systems and Cosmology,*** by Alan Coley
Hardbound, ISBN 1-4020-1403-1, November 2003

Volume 290: ***Astronomy Communication,*** edited by André Heck, Claus Madsen
Hardbound, ISBN 1-4020-1345-0, July 2003

Volume 287/8/9: ***The Future of Small Telescopes in the New Millennium,*** edited by Terry D. Oswalt
Hardbound Set only of 3 volumes, ISBN 1-4020-0951-8, July 2003

Volume 286: ***Searching the Heavens and the Earth: The History of Jesuit Observatories,*** by Agustín Udías
Hardbound, ISBN 1-4020-1189-X, October 2003

Volume 285: ***Information Handling in Astronomy - Historical Vistas,*** edited by André Heck
Hardbound, ISBN 1-4020-1178-4, March 2003

Volume 284: ***Light Pollution: The Global View,*** edited by Hugo E. Schwarz
Hardbound, ISBN 1-4020-1174-1, April 2003

Volume 283: ***Mass-Losing Pulsating Stars and Their Circumstellar Matter,*** edited by Y. Nakada, M. Honma, M. Seki
Hardbound, ISBN 1-4020-1162-8, March 2003

Volume 282: ***Radio Recombination Lines,*** by M.A. Gordon, R.L. Sorochenko
Hardbound, ISBN 1-4020-1016-8, November 2002

Volume 281: ***The IGM/Galaxy Connection,*** edited by Jessica L. Rosenberg, Mary E. Putman
Hardbound, ISBN 1-4020-1289-6, April 2003

Volume 280: ***Organizations and Strategies in Astronomy III,*** edited by André Heck
Hardbound, ISBN 1-4020-0812-0, September 2002

Volume 279: ***Plasma Astrophysics , Second Edition,*** by Arnold O. Benz
Hardbound, ISBN 1-4020-0695-0, July 2002

Volume 278: ***Exploring the Secrets of the Aurora,*** by Syun-Ichi Akasofu
Hardbound, ISBN 1-4020-0685-3, August 2002

Volume 277: ***The Sun and Space Weather,*** by Arnold Hanslmeier
Hardbound, ISBN 1-4020-0684-5, July 2002

Volume 276: ***Modern Theoretical and Observational Cosmology,*** edited by Manolis Plionis, Spiros Cotsakis
Hardbound, ISBN 1-4020-0808-2, September 2002

Volume 275: ***History of Oriental Astronomy,*** edited by S.M. Razaullah Ansari
Hardbound, ISBN 1-4020-0657-8, December 2002

Volume 274: ***New Quests in Stellar Astrophysics: The Link Between Stars and Cosmology,*** edited by Miguel Chávez, Alessandro Bressan, Alberto Buzzoni,Divakara Mayya
Hardbound, ISBN 1-4020-0644-6, June 2002

Volume 273: ***Lunar Gravimetry,*** by Rune Floberghagen
Hardbound, ISBN 1-4020-0544-X, May 2002

Volume 272:***Merging Processes in Galaxy Clusters,*** edited by L. Feretti, I.M. Gioia, G. Giovannini
Hardbound, ISBN 1-4020-0531-8, May 2002

Volume 271: ***Astronomy-inspired Atomic and Molecular Physics,*** by A.R.P. Rau
Hardbound, ISBN 1-4020-0467-2, March 2002

Volume 270: ***Dayside and Polar Cap Aurora,*** by Per Even Sandholt, Herbert C. Carlson, Alv Egeland
Hardbound, ISBN 1-4020-0447-8, July 2002

Volume 269: ***Mechanics of Turbulence of Multicomponent Gases,*** by Mikhail Ya. Marov, Aleksander V. Kolesnichenko
Hardbound, ISBN 1-4020-0103-7, December 2001

Volume 268: ***Multielement System Design in Astronomy and Radio Science,*** by Lazarus E. Kopilovich, Leonid G. Sodin
Hardbound, ISBN 1-4020-0069-3, November 2001

Volume 267: ***The Nature of Unidentified Galactic High-Energy Gamma-Ray Sources,*** edited by Alberto Carramiñana, Olaf Reimer, David J. Thompson
Hardbound, ISBN 1-4020-0010-3, October 2001

Volume 266: ***Organizations and Strategies in Astronomy II,*** edited by André Heck
Hardbound, ISBN 0-7923-7172-0, October 2001

Volume 265: ***Post-AGB Objects as a Phase of Stellar Evolution,*** edited by R. Szczerba, S.K. Górny
Hardbound, ISBN 0-7923-7145-3, July 2001

Volume 264: ***The Influence of Binaries on Stellar Population Studies,*** edited by Dany Vanbeveren
Hardbound, ISBN 0-7923-7104-6, July 2001

Volume 262: ***Whistler Phenomena - Short Impulse Propagation,*** by Csaba Ferencz, Orsolya E. Ferencz, Dániel Hamar, János Lichtenberger
Hardbound, ISBN 0-7923-6995-5, June 2001

Volume 261: ***Collisional Processes in the Solar System,*** edited by Mikhail Ya. Marov, Hans Rickman
Hardbound, ISBN 0-7923-6946-7, May 2001

Volume 260: ***Solar Cosmic Rays,*** by Leonty I. Miroshnichenko
Hardbound, ISBN 0-7923-6928-9, May 2001

Volume 259: ***The Dynamic Sun,*** edited by Arnold Hanslmeier, Mauro Messerotti, Astrid Veronig
Hardbound, ISBN 0-7923-6915-7, May 2001

Volume 258: ***Electrohydrodynamics in Dusty and Dirty Plasmas- Gravito-Electrodynamics and EHD,*** by Hiroshi Kikuchi
Hardbound, ISBN 0-7923-6822-3, June 2001

Volume 257: ***Stellar Pulsation - Nonlinear Studies,*** edited by Mine Takeuti, Dimitar D. Sasselov
Hardbound, ISBN 0-7923-6818-5, March 2001

Volume 256: ***Organizations and Strategies in Astronomy,*** edited by André Heck
Hardbound, ISBN 0-7923-6671-9, November 2000

Volume 255: ***The Evolution of the Milky Way- Stars versus Clusters,*** edited by Francesca Matteucci, Franco Giovannelli
Hardbound, ISBN 0-7923-6679-4, January 2001

Volume 254: ***Stellar Astrophysics***, edited by K.S. Cheng, Hoi Fung Chau, Kwing Lam Chan, Kam Ching Leung
Hardbound, ISBN 0-7923-6659-X, November 2000

Volume 253: ***The Chemical Evolution of the Galaxy***, by Francesca Matteucci
Paperback, ISBN 1-4020-1652-2, October 2003
Hardbound, ISBN 0-7923-6552-6, June 2001

Volume 252: ***Optical Detectors for Astronomy II***, edited by Paola Amico, James W. Beletic
Hardbound, ISBN 0-7923-6536-4, December 2000

Volume 251: ***Cosmic Plasma Physics***, by Boris V. Somov
Hardbound, ISBN 0-7923-6512-7, September 2000

Volume 250: ***Information Handling in Astronomy***, edited by André Heck
Hardbound, ISBN 0-7923-6494-5, October 2000

Volume 249: ***The Neutral Upper Atmosphere***, by S.N. Ghosh
Hardbound, ISBN 0-7923-6434-1, July 2002

Volume 247: ***Large Scale Structure Formation***, edited by Reza Mansouri, Robert Brandenberger
Hardbound, ISBN 0-7923-6411-2, August 2000

Volume 246: ***The Legacy of J.C. Kapteyn***, edited by Piet C. van der Kruit, Klaas van Berkel
Paperback, ISBN 1-4020-0374-9, November 2001
Hardbound, ISBN 0-7923-6393-0, August 2000

Volume 245: ***Waves in Dusty Space Plasmas,*** by Frank Verheest
Paperback, ISBN 1-4020-0373-0, November 2001
Hardbound, ISBN 0-7923-6232-2, April 2000

Volume 244: ***The Universe***, edited by Naresh Dadhich, Ajit Kembhavi
Hardbound, ISBN 0-7923-6210-1, August 2000

Volume 243: ***Solar Polarization***, edited by K.N. Nagendra, Jan Olof Stenflo
Hardbound, ISBN 0-7923-5814-7, July 1999

Volume 242: ***Cosmic Perspectives in Space Physics***, by Sukumar Biswas
Hardbound, ISBN 0-7923-5813-9, June 2000

Volume 241: ***Millimeter-Wave Astronomy: Molecular Chemistry & Physics in Space,*** edited by W.F. Wall, Alberto Carramiñana, Luis Carrasco, P.F. Goldsmith
Hardbound, ISBN 0-7923-5581-4, May 1999

Volume 240: ***Numerical Astrophysics,*** edited by Shoken M. Miyama, Kohji Tomisaka,Tomoyuki Hanawa
Hardbound, ISBN 0-7923-5566-0, March 1999

Volume 239: ***Motions in the Solar Atmosphere,*** edited by Arnold Hanslmeier, Mauro Messerotti
Hardbound, ISBN 0-7923-5507-5, February 1999

Volume 238: ***Substorms-4,*** edited by S. Kokubun, Y. Kamide
Hardbound, ISBN 0-7923-5465-6, March 1999

Volume 237: ***Post-Hipparcos Cosmic Candles,*** edited by André Heck, Filippina Caputo
Hardbound, ISBN 0-7923-5348-X, December 1998

Volume 236: ***Laboratory Astrophysics and Space Research,*** edited by P. Ehrenfreund, C. Krafft, H. Kochan, Valerio Pirronello
Hardbound, ISBN 0-7923-5338-2, December 1998

Missing volume numbers have not yet been published.
For further information about this book series we refer you to the following web site:
http://www.wkap.nl/prod/s/ASSL

To contact the Publishing Editor for new book proposals:
Dr. Harry (J.J.) Blom: harry.blom@wkap.nl